1984
IASS-RUB

Natural Draught Cooling Towers

Proceedings of the 2. International Symposium
Ruhr-Universität Bochum, Germany,
September 5-7, 1984

Editors:
P. L. Gould, W. B. Krätzig, I. Mungan,
and U. Wittek

With 333 Figures

Springer-Verlag
Berlin Heidelberg New York Tokyo 1984

Prof. Ph.-D. Philipp L. GOULD
Dept. of Civil Engineering
Washington University
St. Louis, Missouri 63130 - USA

Prof. Dr.-Ing. Wilfried B. KRÄTZIG
Inst. für Konstruktiven Ingenieurbau
Lehrstuhl III - Statik und Dynamik
Ruhr-Universität Bochum
D-4630 Bochum - FRG

Dr.-Ing. Ihsan MUNGAN
Inst. für Konstruktiven Ingenieurbau
Lehrstuhl I - Massivbau
Ruhr-Universität Bochum
D-4630 Bochum - FRG

Prof. Dr.-Ing. Udo WITTEK
Lehrstuhl für Baustatik
Universität Kaiserslautern
D-6750 Kaiserslautern - FRG

Library of Congress Cataloging in Publication Data.

Main entry under title:
Natural draught cooling towers.
1. Cooling towers – Congresses.
I. Gould, Phillip L. II. International Symposium on Natural Draught Cooling Towers
(2nd : 1984 : Ruhr-Universität Bochum)
TJ563.N38 1984 621.1'97 84-14092

ISBN-13:978-3-642-82347-3 e-ISBN-13:978-3-642-82345-9
DOI: 10.1007/978-3-642-82345-9

Softcover reprint of the hardcover 1st edition 1984

2061/3020 5 4 3 2 1 0

Dedicated to

o. Prof. Dr.-Ing. Dr.-Ing. E. h. Wolfgang Zerna

Preface

Natural draught cooling towers are inevitable for the economic generation of electricity under environmental aspects: In any thermal power plant heated steam drives a turbogenerator, which produces the electric energy. In order to create an effective heat sink behind the turbine, the steam is condensed and recycled into the boiler. This requires a large amount of cooling water which is heated by running through the condensor.

Even in the most efficient fossil fuel thermal (nuclear) power plants, only about 40 (30)% of the generated heat is turned into electric energy. The rest of heat of 60 (70)% is discharged into the environment through the smokestack and the cooling water circuit.

To avoid thermal pollution of natural water resources the heated cooling water should be recooled artificially in a cooling system and reused. Among all technical solutions natural draught cooling towers are effective, economic and widely spread corrective measures. In large power plants only natural draught cooling towers are able to recover the immense quantities of water required for cooling. Thus only these engineering structures are able to balance environmental factors, economic investment and operating costs of the power station with the demands of a reliable electric energy supply.

The present *2. International Symposium on Natural Draught Cooling Towers* has been termed as a state-of-the-art colloquium for all civil engineering problems concerning these structures. After two preceding meetings, the colloquium in Brussels (1975) and the 1. symposium in Paris (1978), the *Ruhr-University Bochum* (RUB) and the *International Association for Shell and Spatial Structures* (IASS) have organized the present symposium in order to bring together from all over the world the most distinguished scientists and engineers working on cooling towers. Theoretical concepts, analytical and numerical methods for a better understanding of the load-bearing behavior of cooling towers will be presented as well as results of experimental investigations. Special structural aspects and construction problems will be discussed and new ways will be searched to increase safety and improve durability of these structures.

The organizers feel auspicious being able to dedicate this symposium to *Professor Dr.-Ing. Dr.-Ing. E.h. Wolfgang Zerna*, who has influenced large parts of the German cooling tower research and practice with his ideas for nearly 20 years. Many large cooling towers would be less safe and less economic without his steady effort to develop appropriate engineering solutions. Many of his academic scholars have adapted his conceptual thinking, some of them contributing to this symposium.

The symposium is gratefully sponsored by the *Ruhr-University, the American Concrete Institute/American Society of Civil Engineers* - Committee 334, the *Deutscher Beton-Verein* and the *Deutsche Forschungsgemeinschaft*. Special supports by *E. Heitkamp GmbH, Hochtief AG, Ph. Holzmann AG, MC-Bauchemie GmbH, Rheinisch-Westfälische Elektrizitätswerke AG* and *Vereinigte Elektrizitätswerke Westfalen AG* are gratefully appreciated. The organizers feel urged to thank all autors for their cooperation and the Springer-Verlag for the willingness to publish these proceedings at the beginning of the symposium. Particular thanks are extended to the staff of *Unikontakt* and to all members of the *Lehrstuhl für Statik* in the *Institut für Konstruktiven Ingenieurbau* of the Ruhr-University. The enormous burden of organizing work carried by both secretaries to the symposium, Professor Dr.-Ing. I. Mungan and Professor Dr.Ing. U. Wittek, has to be accentuated.

Bochum, September 5, 1984

Wilfried B. Krätzig   Philipp L. Gould

Contents

DURABILITY

Impulses of the Research on the Development of Large Cooling Towers

W. ZERNA

Institut für Konstruktiven Ingenieurbau
Ruhr-Universität Bochum
Bochum, Germany

Summary

The rapid development of cooling towers during the last two decades was only possible through fundamental research on their structural aspects, whereas the assessment of the forces induced by wind was the first step. There exist different computation techniques developed to obtain the shell stresses with sufficient accuracy. The buckling behavior could be understood best in buckling tests carried out on models both under axisymmetric loading and in wind tunnel. The shape of the meridian influences the stress state and the buckling behavior substantially. Therefore more emphasis has been devoted to the optimization of the cooling tower shape. With increasing sizes, however, the stiffening of the shell part may be necessary for an optimal design. In this case stiffening rings proved to be most effective. The method of construction allowing the erection of the cooling tower shell with a high accuracy is of vital importance. In the future more attention has to be paid to the durability of the reinforced concrete veil.

Introduction

Natural draft cooling towers are mostly designed as reinforced concrete thin shells supported along the circumference by a system of columns, in general having a height less than 10 % of the total height of the tower. The V- or X-shaped columns are resting on single footings or on a ring foundation, generally being independent of the cold water basin. The fill inside the tower or cooling elements are supported by a system of beams with adequate support posts. Among all these structural units the shell part requires most of the attention in design and construction.

Cooling towers in general and the shell structure in particular belong to the largest structures in civil engineering. Fig. 1 demonstrates the development in the height H of the cooling towers depending on the capacity C of the power plant blocks for which they are designed. The first natural-draught cooling tower having hyperboloidal shape was built in Netherlands in 1914 and was already 35 m high. After half of a century the 100 m limit in the height was reached and the first natural draught cooling tower in Germany was of this size. However, due to dry-type cooling system, the capacity of the power plant was in this case only 150 MW. The development in Germany between the years 1965 and 1982 is shown in Fig. 1 marking the highest cooling tower of each year with its diameter and the corresponding plant capacity.

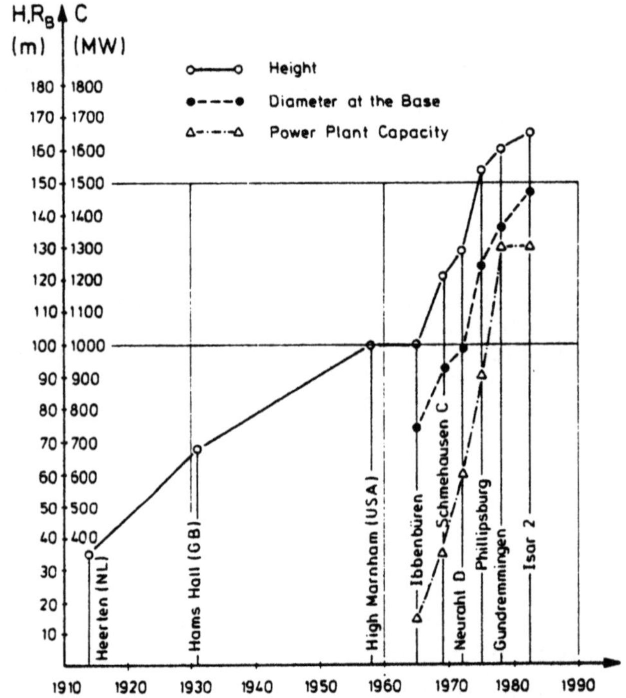

Fig. 1 Development in the Size of the Cooling Towers and Block Capacities

With increasing capacity of the power plants it was necessary to increase both the height and the diameter. Since the second half of the seventies cooling towers having a height of about 160 m and a base diameter of about 140 m are going to be more or less a standard type for nuclear power plants of 1300 MW capacity in Germany. A similar trend of development can be observed also in other western industrial countries. This upsurge during a relatively short time could not be realized without fundamental researchs on these remarkable structures as the following presentation will try to show.

Computation Methods

Two decades ago the membrane solution of the continually supported hyperboloidal shell under axisymmetric dead load and non-axisymmetric wind action obtained using the so-called method of characteristics constituted the standard method of analysis. The disturbances at the stiffened upper and lower boundaries were accounted for by a simplified bending theory for edge disturbancies. The region of the shell immediately above the supporting columns, termed the lower lintel, was designed according to deep beam theory.

Due to rapid development of the computers and computing methods there is today no problem in the analysis of the shell part having the geometric form of a hyperboloid of revolution according to the general bending theory for thin elastic shells. The numerical evaluation of the bending theory may be based on any finite element or finite difference methods and the method of dynamic relaxation as well. For design purposes it is appropriate to assume the reinforced concrete to be isotropic, homogeneous and without cracks. A refinement of the analysis is achieved if the supporting columns too are introduced into analysis. However,in this place it must be pointed out that any of the analysis methods described may be accurate enough, even for large cooling towers, provided that care is given to a sound structural modeling of the cooling tower.

Wind Forces

The problem arising with increasing dimensions of the cooling towers is not to improve the method of analysis but rather the assessment of the forces induced through wind. Besides the dead load, wind is the main loading of a cooling tower and the design of the reinforcement is based on the difference of two stresses of comparable magnitude, namely tension due to wind and compression due to dead load. On the other hand, the buckling of the cooling tower shell is substantially influenced by the compressive wind stresses added to the compression due to dead load. For these reasons to seize the wind effects realistically is of vital importance for the analysis of the cooling tower. This is why a considerable part of the research on cooling towers concentrated during the last decade on this item. Measurements both on aeroelastic models tested in wind tunnel and on real reinforced concrete cooling towers have yielded detailed information on the magnitude and distribution of the wind forces along the circumference and height of the structure. For a group of cooling towers wind tunnel tests are considered inevitable in order to obtain the wind effects on each cooling tower of the group. However, according to measurements on aeroelastic cooling tower models in a boundary layer wind tunnel, cooling towers erected in a distance more than 1,5 times the base diameter or twice of the throat diameter between the cooling tower axes can be handled as singles.

Buckling Behavior

For given base diameter, throat diameter and height the choice of the shape of the meridian and of the wall thickness is the first step in designing a cooling tower shell. The wall thickness is varied along the height taking the buckling resistance into account. The stress state of the cooling tower shell is under wind loading non-axisymmetric. Therefore the final stress state under dead load and wind load is also non-axisymmetric with biaxial stress states changing along both the height and the cirumference. The stability problem of the shell under such a type of loading and due to nonlinearities is rather of limit point type than of bifurcation type. A realistic nonlinear analysis of problems of this type is rather complicated and can therefore not be advocated at the stage of preliminary design. On the other hand according to the theoretical and experimental research carried out, the bifurcation analysis of the cooling tower shell subjected to dead load together with an axisymmetric conjugate loading substituting the wind action proves to be effective in estimating the shell thickness varying along the height. The main idea of this engineering approach rests on the observation done in buckling tests on shells, namely on the local initiation of the buckling process. The local character of the impending buckling is in the case of wind loaded shells obvious, as Figures 2 and 3 taken in wind tunnel show. But this is also the case for axisymmetrically loaded shells due to unavoidable imperfections in the shell geometry and wall thickness or due to variations in the materials properties such as initial shrinkage cracks of concrete and similar defects.

Fig. 2 Wind-Buckling of a Cooling Tower-Model without Upper Edge Beam

Fig. 3 Wind-Buckling of a Cooling Tower Model with a Stiff Upper Edge Beam

The final buckling configuration as a local or periodic one depends mainly on the restraints of the cooling tower shell and on the distribution of the loading along the circumference.

Without any deformation control the first buckling dimple will get always deeper until the strength of the reinforced concrete section is reached and the shell is severly damaged or even collapses. The failure of a reinforced concrete cooling tower shell due to buckling under wind action will take place first along the windward meridian either at the top or near the throat depending on the rigidity of the upper ring beam and of the magnitude and the variation of the wall thickness. In the case that both the upper zone and the region near the throat have a sufficient buckling resistance the areas below the throat at the flanks are the mostly endangered zones through buckling. For this reason all parts of the cooling tower shell has to exhibit a sufficient buckling resistance. For an optimal buckling design this resistance has to be the same along the height of the cooling tower shell. Previous design methods took only care of a global buckling resistance and so did not allow to obtain the optimal variation of the wall thickness. For safety and economy, however, this is a prerequisite in the case of large reinforced concrete cooling tower shells.

Earthquake Resistance

The seismicity of most parts of Germany is restricted to small values. For this reason the question of earthquake resistance for cooling tower shells erected in Germany has been of secondary importance and it is in most cases sufficient to make an approximate earthquake analysis through discretization of the shell into a beam model. On the other hand the seismic loading of the cooling tower is defined as a response spectrum, which is site related. The response spectrum method facilitates the specification of this loading in terms of design spectra, if such spectra are existing or can be evaluated. In other cases the time-history analysis may be used to produce a complete history of the cooling tower response. Of course this last method requires much more computational effort.

Formgiving

The shape of the meridian curve of the cooling tower shell generally is a hyperbola described by

$$R = r_0 + a\sqrt{1+\frac{z^2}{b^2}} \qquad (1)$$

in which R, z and r_o are shown in Fig. 4, r_o being the distance between the axis of rotation and the axis of the hyperbola.

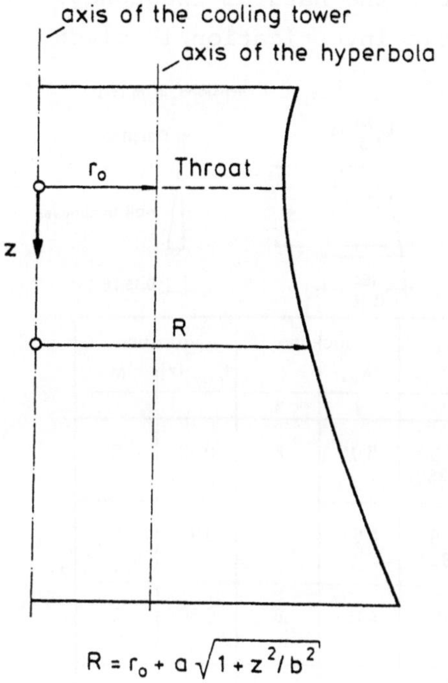

Fig. 4 Geometry of Cooling Tower Shell

For r_o = 0, parameter a is the radius of the cooling tower at the throat. The parameter describing the curvature is b, given by

$$b = \frac{H_T}{\sqrt{\left(\frac{R_B - r_0}{R_T - r_0}\right)^2 - 1}} \qquad (2)$$

in which R_B and R_T = the radii of the cooling tower at the base and at the throat, respectively, and H_T = the vertical distance between the base and the throat.

For given thermodynamic data the values of H_T, R_B and R_T are related with each other and offer only a restricted possibility for variation. On the other hand the offset distance r_o influences not only the stress distribution in the shell, but also the buckling load and the natural frequency. In Fig. 5 the result of a parametric investigation is given.

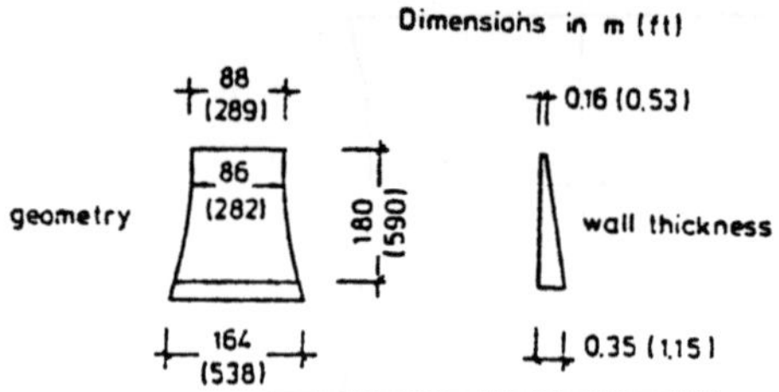

r_o m. (ft)	buckling λ_{Kr}	n	vibration f_{min} [Hz]	m
1	2	3	4	5
-1000 (-3280)	5.04	6	0.95	5
-25 (-82)	4.51	6	0.91	5
0 (0)	4.15	6	0.86	5
+10 (+33)	3.90	6	0.83	5

Fig. 5 Influence of Offset Distance on Buckling and Vibration

The buckling parameter λ_{cr} is defined as the ratio of the critical load q_{cr} and acting load q

$$\lambda_{cr} = \frac{q_{cr}}{q} \tag{3}$$

The load q is the combination of the dead load of the shell with a uniform external radial pressure having the magnitude

1.95 kPa (0.283 psi). This value corresponds to the wind pressure according to DIN 1055 on the windward meridian at the top, including an internal suction of 50 %. The magnitude f_{min} is the lowest natural frequency and n or m is the corresponding harmonic number of the buckling or vibration mode, respectively. By changing r_o from +10 m to -1000 m and keeping the main dimensions of the cooling tower shell (R_B,R_T and H_T) constant the buckling load can be increased about 30 % and the lowest natural frequency about 15 %. The variation of r_o causes some changes in the shape of the meridian between the throat and the base. These changes are however relatively small compared with radii R_T and R_B. For this reason no changes in the thermodynamical efficiency of the cooling tower are expected as r_o is varied. In the future more emphasis should be devoted to the optimization of the cooling tower shell shape, especially for large cooling towers.

Stiffening Means

With increasing dimensions of cooling towers the buckling and dynamic behavior of the shell part becomes of major importance. In order to maintain the prescribed buckling safety the wall thickness of the shell has to be increased to satisfy the buckling criteria with increasing size, the natural frequency drops more rapidly. Therefore it becomes necessary to investigate methods of stiffening the shell in order to improve the stability and vibration resistance of the cooling tower in an economic way. After systematic investigations on different stiffening means it turned out that arranging rings of adequate stiffness and number in appropriate positions proves to be most effective with respect to both buckling and dynamic behavior.

From a certain number of stiffening rings and equidistant meridional ribs having the same concrete volume as the rings, ring stiffeners are nearly three times more effective than meridional ribs if the augmentation of the buckling load after stiffening is considered. On the other hand the lowest natural frequency of the cooling tower shell is augmented considerably after arranging stiffening rings, whereas it remains unchanged in the case of stiffening by means of meridional ribs. After this information the design problem of very large cooling tower shells becomes a problem of optimization.

The wall thickness of the shell can be reduced considerably through arranging stiffening rings in optimal size, number and position. According to buckling tests and numerical parametric studies carried out a number of 4 to 5 intermediate rings proves to be optimal with respect to both buckling and vibration resistance of the shell part. The width of the rings has to be not less than 5 times the local shell thickness. However, the maximum width has to be restricted to about 8 times the shell thickness, because above this width no remarkable augmentation can be observed. The optimal position of the rings can be found out by iteration. In case of more than 4 intermediate rings the equidistant positioning is in general as much as effective as the optimal arrangement.

In Fig. 6 the augmentation of the safety factor against buckling is depicted with increasing number of the intermediate stiffening rings. The cooling tower shell is designed first for a buckling safety factor of γ_B = 5,00 overall. It has an upper edge beam together with the lower lintel shaped through gradual stiffening of the shell wall, as usual. Adding a ring of optimal stiffness at the throat increases the buckling safety only about 1,5 %. A second ring in optimal position between the throat and the lower edge imparts an additional safety of 5,5 %. A considerable augmentation of the buckling resistance is reached first after arranging 3 stiffening rings. However according to Fig. 6 the optimal number of stiffening rings is either 4 or 5. Even through equidistant positioning of the rings in these cases an augmentation of about 50 % or 85 % can be attained, respectively. For a larger number of rings the additional augmentation attainable through each of the additional rings decrease gradually, as Fig. 6 demonstrates.

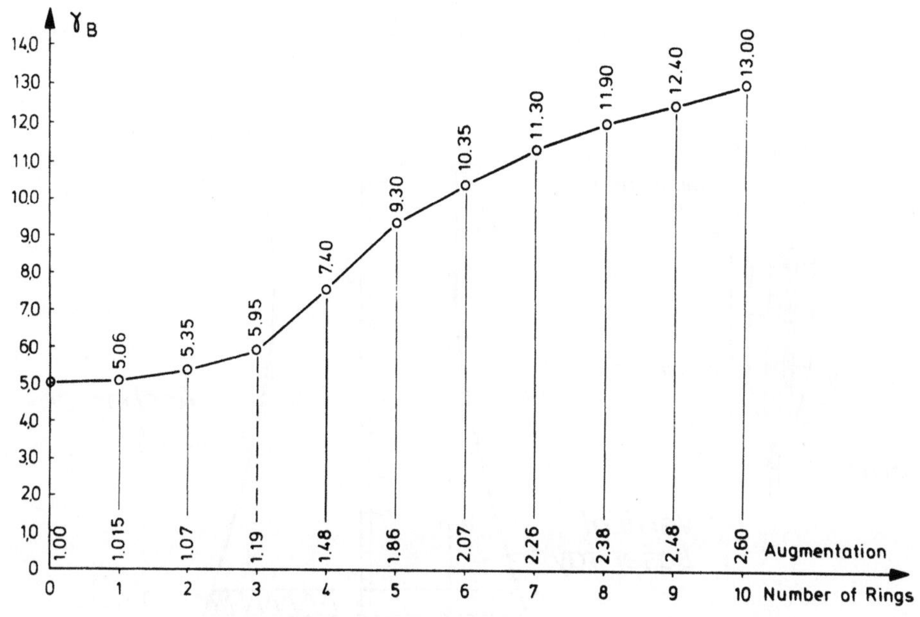

Fig. 6 Augmentation of Buckling Safety for Ring Stiffened Cooling Tower Shell

Method of Construction

The realization of large cooling towers in reinforced concrete has only been possible through construction methods allowing the erection of the cooling tower shell with high accuracy in the shape and wall thickness. Otherwise, both the bending moments and the buckling resistance calculated may be far away from reality. In the last case, it may even become absurd to insist on a high safety factor against buckling, if the ultimate load capacity is exceeded earlier due to usually law amoun of reinforcement. This argument shows the necessity of a construction method minimizing the imperfections in the cooling tower shell. In addition, not only an appropriate method of construction but also its continuous control is necessary.

In Fig. 7, the construction method applied to recent large

cooling towers in Germany is shown schematically.

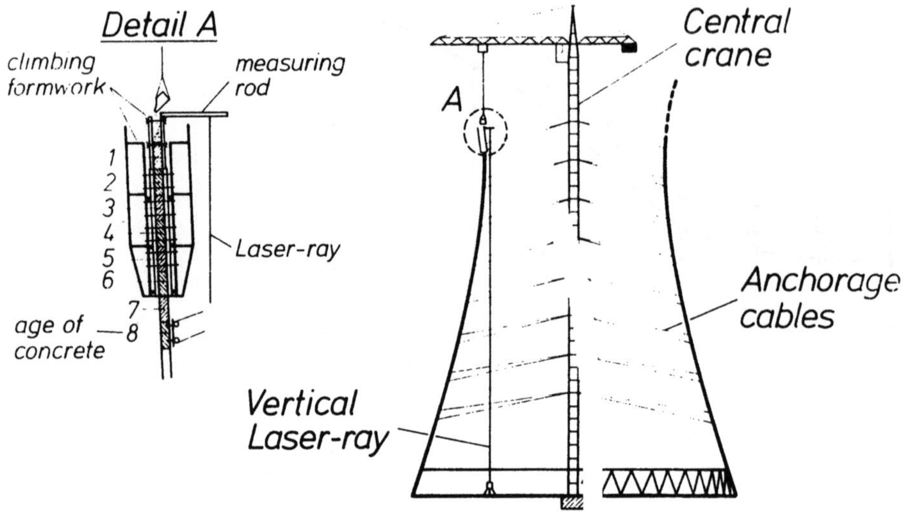

Fig. 7 Method of Construction and Surveying of Diameter of Each Lift

There are two basic ideas behind this method. First, the climbing formwork and the concrete hoisting system are separated from each other. Concrete is hoisted by the central crane which is stabilized by means of cables anchored in the lower parts of the shell. The formwork itself is anchored in parts of concrete which are at least two days old and have a strength of approx.7 MN/m^2. The second important idea behind the method is the shape control and shape correction by means of a vertical laser ray. The inclination of the formwork is adjusted at each lift. After concreting it is controlled and, if necessary, corrected gradually. After hardening of the concrete, the final shape is surveyed once more and the real wall thickness is measured. The results of such a survey are given for one meridian of the cooling tower shell of the nuclear power plant in Mülheim-Kärlich, Germany, in Fig. 8.

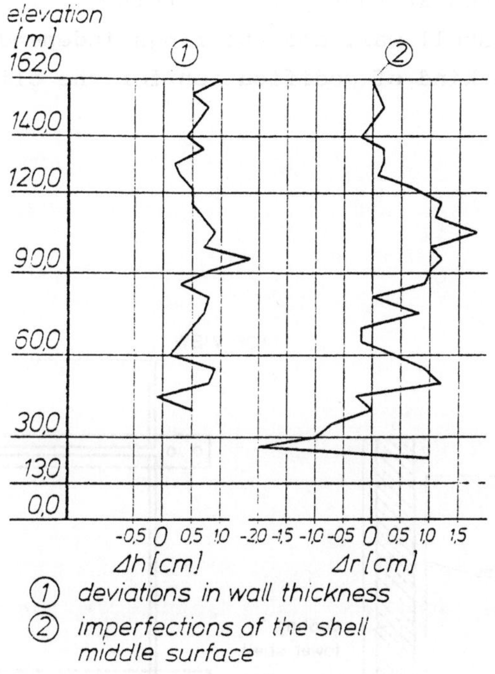

Fig. 8 Imperfections Measured on Cooling Tower Mülheim-Kärlich

After measurements along 48 meridians at more than 100 lifts, the maximum deviation of the wall thickness has been found to be about 10 % and the largest amplitude of imperfection 32 mm. It is obvious that such low values can be tolerated and need not to be considered in design.

The construction of the cooling tower shell stiffened by means of rings needs a modification of the existing formwork systems used for the conventional unstiffened shell. The horizontal rings have to be constructed in such a manner that they do not obstruct the very advantageous use of the climbing formwork. This is possible if the rings can be connected with the shell not continually, but with interruptions. Buckling tests on models showed that the rings are still fully effective if they are only partially connected to the shell. By hanging a platform at the end of the formwork the partial connection between a ring and the shell wall can be established after the climbing process is completed at that part. In this way it is possible to construct the shell wall and the rings independently. In Fig. 9 a possible kind of modification for the climbing formwork is shown.

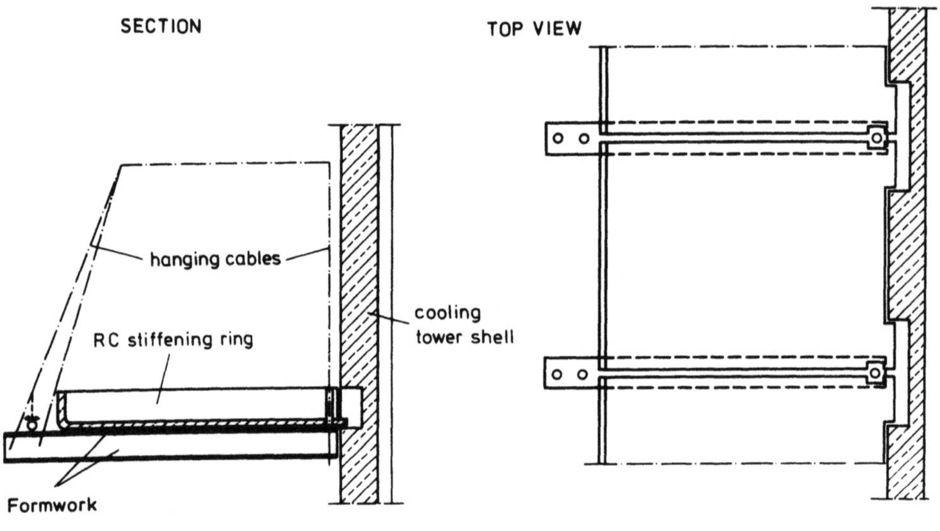

Fig. 9 Modification of Climbing Formwork to Allow Stiffening Rings

Alternative Solutions

As Fig. 1 demonstrates the era of large cooling towers began in Germany during the early 1970s. At that time some alternative solutions were taken into consideration; none of them could however, replace the conventional reinforced concrete shell structure. The idea of constructing the shell part using either precast elements of reinforced concrete or in steel did not prove to be economical. Because in both cases special equipment for mounting is required. In addition, in the first case more material is needed and in the second case maintenance is one of the main problems. Much better than these two variants the cable network version seemed to be a promising solution. The first natural draft cooling tower having a cable network as a veil was erected in 1975 in Germany for the 300 MW nuclear power plant at Schmehausen. This prototype tower has a dry-type cooling system and consists of the following five structural parts:

- Reinforced concrete pylon with foundation,
- ring foundation anchoring the cables,
- an upper compression ring and two intermediate rings with spokes,
- prestressed cable network and finally,
- the covering consisting of aluminium sheets attached to the network.

There are some facts which are not in favour of the cable network cooling tower from the economic point of view. The alteration of the cable cross section along the height is impossible. For this reason the cross section of meridional and inclined cables has to be dimensioned considering the maximum force resulting from superposition of the dead load, wind load and prestressing. In addition, the prestressing has to be transferred into the ground twice, as compression over the reinforced concrete pylon and as tension over the cable network. For this reason a strong foundation for the pylon will be necessary and a ring foundation together with soil anchors for the cables will be inevitable. For cooling towers below 200 m height the cable network structure is more costly than the reinforced concrete shell.

Outlook on the Future

Recent research on reinforced conrete cooling tower shells concentrates mainly on the material properties. The inclusion of the nonlinear material behavior and cracking in computations for ultimate load, buckling and dynamic response is necessary in order to develop a consistent design philosophy and to assess realistic safety factors. The interaction between the load carrying and buckling behavior is one of the main topics investigated for some years.

As a first step the influence of the tangent stiffness of reinforced concrete standing under biaxial stress states was investigated. Considering the biaxial material law given in Fig. 10 for concrete the full curve in Fig. 11 is obtained for the variation of the buckling safety factor under increasing loading.

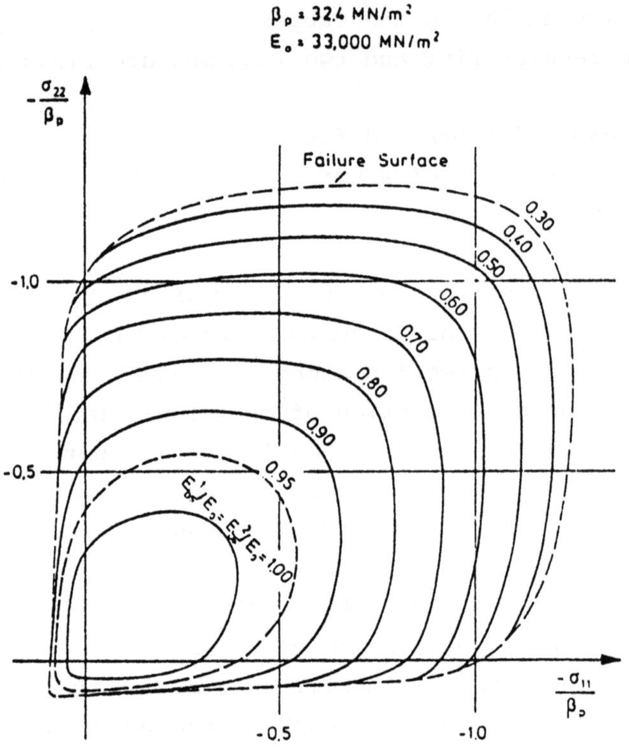

Fig. 10 Biaxial Material Law for Concrete

In this figure LF stands for the Load factor as a multiple of the axisymmetric conjugate loading inducing the worst stress state due to dead load and wind load. The magnitude γ_B is the Buckling Safety Factor defined as the ratio of the bifurcation load and the acting load at each load level LF. The dashed curve in the same figure is valid for the linear elastic approach. It turns out that the bifurcation load decreases to half of its value in the vicinity of the ultimate strength of the cooling tower shell, if instead of the linear elastic behavior the nonlinear material law with decreasing tangent modulus is introduced.

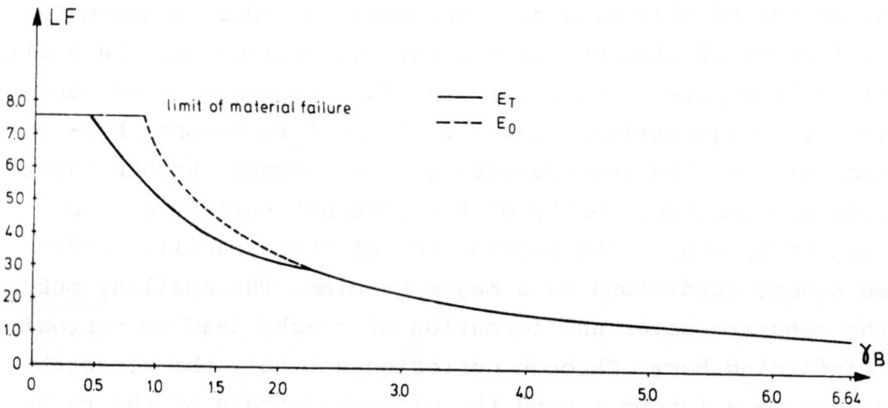

Fig. 11 Effect of Material Stiffness on Buckling Safety Factor γ_B

The buckling safety factor for shell structures in general prescribed in the German Building Code for Reinforced Concrete DIN 1045 as equal to 5 is relatively high and can be reduced in case of cooling tower shells. However, the reliability of the safety margins assessed will highly depend on the accuracy attained during construction. From this point of view it seems to be in place to choose a buckling safety factor which has to be higher than the safety factor against compression failure of concrete, which is taken in DIN 1045 equal to 2,10. Parametric studies in run will help to find out an appropriate value for the safety factor against buckling.

Another aspect investigated at the moment is the influence of the material law on the ultimate load of shells in general and for cooling tower shells in special. Introducing different material laws for reinforced concrete, i.e. varying the ultimate strain of concrete and reinforcement, the tension stiffening and other governing parameters, the ultimate load of cooling tower shells is obtained. The numerical investigations still in run and making use of Semi-Loof Shell Elements will provide indications on the influence of each of the parameters on the load carrying behavior and capacity.

Informations of this kind are necessary in order to estimate the reduction of strength caused through detritation. The cooling tower shell exposes a very large surface to atmospheric conditions like aggressivity causing chemical reactions, biological effects and temperature. Damages caused through these effects are observed mostly of the internal surface at the cooling tower shell. The maintenance of the durability under these severe conditions is a major problem. The spalling out of the concrete cover and formation of cracks lead to corrosion of reinforcing bars. Through corrosion not only the strength drops but to a larger extend the ultimate strain of the reinforcement and in this way the ductility of the whole shell goes down.

Conlusion

The foregoing brief exposition describes the mutual interaction between research and practice. Research on the lines as described here has been performed in many countries. In this contribution, however, research in Germany, mainly done by the Bochum Group, has been given particular attention.

Due to their unusual large sizes combined with likewise unusual small wall thickness reinforced concrete cooling tower shells are among the most sensitive structures of civil engineering. They are a challenge to researchers, practising engineers and contractors as well. The contributions coming from this three different directions which are going to be presented at this symposium underlines the necessity of close cooperation between these disciplines on international level. I wish a big success to the 2nd International Symposium on Natural-Draught Cooling Towers.

1984 IASS-RUB

Recent Structural Developments

Ring-Stiffened Shell Constructions – a Structural Alternative or a Technical and Economical Necessity?

HANS LUDOLF PETERS

E. Heitkamp GmbH, Langekampstraße 36, 4690 Herne 2

1. Introduction

In the last few decades reinforced-concrete natural-draft cooling towers have proved to be the most economical way of evacuating the immense quantities of process heat generated by thermal power plants into the atmosphere.

The geometrical dimensions depend primarily on the cooling procedure chosen, i.e. wet or dry cooling, and moreover on the heat quantities to be evacuated. Assuming that the quantity of heat is a direct function of the electrical power generated by the plant, we find a close relationship between the size of the power plant and the size of its cooling tower.

The evolution of the power plant capacities during the last few decades is a well-known fact: 150, 300, 600 and 750 MW of electrical power for the plants using fossil fuels and 300, 700, 900 and 1300 MW of electrical power for the plants using nuclear fuel. These indicated power ratings were achieved whenever the site was suitable for such plant sizes.

Fig. 1: Natural-draft cooling tower dimensions as a function of the plant capacity

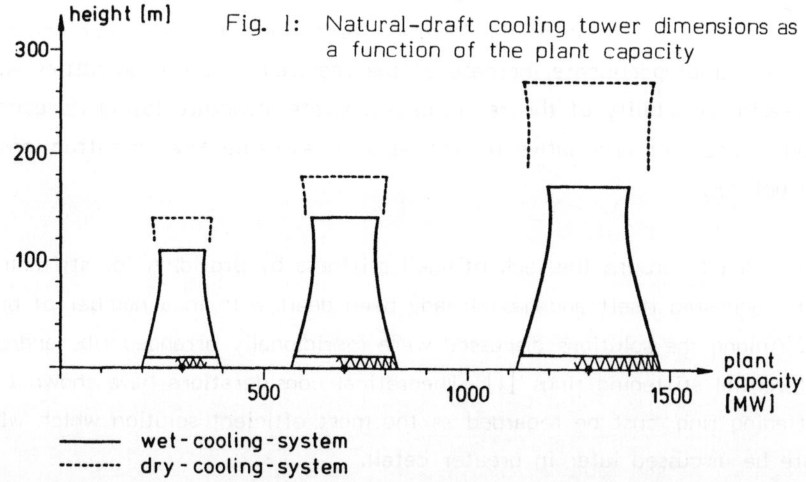

Fig. 1 shows the parallel evolution of power plants and their cooling towers. This figure shows, in the first instance, only wet cooling towers. For cooling towers operating on the basis of the dry-cooling technique far larger dimensions are required which are also outlined in fig. 1.

The design principle followed in the construction of cooling tower shells was as simple as it was economical because the reinforced-concrete shell served the purpose of being the enclosing and the supporting element at the same time.

For cooling-tower shells with a height of up to about 100 m and a corresponding basis diameter of 80 m the wall thickness of 16 cm alone - required anyhow for structural reasons - was sufficient to ensure the required buckling safety as well as a sufficiently high natural frequency. This pre-supposes, however, that the shell is provided with sufficiently stiff lintels at its upper and lower edges and that there is, moreover, a double-axis reinforcement on each face of the shell.

In case of increasing geometrical dimensions the wall thickness chosen for structural reasons is no longer sufficient to ensure the stiffness required with regard to the indispensable buckling safeties and - in case of still larger dimensions - to fulfil the requirement of a minimum natural frequency.

In the past, this lack of shell stiffness was exclusively compensated for by increasing the wall thickness of the shell. The wall-thickness curve for a given height of shell was designed in such a way that the theoretical minimum buckling safeties for each point of the shell were high enough or even higher than required. The buckling criterion thus became the most important parameter for the determination of the wall thickness. As will be shown later, this principle entailed a gradual reduction of the lowest natural frequency. Another consequence was an over-proportionate increase of the required concrete quantities with the increasing possibility of the reinforced-concrete structure losing its economical advantage over alternative techniques such as cable truss constructions or steel structures.

Thus, the idea to ensure the lack of shell stiffness by providing for stiffening elements suggested itself and has already been dealt with in a number of publications. Among the solutions discussed were meridionally arranged ribs and/or circumferential stiffening rings [1]. Theoretical considerations have shown that the stiffening ring must be regarded as the most efficient solution which will therefore be discussed later in greater detail.

This construction principle has already been applied twice by us in practice: in the 170 m-high natural-draft cooling tower of the Doel nuclear power plant in Belgium and for a subsequent rehabilitation measure in the 160 m-high cooling tower of the Mississippi I nuclear power plant of the Mississippi Power and Light Company, USA.

A third tower for the Isar II nuclear power plant is under construction.

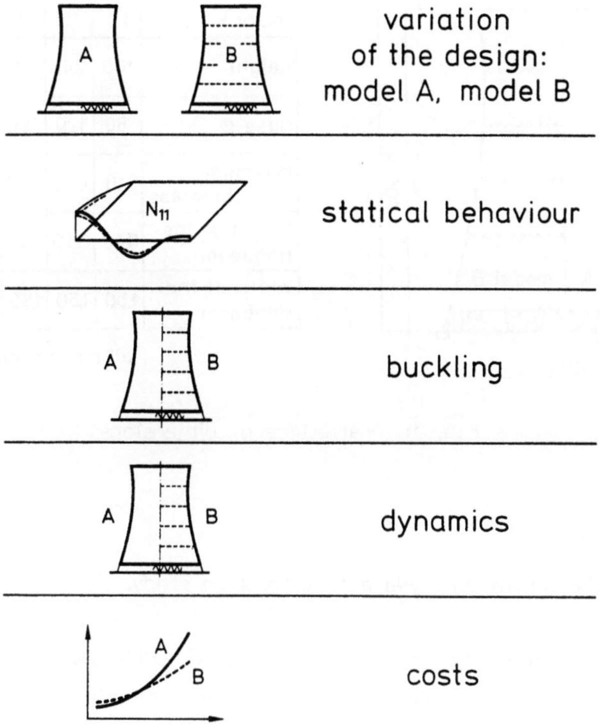

Fig. 2: List of contents

Fig. 2 gives a survey of the subjects to be dealt with in this contribution. At first, we shall discuss the two distinct construction principles and then, in the third and fourth chapter, the differences in the statical, buckling and dynamic behaviour of the construction principles chosen. The last chapter is to show the most economical fields of application of these two alternatives by comparing the related costs.

To demonstrate the relationship between the said alternatives and the geometrical dimensions, a series of tower dimensions will be studied. For simplification purposes we shall make use of a geometrical parameter by which the various cooling towers are distinguished. The procedure enables us to outline dependence relationships which may then serve to facilitate decision-making if a realistic example is to be dealt with.

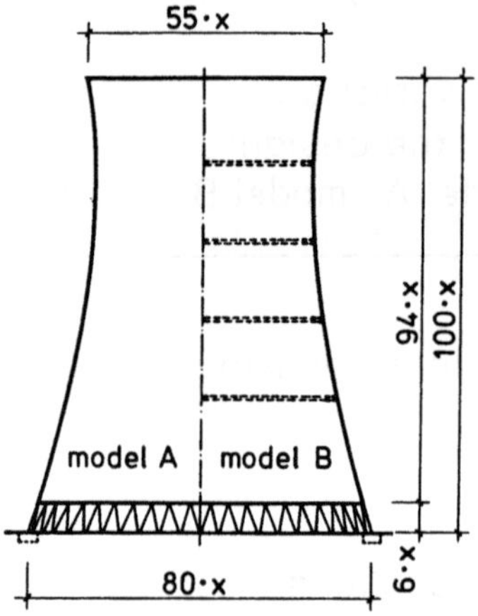

factor x	1,0	1,5	2,0	2,5	3,0
height	100	150	200	250	300
base Ø	80	120	160	200	240
minimum wall-thickness	0,16	0,16	0,20	0,24	0,28
height of the ringbeam	0,40	0,40	0,40	0,40	0,40
width of the ringbeam	1,50	1,50	1,50	3,30	5,20

all dimensions in [m]

Fig. 3: Variations of dimensions

Fig. 3 shows the geometrical variants subject to study.

Height, basis diameter, diameter at upper edge of shell and height of the column framework are distinguished in the different variants by a constant factor X. Minimum wall thickness, height and width of ring have been chosen with due consideration of all statical, structural and technical aspects. The stiffening rings have been arranged equidistantly.

2. Structural alternatives

model A - without stiffening elements
model B - with stiffening elements

The characteristics required for shells have already been outlined in short in the first chapter. The shell structure must be an enclosing and self-supporting element and it must be resistant against all kinds of atmospheric agents. With regard to integrating these immense structures optically into a power plant site they should also offer the possibility of appropriate colouring.

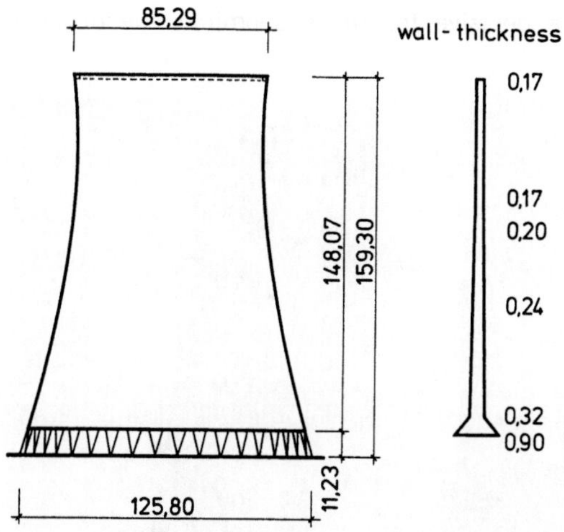

Fig. 4: Example for a model A natural-draft cooling tower: Voerde

The last two criteria can easily be put into practice if reinforced concrete is used as building material. Fig. 4 shows the construction principle referred to in the following text as the model A.

The shell structure consists mainly of the shell itself with a minimum wall thickness of 16 cm. The wall thickness over the height of the shell is chosen just as high as necessary to ensure the buckling safety required. The upper lintel required for statical and dynamical reasons should suitably be executed as a walkway. The dimensions resulting from this requirement are, in most cases, sufficient to obtain a minimum stiffness. The lower lintel must englobe the columns, a fact that automatically leads to a minimum thickness of about 70 cm. The lower part of the wall-thickness curve is then characterized by a continuous reduction of the wall thickness down to the dimensions required due to the buckling criterion.

The main parameters of the shell, such as the lower diameter, the upper diameter, the diameter of throat, the total height and the position of throat, are stipulated by the thermal design engineers. If the parameters mentioned are observed, the generating function can be chosen in accordance with statical requirements as well as with design and construction criteria. The construction criteria, i.e. the maximum gradient of shell, the minimum curvature radius and the ratio of the largest to the smallest diameter, depend to a small extent on the formwork system and are decisive for the economical execution of the structure.

Fig. 5: Natural-draft cooling tower Voerde

Fig. 5 shows a cooling tower of the model A. It was built in 1979-1980 and is part of the Voerde power plant.

Let us now consider the model B. As can be seen from fig. 6 this model is but a variant of the model A. For this reason the main constructional features are the same. Only the ring stiffness has been efficiently improved by the provision of discrete stiffening elements. The optimal arrangement, number and dimensions of the rings must be determined in consideration of their effect on the statical, dynamical and buckling behaviour. It seems suitable to choose the position, number and dimensions of the stiffening rings in such a way that the extension of buckling zones is reduced and that the modes are forced to cross the zero axis at the rings. If the mechanical behaviour can be successfully influenced in this way it will be possible to considerably reduce the wall thickness as compared with the model A. The smaller amounts of concrete, however, must now be paid for by considerably greater demands on the construction techniques employed for the rings. This will have to be taken into consideration when it comes to a comparison of costs in chapter 6. An additional, economical advantage can be gained if the rings can be used, where applicable, as a lighting platform for air traffic warning lights. In such case it would be possible to dispense with the costly walkways.

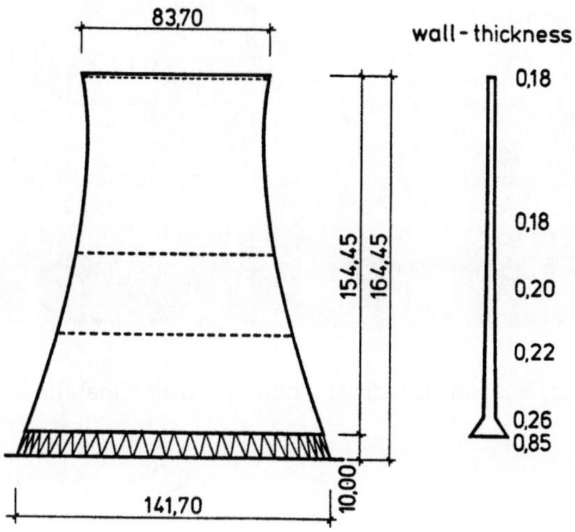

Fig. 6: Example for a model B natural-draft cooling tower: Doel III

Fig. 6 shows the arrangement of the rings on the shell's inner face. Such arrangement is chosen mainly for optical reasons. Moreover, thermo-dynamical calculations have shown that there is hardly any influence on the cooling tower's thermal efficiency. Fig. 7 shows a finished structure offering a view into the cooling tower of the Doel nuclear power plant in Belgium. To this structure only two rings were added so that, for obvious reasons, the resulting effect remained rather limited.

Fig. 7: Natural-draft cooling tower Doel III

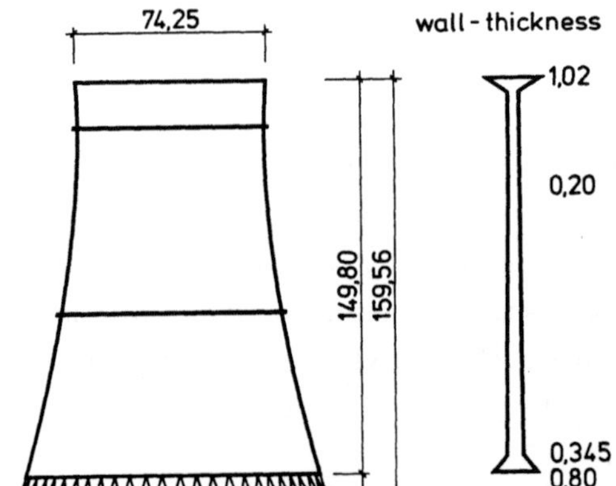

Fig. 8: Example for a model B natural-draft cooling tower: MIssissippi I

Fig. 8 shows a solution with the rings being arranged on the outer face of the shell. The corresponding statical and dynamical effect is similar to the one with the rings inside the shell.

Fig. 9: Natural-draft cooling tower Mississippi I

Fig. 9 shows a cooling tower in the USA belonging to the Mississippi Power and Light Company. This tower had been severely damaged during a tornado by a toppling tower crane. In order to maintain the existing structure it was decided to install the stiffening ring shown in the figure as a rehabilitation measure.

Fig. 10 shows three possibilities of realizing the rings with an economical construction method. The method of in-situ concreting was applied for the natural-draft cooling tower of the Doel nuclear power plant in Belgium. The achievement for a good and trouble-free connection between the shell and the ring may cause problems during the execution as it requires a more or less long interruption of the climbing work at the shell. A method using prefabricated elements has - as far as I know - not yet been put into practice until now. Besides the connection between the shell and the rings also the problem of connecting the individual prefabricated elements in the circumferential direction can be satisfactorily solved only in case of very low tolerances of the prefabricated elements and the shell.

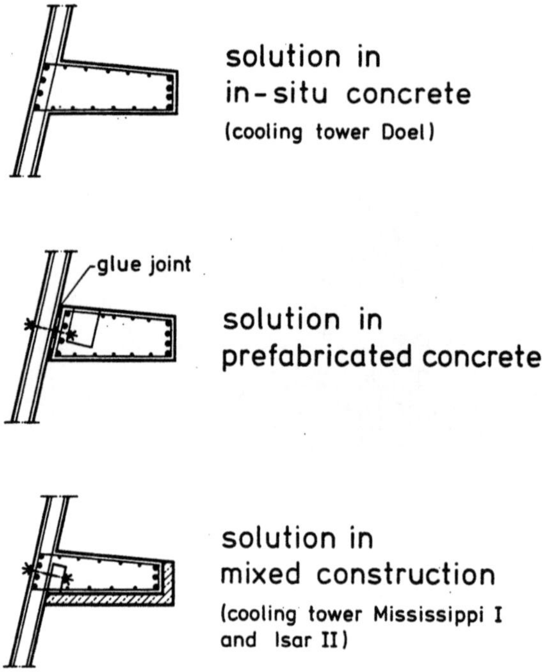

Fig. 10: Possibilities of executing the ringbeams

The least problematic is the mixed-construction method where the prefabricated lost-formwork elements of reinforced concrete are fastened to the shell independently of the climbing process. The connection between the shell and the ring reinforcement is made by bolts. The subsequently placed in-situ concrete ensures the good and shear-resistant bond of the elements to be connected.

The very positive experience gained from the works at the Mississippi I natural-draft cooling tower greatly influenced the decision to apply the same construction method also for the cooling tower of the Isar II plant.

3. Different states of stress in the shell structure of the model A and B towers

The state of stress in a shell is above all determined by the dead load and the wind load. The rotationally symmetrical case of dead load will be modified by the stiffening rings only to a negligible extent. The smaller wall thickness of the model B and the resulting lower dead load only lead to a generally lower level of stresses. As to the case of wind load, however, the modification due to the stiffening elements is quite considerable. The stiffening rings attract ordinary forces in the circumferential direction and there are also high ring bending moments. What seems surprising at first glance are the changes of the membrane state of stress which are a consequence of the anti-deformation effect in the ring direction. This results then in a levelling of the state of stress with a reduction of the stress peaks. Fig. 11 shows some of the important stresses in case of wind load.

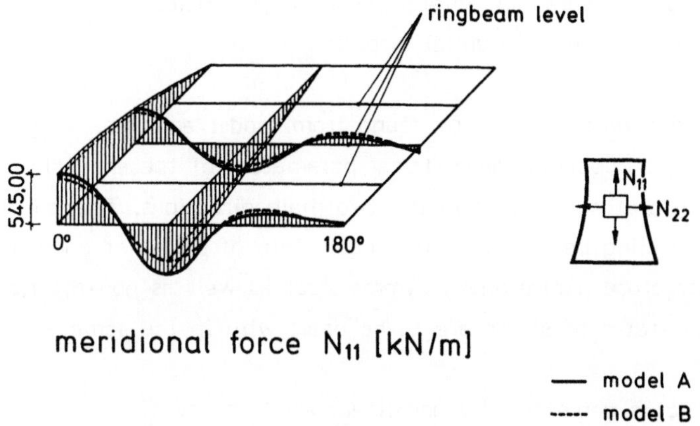

Fig. 11: Membrane forces for wind load

The state of stress indicated was found by means of the linear theory of elasticity. Calculations carried out on the basis of a physically non-linear theory, which takes a realistic material law for reinforced concrete into account, partly leads to considerably deviating membrane forces (cf. [2]). However, as the whole safety philosophy is based in the first instance on the linear theory, this more rigid approach will be left out of consideration.

4. Buckling behaviour

The correct evaluation of the buckling behaviour is decisive in order to harmonize economical aspects with the necessity of having sufficient levels of safety. In the last few years three approaches using different degrees of approximation for the true or realistic buckling behaviour were applied. [3], [4], [5].

a. Application of a linear, classical buckling theory for a rotationally symmetrical model case of load

This rotationally symmetrical load model must deal with the rotationally symmetrical load cases of dead load and inner suction including the non-rotationally symmetrical wind load on the shell's outer face. The realistic buckling value sought for is to be derived from the reducing of the classical buckling value. The reduction factor itself must be found by evaluating appropriate model analyses.

b. Application of a diagram of interaction between critical membrane forces in meridional and circumferential direction

The interaction diagrams can be found from model analyses. They must include the most important geometrical parameters of the shell structure and the different states of stress in the two main directions. This method reduces the buckling behaviour to the local state of stress at each point of the shell structure. Rotationally symmetrical as well as non-rotationally symmetrical states of stress are to be dealt with in the same way.

c. Application of a geometrically non-linear shell theory

This problem can only be solved by means of a very extensive numerical calculation procedure. Theoretically, it can be applied to rotationally symmetrical as well as non-rotationally symmetrical load cases. In practice, however, the very extensive non-rotationally symmetrical calculation is applied here as well in order to establish a factor for the results of a rotationally symmetrical calculation. Consequently, the application as such extends only to a rotationally symmetrical model load case and corrects the result found by means of a previously established reduction factor.

In practice there have also been combinations of the possibilities discussed depending on the precision requirements to be fulfilled. Another often preferred method consists of always starting from the results of a classical buckling theory in a similar way as outlined under a). These results can be found quickly for the rotationally symmetrical load case and offer the possibility of carrying out certain parameter studies in a simple way. A reduction factor can then be found as in b) or even with regard to c). This simple approach was also used for the following considerations.
Fig. 12 gives the classical buckling value of an unstiffened cooling-tower shell. The same cooling-tower shell with an identical wall-thickness curve was then additionally stiffened by means of four rings. The resulting buckling values are

equally shown in fig. 12. If, in a first approach, the reduction factor from the classical buckling safety down to a realistic buckling safety is chosen the same for the stiffened and for the unstiffened shell, we obtain an additional gain of buckling safety. By reducing the wall thickness of the stiffened shell the buckling level of the stiffened shell can now be brought down to the required level.

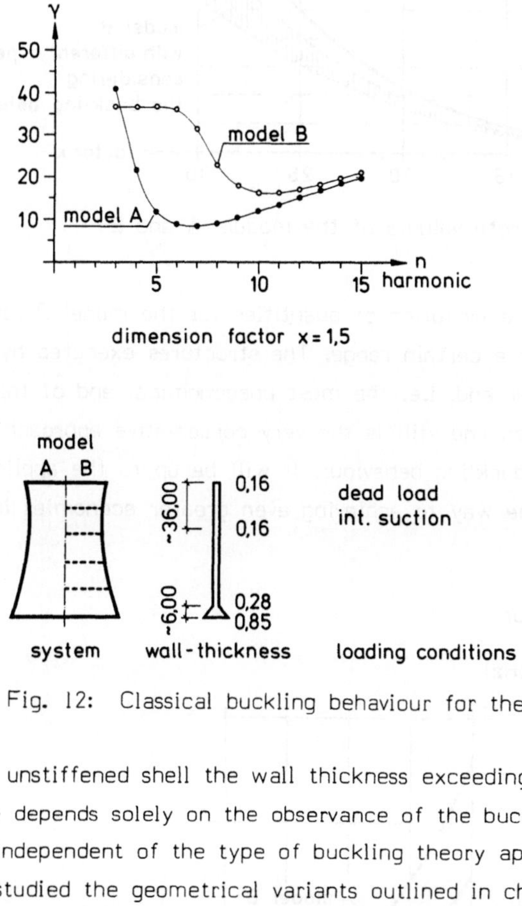

Fig. 12: Classical buckling behaviour for the models A and B

In case of the unstiffened shell the wall thickness exceeding the structurally required measure depends solely on the observance of the buckling criterion. This is, of course, independent of the type of buckling theory applied. In a series analysis we have studied the geometrical variants outlined in chapter 1, determined the required concrete quantities and incorporated the results in the diagram of fig. 13 where the over-proportionate increase of the concrete quantities is clearly shown. This behaviour is indicative of a shell becoming more uneconomical with increasing dimensions. The evolution of quantities for ring-stiffened shells can also be seen in this diagram where the increase is clearly less marked than in the first case. The extent to which such smaller increase of quantities is reflected in the prices will be discussed in chapter 6. In anticipation of this discussion it can, however, already be said that this over-proportionate rise in the quantities alone justifies the necessity of providing for ring-stiffened shells.

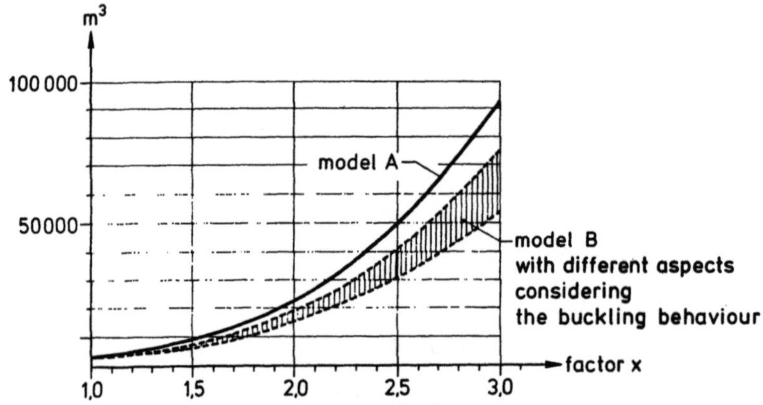

Fig. 13: Concrete volume of the models A and B

As seen from today, the evolution of quantities for the model B can only be indicated as lying within a certain range. The structures executed by us until now always lay at the upper end, i.e. the most uneconomical end of this range. The reason for this has been and still is the very conservative approach applied in the evaluation of the buckling behaviour. It will be up to the application-oriented researchers to show the way to achieving even greater economies in this field.

5. Dynamical behaviour

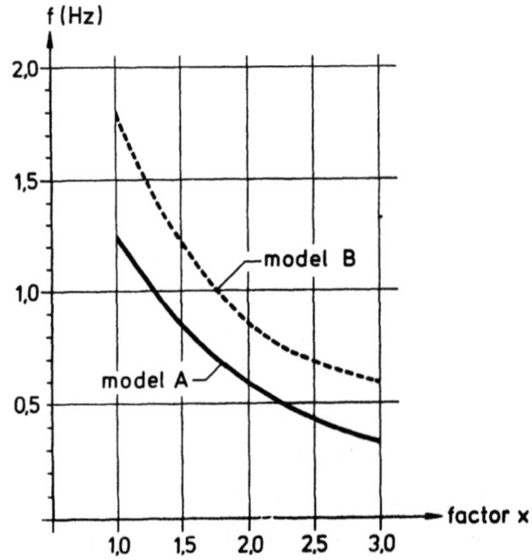

Fig. 14: Lowest eigenfrequency of the models A and B

Considering the main excitation spectra of the natural wind, cooling towers are structures of high natural frequencies. Fully dynamical calculations - regardless of whether these are carried out on a time-history basis or in accordance with a response-spectra method - are possible but generally not required. The procedure takes a dynamical amplification factor for the wind loads into account which depends mainly on the lowest natural frequency and the shell diameter. A prerequisite for this procedure, however, is the existence of a minimum natural frequency that enables us to treat cooling towers with good approximation as structures with high natural frequencies. The lowest natural frequency will be in the order of 0.5 Hz.

If the complete series of different cooling-tower sizes defined in chapter 1 is analyzed and if the results are compared in a diagram, it will be seen that the lowest natural frequencies strongly decline with increasing geometrical dimensions. Special emphasis is put on the fact that for all cooling towers analyzed a minimum buckling safety has always been observed. This diagram clearly shows the technical necessity of considering ring-stiffened shell constructions more closely. Even if one tried to raise the shell stiffness by increasing the wall thickness in order to obtain a higher natural frequency, this would be counteracted by the simultaneous increase of the masses. The lowest natural frequencies of the stiffened shells are therefore indicative of a much more favourable behaviour. The curve shown in fig. 14 contains the decisive hint as to the means by which the natural frequency can be influenced.

6. Cost comparison between the presented shell models A and B

The criteria of buckling safety and minimum natural frequency discussed in chapters 4 and 5 point out two different directions where the application of a ring-stiffened shell is compulsory. In this chapter we shall perform a cost comparison where - for obvious reasons - only cooling towers with a comparable level of buckling safety and a sufficiently warranted dynamical behaviour can be taken into consideration. The calculation of the costs is based on the price level as per January 1984 and the Ruhr area as the chosen site. The simplified calculation accounts only for such cost parameters as concrete, steel, formwork and scaffolding. Other items such as site installation, commissioning of equipment and structural design have been allotted to the main items. The shell structure including the column framework and individual shallow foundations was also considered in the calculation. The natural ground was assumed to have an admissible bearing capacity of 1 N/mm² immediately beneath the structure. No earthworks and no dewatering measures were considered. This applies also to possible coating works of the shell structure.

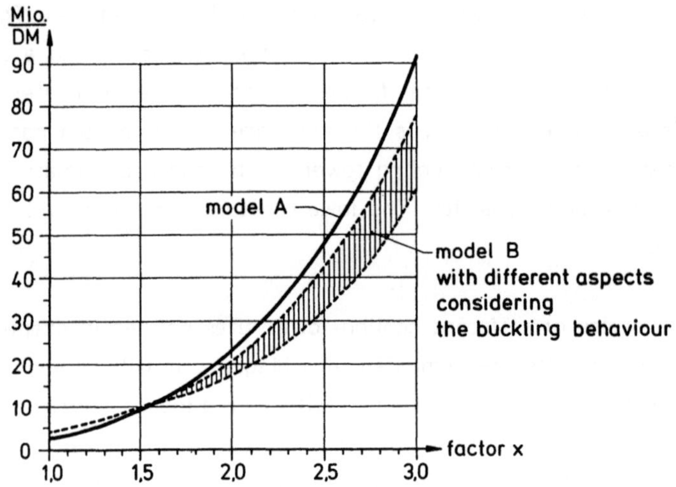

Fig. 15: Cost estimation for models A and B incl. supports and foundation (1/84)

Fig. 15 shows the cost comparison carried out. As already mentioned in former investigations the unstiffened cooling tower is the most economical solution for heights of up to 160 m. From a height of about 160 m onwards the ring-stiffened variant is more economical. It should be remembered that we are faced here merely with a parameter study and that the point of intersection found for the economy is only a first estimate. The definite decision for the shell type to be chosen can only be taken after close consideration of the detailed calculation with all the other manifold parameters of influence.

Once again we are faced here with the wide range of costs in case of ring-stiffened constructions. A more conscious approach to the buckling behaviour of ring-stiffened shell constructions may be helpful in saving a lot of money without reducing the safety to an irresponsible extent.

7. Conclusions

Natural-draft cooling towers of reinforced concrete have been existing now for about 70 years. The geometrical dimensions of the towers built until now generally were such that conventional construction techniques without stiffened shells could be applied. However, in view of the constantly increasing power plant ratings and with the change from the wet to the dry-cooling techniques the dimensional limits beyond which only ring-stiffened constructions are suitable because of otherwise too low natural frequencies are quickly exceeded.

Nevertheless, the question of the most economical construction method becomes acute already with today's structures of about 170 m of height and a basis diameter of about 160 m. Thus, ring-stiffened shells were executed already in three cases and under the author's responsibility which - from the economical and technical point of view - proved fully satisfactory.

The large geometrical dimensions required already today and, possibly, even more so in the future by the plant techniques pose a challenge to the competing concrete, steel structure and cable truss constructions. Concrete can furnish the proof that it is a material able to meet the challenge of even greater dimensions. It will be up to a number of further theoretical studies to pave the way for the continued triumphant advance of these gigantic reinforced-concrete structures.

As a conclusion it can be stated that ring-stiffened shell constructions are not only a structural alternative but, beyond a certain size, also an economical and, for even larger dimensions, a statical and dynamical necessity as well.

References:

[1] Lehmkämper, O.:
Versteifte Kühlturmschalen aus Stahlbeton
Technisch-Wissenschaftliche Mitteilungen Nr. 78-6 des Instituts für Konstruktiven Ingenieurbau der Ruhr-Universität Bochum

[2] Form, J. and Peters, H.L.:
Economical aspects of design and execution of large natural-draft cooling towers
International Symposium on very tall reinforced concrete Cooling Towers, 8.-10. November 1978, Paris

[3] Form, J., Krings, W., Mazur, H., Peters, H.L.:
Berechnung und Ausführung eines ringversteiften Naturzugkühlturms aus Stahlbeton
Beton- und Stahlbetonbau, Heft 9/1980

[4] Form, J., Krätzig, W.B., Peters, H.L., Wittek, U.:
Ringversteifte Naturzugkühltürme aus Stahlbeton
Bauingenieur 1984, in preparation

[5] Zerna, W. and Mungan, I.:
Über das Beulen von Kühlturmschalen mit Versteifungsringen
Beton- und Stahlbetonbau 76 (1981), Heft 2

Large Reticulated Steel Cooling Towers

L. KOLLÁR
Budapest City Council's Architectural
and Town Planning Office,
Budapest, Hungary

Summary

The paper reports on reticulated steel cooling towers of 120 m height, designed in Hungary, and erected in Armenia /Soviet Union/ and in Iran. Structural details and construction methods are presented, and the special reasons for choosing a steel structure will be elucidated. There will also be shown how steel structure can be used to build cooling towers of even larger dimensions.

1 Why steel?

Cooling towers are mostly made of reinforced concrete as thin shells. Given the wide-spread opinion that r.c. structures are, as a rule, more economic than steel ones, at first glance it does not seem justified to build cooling towers of steel. Thus, we first have to investigate what circumstances can render a reticulated steel cooling tower competitive as compared with r.c. ones.

If the tower has to be built in a seismic zone and one has to take into account elevated seismic forces, substantial advantages can be achieved by reducing the own weight of the tower, since seismic forces are /roughly/ proportional to own weight. Harsh climatic conditions, hindering the pouring of the concrete over most part of the year or necessitating costly protection measures call for some kind of prefabrication.

Protection against corrosion and continuous maintenance costs are also considered as economic disadvantages of steel structures. The Hungarian dry cooling system Heller-Forgó /Energiagazdálkodási Intézet, Budapest/, where cooling water circulates in a closed pipeline system, ensures that the steel structure is only exposed to warm, dry air from the inside. From outside it is protected anyway by the cladding. Consequently, no special protection against corrosion is necessary.

There is an additional advantage of static nature of the steel structure which will be treated in Sect 3.

The circumstances mentioned so far all improve the economy of steel structure. In order to become competitive, however, a proper structure and a suitable erection method had also to be developed.

2 Structure and erection method of reticulated steel cooling towers

The development of the steel structure is due to I. Marosi, engineer of the Design Office for Industrial Buildings IPARTERV in Budapest. He chose a comparatively thin reticulated cylinder and stiffened it by three rings in the inside of the tower. These rings, being also reticulated, did not hinder the flow of the air.

For the erection of the structure he chose the principle of manufacturing individual bars in the factory, connecting them on the site to panels, lifting them onto place and assembling them by welding.

The tower had to be able to resist earthquakes also during erection. This requirement was met by using the upper stiffening ring as a working platform, which was always lifted to the upper edge as the tower grew, and which thus served as a permanent upper stiffening ring for the tower being built. The stiffening rings were lifted by hydraulic jacks supported on the assembled panels, like the sliding shuttering of r.c. cooling towers. On the upper stiffening ring an assembly crane of 3,5 t capacity was installed, which moved around the ring and could lift the panels onto place.

3 Static problems

The bar forces of the structure are to be determined by computer, using some standard procedure. However, in order to assess the overall static behaviour of the reticulated tower, we may consider its statically equivalent continuum, which is a cylindrical shell strengthened by stiffening rings [3]. These rings substantially reduce the ring-directed bending moments in the shell, so that this latter receives almost exclusively membrane forces.

The equivalent continuum is also useful to investigate overall stability and vibration of the structure. When analysing buckling behaviour we have to consider that the individual bars of the reticulated structure are trusses themselves, whose deformation due to transverse shear also has to be taken into account. We thus have to deal with a "sandwich shell" [2]. The buckling of the individual bars is to be analysed in the usual way. The interaction between overall and individual bar buckling can be assessed according to [1].

When investigating the dynamic behaviour of the cooling tower under wind gusts, we can make use of the results of analysis of Krätzig [4], who found that no special dynamic analysis is necessary if all eigenfrequencies of the tower are higher than 1 Hz and the tower can carry 1.5 times the basic wind load in the plastic state, i.e. taking its ultimate load bearing capacity into account. This latter requirement necessitates additional reinforcement in concrete towers, since the maximum axial tensile force is obtained as the –comparatively small-difference between the tension due to wind and the compression due to own weight. Thus the increase of the tension due to wind by 50% causes a much larger increase in the resulting axial tensile force. On the other hand, the bar sections of a reticulated steel tower are determined on the basis of maximum axial compressive force in the tower, which is the sum of the compressions due to wind and to own weight. Hence an increase of 50% in the wind load causes a substantially lower increase in total compression, which is, as a rule, less than the difference of yield /or buckling/ stress and the allowable stress. Consequently, the loading case 1.5 times basic wind load generally does not necessitate any excess material in steel towers. This circumstance represents an additional economic advantage of steel cooling towers.

4 Description of erected towers

4.1 Towers built in Rasdan /Armenia, SU/

In the years 1970 to 1975 four steel cooling towers were built in Armenia /SU/, near the town Rasdan /Fig 1/. To our

Fig 1

knowledge these were the first such structures with a height over 100 m built in the world. The power station had a capacity of 4x200 MW, the dry cooling system applied was that of Heller-Forgó. The height of the building site was 1700 m above sea level, in a seismic zone. Maximum and minimum temperatures were +35°C and -32°C respectively. The maximum wind velocity was 32 m/s at 10 m height.

The main load-bearing structure of the tower consisted of a 24-sided reticulated prism with a diameter of 60 m and a height of 120 m. The cooling deltas /System Forgó/ of the air-cooled condensation equipment /System Heller/ were arranged around this prism, so that they had to be protected by a pyramid-shaped roof with a lower diameter of 110 m and a height of 50 m, where this roof joined the prism /Fig 2/. The prismatic main skeleton of the tower was stiffened by reticulated rigid rings at the levels 30, 70 and 110 m. The main skeleton of the tower was assembled of rectangular panels 8 by 10 m consisting of reticulated bars. The individual bars of the panels were trusses themselves, consisting of four angle sections as chords and four sets of bracing bars. The graduation of the bars was accomplished by choosing stronger or weaker angle sections while keeping the contour of the whole bar unchanged. Thus a mass production with the aid of a few master patterns became possible. These individual bars were manufactured in Hungary and transported to Armenia by train. The material used was Steel A 37 B /identical with RSt 37 according to DIN-17100; or with 40 B according to BS 4360/, which has a sufficient resistance against brittle fracture at low temperatures. The steel skeleton was protected against corrosion by painting it twice by KATEPOX, an artificial resin, which also prevents electro-corrosion at points where the aluminium cladding joins to the steel structure.

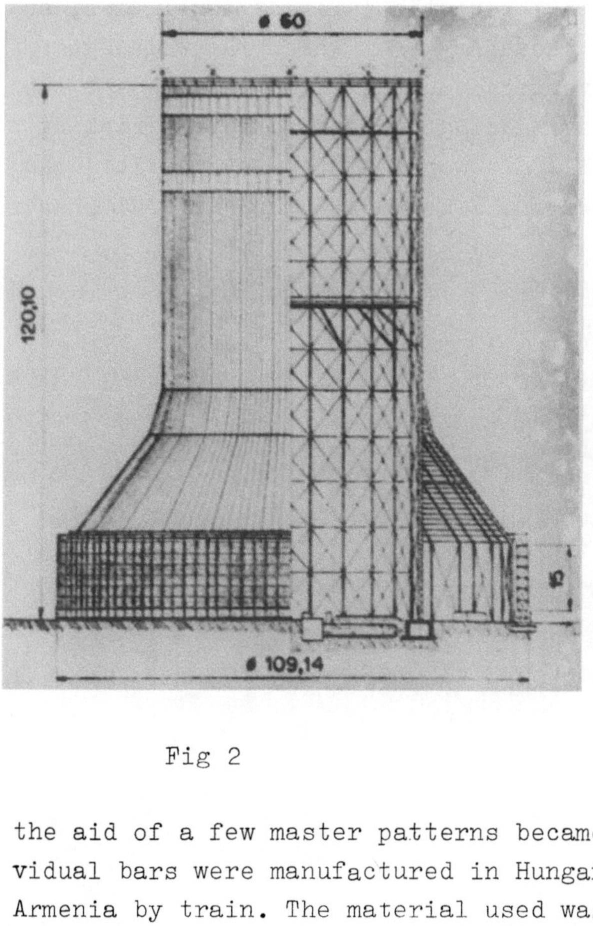

Fig 2

The erection went essentially as described in Sect 2. The two upper stiffening rings were lifted in about six hours by the height of a panel row /10 m/. The 8x10 m panels with cladding were transported into the interior of the tower and lifted onto place by the assembly crane which moved around the stiffening ring. The panels were fastened together by welding some batten plates to the two vertical half bar sections belonging to the neighbouring panels. Due to the double layer of panel bars, an overhanging falsework was necessary from which the welding of the batten plates of the vertical chords could be made and the still missing strip of the cladding replaced. After finishing the assembly, the crane, the falseworks and the hydraulic jacks were lowered by a simple winch. Figs 3 to 7 show several phases of the erection.

Fig 3

The structure of the pyramidal roof over the cooling deltas consists of skew main girders, whose outer ends are supported by solitary foundations along a circle of 106 diameter. There are two such main truss girders joining each edge of the prismatic skeleton. Horizontal purlins, stiffened by wind bracing,

Fig 4

carry the roofing. The structure of this pyramidal roof were lifted onto place by a crane of 50 m hoisting capacity. The lowest stiffening ring was also lifted in pieces by this crane to its final height 30 m. It is placed outside the skeleton, so that it also serves for supporting the main girders of the pyramidal roof.

Fig 5

Fig 6

Fig 7

For one cooling tower in Rasdan there were used for the foundation 4000 m^3 reinforced concrete; for the skeleton 1250 tons of steel; for the cladding 125 tons of corrugated aluminium sheets. The erection times were: earthwork and foundation 3 months; steel structure with cladding 6 months; cooling deltas 2 months; totalling 11 months.

4.2 Cooling towers in Esfahan /Iran/

The experiences obtained during design of the cooling towers in Rasdan made other, more economic solutions possible.

Instead of a prismatic skeleton all along the height and an extra outside pyramidal roof, we applied a pyramidal structure from ground level up to 49 m height, and only from here on a prismatic structure. In this way we substantially reduced the bar forces in the lower part; moreover, instead of two foundation systems, there was only one foundation ring necessary /Fig 8/. The panels became triangular instead of rectangles with diagonals. Consequently, they had to be welded together at the corners only, instead of all along the sides /Fig 9/.

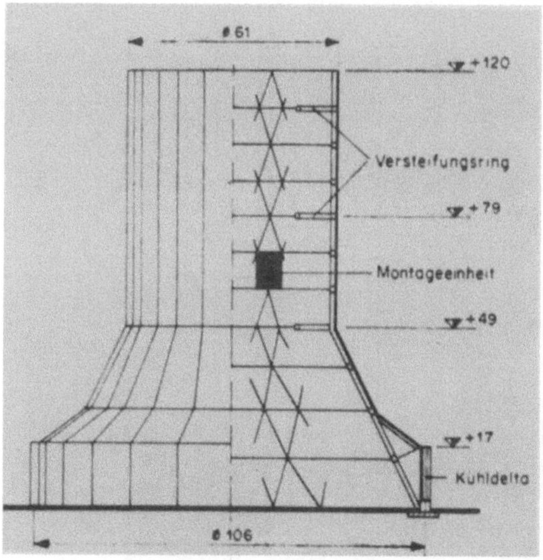

By these two modifications we could have saved about 25% of steel if the wind load had been the same as in Rasdan. However, since we had to take a higher wind load into account here, we saved 15% of steel and 48% of reinforced concrete as compared with the cooling towers in Rasdan. In addition, the labour necessary on site was also reduced by 30%, and there was no need of welding or cladding work outside the tower.

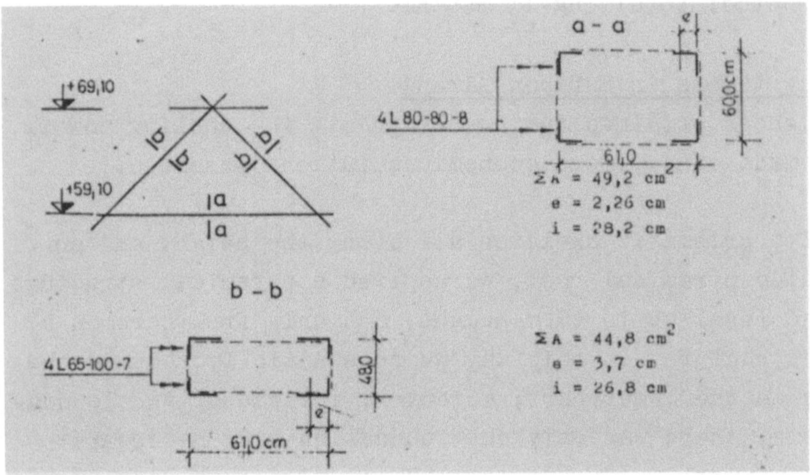

Fig 9

Due to the lower, pyramid-shaped part of the skeleton, the stiffening rings could not be lifted here step by step, simultaneously with the assembly of the panel rows, but had to be lifted directly to the upper edge of the pyramid after this has been assembled. The pyramid was mounted with the aid of a

Fig 10

crane with 50 m hoisting capacity and of a skew assembly platform of 49 m height which could roll around the circumference. The erection of the prismatic part of the tower was carried out in a similar way as with the Rasdan towers. In Figs 10 to 12 we show some phases of the erection of the Esfahan cooling towers.

Fig 11

Fig 12

5 Further developments

At present, further cooling towers are being designed for Iran with the height of 160 m and the diameter of 80 m. With these dimensions we reached the upper limit that can be realized by the structural solution described earlier. That is, in order to increase the load bearing capacity of individual bars we would in vain choose a larger side length for the surrounding cross section rectangle. The slenderness ratio of the bar trusses was about 60 even in the case of the towers described in Sect 4, so that no substantial increase of strength could be achieved by pulling them further apart. We can only increase the cross section of the angle sections. With these towers of 160 m height some bars needed the maximum available angle sections. Thus, to increase the tower height even further, some other bar buildup should be chosen.

References

1. Forman, S.E.–Hutchinson, J.W.: Buckling of Reticulated Shell Structures, Int. Journ. Solids Struct. 6 /1970/ 909-932.

2. Kollár, L.–Dulácska, E.: Buckling of Shells for Engineers, John Wiley and Sons, Chichester/New York–Akadémiai Kiadó, Budapest, 1984.

3. Kollár, L.–Hegedüs, I.: Analysis and Design of Space Frames by the Continuum Method. Elsevier, Amsterdam–Akadémiai Kiadó, Budapest, to be published in 1985.

4. Krätzig, W.B.: Grosse Naturzug-Kühltürme aus Stahlbeton – Derzeitiger Stand und zukünftige Entwicklung. VGB Kraftwerkstechnik 1970. 191-197.

Design and Construction of Cooling Towers – Recent Developments – Constructional Problems

HANS DAMJAKOB

BALCKE-DÜRR AG
HOMBERGER STRASSE 2
D-4030 RATINGEN 1

Summary

Design of cooling towers means the cooperation of engineers from various disciplines in finding an adequate solution to a cooling problem. The solutions found can be very different depending on the various requirements and limiting conditions as stipulated in the specification. On the basis of current projects a survey is given on recent developments in cooling tower design and construction covering both mechanical draft and natural draft cooling towers: Multi-Fan cooling towers, cooling towers with integrated desulfurizing plant, wet cooling towers with special founding solution, dry cooling towers with large shell diameters, forced draft wet/dry cooling towers.

Fig. 1. Natural draft cooling towers under construction at Lethabo Power Station site, RSA.

General

From the point of view of cooling tower companies the d e s i g n o f c o o l i n g t o w e r s means the cooperation of engineers from various disciplines such as thermodynamics, flow mechanics, mechanical, electrical, acoustical, chemical, structural and civil engineering, i.e. cooperation in finding a solution to a cooling problem towards the unique and specific requirements of each power generating plant. This paper is concerned only with cooling towers for power stations.

All cooling towers described in this paper are currently under construction (1984) or have recently been completed. In the case of all the projects the competence and responsibility for both design and construction including the civil part are in one hand which satisfies all the requirements of the complex heat exchanger, cooling tower. For some of the cooling towers described feasibility studies were carried out in a preliminary design phase.

The solutions with regard to the cooling system to be chosen can be very different depending on the various specific requirements and limiting conditions as laid down in the specification. In this sense the cooling towers presented below represent r e c e n t d e v e l o p m e n t s in cooling technology.

Fig. 2. Construction of diagonals for the cooling towers at Lethabo Power Station site, RSA. The columns are cast in situ together with the shell rim using a mobile formwork system.

Multi-Fan Cooling Towers

The decision as to whether a natural draft or mechanical draft cooling tower is to be used depends on the values given in the specification for power consumption and investment cost and the local conditions. Regarding the cooling tower shown in Fig. 3 the decision was taken in favour of mechanical draft cooling towers. The fans are arranged over the fill area (counterflow principle) to provide equal fan coverage and uniform air flow for optimal thermal performance. Its main dimensions are a diameter of 74 m and a height of 19 m. The tower is totally of precast concrete.

Fig. 3. Zurn Round Multi-Fan Mechanical Draft Cooling Tower. Zurn cooling towers are constructed in the USA under licence from Balcke-Dürr AG.

Cooling Tower with Integrated Desulfurizing Plant

A study carried out in 1978 describes the feasibility of installing a flue gas desulfurizing plant within a cooling tower as shown in Fig. 4. The desulfurized flue gases shall be mixed with the cooling tower exhaust air and discharged into the atmosphere. The power station concerned does not have a chimney. Due to the fact that the CO_2 content of the cooling tower exhaust air/desulfurized flue gas mixture is considerably higher than that of normal air, at least the inner surface of the shell has to be protected so as to

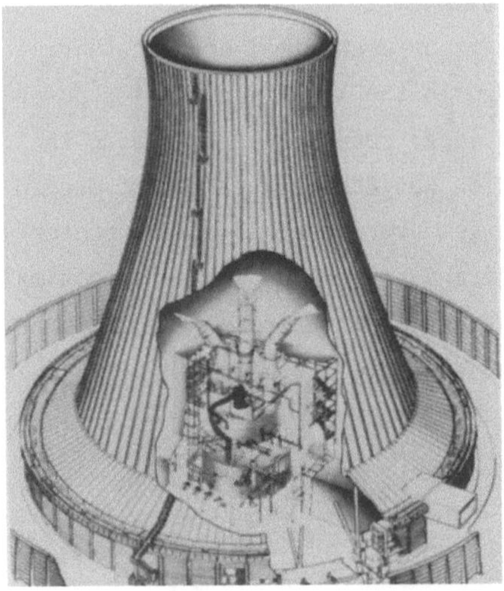

Fig. 4. Natural draft wet-type cooling tower with integrated flue gas desulfurizing plant (System Saarberg/Hölter) for the 230 MWe Völklingen Prototype Power Station of Saarbergwerke AG, FRG.

Fig. 5. Model used to optimize the mixing of cleaned flue gas with cooling tower exhaust air with the Balcke-Dürr gas mixing system.

keep the carbonatisation of the concrete within acceptable limits. On the basis of the above feasibility study the cooling tower with integrated desulfurizing plant was designed and built and put into operation in 1982. The following products of Messrs. MC-Bauchemie, FRG, were used to protect the shell concrete: Immediately after stripping of shuttering MC 1277 WV for the inner surface and MC 702 for the outer surface. After completion of the shell MC DUR VS was applied to the inner surface and MC 702 to the outer surface using a hanging scaffolding system.

Apart from other technical concepts alternative solutions will be realized in such a way that the desulfurizing plant will be located outside the cooling tower the desulfurized flue gas being directed into the cooling tower via channels which pass through the tower shell. The appropriate openings in the shell are approximately 100 m^2.

The decision in favour of a particular concept is dependent on the evaluation of all local conditions concerned. In any case it is important to achieve an efficient and carefully directed mixing of the different gas flows which can be studied on a model as illustrated in Fig. 5.

Wet Cooling Tower with Special Founding Solution

Special soil conditions have to be taken into account for the natural draft cooling towers shown in Fig. 1 and Fig. 6. Each shell is founded on 36 large bored piles. Due to the high dynamic wind pressure of 2910 N/m^2 these piles have not only to be designed for compression but must also be capable of taking high tension forces. In the upper pile area there are layers of soil which will expand by up to 100 mm in the course of time due to absorption of surface water. Thus the piles are subjected to additional tension forces as a result of such heaving of the ground.

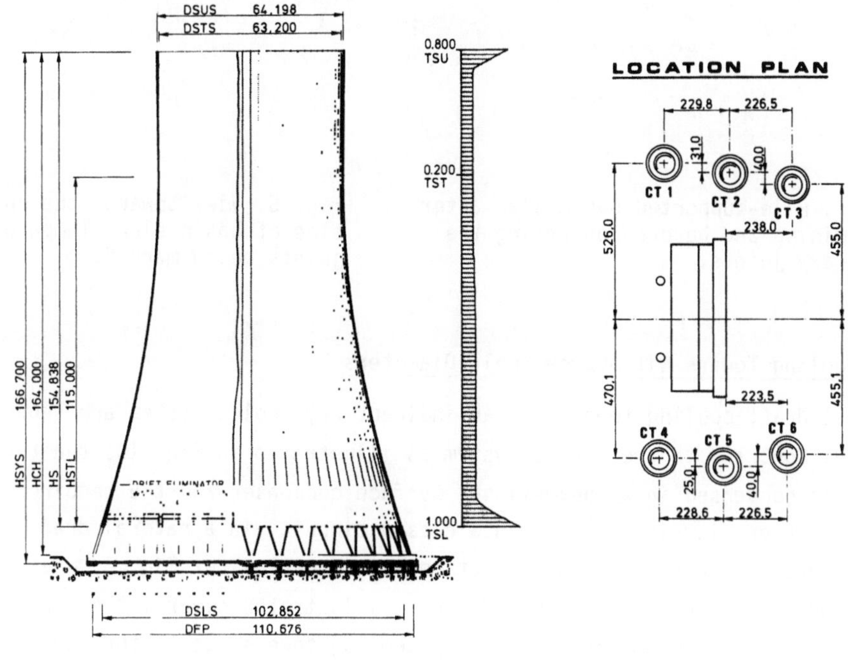

Fig. 6. Natural draft, wet-type counterflow cooling towers with pile-supported shell and basin slab for the 6 x 600 MWe Lethabo Power Station, RSA.

Vertical shell piles had to be chosen instead of battered piles which would be the normal and more economic solution. The basin slab is connected to the ring beam foundation in order to absorb the balanced horizontal/radial forces due to deadload of the shell and to distribute the unbalanced horizontal forces due to wind load over a large number of piles. Both the ring beam foundation and the basin slab will be underminded shortly after concreting has taken place in order to keep the foundation and the basin slab free from soil pressure due to heaving of the ground. In this connection please refer to Fig. 7. The temporary joints as shown in Fig. 8 will be concreted before a certain shell height is reached.

Fig. 7. Pile-supported basin slab after undermining and before concreting the temporary joints.

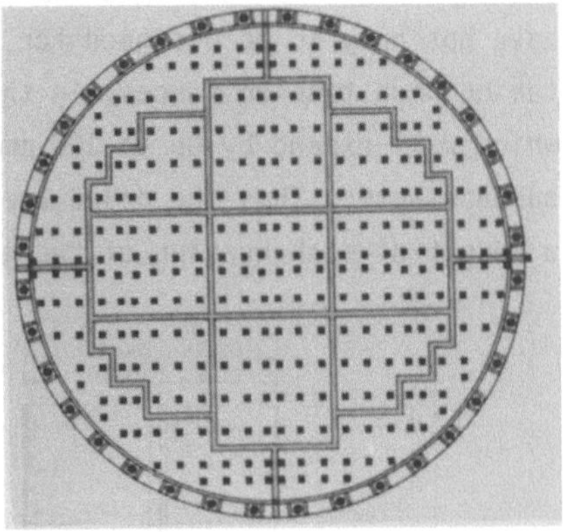

Fig. 8. View towards the underside of basin slab. Temporary joints being marked.

Dry Cooling Towers with Large Shell Diameters

Natural draft cooling towers for an indirect dry cooled system are shown in Fig. 9. Contrary to the direct system as illustrated in Fig. 10, the turbine steam is condensed in a conventional surface condenser and the required cooling water flow is recooled in a closed circuit via a natural draft cooling tower. A comparison with the wet cooling tower as per Fig. 6, whereby both cooling towers serve a 600 MWe unit, illustrates clearly the considerably larger shell diameters of the dry cooling towers due to the thermodynamic conditions. The dynamic wind pressure for which these cooling towers have to be designed is 3800 N/m^2. The hyperbolic reinforced concrete shell was found to be the adequate solution to the structural problem.

Alternative concepts for large shell structures are available as for example the cable-net cooling tower as per Fig. 11 or steel framework structures. The decision as to which concept, i.e. concrete shell or the above alternative concepts, is applied depends on the evaluation of all relevant criteria, e.g. extreme seismic conditions.

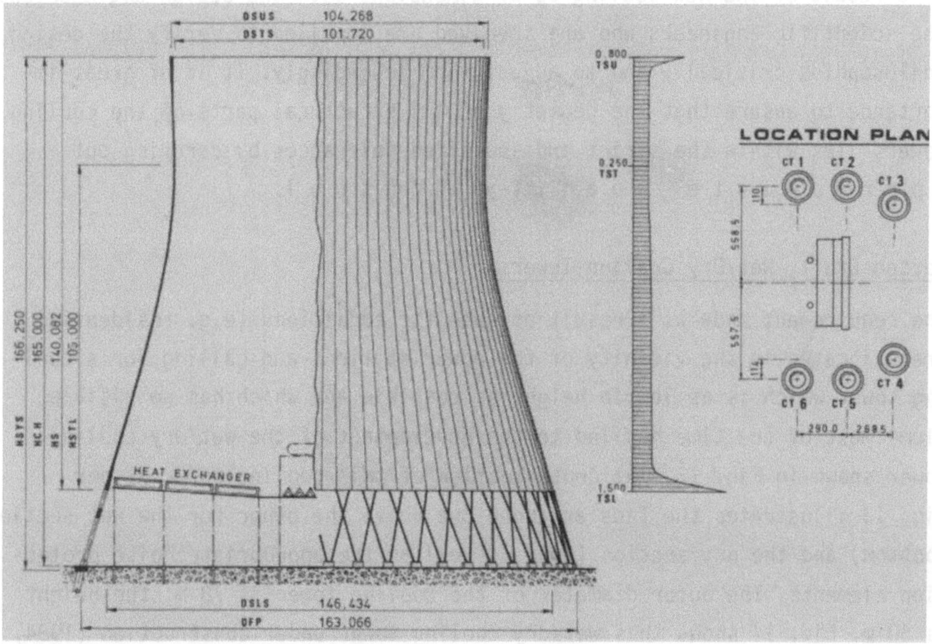

Fig. 9. Natural draft dry cooling towers for the indirect dry cooled system of the 6 x 672 MWe Kendal Power Station, RSA.

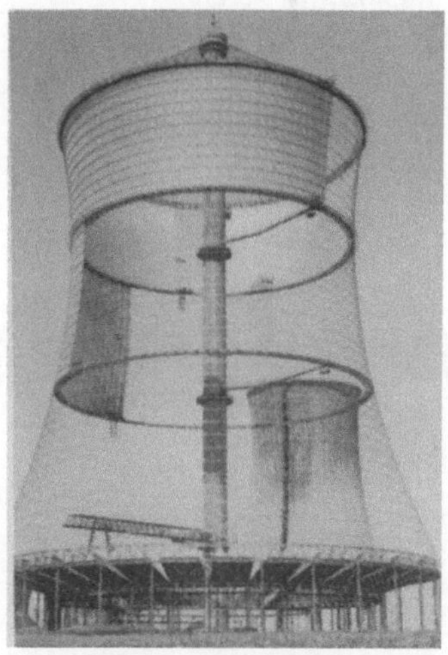

Fig. 10. Air-cooled condenser (direct system) for the 4 x 150 MWe Mashad Power Station, Iran.

Fig. 11. Cable-net natural draft cooling tower for the Schmehausen Nuclear Power Station, FRG under construction in 1976. Height of pylon 181 m, height of shell 146 m, diameter at the base 141 m.

As a result of the new scaling up of dimensions both the design engineer and the scientific engineers who are involved are impelled to verify the design philosophies critically and to adjust them accordingly. It is of great importance to ensure that the geometry of all structural parts of the cooling towers lies within the strict and specified tolerances by carrying out appropriate quality control.

Forced Draft, Wet/Dry Cooling Towers

The requirement made as a result of specific conditions (e.g. residential areas located in the vicinity of the power station) and calling for a cooling tower which is as low in height as possible and which has no visible plume most of the time has led to the development of the wet/dry cooling tower shown in Fig. 12. The cross-section of this cooling tower as per Fig. 13 illustrates the fans arranged one above the other for the wet section (bottom) and the dry section (top) as well as the appropriate noise protection elements. The outer diameter of the cooling tower is 78 m; the height is 50 m. Fig. 14 shows this wet/dry cooling tower under construction (1984).

Fig. 12. Model of the 420 MWe Altbach V Power Station, FRG, showing the Balcke-Dürr forced draft, wet/dry cooling tower on the left.

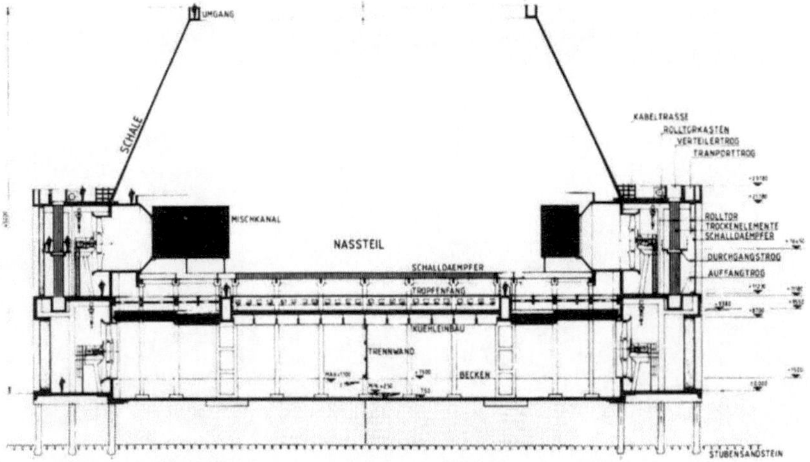

Fig. 13. Cross-section of the wet/dry cooling tower serving the 420 MWe Altbach V Power Station, FRG.

Fig. 14. Wet/dry cooling tower Altbach V under construction.

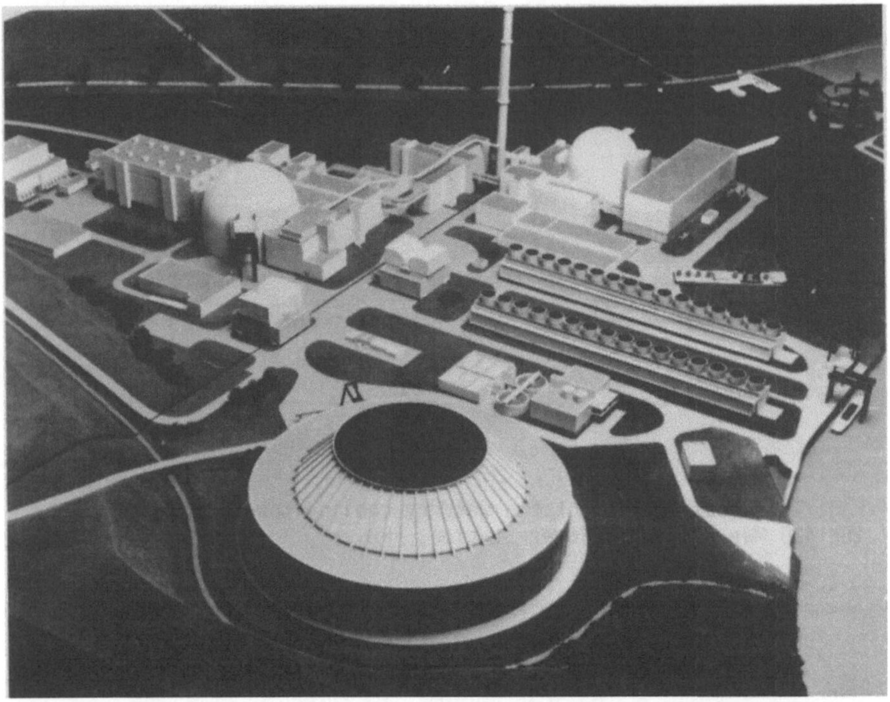

Fig. 15. Model of the nuclear power plants GKN I (right) and GKN II (left), FRG. The Balcke-Dürr forced draft, round, wet/dry cooling tower for the 1300 MWe GKN II unit is shown at the bottom of the picture.

Fig. 15 shows a further development on this wet/dry cooling tower with an outer diameter of 160 m and a height of 50 m. Construction on site has started in 1984.

Conclusion

As can be seen from the examples presented above, optimal cooling systems can be developed for any specific requirement and location. Significant impulses for new developments result from the increasing demand for environmental protection, increase in capacity and thermal efficiency as well as reduction in operating cost. The large shells for natural draft cooling towers show the necessity for continuous scientific research on all structural parts concerned. It is also evident that many different technical and scientific disciplines are involved in finding an optimal solution to a cooling problem. This requires both close cooperation of all partners and efficient coordination.

Improvement of the Structural Behavior of Cooling Tower Shells by Ring-Stiffeners

U. Eckstein*, Ç. Eller*, R. Harte*, Z. Sanal*, W.B. Krätzig* and U. Wittek**

* Institut für Konstruktiven Ingenieurbau - Statik und Dynamik
Ruhr-Universität Bochum, Germany

** Lehrstuhl für Baustatik
Universität Kaiserslautern, Germany

Summary

Increasing capacities of power plants as well as the intensified use of dry cooling systems have induced cooling tower dimensions of such a degree that the installation of additional ring-stiffeners must be regarded as a suitable alternative *recent structural development*. Such horizontal ring-stiffeners can be located at the inner or at the outer surface or at both sides of the shell. The intension is to improve the stresses, the stability and the dynamic behaviour of the structure. Furthermore their use as additional service stages leads to an enhanced operational utility.

In the present paper essentially the mechanical influence of ring-stiffeners to the *buckling and dynamic behaviour* shall be investigated from a theoretical-numerical view-point. The improvement of the structural behaviour of cooling towers with 1, 2, 3 and 4 stiffening rings will be quantified and recommendations for the design procedure are given.

1. Introduction

The investigations are concerned with a typical dry cooling tower of an 1300 MW power plant, the geometry of which is given in fig. 1. The unstiffened shell has the usual upper edge beam and a continuous enlarging wall-thickness at the bottom. The minimum wall-thickness is 20 cm. Four versions of ring-stiffening will be investigated:

version 1 :	1 ring-stiffener at	80.00 m
version 2 :	2 ring-stiffeners at	120.00 m
		70.00 m

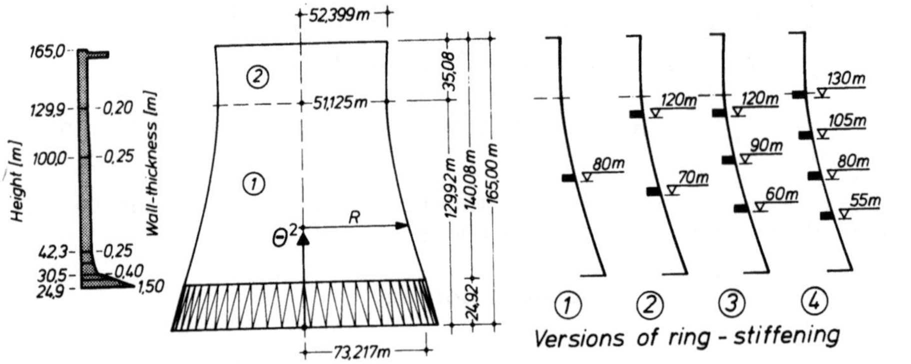

$$R = DR + \frac{A}{B}\sqrt{B^2 + (129{,}92 - \Theta^2)^2}$$

section	DR	A	B
①	33,077477	18,047522	52,853802
②	50,724535	0,400465	8,640975

Material:
Young's modulus $E = 3{,}0 \cdot 10^7\, kN/m^2$
Poisson's ratio $\nu = 0{,}2$

Fig. 1. Cooling tower geometry

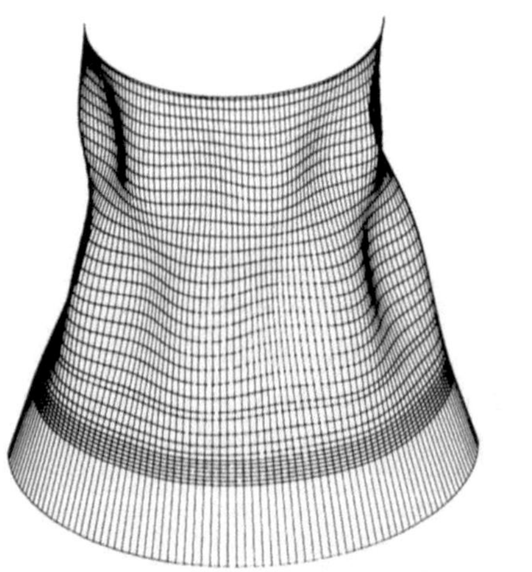

Fig. 2. Buckling pattern of the unstiffened cooling tower

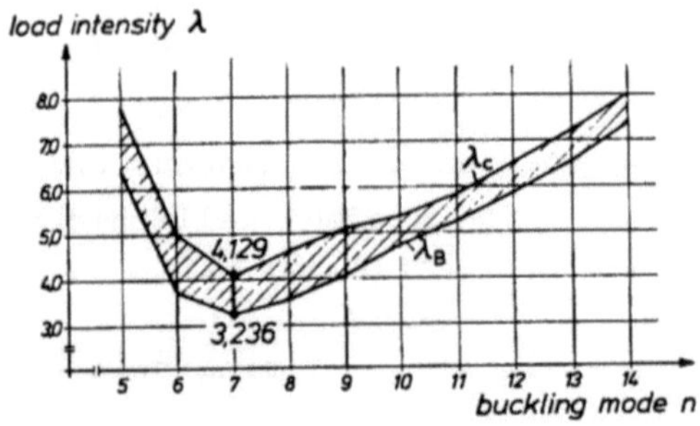

Fig. 3. Load factor of the unstiffened cooling tower

version 3 : 3 ring-stiffeners at 120.00 m
90.00 m
60.00 m

version 4 : 4 ring-stiffeners at 130.00 m
105.00 m
80.00 m
55.00 m .

The ring-stiffeners are installed at the inner surface of the shell with monolithic bond and have a cross-section of 40 cm/ 120 cm, that means the height of the rings in horizontal direction is about 6 times the wall-thickness.

The numerical realization is done by the finite element method using 35 conical *axisymmetric shell elements* [12] with 8 degrees of freedom to discretize the whole cooling tower. The displacement field is approximated by Fourier-series in the circumferential direction and by cubic polynominals along the meridian. The ring-stiffeners are discretized as axisymmetric beam elements.

The elastic properties of the supporting column system, which consists of 58 pairs of round columns with a diameter of 1.26 m and a height of 26.92 m, are transformed into an anisotropic axisymmetric shell element with comparable flexibility by converted thicknesses for the shear stiffness $h_S = 1.5$ cm , for the membran stiffness $h_M = 27.0$ cm and for the bending stiffness $h_B = 69.5$ cm , both in meridional direction.

All numerical calculations have been carried out with the modular software-system FEMAS - *Finite Element Modules for Arbitrary Structures* [11].

2. Buckling behaviour

2.1 Overall buckling

The resistance of cooling towers to buckling essentially depends upon the geometrical shape of the shell, its wall-thickness, locations and stiffnesses of the ring-stiffeners, the flexibility of the supporting column system and last not least upon deviations from the perfect shape. As such *geometrical and material imperfections* are not known, when the shell is designed, the buckling safety has not to be related to the snap-through load of the imperfect shell, but to the bifurcation load of the perfect shell.

A complete *geometrically nonlinear calculation* [2, 13] of cooling towers under the load combination of [1]

axisymmetric dead weight	g	see fig. 6 in [13]
axisymmetric internal suction	s	
non-axisymmetric wind load	w	

demonstrates that the nonlinear deformations up to bifurcation are sufficiently small so that their influence on the buckling behaviour can be neglected completely. Therefore the problem reduces to the *classical linear stability analysis*, characterized by the linear eigenvalue problem of the discretized shell:

$$[\underline{K}_e + \lambda \underline{K}_g]\ \underline{\dot{V}} = \underline{0} \rightarrow \det\ [\underline{K}_e + \lambda \underline{K}_g] = 0 \rightarrow \lambda_c,\ \underline{\dot{V}}_c \qquad (1)$$

$\underline{K}_e$ - the elastic stiffness matrix,

$\underline{K}_g$ - the geometrical stiffness matrix,

$\underline{\dot{V}}$ - the vector of the nodal displacements.

Solution of this homogeneous algebraic equation system leads to the smallest eigenvalue λ_c as the *critical load factor* of $g + w + s$ and to the corresponding eigenfunction $\underline{\dot{V}}_c$ as the *buckling mode* at the bifurcation point.

Furthermore it is possible to substitute the nonaxisymmetric wind load $w(\theta^1, \theta^2)$ by an axisymmetric reduced value of the wind pressure in the windward face: $\alpha \cdot w\ (\theta^1 = 0, \theta^2)$. This simplification, which is essential for the design procedure, enables us to discretize the shell by axisymmetric elements with harmonic functions for the displacement values in the circumferential direction. Then we obtain a series of eigenvalues $\lambda(n)$, depending upon the number n of buckling waves in the circumferential direction, the smallest of which represents the critical load intensity factor $\lambda_c(n_c)$; the corresponding eigenvector $\underline{\overset{+}{V}}_c\ (n_c)$ yields the buckling pattern along the meridian.

First of all such an *overall* buckling analysis may be demonstrated for an unstiffened cooling tower, the basic version of fig. 1. Fig. 2 shows the buckling pattern and fig. 3 the load factor $\lambda(n)$, normalized with respect to the axisymmetric load combination

$$g + \alpha w + s\ , \qquad \alpha = 0.6\ .$$

Then the smallest eigenvalue occurs for 7 buckling waves in the circumferential direction with a value of $\lambda_c = 4.129$.

The neglect of all curvature-depending membrane terms in the elastic stiffness matrix $\underline{K}_e$ leads to the *concept of reduced-membrane-buckling* [3], where the stability is only guaranteed by bending action of the buckled configuration. The corresponding critical load factor $\lambda_B = 3.235$ has to be looked upon as the lowest bound of buckling load for a geometrically imperfect shell in the sense of a conservative estimation [3].

Now the installation of *additional ring-stiffeners* changes the overall buckling behaviour decisively. It can be recognized from figs. 4-7, where the ring-stiffeners have been located in the range of large normal displacements $\overset{+}{V}_3$ of the unstiffened shell, that the wave-lengths of the deformation patterns in the buckled configuration become smaller with increasing number of ring-stiffeners. Because of their great horizontal bending stiffness they enforce *nodal lines* in the buckling mode shapes, as can be seen in fig. 8, with the

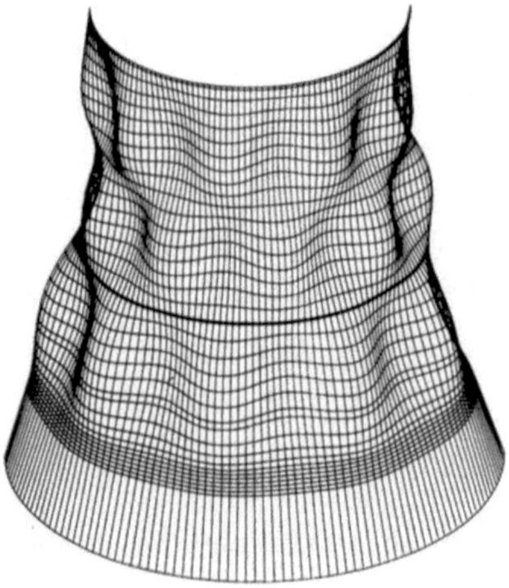

Fig. 4.

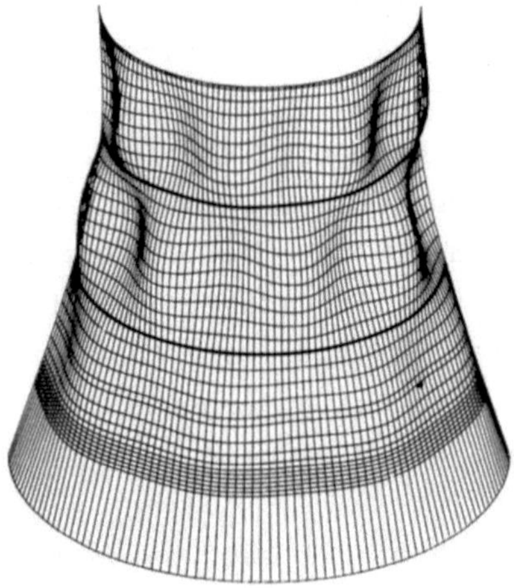

Fig. 5.

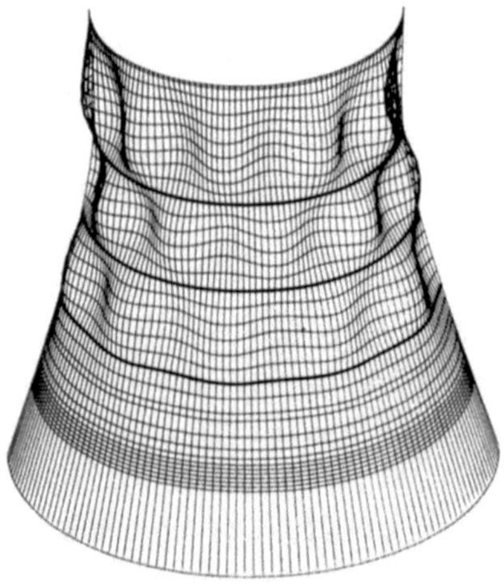

Fig. 6.

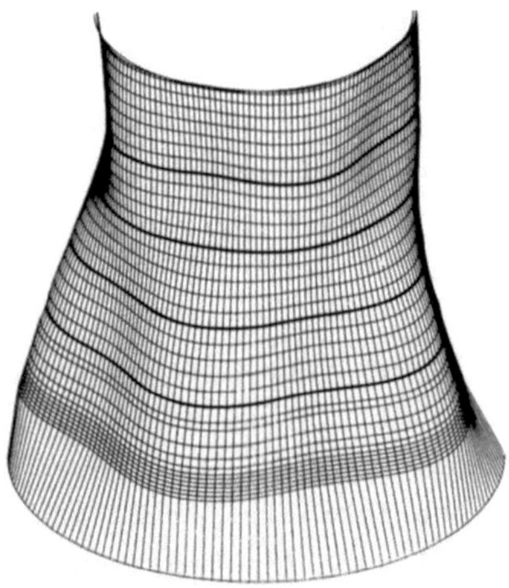

Fig. 7.

Buckling patterns of the cooling tower with 1, 2, 3 and 4 ring-stiffeners, respectively

exception of version 4. Furthermore fig. 8 demonstrates the elastic flexibility of the shell's bottom lintel due to the consideration of the supporting column system.

The real effectiveness of ring-stiffening is expressed by a *drastical augmentation of the buckling load*. As fig. 10 demonstrates this augmentation relative to the unstiffened shell amounts 44% for version 1, 65% for version 2, 107% for version 3 and 170% for version 4. On the other hand these classical buckling loads are reduced to about 80% by the *imperfection sensitivity* of the shell.

Another remarkable phenomenon of ring-stiffened shells can be recognized from fig. 9, where the load factors λ are shown depending upon the harmonics n in circumferential direction. The unstiffened shell has a relatively definite buckling minimum $\lambda_c(n_c)$, whereas the ring-stiffened versions tend to admit different buckling modes with nearly equal load intensities. This *multi-mode-buckling* means with respect to imperfection sensitivity that a random system variable, i.e. the actual imperfection pattern, causes a *greater* probability for the decrease of the buckling load to the value λ_B. This circumstance requires particular care in constructing the shell as accurate as possible.

2.2 Local buckling of an unstiffened segment of the shell

In order to deal with the problem of local buckling of a shell segment considering interaction with the ring-stiffeners the unstiffened segments between two ring-stiffeners have to be investigated. This seems to be necessary, because in the case of cooling tower shapes as given here there is *no decisive hyperbolic curvature* in these segments any more. The investigated segments have immoveable edges in radial direction, which are loaded by the stresses of the whole structure from $g + 0{,}6w + s$ and the segment itself is loaded by the surface loads from $g + 0{,}6w + s$. For the middle segment of version 2 we receive a buckling mode $n=10$ with critical load factors

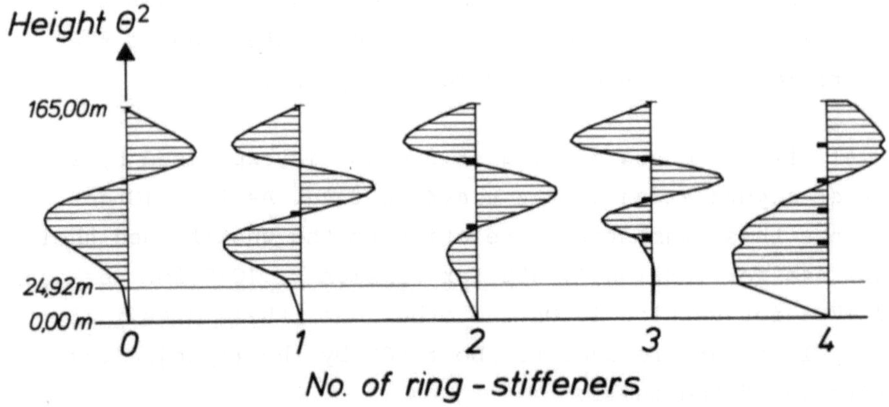

Fig. 8. Meridional buckling mode shapes of the unstiffened and stiffened versions

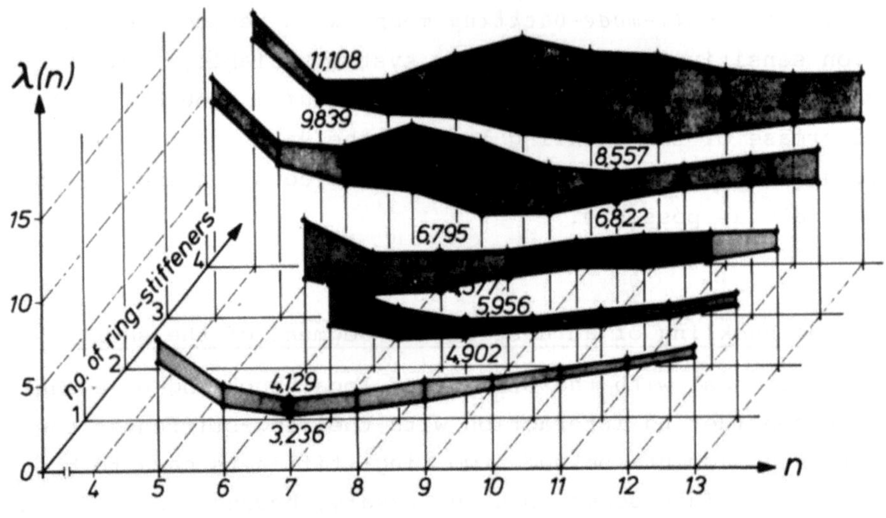

Fig. 9. Load factors λ(n) of the unstiffened and stiffened versions

$$\lambda_C = 9.560 \ ,$$

$$\lambda_B = 6.121 \ .$$

The corresponding buckling pattern is shown in fig. 11. Additionally it is necessary to prove that the critical load of the single uncoupled ring-stiffener under radial compression for one special harmonic n has to be greater than the corresponding buckling load of the unstiffened shell.

2.3 Buckling in the state of construction

Naturally stiffened cooling towers *under construction* are more resistant to buckling than unstiffened ones, which preferably can be investigated by means of the overall buckling concept, too. For version 2 with 2 ring-stiffeners the critical load factors have been calculated just before installation of the ring-stiffeners and the upper edge beam respectively. The results are shown in fig. 12. For the critical state of construction at a height of 164.00 m without upper edge beam the buckling pattern is presented in fig. 13.

2.4 Local buckling criterion

Besides the concept of overall buckling with critical load factors the buckling safety of a ring-stiffened shell may also be estimated by a *local stability criterion*, where suitable interaction-diagrams for the critical stresses of unstiffened shells are provided [6]. The ratio between the local buckling safety according [6] and the global safety factor of the classical bifurcation load (1) for the unstiffened shell respectively leads to a *reduction factor*, which is supposed to be valid for ring-stiffened shells, too.

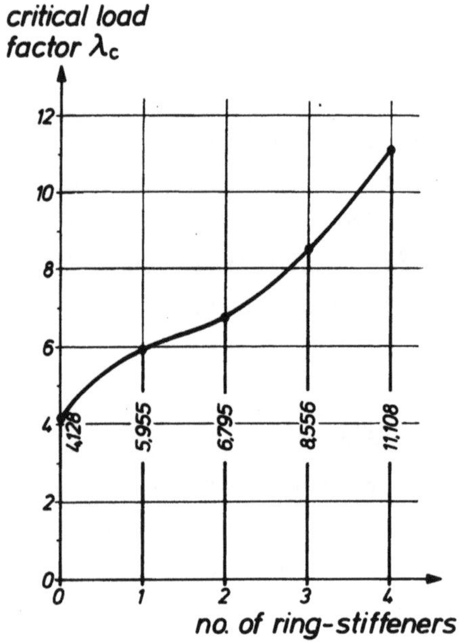

Fig. 10. Augmentation of critical load factors by ring-stiffening

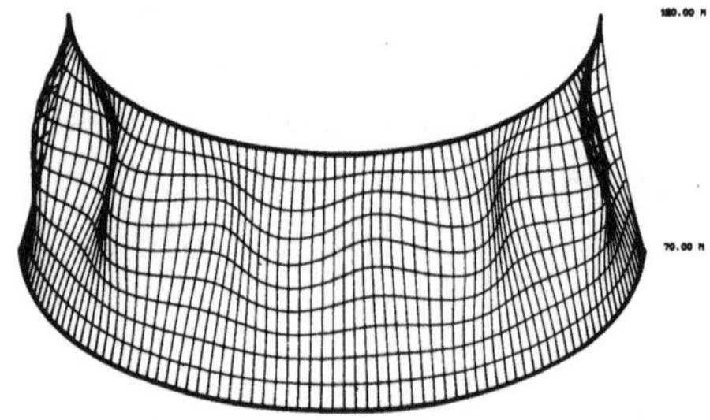

Fig. 11. Buckling pattern of the middle segment of the cooling tower with 2 ring-stiffeners (version 2)

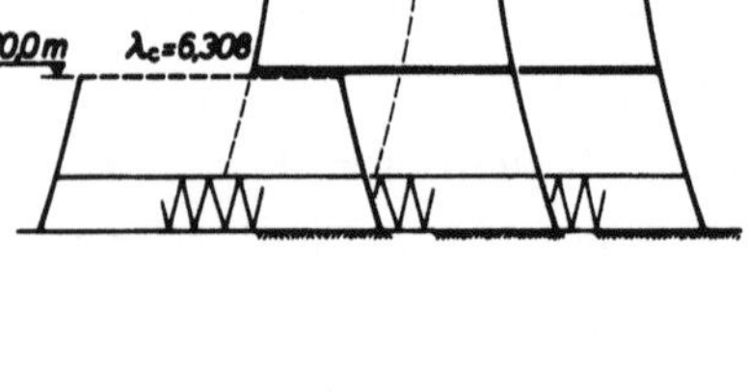

Fig. 12. Critical load factors λ_c for different states of construction (version 2)

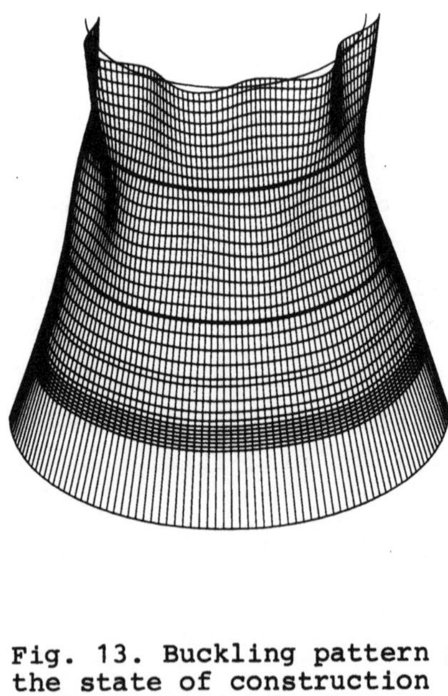

Fig. 13. Buckling pattern in the state of construction without upper edge beam (version 2)

3. Dynamic behaviour

3.1 Fundamentals of kinetic stability

The dynamic behaviour of cooling tower shells under time-varying natural wind load can be treated - from the actual state of the art [1, 8] - by a *quasi-static analysis*, at which the resonant part of the structural response is considered by a *dynamic magnification factor* φ [8]. Therefore the calculation of the lowest natural frequency obtains decisive importance. The constructional variant of a ring-stiffened cooling tower in most cases implies a further rise of the natural frequency, which is already high compared with wind-excitation.

The dynamic behaviour of the shell, which is prestressed by time-invariant loading (dead weight, internal suction, quasi-static wind), can be described by the *classical eigenvalue problem of kinetic stability* [9, 10]

$$[(\underline{K}_e + \underline{K}_g) - \omega^2 \underline{M}] \, \underline{\overset{+}{V}} = \underline{O} \tag{2}$$

where ω denotes the circular frequency and $\underline{M}$ the mass matrix.

For unloaded shells $(\underline{K}_g = \underline{O})$ we directly receive the eigenvalue problem of *free vibration* with the characteristic equation

$$\det [\underline{K}_e - \omega^2 \underline{M}] = 0 \, , \tag{3}$$

from which the *natural frequencies and the modal matrix of eigenvectors* of the vibrating shell are obtained. Summation of the elastic stiffness matrix $\underline{K}_e$ and the initial stress matrix $\underline{K}_g$ yields the natural frequencies and corresponding modes as functions of the static preload.

3.2 Free vibration behaviour

Fig. 14 shows the lowest natural frequencies for different harmonics n of the unstiffened and ring-stiffened cooling tower versions, at which, according to the buckling analysis the supporting column system has been considered. The lowest natural frequency of the unstiffened shell is min f = 0.714 Hz (n=5) and on the contrary for the stiffened version 4 min f = 0.780 Hz (n=4) . With simply supported bottom lintel the natural frequencies increase to min f = 0.952 Hz (n=7) for the unstiffened shell and min f = 1.167 Hz (n=5) for the stiffened one, respectively. Different from the buckling analysis the ring-stiffeners do not enforce nodal lines in the vibration modes, as can be seen from fig. 15 and 16.

The influence of the ring-stiffeners on the frequencies is *not definitely increasing*, because their additional masses will slightly lower the frequency-values again. Thus it must be stated that the positive effect of ring-stiffening to the natural frequencies is rather poor.

3.3 Load-frequency interaction

The complete evaluation of the kinetic stability analysis according to (2) is shown in fig. 17 and 18. The load-frequency graphs demonstrate the influence of the *static preload* λ (g + w + s) on the free vibration behaviour. For the load intensity $\lambda = 1$ hence the lowest natural frequency of the unstiffened shell decrease by 8% from min f = 0.714 Hz to min f = 0.657 Hz . For the stiffened shell version with 4 ring-stiffeners the frequency-decreasing influence amounts only 4%.

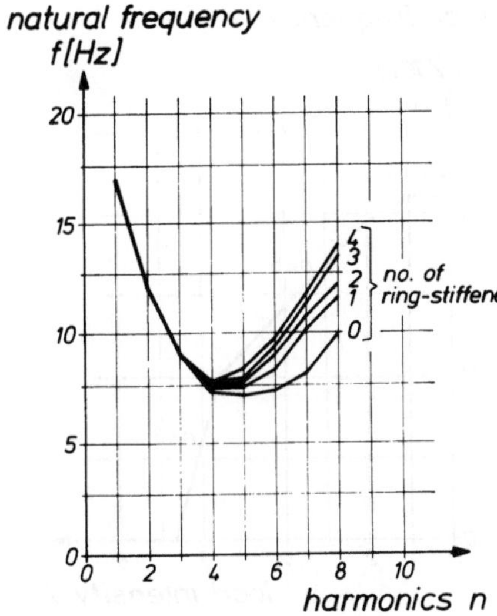

min f (Hz)	n	ring-stiffeners
0,714	5	0
0,753	5	1
0,757	4	2
0,770	4	3
0,780	4	4

Fig. 14. Natural frequencies of the unstiffened and stiffened versions

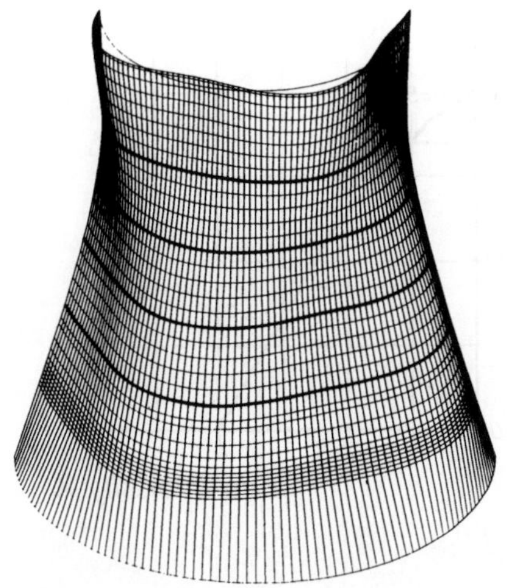

Fig. 15. Vibration pattern of the cooling tower with 4 ring-stiffeners (version 4)

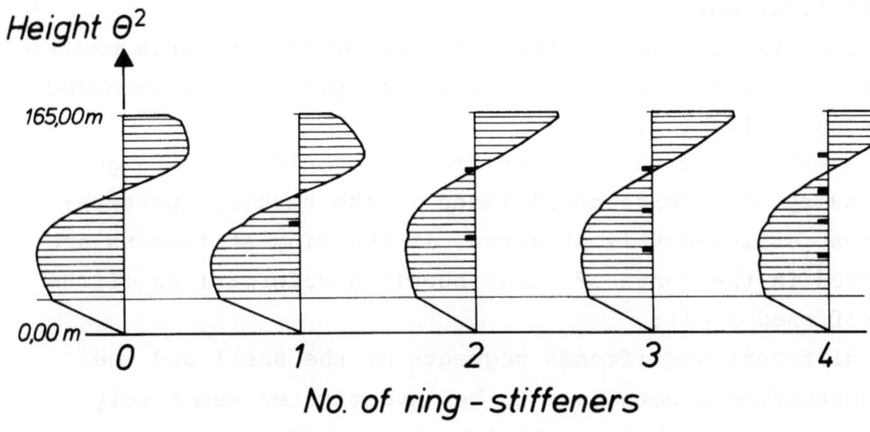

Fig. 16. Meridional vibration mode shapes of the unstiffened and stiffened versions

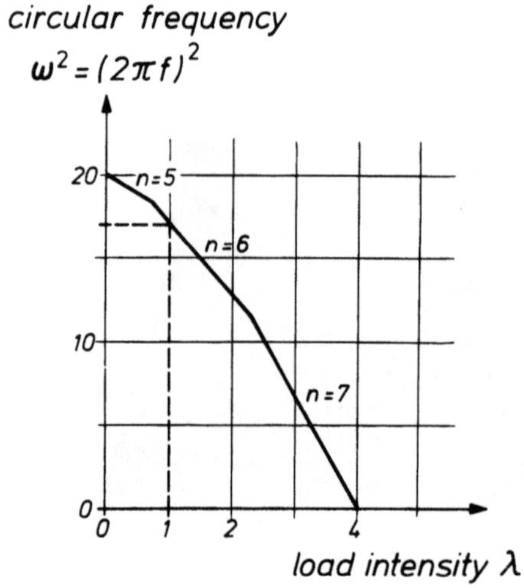

Fig. 17. Load-frequency interaction of the unstiffened cooling tower

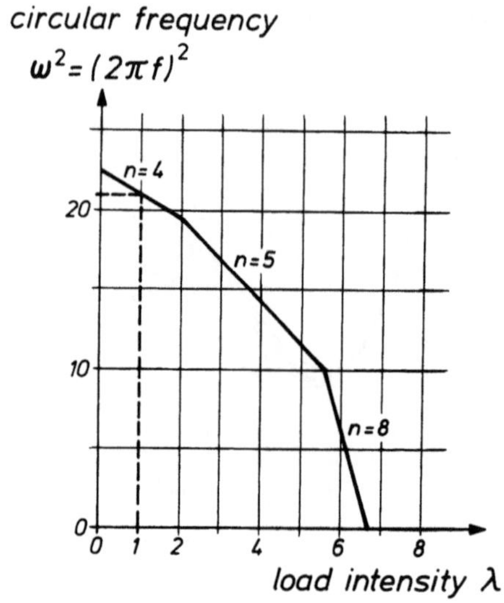

Fig. 18. Load-frequency interaction of the cooling tower with 2 ring-stiffeners (version 2)

4. Conclusions

The structural behaviour of cooling towers can be improved by the installation of ring-stiffeners with respect to the following specifications:

- the linearized critical buckling factor for the axisymmetric load $g + \alpha w + s$ (here: $\alpha = 0.6$) is perceptibly enhanced by ring-stiffeners;
- the stiffness of the ring-stiffeners should be as large as necessary to enforce nodal lines in the buckling pattern;
- highest efficiency is obtained, if the ring-stiffeners are located in the range of large buckling deformations of the unstiffened shell;
- the different unstiffened segments of the shell and the ring-stiffeners both have to be investigated separately with respect to their partial buckling safety;

- the buckling safety of critical states of construction has to be proved by means of an overall buckling analysis;
- geometrical imperfections may reduce the classical buckling load of the perfect shell by about 20% (for the shells investigated here);
- ring-stiffened cooling tower shells tend to multi-mode-buckling;
- the natural frequency can be enhanced by installation of ring-stiffeners, but the efficiency is relatively small compared with the improvement of the buckling behaviour;
- the stiffness of the ring-stiffeners in general is not sufficient to enforce nodal lines in the vibration pattern;
- the influence of static preloading on the natural frequency is small and may be neglected in general;
- especially in the case of high column systems their flexibility has to be taken into consideration both for buckling and vibration analysis.

All of the presented calculations are based on the *theory of elasticity*. The nonelastic behaviour of reinforced concrete including tension-stiffening, cracks and bond-slip have not been considered. Thus *high safety factors* are still indispensable for the proof of the buckling safety as well as for the assessment of the dynamic behaviour.

5. References

1. BTR - Bautechnik bei Kühltürmen, Teil 2: Bautechnische Richtlinien. VGB-Verlag, Essen 1980.

2. Eckstein, U.: Nichtlineare Stabilitätsberechnung elastischer Schalentragwerke. Technical Report of the Institut für Konstruktiven Ingenieurbau, Ruhr-Universität Bochum, 1983, No. 83-3.

3. Wittek, U.: Beitrag zum Tragverhalten der Strukturen bei endlichen Verformungen unter besonderer Beachtung des Nachbeulmechanismus dünner Flächentragwerke. Technical Report of the Institut für Konstruktiven Ingenieurbau, Ruhr-Universität Bochum, 1980, No. 80-1.

4. Form, J., Krings, W., Mazur, H., Peters, H.L.: Berechnung und Ausführung eines ringversteiften Naturzugkühlturms aus Stahlbeton. Beton- und Stahlbetonbau, Heft 9, 1980.

5. Zerna, W., Mungan, I.: Über das Beulen von Kühlturmschalen mit Versteifungsringen. Beton- und Stahlbetonbau, Heft 2, 1981.

6. Zerna, W., Basar, Y., Mungan, I., Tiemann, M.: Beuluntersuchungen an hyperbolischen Rotationsschalen. Forschungsbericht des Landes Nordrhein-Westfalen Nr. 2439. Opladen: Westdeutscher Verlag, 1974.

7. Lehmkämper, O.: Versteifte Kühlturmschalen aus Stahlbeton. Technical Report of the Institut für Konstruktiven Ingenieurbau, Ruhr-Universität Bochum, 1978, No. 78-6.

8. Jahrbuch Bautechnik im Kraftwerksbau. Kapitel VIII Kühltürme. Vulkan-Verlag Essen. 2. Ausgabe 1982/83.

9. Krätzig, W.B., Sanal, Z., Wittek, U.: Zum Einfluß oberer Randaussteifungen auf das dynamische Verhalten von Kühlturmschalen. Heft 3, 1984.

10. Wittek, U.: Die Stabilitätsgleichungen der Schalentheorie und ihre Anwendung zur Beurteilung des Schwingungs- und Stabilitätsverhaltens der Rotationsschalen. Konstruktiver-Ingenieurbau in Forschung und Praxis. Werner-Verlag, Düsseldorf, 1976.

11. Beem, H., Brink, K., Lechtleitner, B.: FEMAS - Finite Element Moduln allgemeiner Strukturen, Internal Report of Lehrstuhl III, Institut für Konstruktiven Ingenieurbau, Ruhr-Universität Bochum, 1981.

12. Eckstein, U., Krätzig, W.B., Wittek, U.: Finite-Element-Berechnungen zur Grenztragfähigkeit der Rotationsschalen, Technical Report of the Institut für Konstruktiven Ingenieurbau, Ruhr-Universität Bochum, 1980, No. 80-4.

13. Eckstein, U., Harte, R., Krätzig, W.B., Wittek, U.: Nonlinear Elastic Stability of Cooling Tower Shells, in: Proc. of the 2nd International IASS-Symposium on Natural-Draught Cooling Towers, Ruhr-Universität Bochum, Springer-Verlag, 1984.

Review and Feedback of Experience Gained over the Last Fifty Years in Design and Construction of Natural Draught Cooling Towers

P B BOSMAN B.Sc., M.I.C.E. F.S.A.I.C.E.

WLPU CONSULTANTS
35/41 STATION ROAD
ASHFORD, KENT, TN23 1PP
ENGLAND

Summary

A review of natural draught cooling towers and their development from the early 1930's to the present is given. Experience gained with these older towers, some of which are still in operation, is described and attention is drawn to aspects of construction and design which have proved to be unsatisfactory. Examples of towers with structural defects, which have led to current design philosophy, are described. Observations on other aspects of natural draught cooling towers design, such as foundations, packing, packing supports and lightning protection, are given.

Historical Review

Over the past 50 years we have seen the growth of natural draught cooling towers from units of 20 MW and 40 MW (1) to units handling 900 MW of heat load and Fig. 1, representing more than 150 cooling towers, illustrates the increase in tower size against this time scale.

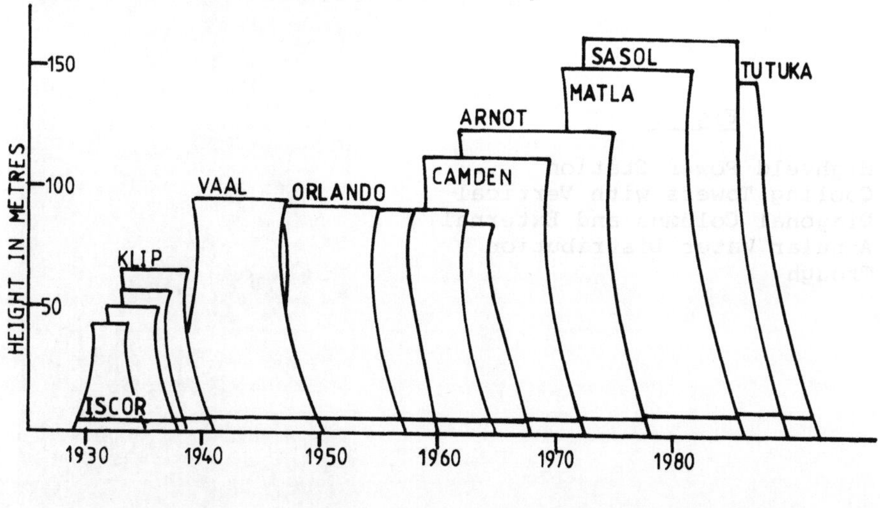

Fig. 1 : Increase in tower size 1932 to 1984

While methods of design with the availability of digital computers and finite element analysis techniques have made tremendous strides, particularly in the post Ferrybridge (2) era, the actual material, with but few exceptions, used in the construction of natural draught cooling towers, has remained essentially the same, namely reinforced concrete.

Fig. 2

Klip Power Station
10 - 40 MW Towers

Klip Power Station (Fig.2), having twelve 33 MW generating sets served by ten 40 MW natural draught cooling towers was typical of Power Stations constructed during the 1930's era. This power station remained one of the largest coal fired power stations in the Southern Hemisphere until the 1950's.

Fig. 3

Highveld Power Station
Cooling Towers with Vertical
Diagonal Columns and External
Annular Water Distribution
Trough

With the increasing size of generating sets, cooling tower design and construction was extrapolated from the comparatively small size of the 1930's era to the intermediate size existing at the time of the Ferrybridge disaster.

It was also during this period that non-hyperbolic cooling towers of the conical/toroid type were introduced. Some designers departed from the more commonly accepted raking diagonal column geometry with vertical diagonal column arrangements (Fig.3).

Early natural draught cooling towers were all constructed from an internal timber scaffold erected as the work progressed (Fig.4). However, in 1949, a new method of

Fig. 4 : Klip Towers during construction 1934

construction was introduced (3), utilising a central mast with revolving bridge and climbing formwork supported off the shell concrete (Fig. 5). Minimum shell thickness of 125 mm was considered necessary for supporting the formwork and working platforms off the green concrete.

Fig. 5 : Mast and bridge used for tower construction from 1949

Shell Reinforcement

On cooling towers of the hyperbolic shape, constructed up to the 1960's, it was common practice for the vertical reinforcement to be placed diagonally along the lines of the shell generators. Initially, reinforcement was provided on both faces of the shell and this resulted in 6 layers of reinforcement being used.

This arrangement was later modified by placing the diagonal vertical reinforcement in one direction on the outside face and in the other direction on the inside face thereby reducing the number of layers of reinforcement in the shell from six to four. The Klip cooling towers provide a good example of shells constructed with 6 layers of reinforcement. Although minimum shell thickness on these small towers was 125 mm, difficulty was experienced in placing the concrete which was hand puddled and not vibrated. Considerable disintegration and deterioration of the concrete in the shells is now evident. Large, as well as numerous small holes have appeared in the shells.

With minimum shell thickness of 125 mm, it was accepted practice to provide minimum cover and the first generation of towers were constructed with only 12 mm of cover to the reinforcement on both faces. Instead of increasing shell thickness, some designers adopted reinforcement placed centrally within the shell (Fig. 6) and in some instances shell thickness was reduced to a mere 100 mm. As a result of the above practice we have today a legacy of some 63 natural draught cooling towers constructed up to the late 1950's that were built with minimum cover to the reinforcement.

Fig. 6

Reinforcement placed centrally within the shell.
Bulawayo Cooling Towers 1945/53

These towers display extensive spalling and corrosion of the reinforcement. This experience has led us to adopt minimum cover of 40 mm and 35 mm to the outside and inside faces of the shell respectively.

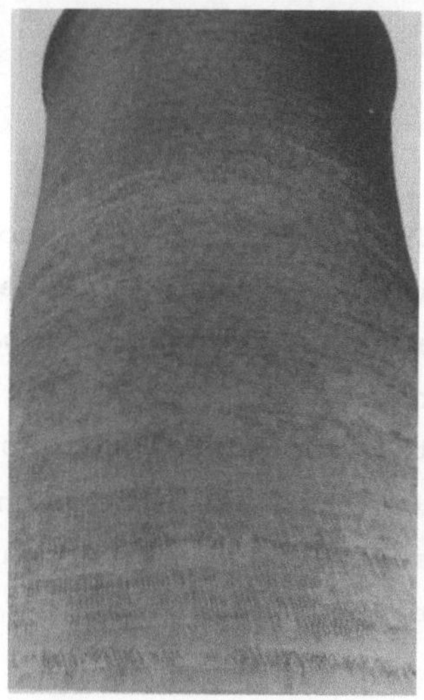

Fig. 7

Deterioration of outside surface of cooling tower shell

Our reasons for the adoption of a greater cover to the

outside face of the shell are, firstly, that chemical attack on the concrete, in the form of carbonation, occurs on the outside surface. It has been our experience that the moist environment inside a shell is not detrimental to the long term performance of concrete. On early cooling towers, it was considered necessary to protect the inside concrete surface of the shell with bitument paint. This practice, however, was abandoned when it was realised that the coating did not last more than a few years and subsequent experience has in fact shown that it was quite unnecessary. Even on our oldest cooling towers, constructed with minimum cover to the reinforcement, spalling of concrete and corrosion of reinforcement on the inside surface has been a fraction of that, that has occurred on the outside surface of the same shell (Fig.8).

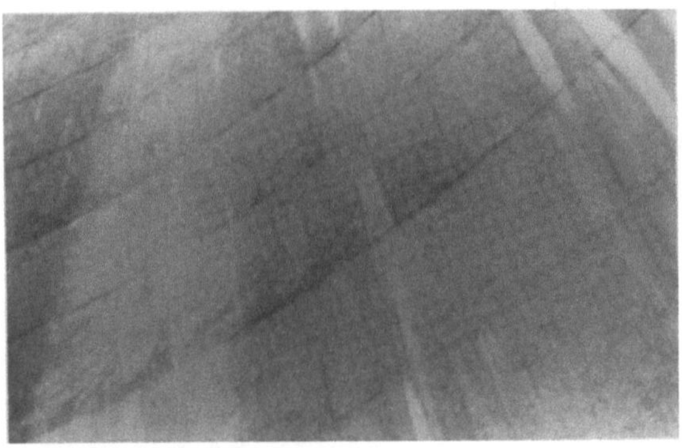

Fig. 8

Inside surface of the same shell as shown in Fig. 7.

The second reason concerns the placing of concrete between inward leaning closely spaced forms. It is difficult to achieve adequate compaction of the concrete against the outer overhanging shutter. If insufficient cover is provided and the horizontal reinforcement is placed on the outside of the vertical reinforcement, there is a tendency for aggregate separation to occur above and below the horizontal reinforcing bars.

The formation of voids under the horizontal bars can occur and this results in horizontal hairline cracks propagating from the void to the surface (Fig.9), leading to corrosion of

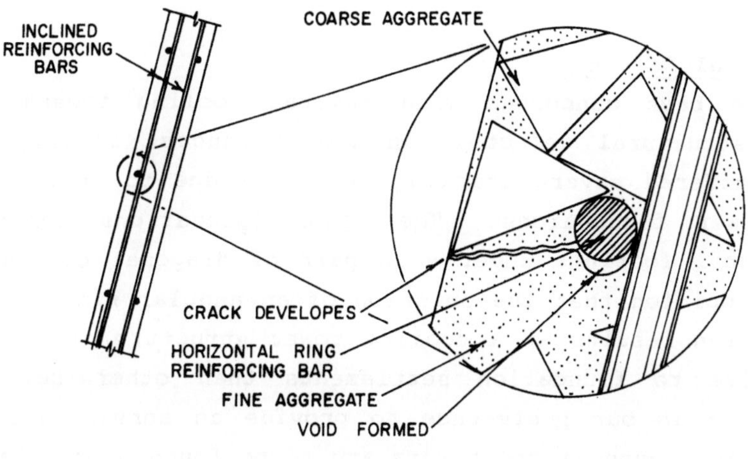

Fig.9

the reinforcement and to spalling of the concrete (Fig.10). Aggregate separation can be identified by a slight change in the colour of the concrete above and below the horizontal reinforcement.

Fig. 10

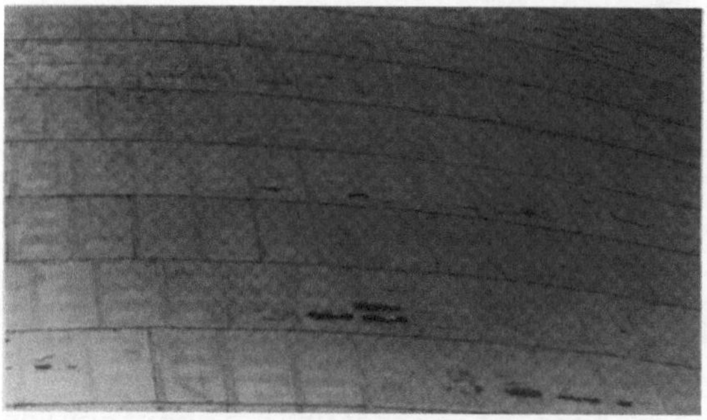

Techniques for carrying out repairs on cooling tower shells have been developed. The major problem of access to the shell surface can be overcome with the use of a gondola suspended from a wheel mounted dolly straddling the top ring beam of the shell. Repairs to cracks and spalled areas have been undertaken using epoxy mortars or guniting. On some towers, repairs have been considered uneconomical and, as a result, we are now monitoring the gradual deterioration of these towers.

Structural Defects

We have been concerned with several cooling towers having major structural defects. In two instances, cooling towers have suffered severe cracking (Fig. 11) due to settlement of the tower foundations. The tower foundations consist of individual footings under each pair of diagonal columns. It is our opinion that the omission of an annular ring beam type foundation results in a cooling tower structure that is more sensitive to foundation settlements than otherwise. As a result it is our preference to provide an annular ring beam foundation, even if the towers are to be founded on piles.

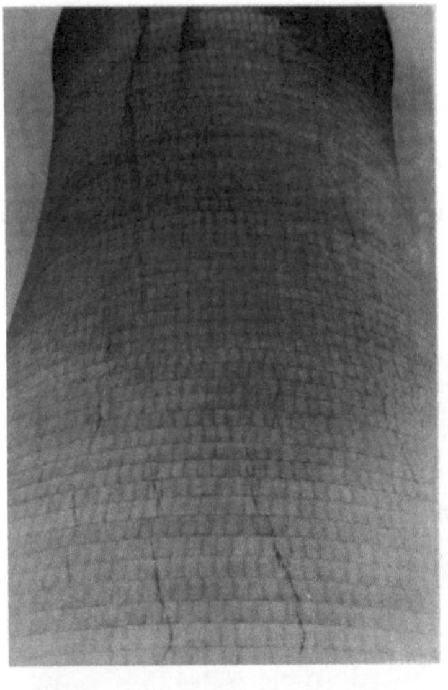

Fig. 11

Severely cracked cooling tower shell as a result of foundation settlement

Another form of structural imperfection that can occur during construction of a cooling tower shell is the misalignment of the shell, i.e. shell bulges. We believe that the method adopted for setting out the shell plays an important role in the ability of the contractor to construct the shell within tolerance.

Numerous shell setting-out techniques are in use, but a method which we favour consists of the permanent installation of an optical plumb at the centre of the tower. This is actually located within the tower crane. At the level of construction, a central target is located by means of adjustable suspension guys fixed to the shell. For the setting out of each lift, the target is accurately centred over the optical plumb and thereafter serves as a central taping point for setting each and every shutter around the shell. A spring tensioned stainless steel wire is used for taping to overcome the problem of sag and wind effect.

Forms are adjusted to their correct position before casting. We do not favour methods which rely on the establishment of the shell position after casting, as successive over correction for measured out of tolerance can occur resulting in shell misalignment.

Where serious misalignment occurs, the best remedy is to cut the shell back. However, under certain circumstances this may not be possible and a solution which we have successfully used is to cast a new inner and outer shell of varying thickness, to eliminate excentricity, over the original shell (Fig. 12). The new inner and outer shells being tied together with shear connectors bolted through the original shell.

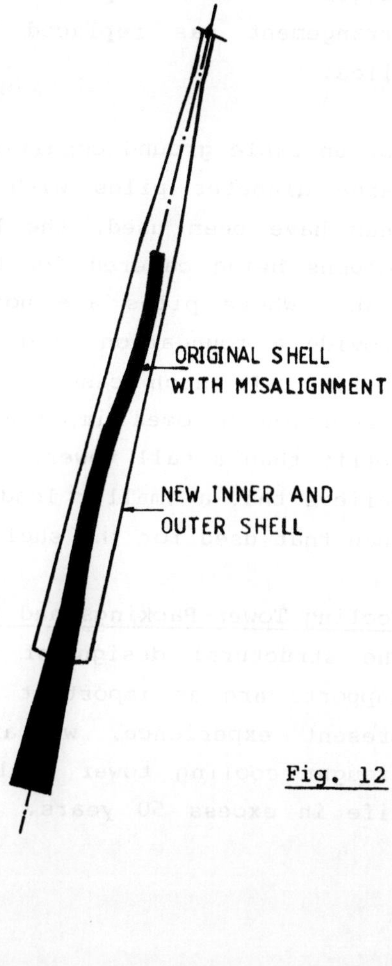

Fig. 12

Design requirements for horizontal reinforcement are generally based on the perfect shell for which nominal reinforcement of, say, 0,25% suffices. However, in practice, the perfect shell does not exist and as a further protection against shell misalignment, we believe that minimum horizontal reinforcement should not be less than 0,35% in the lower two-thirds and 0,6% in the top third of the shell. It is also our practice to specify that all reinforcement used in the shell shall be deformed high tensile and that the spacing of reinforcement shall not be greater than 250 mm.

Foundations

As mentioned, it is our policy to construct all our towers with annular foundation ring beams. Where piled foundations were considered necessary on earlier towers, pile groups arranged in the form of a tripod were employed. However, as larger diameter piles became available, this tripod arrangement was replaced by single large diameter raking piles.

For unstable ground conditions such as heaving clay, vertical large diameter piles with voids under the foundation ring beam have been used, the horizontal forces from the raking columns being catered for by means of a stressed pond floor slab. Where piles are not necessary, it is our policy to provide a foundation ring beam of sufficient mass to resist uplift. In such cases, the selection of tower shape and proportion becomes important, a squatter tower having less uplift than a tall tower. However, for foundation design, we believe that a smaller load factor, applied to the wind load, than that used for the shell design, is justified.

Cooling Tower Packings and their Supports

The structural design of cooling tower packings and their supports are as important as the shell structure. With our present experience, we are today confident that natural draught cooling tower shells will achieve a useful working life in excess 50 years. However, the same cannot be said for many current packings where the emphasis has been on higher thermal performance at lower costs.

With the introduction of plastics with their low mass, the tendency has also been to design lighter pack structures, in which the packing is often suspended rather than supported.

Many packings, originally developed for induced draught tower application, are now used in natural draught cooling towers without an appreciation of the more severe conditions imposed on a packing system in a natural draught tower environment. Problems which we have encountered with packing systems include clogging through the build-up of silt and algae, damage through wind buffeting and ice loading and water overload. Besides these problems, deterioration of the materials used for the packing has also occurred. Of all the packing systems used in cooling towers, we consider that asbestos cement has proved to be the most suitable and it is indeed unfortunate that this material is no longer permitted in many countries for health reasons.

With increasing involvement in cooling tower maintenance, we find that cooling towers are generally neglected. Many Utilities do not regard their cooling towers as items of operating plant. Unfortunately the maintenance free cooling tower does not exist and although maintenance on a well designed natural draught cooling tower should be minimal, if it is totally neglected, a gradual deterioration in the tower and its performance results.

The inside surfaces of a natural draught cooling tower shell provides the ideal environment for algae growth. As the algae is regularly shed from the shell surface it is important that this material can pass through the drift eliminators and packing. Almost total blockage of drift eliminators can occur after 15 to 20 years of operation. Build up of algae and silt in high performance film packs have also been observed after 5 to 10 years of operation.

Their condition after 20 years of operation is therefore of concern.

With rising fuel costs, refurbishing and uprating the performance of older cooling towers is on the increase. While the performance of the older towers can be significantly uprated through the use of new high performance packings, the structural requirements of packing design must not be overlooked and this is a major consideration in our own research into packing systems.

Lightning Protection

Very few of the cooling towers with which we have been associated have been provided with lightning protection. For those that have, the reason has generally been that the cost is relatively minor if the installation is carried out at the time of construction.

Although most of the cooling towers are located in regions of high electrical thunderstorm activity we do not know of any tower that has actually been damaged by lightning. On some older towers pieces of concrete from the top cornice are missing. This minimal damage may be the result of lightning strikes. However, corrosion of the reinforcement leading to spalling is just as likely to have occurred.

The Next Fifty Years

During the past few years, we have seen the emergence of the next era of natural draught cooling towers. Towers of 165 m height with base diameters of 145 m are already being constructed and designs of even larger towers are being proposed. With some Utilities nervously contemplating this jump in size, rigorous specifications have been proposed resulting in substantially thicker shells and substantially greater volumes of concrete.

We have the confidence and ability to design and construct cooling towers of this size. However, should we not pause and think of fifty years hence. What will we do with these structures when the useful life of the plants they serve has drawn to an end? We are facing this problem today, with cooling towers that are only a fraction of the sizes now contemplated. Unfortunately, these structures bear little resemblance to other ancient structures which we can marvel at today. It is therefore very doubtful that future generations will want to be saddled with enormous deteriorating empty concrete shells that have no other useful purpose. It seems a great shame that the only future we can contemplate therefore appears to be one of demolition and if this is the case, are we wise to be considering towers of even greater sizes than those being built today?

REFERENCES

1. Gueritte T.J.; Modern Tendencies in Cooling Tower Design, London Technical Group of the Electrical Power Engineers' Association. November 1937.

2. Central Electricity Generating Board; Report of the Committee of Inquiry into Collapse of Cooling Towers at Ferrybridge Monday 1 November 1965.

3. Watermeyer T.C.; Notes on Natural Draught Cooling Towers South African Institution of Civil Engineers Jubilee Convention 1953.

The Ring-stiffened Shell of the Natural Draft Cooling Tower of the ISAR II Nuclear Power Plant

JÜRGEN FORM

E. Heitkamp GmbH, Langekampstrasse 36, 4690 Herne 2, West Germany

Summary

The natural-draft cooling-tower of the ISAR II nuclear power plant is one of the largest cooling towers in the world. The bid specifications provided for an unstiffened cooling-tower shell. For the execution, however, it has been decided to carry out a shell with three additional stiffening rings. The present contribution deals with the statical and dynamical calculations performed in the course of the execution planning. In one of the chapters I shall also deal with the execution and, in particular, with the working technique employed for the construction of the rings.

I. Introduction

In November 1981 KWU Offenbach issued an invitation to tender in their capacity as general contractors for the works on the natural-draft cooling tower of the 1300 MW nuclear power plant ISAR II near the town of Landshut in Eastern Bavaria. The Owner of the plant is a group of utilities consisting of the power supply companies Bayernwerk AG (leader firm), Isar Amper Werke AG, Stadtwerke München and Energieversorgung Ostbayern AG. The works were awarded to GEA Kühlturmbau und Luftkondensation GmbH with the understanding to sub-contract with E. Heitkamp GmbH for the building works.

The natural-daft cooling-tower ISAR II is a counter-current wet cooling tower with a water through-put of max. 259.200 m³/h. With its total height of 165,50 m and a diameter of 145,50 m at lower edge of shell (cf. fig. 1) it is the largest reinforced-concrete structure of its kind built to date.

Rentability studies [1], [2] had shown that a ring-stiffened cooling-tower shell constitutes a technical and economical alternative in case of the dimensions under review. This is confirmed by the numerous calculations and cost analyses supported by the experience made during the construction of the ring-stiffened cooling-tower shell Doel III and Mississippi I.

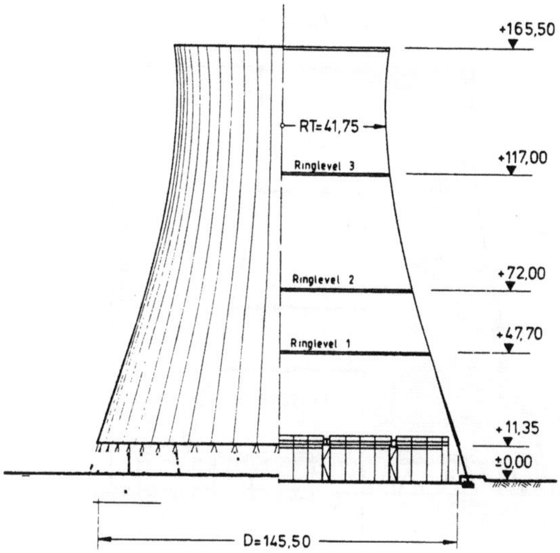

Fig.1. Natural-draft cooling tower ISAR II - Main dimensions

Although the bid specifications for the ISAR II cooling tower provided for an unstiffened shell it was decided to carry out a shell stiffened with three rings. The present contribution will therefore show the particularities of the statical and dynamical calculations, the stability studies, the construction and the building technique of the ring-stiffened cooling-tower shell ISAR II.

2. Calculations

2.1 Shell geometry

It is well known that the state of stress of shell structures is heavily dependent on the curvature of their center surface [3]. Only minor changes in the geometry - which, for cooling towers, are unimportant from the point of view of thermo-dynamics and optically hardly visible - may lead to considerable shifts of stresses and thus decisively influence on the material consumption.

With regard to the optimization it was therefore important to fix the shell geometry of the ISAR II cooling tower in such a way as to find out the most economical shape for the concrete and steel consumption by simultaneously observing the main input data supplied by the thermal engineers (cf. fig. 2).

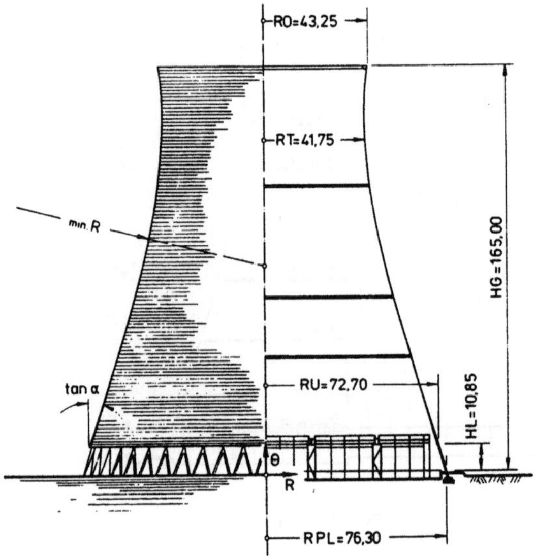

Fig.2. Input data supplied by thermal engineers

Moreover, it was important not to exceed certain limit values which are to guarantee a trouble-free climbing of the scaffolds along the shell geometry (cf. fig. 2). These values are in first respect tne smallest curvature radius R_{min}, and the greatest inclination realizable with the scaffolding at lower edge of shell tan α max.

The load cases considered in the studies were - in accordance with the Civil Engineering Guidelines for cooling towers [4] - the dead load, the wind, the internal suction and the temperature. The result of this optimization was a meridional curve composed of two hyperbola branches. The corresponding equations are shown in fig. 3.

Whereas it was still necessary during the tender phase in autumn 1981 to calculate each variant of the shell geometry individually and to evaluate it from the point of view of cost calculation, we have now elaborated a program in our company enabling us to calculate in the shortest possible time as many variants of the cooling-tower shell together with the column framework as required, from the calculation of the internal forces to the determination of the quantities and to the cost estimation. By means of this program the details of which are described in [5], the geometry chosen for the ISAR II cooling tower was examined. The results of such examination are shown partly in fig. 4.

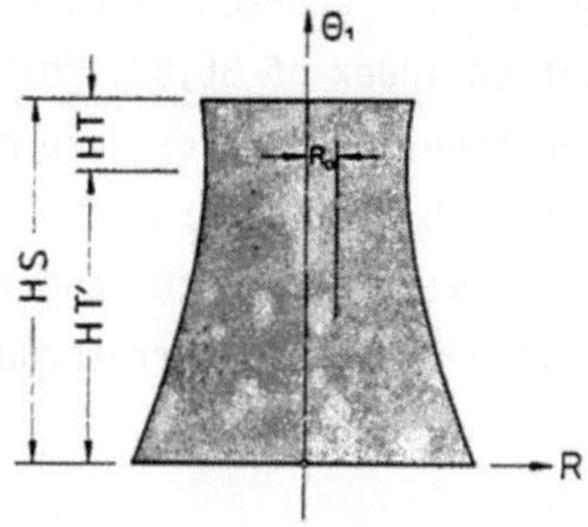

area below the throat ($0 \leq \Theta_1 \leq 124{,}15$):

$$R_{(\Theta_1)} = 28{,}061 + \frac{13{,}689}{40{,}00}\sqrt{(\Theta_1 - 124{,}15)^2 + 40{,}00^2}$$

area above the throat ($124{,}15 \leq \Theta_1 \leq 154{,}15$):

$$R_{(\Theta_1)} = 41{,}343 + \frac{0{,}407}{6{,}90}\sqrt{(\Theta_1 - 124{,}15)^2 + 6{,}90^2}$$

Fig.3. Function of the shell's center surface

The "best geometry" established by the program is shown in the lower part of the figure: two hyperbolas (HYPE) with the intersection HT = 31,00 m.

SHELL + COLUMN - FRAMEWORK
EVALUATION INDEX

HT =	28.00	29.00	30.00	31.00	32.00
YU = 25.00	0.00	0.00	0.00	0.00	0.00
YU = 26.00	37.02	0.00	0.00	0.00	0.00
YU = 27.00	37.16	36.90	36.82	0.00	0.00
YU = 28.00	37.27	37.13	36.95	36.82	0.00
YU = 29.00	37.47	37.30	37.15	37.04	36.89
YU = 30.00	37.67	37.50	37.36	37.23	37.11
YU = 31.00	37.92	37.78	37.63	37.48	37.37

BEST GEOMETRY

TYP	Y	A	B	C	HT
HYPE	41.390	0.360	6.440	0.000	31.000
HYPE	28.000	13.750	39.813	0.000	31.000

Fig.4. Optimization of shell and supporting framework

In the upper matrix this geometry - characterized by HT = 31,00 and YU = 28,00 - is given the evaluation index of 36,82. This figure is obtained by multiplying the concrete and steel quantities determined for the shell and the column framework with price-bound factors. The ISAR geometry is immediately adjacent to this value with HT = 30,00 and YU = 28,00. Its evaluation index is 36,95 which is a very good correspondence (0,35 % of deviation).

2.2 Position and dimensions of the stiffening rings

The position and the dimensions of the stiffening rings should be chosen such that the stability and vibration behaviour of the shell can be influenced in the best possible way [6].

For the stability behaviour this condition is fulfilled if the rings are positioned at those points where the relevant buckling modes of the unstiffened shell would produce the greatest deformations. If the relevant buckling modes of the ring-stiffened shell can be forced to cross the zero axis at these points this is indicative of the ring dimensions having been chosen in the best possible way.

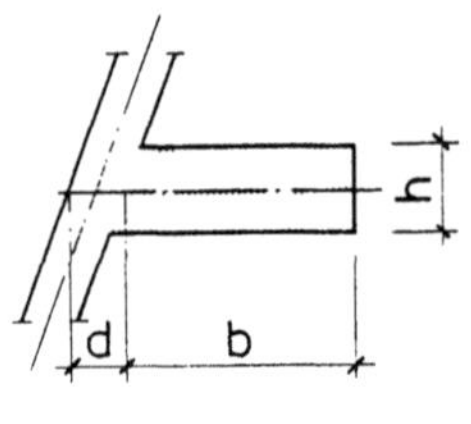

	b	h	d	b/d
	[m]			
Ring No. 1	1.39	0.40	0.22	6.32
Ring No. 2	1.40	0.40	0.20	7.00
Ring No. 3	0.91	0.40	0.18	5.06

Fig.5. Ring dimensions

In [6] it is recommended to choose the ring width 'b' (cf. fig. 5) 5 to 8 times as high as the local thickness of the shell 'd', whereas the ring height 'h' can be chosen in acc. with structural aspects.

It is not so easy, however, to influence the shell's vibration behaviour positively by means of stiffening rings as there are several influences with partly opposite effects to be taken into consideration. Whereas the eigenfrequencies are reduced on the one hand by the additional ring masses, they are, on the other hand, increased due to the higher bending stiffness resulting from the rings.

A very decisive influence on the vibration behaviour of the shell is exercised by the eigenfrequencies of the decoupled rings. As has been shown in [7] the eigenfrequency of the coupled shell-ring system can be considerably raised only if the eigenfrequency of the free ring is higher for transverse modes at the ring level than that of the unstiffened shell. Only in such case is it possible to obtain the desirable frequency shift.

In case of the ISAR II cooling tower, however, choosing the position of the stiffening rings for optimality was possible only to a limited extent. The two lower rings had to be provided for at the level of the air-traffic warning lights as they are intended for use as a maintenance walkway after completion, whereas the location of the upper ring could be chosen freely. The studies performed and described in the following chapters have shown, however, that it is possible - despite these restrictions of position (cf. fig. 1) - to achieve higher buckling safeties and an increased lowest eigenfrequency.

2.3 Stability studies

The stability behaviour of unstiffened shells is actually justified by means of the procedure recommended in [4]. With the interaction diagrams contained in [4] the wall-thickness curve of the cooling-tower shell is to be designed in such a way that its buckling safety at each point is greater than or equal to 5 as required in DIN 1045, section 24.3 and in [4], section 1.3.4. For ring-stiffened shells no such interaction diagrams have been published until now.

In order to be able to evaluate the buckling behaviour of the ISAR II ring-stiffened shell from an engineering point of view it was therefore decided to resort to a procedure that had already been applied for the dimensioning of the ring-stiffened shell of the Doel III cooling tower [7]. This procedure is based on the assumption that the realistic buckling safety of the ring-stiffened shell can be deduced from the classical buckling safety of the ring-stiffened shell by allowing for a reduction factor, such that:

$$\bar{\nu}_R = C \times \bar{\nu}_{kl}$$

where

$\bar{\nu}_R$ = realistic buckling safety of the ring-stiffened shell

$\bar{\nu}_{kl}$ = classical buckling safety of the ring-stiffened shell

C = reduction factor

The reduction factor is determined from the ratio of the realistic buckling safety of the unstiffened shell in acc. with [4] to the global buckling safety obtained from the classical theory of stability also for the unstiffened shell with:

$$C = \frac{\nu_R}{\nu_{kl}}$$

where

ν_R = realistic buckling safety of the unstiffened shell

ν_{kl} = classical buckling safety of the unstiffened shell

It is assumed that this reduction factor is valid for the ring-stiffened shell as well.

All stability studies were based on the following load cases:

D e a d l o a d 'g' with a calculation weight for reinforced concrete of 25 kN/m³.

W i n d l o a d 'w' with a height-dependent dynamic pressure for wind zone III in acc. with [4] (cf. fig. 6). The dynamical amplification factor to be applied in acc. with [4], fig. 7, is φ = 1,02 on the assumption of a lowest eigenfrequency of abt. 1,00 Hz. The distribution of pressure in the circumferential direction led to the distribution curve K 1.1 in acc. with fig. 1 in [4] and depending on the roughness produced by the formwork system used.

I n t e r n a l   s u c t i o n 's', to be chosen in acc. with [4], section 1.1.2 f. in the vertical and circumferential direction as a constant value

$$s = -0,5 \times q_E(\Theta = 165,50) = -0,695 \text{ kN/m}^2$$

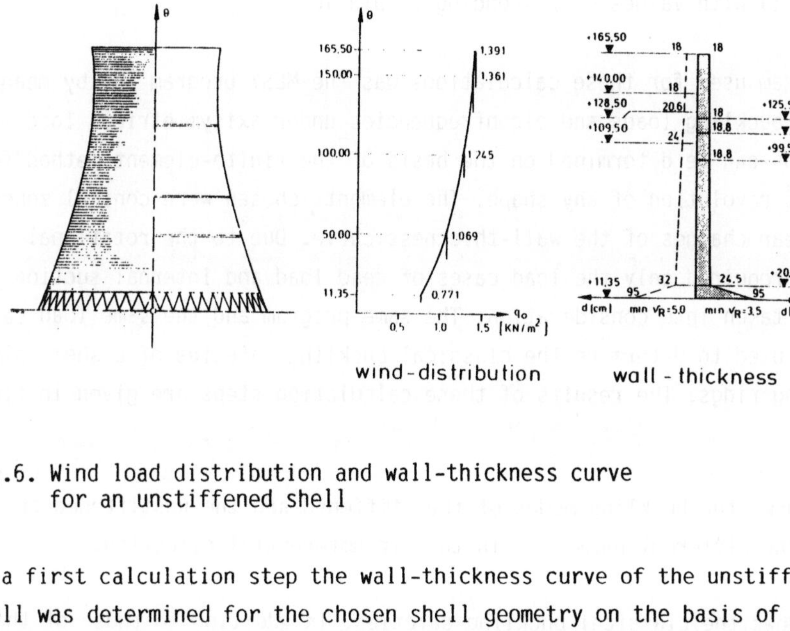

Fig.6. Wind load distribution and wall-thickness curve for an unstiffened shell

In a first calculation step the wall-thickness curve of the unstiffened shell was determined for the chosen shell geometry on the basis of the load cases given above. The design criterion was a lowest realistic buckling safety of ν_R = 3,5 to be determined in acc. with [4]. The calculation of the pertaining internal forces was carried out in acc. with the membrane theory and in application of the characteristics procedure [8].

For the purpose of comparison fig. 6 shows the respective calculation results together with the wall-thickness curve that would have been required if the shell had been executed without stiffening rings. It can be clearly seen that the shell would have had to be executed over large sections with considerably higher wall thicknesses and correspondingly higher amounts of material.

In a second calculation step the buckling safeties are determined in acc. with the classical stability theory for the shell designed with $\nu_R = 3{,}5$. In order to allow, at least approximately, for the quite considerable influence of the column framework it will be considered in the calculation as a substitute shell with a reduced modulus of elasticity and a reduced density. The transition to the foundation will be represented by a non-displaceable support allowing for free rotation in the meridional direction. The material assumed is an isotropic and fully elastic material (concrete in state I) with values corresponding to DIN 1045.

The program used for these calculations was the MESY program [9] by means of which buckling loads and eigenfrequencies under axisymmetrical load conditions can be determined on the basis of the finite-element method for shells of revolution of any shape. The elements chosen were conical zones with linear changes of the wall-thickness curve. Due to the rotational symmetry required only the load cases of dead load and internal suction could be taken into consideration. The same program and the same load cases are then used to determine the classical buckling safeties of a shell plus stiffening rings. The results of these calculation steps are given in figs. 7 to 9.

Fig. 7 shows the buckling modes of the stiffened and the unstiffened shell and for the different harmonics in the circumferential direction.

Fig. 8 shows the classical buckling safeties. In addition to those of the unstiffened shells with a lowest realistic buckling safety of 5 (curve 1) and of 3,5 (curve 2), we find here the classical buckling safeties for a twofold (curve 3), a threefold (curve 4) and a fourfold ring-stiffened shell (curve 5). These shell variants were also taken into consideration during the preparation of the tender and this for the following reasons: firstly in order to find out whether the twofold ring-stiffening, such as it had already been executed for the Doel III cooling tower, would not also have been sufficient for ISAR II, and secondly to evaluate the extent by which the level of safety of a threefold stiffened shell could have been raised by simply adding a fourth stiffening ring to the shell.

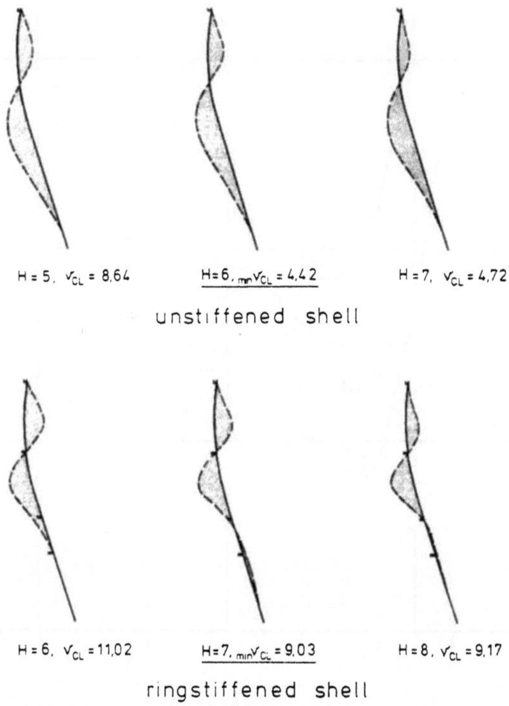

Fig.7. Buckling modes of the shell variants

Fig. 9 is a summary of the most important values giving the results of the buckling studies, the reduction factors and the realistic buckling safeties. For the purpose of comparison we have once again added the corresponding values for the Doel III cooling tower. Particular attention is drawn to the 3rd column for these are the values that would have been attained if the three stiffening rings could have been placed in their optimal positions. The lowest realistic buckling safety would then have risen from $\overline{\nu}_R$ = 7,13 to $\overline{\nu}_R$ = 8,10 for the load combination of dead load + wind + internal suction. As can be seen from fig. 10, it would, however, not have been possible any more in such case to use the rings as maintenance platforms for the air-traffic warning lights.

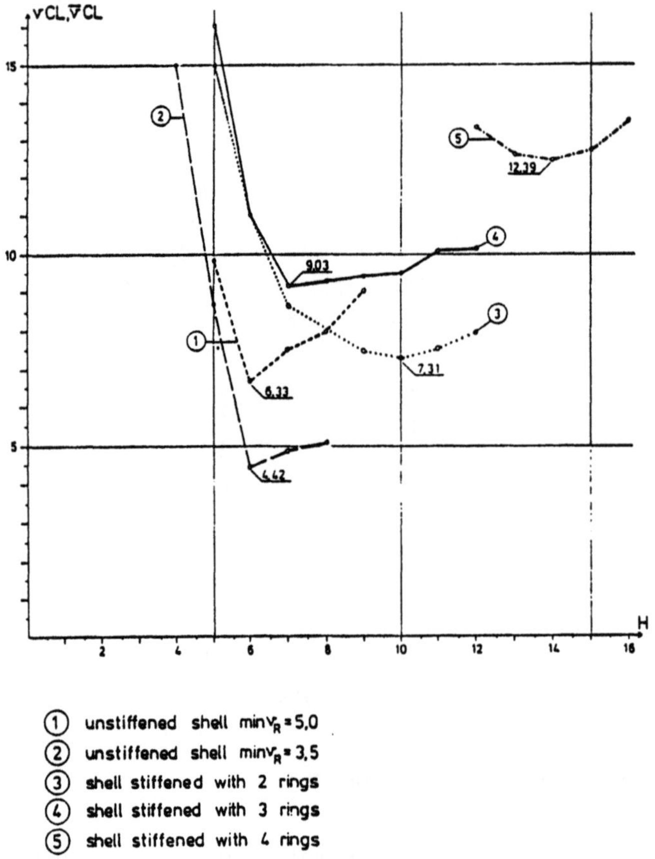

Fig.8. Classical buckling safeties of the shell variants

2.4 Dynamic studies

Cooling towers whose main load element is constituted by the natural wind are structures of high eigenfrequencies with regard to the wind's predominant excitation frequencies. In their design, the effects of the natural wind are therefore considered as quasi-statical loads with the resonant part of the structure's response being allowed for by a dynamical amplification factor that depends mainly on the lowest eigenfrequency and the shell's diametre [4]. Thus, with rising dimensions of cooling towers the dynamical studies are becoming more and more important. According to [10] the lowest eigenfrequency should be near 1 Hz. The dynamical calculations equally carried out with the MESY PROGRAM have shown that this would not have been the case for the ISAR II cooling-tower shell the lowest eigenfrequency of which was 0,74 Hz and 0,69 Hz respectively.

		Isar II				Doel III
		shell stiffened with				
		2 rings	3 rings	3 rings optim.	4 rings	2 rings
1	real. buckling safety ν_R shell unstiffened for d+w+s		3,50			3,50
2	real buckling safety ν_R shell unstiffened for d+s		3,83			3,94
3	class. buckling safety ν_{CL} shell unstiffened for d+s		4,42			5,31
4	class. buckling safety $\bar{\nu}_{CL}$ shell ringstiffened for d+s	7,31	9,03	10,25	12,39	10,69
5	reduction factor $C = \frac{\nu_R}{\nu_{CL}}$		0,87 (0,79)			0,74 (0,66)
6	real. buckling safety $\bar{\nu}_R$ shell ringstiffened for d+s	6,36	7,86	8,92	10,78	7,93
7	real. buckling safety $\bar{\nu}_R$ shell ringstiffened for d+w+s	5,02	7,13	8,10	8,52	7,04

Fig.9. Results of the buckling studies

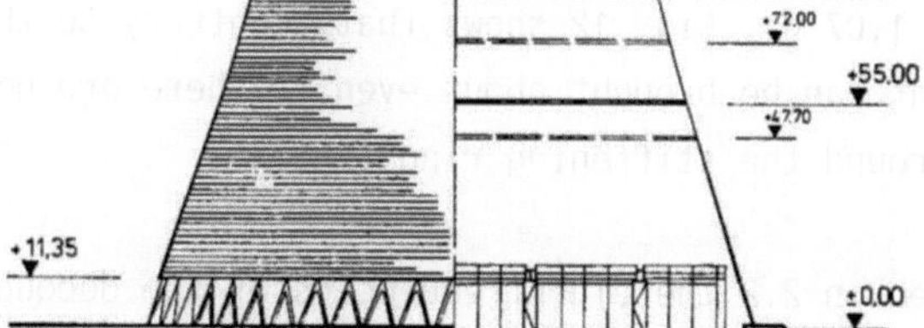

Fig.10. Optimized ring levels

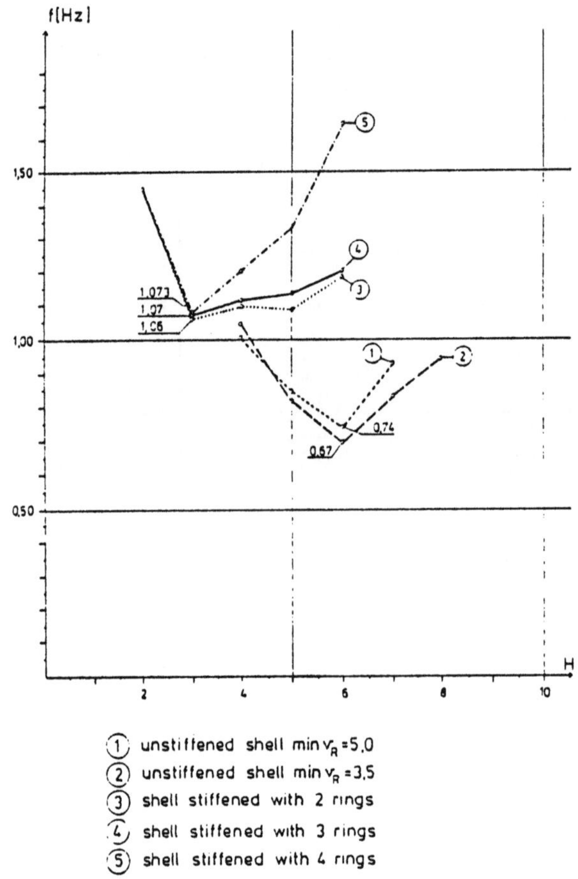

Fig.11. Eigenfrequencies of the shell variants

However, a marked improvement was brought about by the stiffening rings. As shown in fig. 11 the lowest eigenfrequency of the threefold stiffened shell has risen to 1,07 Hz. Fig. 12 shows that, contrary to the stability, a marked improvement can be brought about even if there are no nodes of the eigenfrequencies around the stiffening rings.

As already mentioned in 2.2 the eigenfrequencies of the decoupled rings have a considerable influence on the shell's vibrational behaviour. Fig. 13 demonstrates that the rings will produce a notable frequency shift of the coupled system only after the eigenfrequency of the free ring exceeds the frequency of the unstiffened shell for the circumferential wave of the same order.

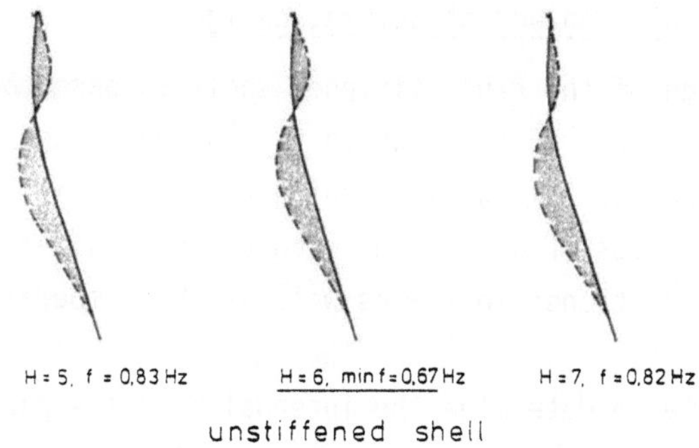

Fig.12. Modes of the shell variants

waves in circumference direction	unstiffened shell	ringstiffened shell	free ringeigenfrequencies ring No.1	ring No.2	ring No.3
2	1,49	1,47	0,48	0,46	0,35
3	1,07	1,07	1,35	1,29	1,00
4	1,04	1,11	2,59	2,48	1,92
5	0,82	1,12	4,19	4,01	3,11
6	0,69	1,20	6,15	5,88	4,57
7	0,81	1,30	8,47	8,09	6,29
8	0,92	1,51	11,14	10,64	8,27

Fig.13. Table of eigenfrequencies

2.5 Statical calculation and structural design

The statical design of the ring-stiffened shell is based on the equations for the shell's center surface shown in fig. 3 and on the wall-thickness curve with the three additional stiffening rings such as resulting from the preliminary studies (fig. 6). The load cases considered in the calculation were those mentioned in 2.3 as well as the temperature conditions.

In order to be able to determine the internal forces - particularly at the ring levels - with sufficient precision the cooling-tower shell was calculated in acc. with the bending theory by means of a shell program with axisymmetrical elements [11]. The calculation model comprised the shell with the upper and lower lintel, the stiffening rings and the column framework in acc. with the simplified modelling already explained in 2.3. The fact that the resting of the shell on the columns is discontinuous in reality has been additionally accounted for by a plate-theory calculation.

The results of these calculations have shown that even the ring-stiffened shell reacts to the stresses to which it is exposed mainly with membrane forces. Additional ring and meridian bending moments due to wind and temperature loading are produced only in the area of the upper and lower lintels and near the stiffening rings.

The dimensioning was carried out in acc. with the rules summed up in [4], section 1.3. The required cross-section of the meridional reinforcement in the lower part of the structure is determined by the load resulting from dead load + increased wind load, whereas only the minimum reinforcement was required for practically all the remaining parts of the structure and for the rings.

The criteria for the structural design of ring-stiffened shells hardly differ from those applicable to unstiffened shells. The recommendations given in [4], section 1.4, for the structural design of cooling-tower shells can thus be applied in full also to ring-stiffened shells.

The same refers also to the experience gained in practice. Due to the permanent contacts with the building site and the resulting change of ideas the practical experience made on site can be usefully combined with the structural engineering work so that the shell may be built not only to the high quality standards required but also with the corresponding economical efficiency.

The two following construction principles which contribute to the trouble-free and faultless placing of the shell reinforcement may serve as an example.

It has proved particularly advantageous to fix the staggering of the meridional reinforcement already in the first course and to adapt it to the concreting sections. It is obvious that such staggering must then be observed over the total height of shell if the connecting bars are chosen in lengths corresponding to a multiple of the concreting sections plus overlapping and if they are to begin at the concrete construction joints. The relatively complicated adjustment of the overlapping lengths is thus no longer necessary. It is equally advantageous to keep the number of the bars to be placed for the meridional reinforcement constant over the total height of shell. By placing the required number of bars once at the lower edge of shell possible omissions, confusions and disturbances in the distribution of forces around splices are excluded. The constantly changing bar spacing over the tower height and the graduation of the bar cross-sections allow a good adaptation to the required steel cross-section.

3. Construction of the shell

The working technique adopted for the construction of the ring-stiffened cooling-tower shell is guided by the effort to stick as far as possible to those building methods for "conventional" shells that have proved successful in the past.

Those changes, however, that are inevitably required due to the new design variant are to be conceived in such a way that the economical efficiency attained in the construction of ring-stiffened shells is not affected but rather assisted by the additional advantages of a reduced consumption of material. It has therefore been decided to build also the shell of ring-stiffened cooling towers with the help of conventionally used climbing scaffolds.

They consist of pairs of climbing towers fastened opposite each other on the inside and the outside of the shell with working platforms at two different levels arranged between them. The upper platform is used for placing the formwork and the reinforcement and for concreting whereas the lower platform is used as a working platform for stripping and curing of the newly placed concrete. To allow a vibration-free hardening of the fresh concrete the climbing scaffold is completely independent of the transfer formwork. It consists of two sets of forms with a height of 1 m each. The lower set is already fastened to the hardened concrete and supports the upper set as required by means of formwork brackets. The brackets are designed so as to offer at the same time the possibility of adjusting the constantly changing inclination of the shell. Transfer formwork and climbing scaffolding allow a work progress of at least 1 m a day. For the stiffening ring as such it was now required to develop a building method that adapts itself smoothly to the climbing procedure and that interferes with the latter as little as possible.

The in-situ concrete method already applied in the construction of the Doel III cooling tower has proved a definitely feasible solution. Details can be found in [7].

However, the positive experience gained with the mixed building method in the execution of the rings of the Mississippi I cooling tower led to the decision to use the same procedure also for the ISAR II tower.

The mixed construction is based on the principle of adding the rings to the shell without using the climbing scaffolds as a working platform. For this purpose a prefabricated element is used (cf. fig. 14) which - after having been fastened to the shell at the respective level - can be employed as a working platform and, at the same time, as lost formwork for the construction of the stiffening rings themselves!

This leads to the following working procedure:

- construction of the shell by means of the climbing scaffolds;
- manufacture of the prefabricated elements;
- insertion of the prefabricated element suspension bolts into threaded sleeves which are installed in the course of the climbing procedure (cf. fig. 15);
- lifting of the prefabricated elements to their respective position by

means of a mobile crane and fastening of these elements at the bolts;

- bending of the shell connection bars and placing of the ring reinforcement;
- casting of the stiffening ring.

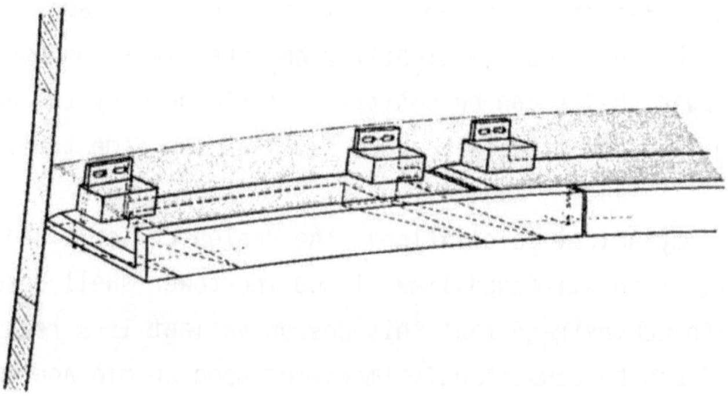

Fig.14. Prefabricated element for the construction of the stiffening rings

Fig. 15 shows the suspension, the prefabricated element and the in-situ concrete section.

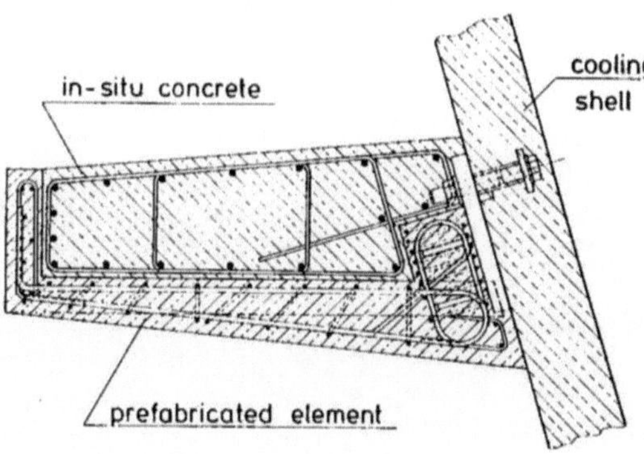

Fig. 15. Stiffening ring cross-section

Although the execution works on the ISAR II tower are still lying ahead of us we are sure that the smooth and efficient work progress achieved with this method at the Mississippi I cooling tower will also be possible for ISAR II.

4. Summary

Rising dimensions of cooling towers lead to lower buckling safeties and lower eigenfrequencies of their shells. Already some time ago, theoretical considerations had shown how the stability and the vibration behaviour of large cooling tower shells can be positively influenced by the provision of stiffening rings and how material can be saved at the same time.

The statical and dynamical calculations, the design criteria and the building method of the ring-stiffened ISAR II cooling-tower shell described in this contribution demonstrate that this design variant is a real alternative which has been found by consistently improving upon an old and successful construction method.

The ring-stiffened reinforced-concrete natural-draft cooling tower will therefore continue to play an important role as a safe and economical structure also for the cooling-tower dimensions of the years to come.

References

[1] Peters, H.L.: Neue Entwicklungen beim Bau von Naturzugkühltürmen. Wechselbeziehungen zwischen Bauausführung und Konstruktion. Vorträge Betontag 1977. Wiesbaden: Deutscher Betonverein e.V. 1977.

[2] Form, J. und Peters, H.L.: Economical aspects of design and execution of large natural-draft cooling towers. International Symposium on very tall reinforced concrete Cooling Towers, 8. - 10. November 1978, Paris.

[3] Krätzig, W.B.: Kühlturmformen besonderer Wirtschaftlichkeit. Konstruktiver Ingenieurbau/Berichte, Heft 1, Vulkan-Verlag, Essen, 1970.

[4] BTR-Bautechnik bei Kühltürmen, Teil 2: Bautechnische Richtlinien. VGB-Verlag Essen 1980.

[5] Krings,W.: Wirtschaftliche Berechnung des optimalen Entwurfs beliebiger Rotationsschalen mit der Membrantheorie für Naturzugkühltürme (in Vorbereitung).

[6] Zerna, W., Mungan, I.: Über das Beulen von Kühlturmschalen mit Versteifungsringen. Beton- und Stahlbetonbau, 76 (1981), S. 33 - 36.

[7] Form, J., Krings, W., Mazur, H. und Peters, H.L.: Berechnung und Ausführung eines ringversteiften Naturzugkühlturmes aus Stahlbeton. Beton- und Stahlbetonbau 75 (1980), S. 205 - 212.

[8] Krätzig, W.B.: Schnittgrößen und Verformungen windbeanspruchter Naturzugkühltürme. Beton- und Stahlbetonbau 61 (1966). S. 247 - 255.

[9] Hartmann, B., Schrader, K.H., Winkel, G.: MESY. Ein Programmsystem zur Untersuchung von Tragwerken. Konstruktiver Ingenieurbau Berichte, Heft 22, Essen: Vulkan-Verlag 1976.

[10] Jahrbuch Bautechnik im Kraftwerksbau. Kapitel VIII Kühltürme, Vulkan-Verlag Essen. 2. Ausgabe 1982/83.

[11] Doboviesek, B.: Berechnung prismatischer Schalen. Der Bauingenieur 47 (1972), S. 393 - 400. Doboviesek, B.: Berechnung des statischen Zustandes der Rotationsschalen und deren Sektoren. Der Bauingenieur 48 (1973), S. 109 - 115.

A New Type of Cooling Tower with Vertical Columns Disconnected from Shell Designed and Built in France

R. BORDET, L. CAUDRON, M. DIVER, Y. GROVALET

E.D.F. - SEPTEN - Cêdex n° 8 - 92080 PARIS LA DEFENSE - FRANCE

SUMMARY : Research has been done by E.D.F., concerning a new type of cooling tower, the traditional diagonal columns being replaced by vertical columns. Bearing pads can be placed between columns and lower lintel of shell.
The aims of the new type of cooling tower are :
- to improve thermal performance
- to cope, by jacking, with differential settlements, when bearing pads are provided.

1. SITUATION AT THE TIME THIS PAPER IS WRITTEN

Though the decision to use the new type of cooling tower is already taken, howewer at the time this paper is written (March 1984), final calculation and drawings are not yet available, design being in progress.
It is believed that the first cooling tower with vertical columns will be built maybe end 1984 or the beginning of 1985 at Chooz in northern France. (fig 1.)*

2. HISTORY

The idea to dissociate shell from columns came some years ago when engineers helplessly contemplated a high French cooling tower distorting under the unexpectedly important action of differential settlement. The settlement came to end before any dangerous effect was noted. But the feeling of powerlessness traumatised many specialists and led to the idea that a new type of cooling tower had to be studied, with the possibility of human intervention in case of non-homogeneous soil.

* reproduced with the agreement of HAMON, PARIS.

It was during research on models that what is probably the most important advantage of the new type of cooling tower was realised, that is better thermal performance due to improved air flow between columns. Incidentally it was observed that the new cooling tower had a better behaviour concerning Euler buckling.
From economic point of view, the cost of traditional and new cooling tower is pratically the same. In fact, the new cooling tower on vertical columns being a prototype, its thermal and structural advantages are not fully taken into consideration as a measure of safety. It is possible that in future, if such new cooling towers are to be built, their cost will be smaller than traditional structures for a pratically identical degree of safety.

3. STRUCTURE ON ELASTIC SOIL

The traditional cooling tower is replaced by two independant structures :

- the shell, becoming a self-standing structure, with the lower lintel being, in a principal variant, stiffened by an annular beam ;
- the substructure, formed by vertical columns, resting on soil.

Between the two structures, bearing pads may be provided.
It is supposed that differential settlement practically comes to end after limited time (e.g.5-10 years). The two structures may be linked together after this period.
Fig 2 and 3 show lower part of such a cooling tower on vertical columns. Distance between columns is about 18 m. Height of cooling tower about 175 m.
Fig 4 shows Euler-buckling deformation of structure indicated in fig. 2 and 3. Factor of safety for (Own weight + Normal French Code wind) = 1

Period of vibration is 1.1 to 1.2 seconds, related to elasticity of soil.

4. STRUCTURE ON RIGID SOIL

In that case physical disconnection between shell and substructure is no longer useful (fig.5).

5. SECTION OF COLUMS

The columns are to be designed in such a way that thermal performance shall be improved at its uppermost. A profile of column is shown fig 6.

6. CONCLUSION

It is disappointing that because of the timing of the symposium this paper has to be sent for printing before all research and design are concluded.
Nevertheless, researchs done until now give the impression that the new type of cooling tower, if and when built, has a good case concerning stability and performance when compared with traditional structures.

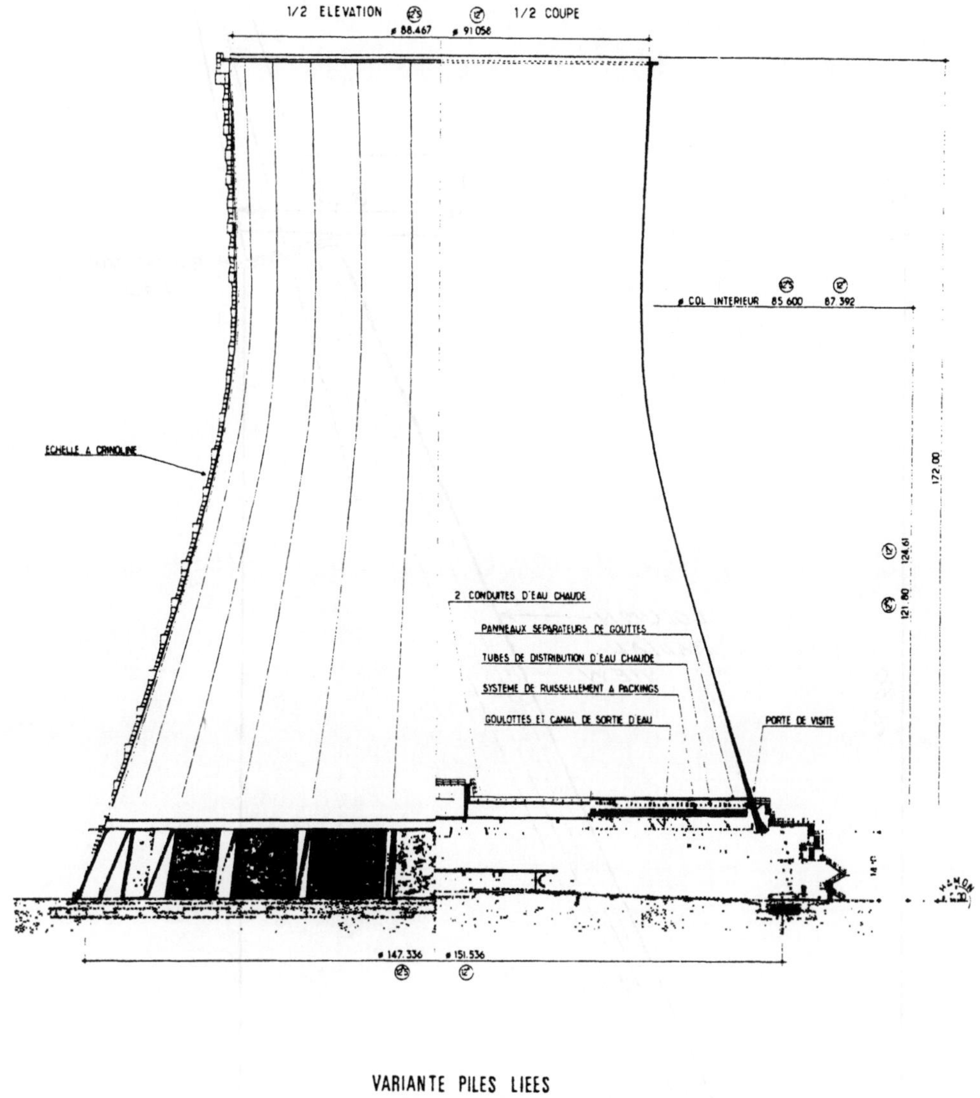

Fig. 1. Cooling Tower with Vertical Columns

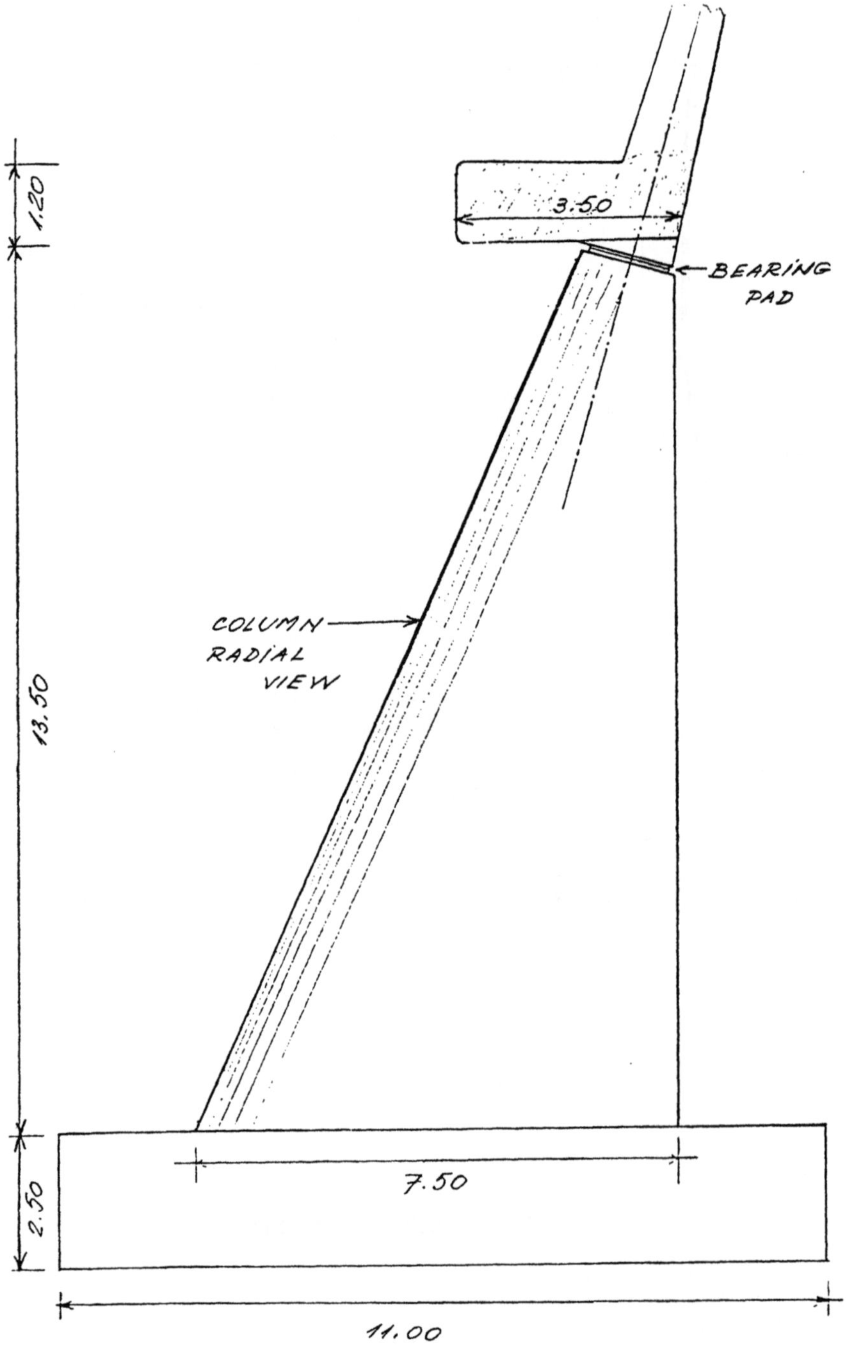

Fig. 2. Columns of the Cooling Tower, Side-view

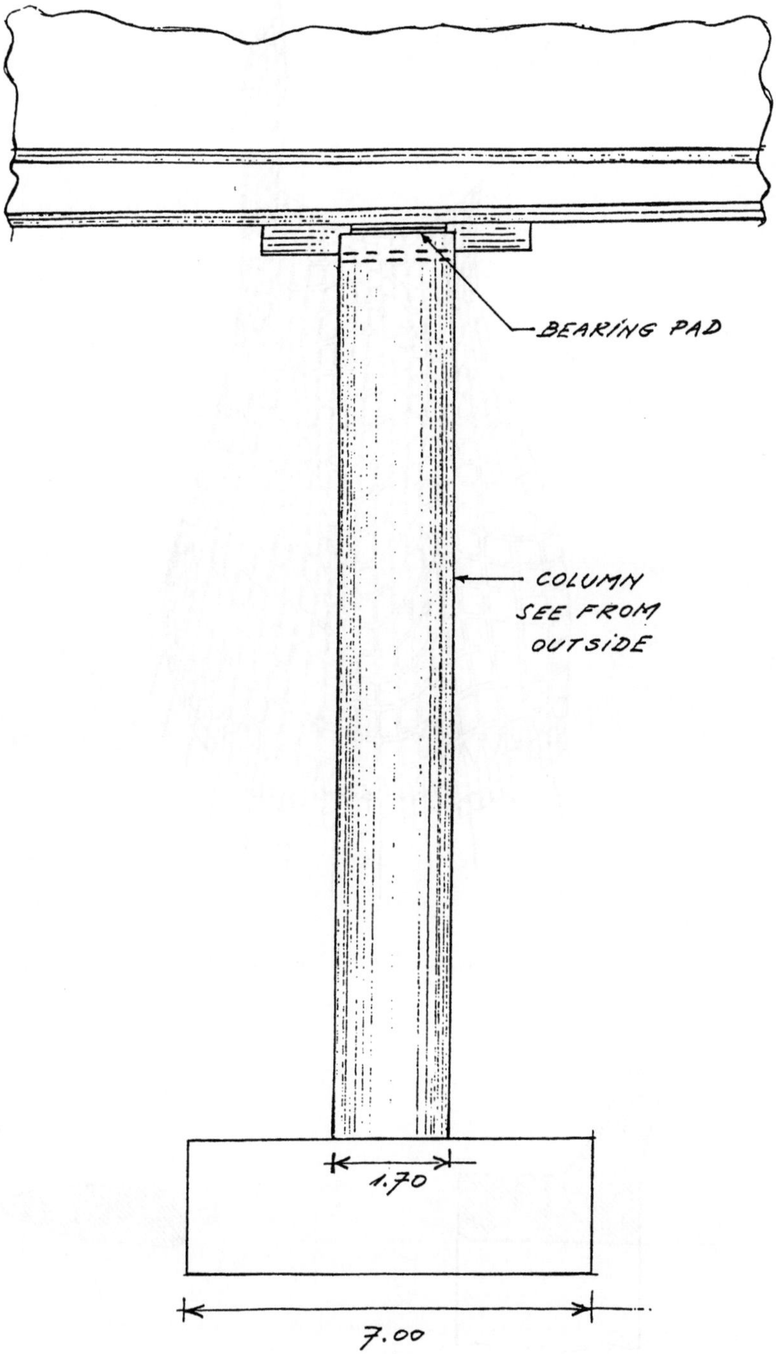

Fig. 3. Columns of the Cooling Tower, Front-view

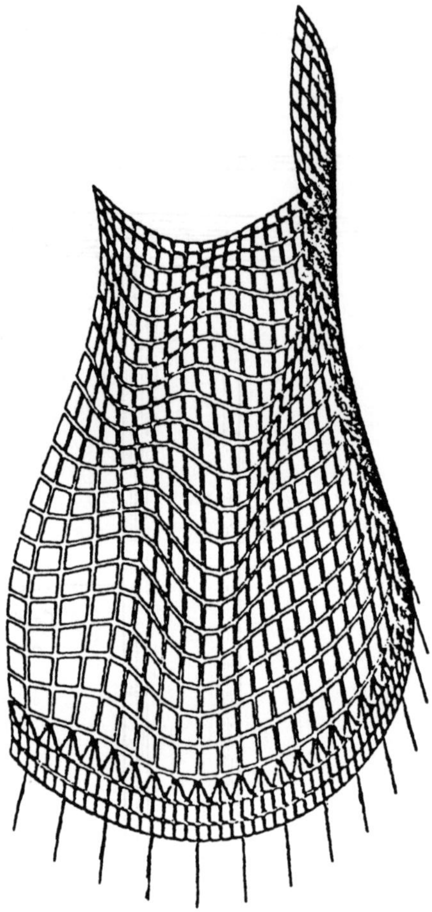

Fig. 4. Buckling Mode

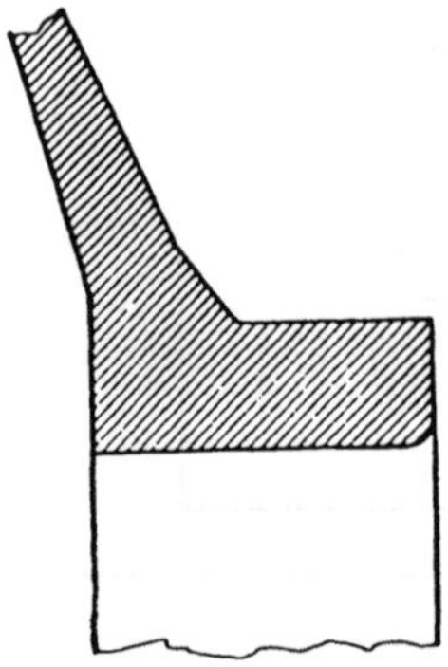

Fig. 5. Connection between Shell and Columns

Fig. 6. Profile of Columns

1984
IASS-RUB

Wind Loading

Reliability of Current Design Methods for Wind-Induced Stresses

H.-J. NIEMANN

Institut für Konstruktiven Ingenieurbau
Ruhr-Universität Bochum

1. Introduction

After nearly two decades of research related to the wind action on cooling towers it may be questioned, whether there is anything left worth mentioning. As a matter of fact, cooling towers are extremely sensitive to wind loads, static and fluctuating, and a thorough knowledge of the load and the structural response is a demand of reliability and economy at the same time. The static component, generally defined as a temporal average taken over 10 min or so, is rather well approximated in an engineering sense, although from the point of view of science the influence of wind shear is still not very well understood. A strong fluctuating load component is introduced by wind gustiness and by vortices separating from the tower surface. This component contributes strongly to the total response peaks at a rate of at least 50 %. We have to deal with a random excitation, varying statistically in time and space. The response may be calculated in the frequency domain by spectral analysis. This method has been introduced for cooling towers by Abu-Sitta (1974, ref.1) and applied for a systematic analysis by the author (1976, ref. 2). Although well established in principle, it is not used for practical design, mainly due to the large amount of calculation involved. Instead, the gust effect is accounted for by a design wind speed, which includes the fluctuating response in the sense of an equivalent static load. This approach is quite usual in many design codes. It is also the background for the VGB-recommendations, ref.3. The basic limitations of a concept based on the gust-response factor are obvious: the dynamic factor can include just one response quantity exactly. The others are obtained with different degrees of conservative or inconservative approximation. For a cooling tower there are roughly speaking four design situations:

- the maximum tensile stress for the design of reinforcement in both, meridional and circumferential direction;
- the maximum compressive stress at the tower sides for the consideration of compressive strength and buckling safety;
- the maximum shear at the base of the shell for the design of the columns;
- bending moments for the design of bending stiffness.

The design wind load is modelled to reproduce the meriodinal tensile stress. The gust response factors of the other design stresses are different and so would be appropriate design loads.

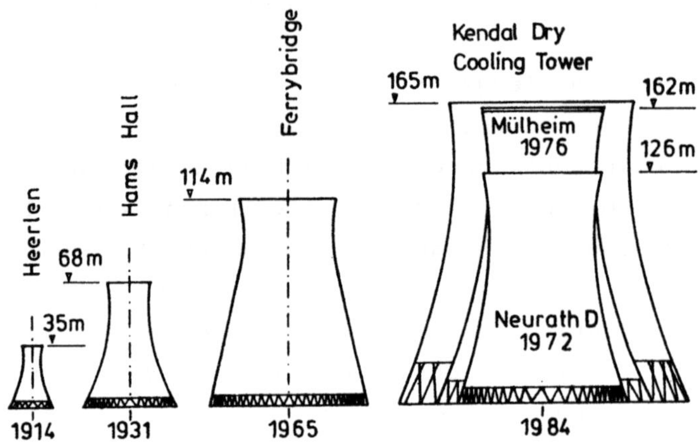

Fig.1. Developement of cooling tower dimensions

The deficiencies of the equivalent static load concept are balanced by a set of provisions such as minimum reinforcement, high buckling safety etc. which have to be observed within this design concept. Many cooling towers have been designed successfully and experience has proved that a satisfactory level of reliability is obtained. Now, a new generation of cooling towers has come up, characterized by a ratio of height to mean diameter of about 1.3, with heights of 165 m, and a much smaller curvature than the preceeding, see fig. 1. Furthermore, cooling towers are designed as single towers in general, if some mininum distance from neighbouring buildings is observed. Experimental evidence indicates, that in spite of this regulation interference effects produce additional stresses. These aspects give rise to the question whether the running concepts are still applicable or need some refinement.

2. Gust response of single towers

2.1 Design wind load based on gust-response factor

The response of a cooling tower to a gusty wind fluctuates randomly. The statistical distribution of the amplitudes is Gaussian. Therefore, the mean value, $\bar{Y}$, averaged over 10 min, and the mean square, σ_Y^2, are sufficient to describe the statistical properties completely. The peak response occuring in a time interval of 10 min duration of an extreme storm is an appropriate design quantity. It is composed of the mean, static part and the fluctuating component in terms of the root of the mean square σ_Y (rms-value or standard deviation):

$$Y_p = \bar{Y} + g \cdot \sigma_Y \tag{1}$$

The peak factor, g, is in the order of 3 to 4, depending on the frequency content of the process. The ratio of peak response to mean response is termed gust-response factor GRF:

$$GRF = \frac{Y_p}{\bar{Y}} = 1 + g \cdot \frac{\sigma_Y}{\bar{Y}} \tag{2}$$

The GRF may be used to determine a design wind load in terms of an equivalent static load. In the VGB-guidelines, it is based on the total wind force:

$$W_D = GRF \cdot \bar{W} \tag{3}$$

Obviously, a main requirement is to ensure a sufficient tensile resistance. Therefore, the gust response of the meridional tensile force is used. For open terrain (profile exponent of mean wind $\alpha = 0{,}16$) the design velocity pressure q_D obtained from equ. (3) is

$$q_D(z) = 2{,}3\ \bar{q}_{10} \left(\frac{z}{10}\right)^{0{,}234}, \tag{4}$$

which includes a gust-response factor of roughly 1,9 (z = height above ground in m, $\bar{q}_{10}$ - velocity pressure of mean wind at reference height z = 10 m). In fig. 2, the result of the design load calculation is compared to the real behaviour of the meridional force. At each point at the circumference, fluctuations are superimposed on the mean response, with peaks occuring non-simultaneously. The envelopes of the positive and negative peaks are needed to design for tensile and compressive strength. The calculation with a static design wind coincides with the positive peak at the windward side. It underestimates the negative peak at $\Theta = 70°$. Regarding the gust response factor, it is greater for the compressive meridional force than for the tensile force. Before inspecting the limitations of the design wind load concept in more detail, a method is needed for the calculation of gust response.

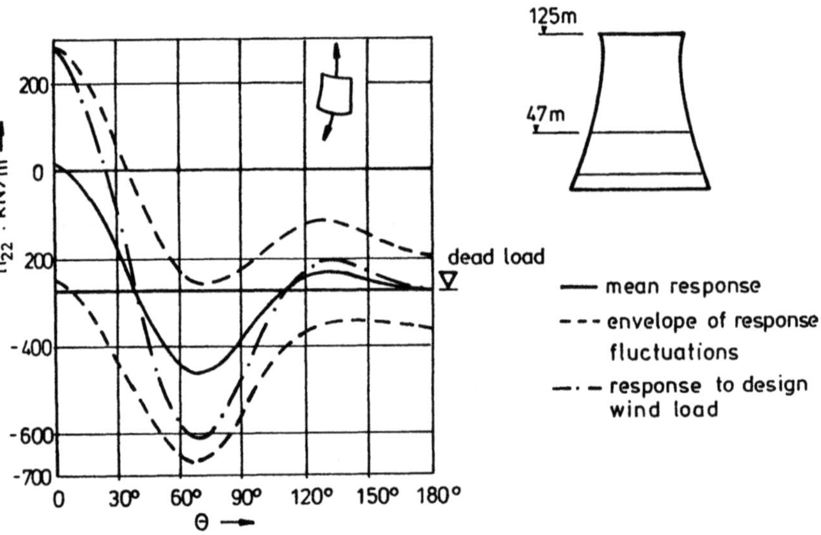

Fig. 2. Peak meridional forces

2.2 Quasi-static calculation of response variance

Excitation from wind turbulence is distributed over a broad range of the frequency. Most energy is concentrated at low frequencies. Only a small proportion produces resonance with the relatively high eigenfrequencies of cooling towers. The total variance of the response σ_y^2 may be split up into two parts: the low frequency component is produced without dynamic magnification and called quasi-static, σ^2_{YQ}. The high-frequency component σ^2_{YR} stems from resonance. The sum of both equals the total variance:

$$\sigma^2_Y = \sigma^2_{YQ}+\sigma^2_{YR} \tag{5}$$

Normally, the contribution of resonance to the total variance is less than 20 %, which means that the quasi-static rms-response comes up to more than 90 % of the total rms-response. Therefore, the knowledge of the quasi-static response will be largely sufficient.

The calculation of the total variance is rather complicated partly because the spectral distributions of the wind pressures cannot be established in an easy manner. Contrarily, the quasi-static response can be calculated much easier. The mechanical admittance in the low frequency range is simply the static flexibility. The fluctuating loads may be introduced in terms of pressure variances and correlations, both of which can be measured with great precision. The matrix of quasi-static response variances for a linear behaviour of the structure is obtained from

$$\underline{\sigma}^2_{YQ} = \underline{H}_0 \cdot \underline{\sigma}^2_p \cdot \underline{H}^T_0, \tag{6}$$

where $\underline{H}_0$ - flexibility matrix, T = transposed,

$\underline{\sigma}^2_p$ - covariance matrix of fluctuating loads.

The diagonal terms $\sigma^2_{p,ii}$ are the single point pressure variances, $\sigma^2_{p,ik}$ are the pressure covariances, which may be expressed by the correlation coefficient ρ_{ik}:

$$\sigma^2_{p,ik} = \rho_{ik} \cdot \sigma_{p,ii} \cdot \sigma_{p,kk} \tag{7}$$

Three sets of data are now necessary for the wind load,

- the pressure coefficients c_p as usual, for the static load,
- the r.m.s.-pressure coefficients c_{rms}, and
- the correlation coefficients ρ.

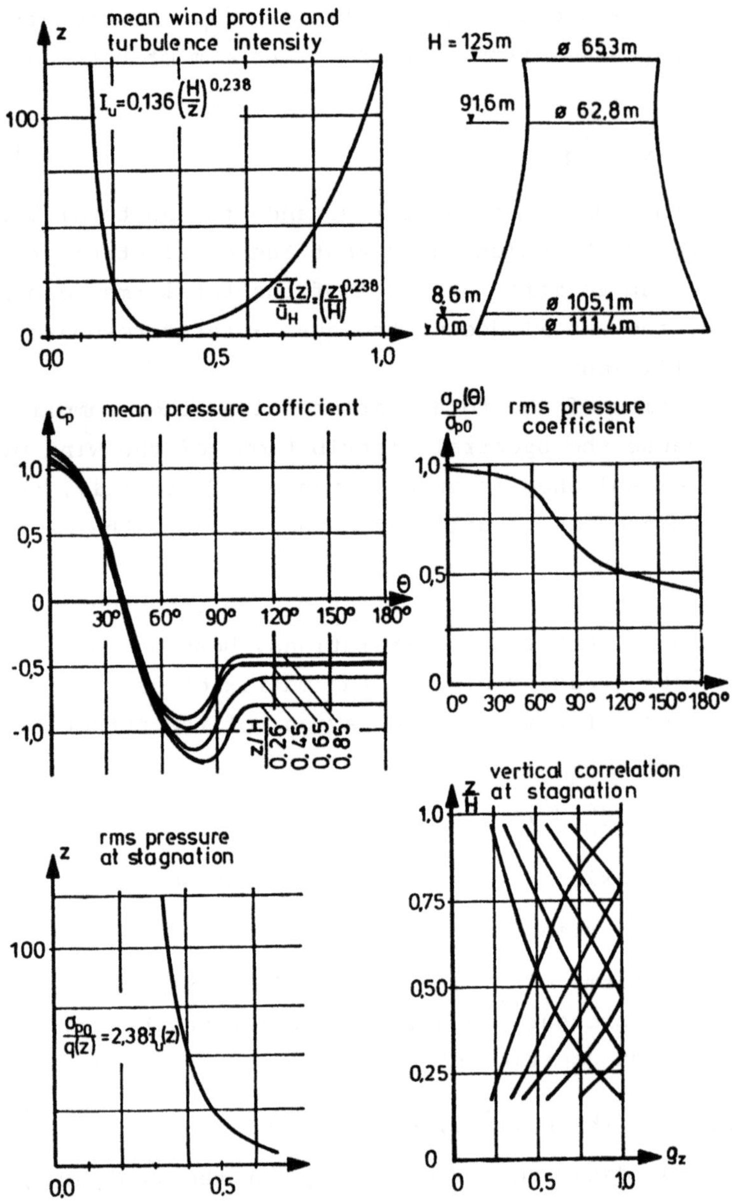

Fig. 3. Mean and rms wind pressures

The rms-pressure coefficients and correlations may easily be included in codes. Eventually, resonance factors have to be added to account for resonant contribution. The load data for a typical cooling tower are shown in figrs. 3 and 4. They were measured in a boundary layer wind tunnel at a scale of 1:400.

Wind profile $\bar{u}(z)$ and turbulence intensity $I_u(z) = \sigma_u(z)/\bar{u}(z)$ correspond to a moderately rough terrain. The pressure coefficients c_p at a height z are referred to the velocity pressure at the same height. It is typical in a wind profile, that the pressure distributions vary with height: They increase with increasing height. The rms-pressure at stagnation, σ_{po}, is referred to the mean velocity pressure. This coefficient is approximately proportional at each height to turbulence intensity I_u. A factor of 2,38 was found in these measurements. The value is rather high compared to 1,8 an 2,0 stated elsewhere. This is an important factor. It relates the fluctuating load to the mean and the rms-response is directly proportional to it. The correlation of the pressure fluctuations at stagnation, ρ_z, is similar to the correlation of the velocity fluctuations. It is positive everywhere. At other meridians, the correlations show the same typical features with some differences concerning the

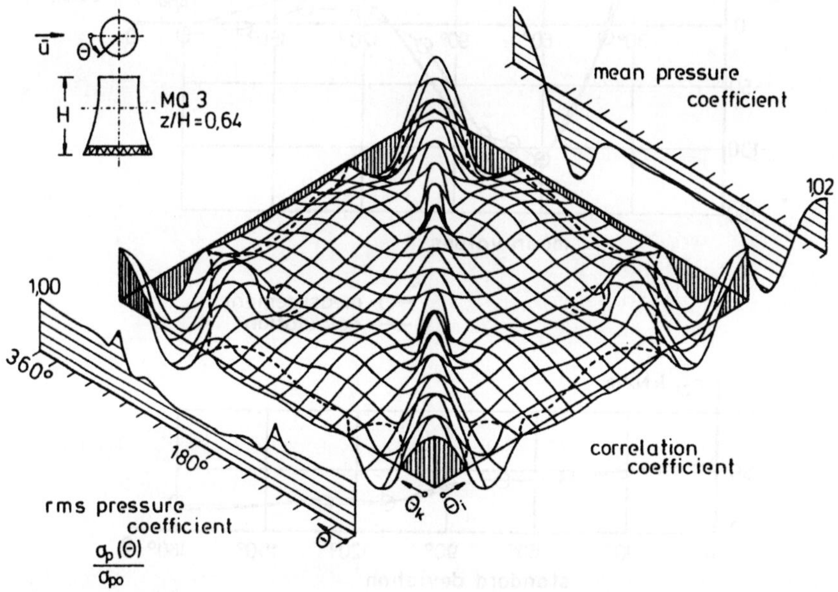

Fig. 4. Circumferential pressure correlation

rate of decrease. The distribution of rms-pressures $\sigma_p(\Theta)$ along the circumference is related to σ_{po}. It is nearly independent of height z above ground. Fig. 4 contains the correlation ρ_Θ along the circumference. The reference points are found on the diagonal.Reference points in the vicinity of stagnation show strong negative correlation with points in the range of maximum suction. Reference points at maximum suction at one side have positive correlation with points at the other. The circumferential correlations are rather independent of height. For the calculations, the correlation of points separated vertically

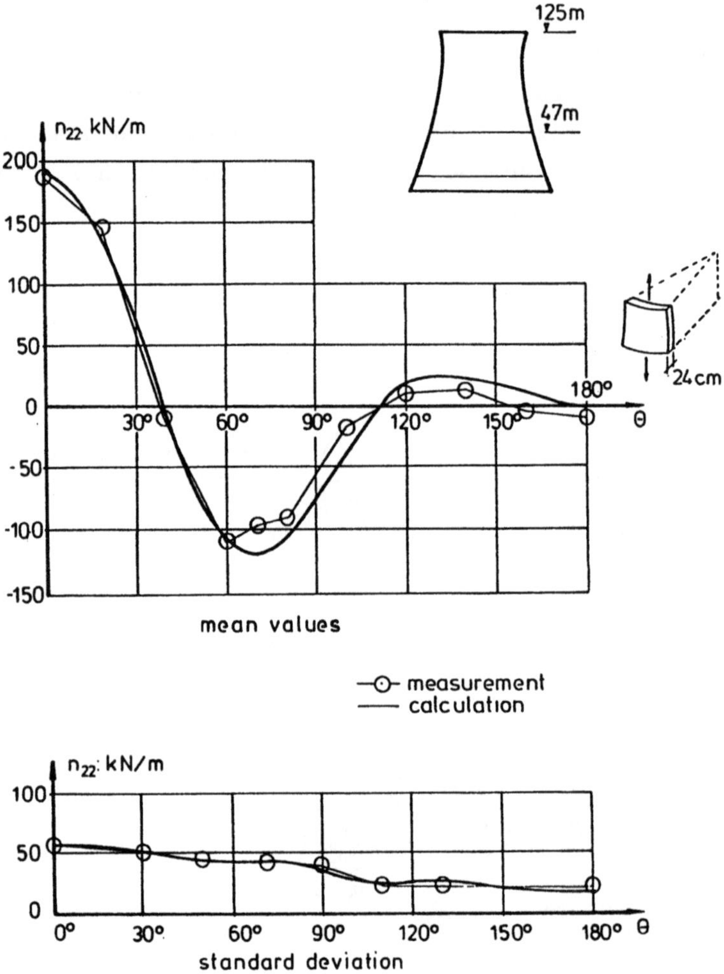

Fig. 5. Comparison of mean and rms meridional forces obtained from measurements and quasi-static calculations

and horizontally $\rho(z_1,z_2,\Theta_1,\Theta_2)$ is needed. It may be approximated by

$$\rho(z_1,z_2,\Theta_1,\Theta_2) = \rho_z(z_1,z_2)\cdot\rho_\Theta(\Theta_1,\Theta_2). \tag{8}$$

Eventually, this approach can be improved, based an additional experiments. Calculations of the mean and rms-response were performed with this set of loading data. The computer program MESTRA was a very helpful and reliable tool. For experimental verification, an aeroelastic model was tested in the boundary layer wind tunnel. The model is constructed in such a way, that elastic and dynamic similarity is ensured. The lowest model eigenfrequency is f_M = 186 Hz corresponding to a full-scale f_F = 1,12 Hz with a frequency scale of 166. The internal forces are obtained from measured strains. In fig. 5, the mean and rms-values of measured and calculated meridional forces are compared. The agreement is excellent. It is even better for the shear force but not quite as good for hoop forces and bending moments, partly due to experimental errors but also due to deficiencies of the mechanical model applied in the calculations. The suitability of the quasi-static method proposed here is obvious.

2.3 Gust response factors.

The gust response factors (GRF) of different design forces and bending moments are determined for the cooling tower from fig.3. The following maxima or minima with respect to circumference are considered, s.fig. 6:

Meridional tension max n_{22} occurs at stagnation $\Theta = 0°$. The gust response factor is about 2,2 and rather constant in the lower parts of the shell. The usual calculation with a design wind load well approximates the real peak response. In this region maximum reinforcement is necessary. In the upper part the gust effect increases relatively and is underestimated by the design wind load. Here, a minimum of reinforcement is required to cope for additional gust effects.

Meridional compression min n_{22} occurs at about $\Theta = 70°$. The GRF is somewhat larger than included in the design wind. This force is mainly important for determining buckling safety. Regarding the high safety factor of 5 and the time required to initiate buckling, the underestimation of peak compressions does not seem to be critical.

Shear force max $|n_{12}|$ occurs at $\Theta \cong 40°$. Its approximation by the design wind is best.

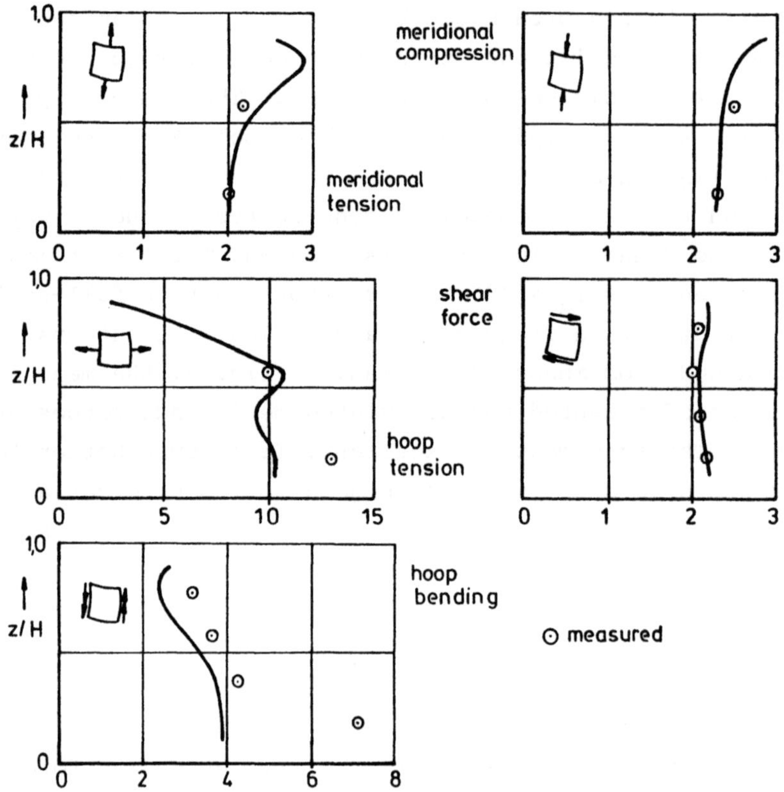

Fig. 6. Gust response factors

All of these design forces in a single tower are rather well approximated by the design wind with the exception of the meridional tensile force in the upper part of the shell. Here, a sufficient tensile strength has to be ensured by a minimum reinforcement. Calculations for a much larger tower with a height of 165 m showed that similar conclusions could be drawn. Contrarily, the peaks of tensile hoop force and bending moments are largely underestimated applying the design wind load. This approach is improper to model the structural response in these cases.

For the hoop force n_{11}, the cause of the large amount of gust response may be read from fig. 7: The internal pressure p_i produces circumferential compression, superimposed on the tension due to external pressure. The difference is small and so GRF is large. Clearly, a calculation with the design wind

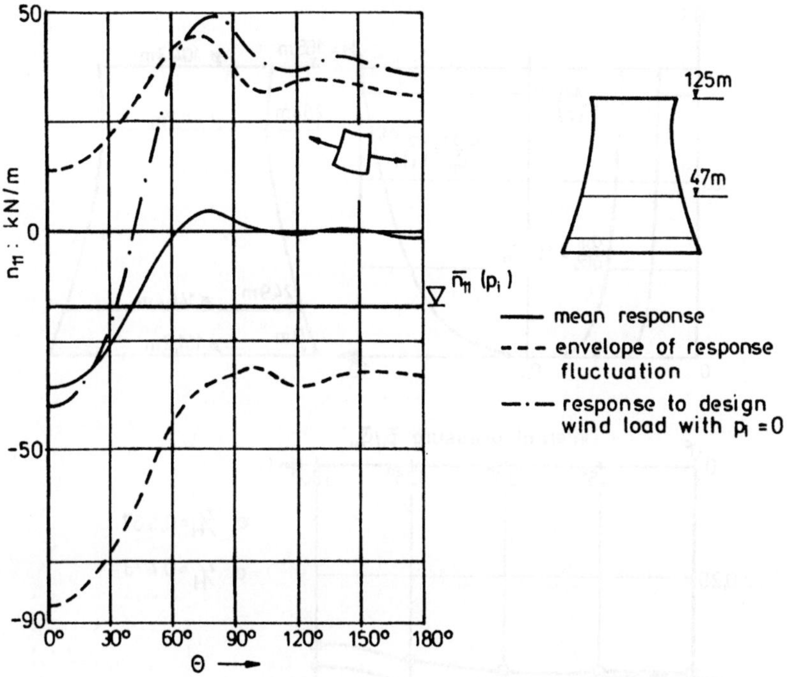

Fig. 7. Peak hoop force due to external and internal wind pressure

based on GRF = 2,2 does not reproduce peak tension. If, however, hoop tension is calculated for external pressure alone, $p_i = 0$, then the peak tension will be well approximated. Contrarily, internal pressure should be included to determine the maximum of hoop compression. Additional investigations are necessary to check whether this procedure is valid in general.

The mean internal pressure is not distributed uniformly around the circumference, s.fig. 8, particularly for a large crest diameter. An additional resulting wind force is introduced. The pressure fluctuations are small at the base and increase with height. At the crest, the rms-values are similar to external pressure at stagnation. These effects can easily be incorporated in a quasi-static calculation as described in section 2.2.

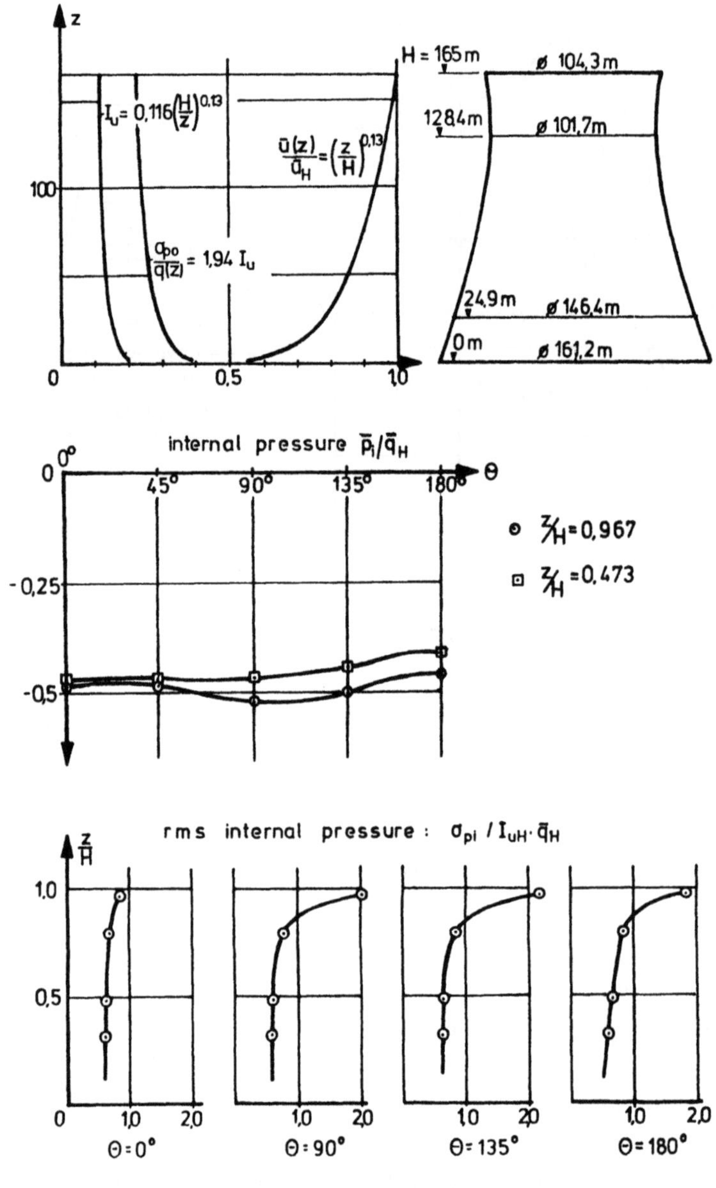

Fig. 8. Mean and rms values of internal pressure

2.4 Bending.

Bending moments occur at the upper and lower boundaries of the shell and eventually at stiffening rings as a disturbance of the membrane state of the shell. They are usually taken into account in the design based on a static design wind calculation.

The real peak values due to turbulence effects are underestimated approximately by a factor of 1,2. Bending occurs also outside of the regions of bending disturbance, due to the non-uniformity of the wind load distribution. These are not considered for the design in general, since the design wind calculation show a small proportion of bending stress. But the more sophisticated calculations as well as strain measurements reveal that bending increases in a turbulent wind, which may be explained by the additional non-uniformity of the load distribution introduced by turbulence. The bending moments are increased approximately by a factor of 2 compared to the design load calculation. Still, the proportion of bending stress remains small in meridional direction and will generally not be important for the reinforcement. Contrarily, bending in hoop direction is more critical. The amount of bending is closely linked to the bending stiffness. For an estimation it is assumed that bending stiffness is not reduced by cracks. Then, peak bending stresses are about twice the peak stresses due to the hoop membrane force in the 125-m-tower. In a 165 m tower, the proportion is even up to 3. The latter result indicates, that towers with large diameters are subjected to more intense bending. The effect of turbulence on bending needs further investigations. A realistic result will be obtained applying the quasi-static calculation of response variance. With the assumption of uncracked concrete this result will be somewhat conservative.

3. Interference effects

In any practical situation, the wind flow approaching a cooling tower will be disturbed by surrounding buildings, either by further cooling towers or by other power plant buildings or by both. There are mainly four effects on the wind load:

- The wind speed may increase locally introducing a local increase of the mean load.
- The distribution of the mean wind pressures is no longer symmetric which may result in higher values of certain design stresses.
- The amount of turbulence will increase producing higher rms-

response, even if the mean response may be diminished.

- The frequency distribution of turbulence in the undisturbed flow is altered in such a way that additional excitation occurs in the high frequency range, where the eigenfrequencies are found. As a result, resonance will increase.

Actually, it is rather difficult to establish general rules to cope for interference effects, since the increase of stresses depends very much on the individual situation. Instead, the increase of shell stresses in some case studies is discussed, giving an indication of the order of magnitude which has to be expected. In any case it will be necessary to refer to experimental investigations. Either the wind loads may be measured and the resulting response be calculated or the response may be measured directly in terms of strains. The first method requires a large amount of static and dynamic calculations until the largest value of a design stress is found out. The latter method needs more sophisticated experimental methods but the final result is obtained quickly and reliably. The following findings are based on strain measurements using aeroelastic models. The wind flow was modelled in a boundary layer wind tunnel.
The first example is a systematic investigation of a group of two identical towers. Details are described in ref. 4. The tower distances are somewhat smaller and larger than required in the VGB-guidelines (2·throat diameter d_T or 1,5·base diameter d_B). For certain wind directions, the peak membrane forces will be increased compared to the single tower. The rate of increase, the interference factor f_G, of some response Y,

$$f_G = \frac{(\bar{Y}+g\cdot\sigma_y)\,\text{group}}{(\bar{Y}+g\cdot\sigma_y)\,\text{single tower}} \qquad (9)$$

is indicated in fig. 9 for a level 10,5 m above the base of the shell. All factors are smaller for the large distance, where they might be disregarded for practical design excepting hoop tension. For the small distance, an overall increase of 15 % would be necessary, again excepting hoop tension. Typically, the meridional tensile forces max n_{22} increase for a wind direction where the inspected tower is hit by the streamlines, separating from

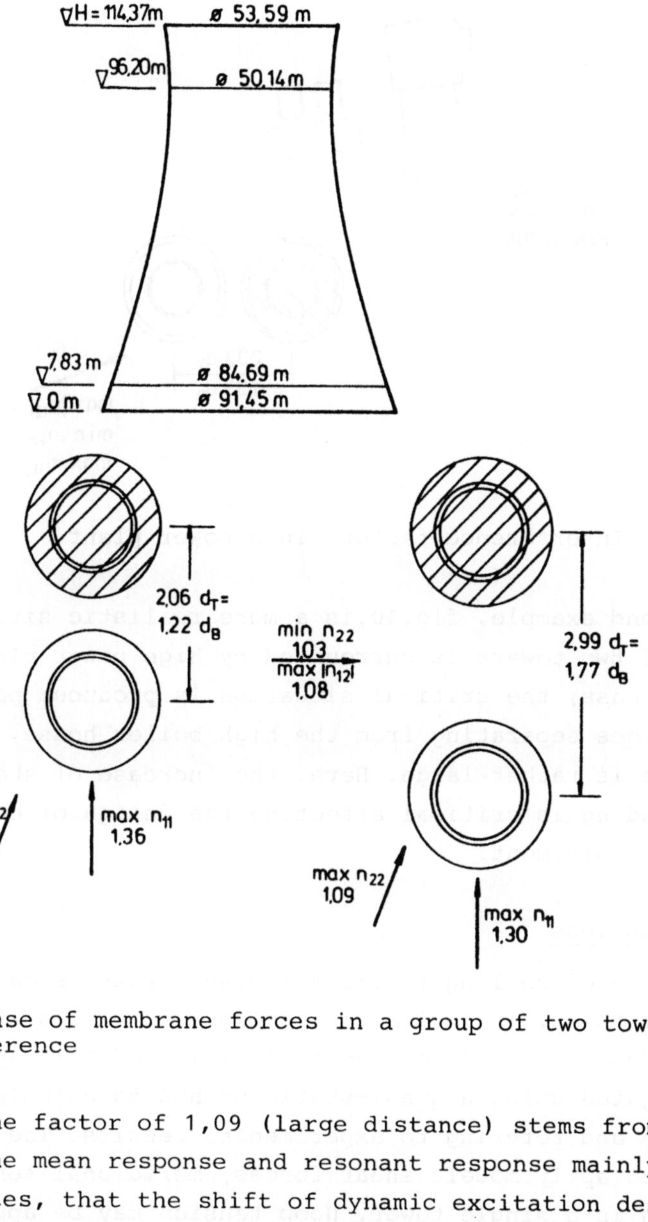

Fig. 9. Increase of membrane forces in a group of two towers due to interference

the second. The factor of 1,09 (large distance) stems from an increase of the mean response and resonant response mainly. The latter indicates, that the shift of dynamic excitation described above has a strong persistence and cannot be counteracted by a larger distance between towers. Hence, towers with a tendency to resonant vibrations will be affected more strongly by interference. Hoop tension is increased considerably by 30 % even for the large distance. Similar to the single tower, it may be underestimated by current design practice. Hoop reinforcement should be designed with particular care to avoid vertical cracks.

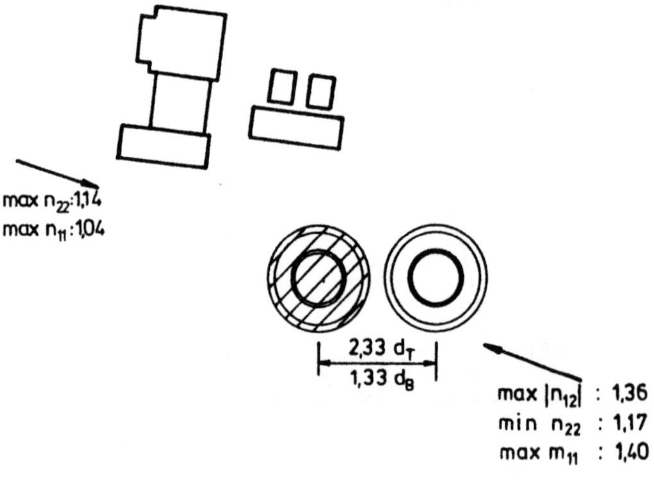

Fig. 10. Interference factors in a power plant

The second example, fig.10,is a more realistic situation. A group of two towers is surrounded by high power plant buildings. In this case, the critical situation is produced partly by the streamlines separating from the high boiler house, although its distance is rather large. Here, the increase of shear force and hoop bending is critical affecting the design of columns and hoop reinforcement.

4. Conclusions

The design of cooling towers for wind forces is carried out for a single tower, including the response to wind gustiness in a static design wind load. The feasibility of this approach is investigated using a quasi-static method to calculate the rms-response and refering to experimental results. The static design wind load aptly models shear forces, meridional tension and compression in a single tower. Hoop tension may be approximated neglecting internal pressure in the calculation. If hoop bending has to be considered, the peak bending stresses obtained from the design wind calculation should be increased by a factor of approximately 2. Bending effects due to turbulence are larger in towers with large diameters.

Interference with surrounding buildings introduces additional stresses. The increase depends very much on the individual situation. Wind tunnel tests in a boundary layer wind tunnel using aeroelastic models are praticularly helpful in such cases. Argueing from two case-studies, a rough estimate would be to increase the loads by 20 % for the design of columns, for meridional tension and compression and by 40 % for hoop tension, hoop bending, and shear force. Generally speaking, it should be avoided, if possible, in the lay-out of power plants that the critical wind directions coincide with the direction of extreme wind speeds.

5. Acknowledgments

Part of the work was carried out within a special research programm "Dynamics of Structures" of the Deutsche Forschungsgemeinschaft at the Ruhr-Universität Bochum. The support is gratefully acknowledged. The author is indebted to M. Kasperski and J. Ruhwedel who assisted in the calculations and the experimental work.

6. References

1. Hashish, M.G., and Abu-Sitta, S.H.: The Response of Hyperbolic Cooling Towers to Turbulent Wind. ASCE J. Struct. Div., 1974.

2. Niemann, H.-J.: Dynamische Windwirkung bei Naturzugkühltürmen auf der Grundlage eines statistischen Sicherheitsbegriffs. Habilitationsschrift, Bochum 1976.

3. Bautechnik bei Kühltürmen, Teil 2: Bautechnische Richtlinien (BTR). VGB-Kraftwerkstechnik GmbH, Essen 1979.

4. Niemann, H.-J., and Ruhwedel, J.: Statische und dynamische Beanspruchungen von Naturzugkühltürmen bei Einzel- und Gruppenanordnung. Konstruktiver Ingenieurbau/Berichte, H. 35/36, 1981.

Wind Loading of Hyperbolic Cooling Towers without Meridional Ribs

Lu Wenda[1], Zhou Liangmao[2], Jin Minzhong[3]

1. Institute of Applied Mathematics and Mechanics, Shanghai University of Technology, Shanghai, China
2. Xian Institute of Thermal Engineering, Xian, China
3. Xian Institute of Metallurgy and Construction Engineering, Xian, China

Abstract

In the present paper, the wind pressure measurements carried out at the 90 m full-scale cooling tower without meridional ribs in Maoming are given. The results from full-scale tests are in excellent conformity with the model results. Through statical analysis, it has been found that to use the simplified height constant pressure distribution in determining the static design wind load for a smooth walled cooling tower is reasonable. According to the information of full-scale pressure measurement an approximate physical model of power spectral and cross-spectral density functions are also presented. The results of numerical calculation of dynamic response of normal displacement to turbulent wind bassed on this model agree quite well with the results from full-scale tests on Maoming tower by means of a ranging laser.

1. Introduction

In November 1965, three cooling towers with a height of 113 m at Ferrybridge Power Station suddenly collapsed under a storm. People were awakened by a series of collapses of towers and believed that the main loading case for hyperbolic cooling tower of reinforced concrete is produced by wind forces. Wind flow near the earth's surface is turbulent, that is gusty, and unsteady. During a storm such turbulence produces drag wind forces acting on the surface of the tower shell that are not only random with respect to time but are random with respect to both vertical and circumferential spatial coordinates. The

* Since 1984 I have been working at the Institute of Mathematics and Mechanics, Shanghai University of Technology

essential feature of the so-called "gust response factor" is the separation of the mean and fluctuating components of wind load which has been made to place the evalution on wind loads on a more rational basis introduced by Davenport. In the present paper, we suggest a simplified design mean wind pressure formula which is valid for "smooth" surface cooling towers. Meanwhile, using the finite element method based on a simplified physical model of fluctuating pressure spectral density measured on Maoming tower we investigate the dynamic response by means of statistical method. The numerical calculation of gust response of normal displacement to turbulent wind agree quite well with the result of displacement measured by laser. In the present paper, the reliability of hyperbolic cooling tower is also studied.

2. Mean Wind Load

The hyperbolic cooling tower at the Maoming Power Station, Guangdong, China, is 90 m high, without meridional ribs on the external shell surface. Its geometry and dimensions are given in Fig. 1. The site of Maoming tower and its surrounding are sketched in Fig. 2. The circumference of the shell at 50 m and 72 m height are chosen as the measurement level for full-scale study. Fifty-eight pressure transducers were arranged for the outer surface and, at the 50 m height level, eight transducers to measure the pressure on the inner surface were provided, except above mentioned, in the meridional directions of the windware and wake region, at the 88, 36, 18 m height above ground six transducers to measure the pressure on the outer surface were also provided. In full-scale measurement, the pressure difference $\Delta p(\Theta, z)$ given by transducer will be taken as the pressure on shell surface. In the next place, the maximum pressure difference $\Delta p(z)_{max}$ is going to be taken as the dynamic pressure of the windware at the height z. Thus pressure coefficient C_p is given by

$$C_p(z, \Theta) = \Delta p(z, \Theta)/\Delta p(z)_{max} \quad (1)$$

From Fig. 3 it will be seen that the results from full-scale tests are in excellent conformity with the model results [4]. From the full-scale measurement on Maoming tower we have obtained the confidence interval of the coefficients of pressure ascent ΔC_p with respect to 95% confidence probability as follows:

$\Delta C_p = 1.22 - 1.27$, at the 50 m height level

$\Delta C_p = 1.21 - 1.29$, at the 72 m height level

The mean wind profile could be approximated by a power law with two profile exponents $\alpha = 0.2$ and $\alpha = 0.3$. The corresponding turbulence intensity $I_v = \sigma_v/\bar{V}_{10} = 25\% - 33\%$; Reynolds number $Re = 4.2\text{-}4.9 \cdot 10^7$ (calculation with respect to the mean wind speed of corresponding height level). It will be seen that the coefficient ΔC_p is approximately constant over height, i.e.

$$\Delta C_p = C_{pb} - \min C_p = \text{const.} \qquad (2)$$

This conclusion has been point out by Cesar Farell [7] and Niemann [8]. Instead of $\min C_p$ the coefficient ΔC_p can be established as the characteristic value of the mean wind pressure distribution. Fig. 5 shows the effect of roughness on pressure-rise coefficients which are given by wind-tunnel tests and full-scale measurements. It should be pointed out that although the Maoming tower has no ribs, its concrete shell surface is far from a "smooth" surface. Take the average of height of the ribs to be 5 mm, the relative roughness K/D of "equivalent" ribs will be $1.16 \cdot 10^{-4}$. It will be seen that the ΔC_p values should be ascended with descending surface roughness in the range of Reynolds number independence. The average limit value of ΔC_p with respect to the tower of "smooth" surface is about 1.25. According to the results measured on full-scale the variation of the wake pressure coefficient with height z is given as follows:

$$C_{pb}(z, \theta) = -0.32\ (z/H)^{-2\alpha} \qquad (3)$$

The results show that the wake pressure ascends with the height z of tower, hence the suction peak ascends also with the height z . Fig. 4 shows the variations of $\min C_p$ and C_{pb} with height z obtained from full-scale measurement. It will be seen that the variations are of significance, whereas they are similar to the results obtained from model tests by Pröpper [9] in simulated atmospheric turbulence, and by Cowdrey, O'Neill [10] in turbulent shear flow. From Equ. (2), the variation of the suction peak coefficient for "smooth" surface cooling towers can be given approximately as follows:

$$\min C_p(z) = -0.32 \ (z/H)^{-2\alpha} - \Delta C_p \tag{4}$$

(1). Evalution of Simplified Design Wind Pressure

In the great majority of the design of cooling towers, it is assumed that the mean wind pressure coefficient is a simplified height-constant pressure distribution $C_p(\theta)$. In order to consider the effect of the wind profile on the mean pressure, in general, the following simplified design wind pressure is used:

$$p(z, \theta) = C_p(\theta) \ q(z) \tag{5}$$

in which $q(z) = \frac{1}{2}\rho \ \bar{V}^2(z)$, but the mean pressure coefficient $C_p(z, \theta)$ varies with height z . Then the realistic wind pressure will be

$$p(z, \theta) = C_p(z, \theta) \ q \ (z) \tag{6}$$

Based on the full-scale measurements on Maoming cooling tower as well as by reference to the results of model tests in wind-tunnel [4, 9, 10], the approximate distribution of pressure coefficients can be obtained as follows:

$$\begin{aligned} C_p(z, \theta) &= 1 - (1 - \min C_p) \sin (90\cdot\theta/75)^{\gamma}, & \theta \leq 108^{\circ} \\ C_p(z, \theta) &= -0.32 \ (z/H)^{-2\alpha} & \theta > 108^{\circ} \end{aligned} \tag{7}$$

in which the maximum pressure max C_p is assumed approximately to equal one, that is max $C_p = 1$, and the circumferential angular coordinates of zero crossing as well as suction peak are independent of height z. In Fig. 3 the pressure distribution given by Equ. (7) is in excellent conformity with the model results [4]. The distribution of mean wind pressure coefficient is approximate to turbulence intensity be 25% - 33%. The membrane forces, axial tensile and compressive forces of the columns of Maoming tower calculated by formulas (5) and (6) are given in Table 1, which are compared with the results by using formula (5).

Distribution types of wind pressure	max N_φ (T/M)	max N_θ (T/M)	max $N_{\theta\varphi}$ (T/M)	Axial forces of the columns	
				Maximum tensile (T)	Maximum compressive (T)
810827-1 records	1.00	1.00	1.00	1.00	1.00
C_p (50, θ)*	1.04	1.03	1.04	1.06	1.02
C_p (z, θ)*	1.04	1.04	1.04	1.06	1.04

*C_p (50, θ), C_p (z, θ) are determinated by formula (7).

Table 1. The Effect of Different Formulas of Design Pressure on the Membrane Forces of Tower Shell and Axial Forces of Columns

It will be seen that the membrane forces and the axial forces of the columns at the bottom of tower shell calculated by using once measurement on full-scale are five percent less than that by using the formula (7). Whereas, in the case of simplified design wind pressure the axial compressive forces are underestimated by about 3%. Therefore the assumption of the distribution of the mean wind pressure coefficient independance with height z can be used in design practice. The $C_p(50, \theta)$ determinated by formula (7) may be used as the $C_p(\theta)$ in the simplified procedure (5).

(2). Evalution of Different Curves of Mean Wind Pressure

The maximum membrane forces, meridional bending moment and the maximum axial tensile as well as compressive forces induces by different types of mean pressure curves are given in Table 2 which are compared with the wind-tunnel results. Tabel 3 shows the variation of the meridional reinforcement ($\Theta = 0^{\circ}$) obtained from different distribution of mean pressure with height z. It will be seen that in the same design condition the curve (III) requires greater reinforcement from bottom lintel until 54 m height of tower, curve (II) requires a little less reinforcement at 10 - 54 m height, but above the throat level all curves only require the reinforcement according to the minimum reinforced ratio. In reference [11], we have discovered that the effect of meridional geometric imperfection on the natural frequency in the cooling tower is significent. The lowest natural frequency will be reduced by about 7% - 15%, because the dominant frequency of turbulent gust wind is lower, so that the dynamic response induced by gust wind will increase. In the case of an allowable slope of geometric imperfection $(2\Delta/H_i)_{max} = 1.34\%$ suggested by reference [11] for imperfect wave length $H = 4$ m - 10 m the meridional reinforcement of perfect shell must be to have an overstress by factor $\alpha^* = 0.2 - 0.3$. If the curve of mean wind pressure (I) is used, then the curve (II) is corresponding to take account of an overstress by a faktor $\alpha^* = 0.2$. Hence, in the normal construction condition, for example, the deviation of horizontal radius is 2.7 cm ($H_i = 4$ m) or 6.7 cm ($H_i = 10$ m), the curve (I) can give an enough overstress of meridional reinforcement.

Distribution of wind pressure	Roughness type of surface	Bottom of tower shell				Axial force of column		min C_p
		max N_φ (T/M)	max N_θ (T/M)	max $N_{\theta\varphi}$ (T/M)	max N_φ (T)	Maximum tensile (T)	Maximum compre-ssive (T)	
Wind-tunnel test (I)	S32	1.00	1.00	1.00	1.00	1.00	1.00	-1.56
$C_p(50,\theta)$ (II)	smooth	1.08	1.08	1.08	1.15	1.06	1.07	-1.62
Ribich (III)	smooth	1.05	1.10	1.09	1.81	1.05	1.09	-1.90
ACI-ASCE (IV)	smooth	1.11	1.12	1.12	1.46	1.09	0.99	-1.70

Table 2. The Effect of Different Distributions of Wind Pressure on the Membrane Forces of Tower Shell and the Axial Forces of Columns

Distribution of wind pressure \ Ratio \ Height level z (m)	5.8	6.8	10	20	37	54	72	80	90
C_p (50, θ) (I)	1.00	1.00	1.00	1.00	1.00	1.00	1.00	1.00	1.00
Ribich (II)	1.05	1.05	0.94	0.87	0.86	0.73	1.00	1.00	1.00
ACI-ASCE 334 (III)	1.10	1.05	1.03	0.95	1.03	1.04	1.00	1.00	1.00

Table 3. Relative Ratio of Meridional Reinforcement (θ = 0°)

3. Gust Response

The RMS of gust response of normal displacement is given (2) as follows:

$$\sigma_w(s,\theta) = [\,2\int_0^\infty S_w(s,\theta\,;f)\,dF\,]^{1/2}$$

$$= \{\sum_{n=0}^{\infty}\sum_{m=0}^{\infty}\sum_{r=1}^{\infty}\sum_{l=1}^{\infty} 2\,[\int_0^\infty S_{q_r^n q_l^m}(f)\,df\,\cos m\theta\,\cos n\theta$$

$$+ \int_0^\infty S_{q_r'^n q_l'^m}(f)\,df\,\sin m\theta\,\sin n\theta]\,\mu_{3r}^n(s)\,\mu_{3l}^m(s)\}^{1/2}$$

in which the coupled terms between symmetrical and antisymmetrical deformation are neglected. The relation between cross-spectral density of generalized coordinates and cross-spectral density of generalized loads are given as:

$$S_{q_r^n q_l^m}(f) = \frac{\chi_{rn}^{*}(jf)\,\chi_{lm}(jf)}{16\pi^4 f_{rn}^2 f_{lm}^2 M_r^n M_l^m} S_{p_{rn} p_{lm}}(f) \quad ;$$

$$S_{q'^n_r q'^m_l}(f) = \frac{\chi_{rn}^{*}(jf)\,\chi_{lm}(jf)}{16\pi^4 f_{rn}^2 f_{lm}^2 M_r^n M_l^m} S_{p'_{rn} p'_{lm}}(f) \quad ;$$

$$\chi_{rn}^{*}(jf) = X_{rn} + jY_{rn} \; ; \qquad \chi_{lm}(jf) = X_{lm} + jY_{lm} \; ;$$

$$X_{rn} = \frac{1 - f^2/f_{rn}^2}{[(1-f^2/f_{rn}^2)^2 + 4f^2/f_{rn}^2\zeta_{rn}^2]} \; ; \quad Y_{rn} = \frac{2\zeta_{rn} f/f_{rn}}{[(1-f^2/f_{rn}^2)^2 + 4f^2/f_{rn}^2\zeta_{rn}^2]} \; .$$

Replacing rn into lm we obtain the expressions of X_{lm} and Y_{lm} from X_{rn} and Y_{rn} . The cross-spectral density of generalized load is defined as:

$$S_{p_{rn} p_{lm}}(f) = \int_0^{2\pi}\int_0^{2\pi}\int_0^{S_0}\int_0^{S_0} S_p(r_A, r_B; f)\,\mu_{3r}^n(S_A)\,\mu_{3l}^m(S_B)\cos n\theta_A \cdot \cos m\theta_B\, r(S_A)\, r(S_B)\, dS_A\, dS_B\, d\theta_A\, d\theta_B \; .$$

Using sinusoidal function instead of cosine function in the $S_{p_{rn} p_{lm}}(f)$, the $S_{p'_{rn} p'_{lm}}(f)$ will be obtained. The gust response factor is defined as follows:

$$G_w = 1 + g\,\frac{\sigma w(s,\,\Theta)}{\bar{w}(s,\,\Theta)} \tag{8}$$

in which

$$g = \sqrt{2\ln\nu T} + \frac{0.577}{\sqrt{2\ln\nu T}}$$

is called peak factor; T = duration of the gust wind record;

$$\nu = \left[\int_0^{\infty} f\, S_w(f)\, df \Big/ \int_0^{\infty} S_w(f)\, df\right]^{1/2}$$

For a lightly damped system, ν = the lowest natural frequency of the structure. $\bar{W}(S, \Theta)$ is called the normal displacement response induced by mean velocity. It can be obtained by means of static procedure or we can also use the mode-superposition procedure to obtain the static mean response of the tower. From Eqn. (8), the normal displacement response of hyperbolic cooling tower induced by gust wind is given as:

$$W(S, \Theta) = \bar{W}(S, \Theta) + g\,\sigma_W(S, \Theta) \tag{8a}$$

If we want to obtain the RMS of stress resultants of cooling tower, the corresponding mode shapes are used in place of $\mu^n_{3r}(S)$ in Eq. (8).

(1). Power Spectral and Cross-spectral Density Functions of Fluctuating Pressure

In Fig. 7 the power spectral densities of fluctuating pressure given by No. 8007-8 pressure records at the throat level with different circumferential angles are obtained by means of FFT method [5]. Spectral peak is appeared near to the $f = 0.016$ Hz. In wake region, vortex shedding frequency $f_{st} = 0.053$ Hz . It must be pointed out that the power spectral density of pressure varies with height z obeyed exponent law as:

$$S_p(s, \Theta; f) = \left(\frac{Z(s) + Hd}{10}\right)^{\alpha_1} S_{p10}(f) \tag{9}$$

in which Hd = height level of the throat of cooling tower; $\bar{K}$ = ground roughness coefficient. The values of α_1 are given as follows:

Ordinal number of pressure record	Exponent of pressure power spectrum α_1	Profile exponent of wind speed α
8007 - 8	1.273	0.30
810827	0.833	0.20

It will be seen that along with the increment of ground roughness in the windware the wind profile exponent α increases in same ratio with the spectral exponent α_1. Fig. 8 shows the cross-spectral density of fluctuating pressure at the points between windware and wake region. It will be seen that the quadrature spectral densities are all equal to zero, and the co-spectral densities appear minus zone. That is to say, if it appears as pressure in the windware, suction will appear as in the wake region which both have the same phase. The maximum minus value of co-spectral density is equal to 0.18 near $f = 0.05$ Hz. From what is said above, the time-spatial correlation of fluctuating pressure at different points on the surface of tower shell can be assumed as follows:

In the windware:

$$S_p\ (r_A,\ r_B;\ f) = C_R\ (\Theta_A,\ \Delta\Theta)\ R_f\ (\Delta\Theta;\ f)\ R_V\ (s_A,\ s_B;\ f) \cdot \sqrt{S_p\ (s_A,\ \Theta_A;\ f)\ S_p\ (s_B,\ \Theta_B;\ f)} \tag{10}$$

In the wake region:

$$S_p\ (r_A,\ r_B;\ f) = R_V\ (s_A,\ s_B;\ f)\ \tilde{R}_r\ (\Theta_A,\ \Theta_B;\ f) \cdot \sqrt{S_p\ (s_A,\ \Theta_A;\ f)\ S_p\ (s_B,\ \Theta_B;\ f)}$$

in which

$$R_{VW}\ (s_A,\ s_B;\ f) = e^{-\beta_1 \tilde{f}_1}$$

$$R_{VL}\ (s_A,\ s_B;\ f) = e^{-\beta_1 \tilde{f}_1} \tag{11}$$

the lower indices VW and VL represent the meridional directions in the windware and wake region. In the windware and wake region of circumference

$$\tilde{R}_f\ (\Theta_A,\ \Theta_B;\ f) = C_R\ (\Theta_A,\ \Delta\Theta)\ R_f\ (\Delta\Theta;\ f) = C_R\ (\Theta_A,\ \Delta\Theta)\ e^{-\beta_2 \tilde{f}_2}$$

$$R_r\ (\Theta_A,\ \Theta_B;\ f) = R_r\ (\Delta\Theta;\ f) = e^{-\beta_3 \tilde{f}_2} \tag{12}$$

$\tilde{f}_1 = f\Delta s/\bar{V}(Z_G)$; s = distance between two points along meridional direction; $\tilde{f}_2 = fr\Delta\theta/\bar{V}(Z_G)$; r = radius of horizontal circle; $\Delta\theta$ = centre angle of a circle between two points at the same height level; $\bar{V}(Z_G)$ = mean wind speed at the gradient height; Z = 430 m. The values of coefficients β_1, β_2 and β_3 are given as follows.

If the effect of circumferential angle on power spectral density can be neglected, then

$$S_p(s, \theta; f) = \left(\frac{Z(s) + Hd}{10}\right)^{\alpha_1} S_{p10}(f) .$$

It will be seen that the vertical coherence function along the meridional curve may use a same characteristic function to express in the windware and wake region; but the circumferential

Ordinal number of pressure record	Profile exponent of wind speed α	Turbulence intensity $\sigma_v/\bar{V}_{10}$	Reynold number Re	Windware		Wake	
				β_1	β_2	β_1	β_3
8007 - 8	0.30	24,5 %	$5.49 \cdot 10^7$	8.3	16.5	7.7	24.8
810827	0.20	23,9 %	$4.90 \cdot 10^7$	11.7	16.3	10.7	26.4

coherence function bears a relation to the horizontal location, distance between two points and frequency. Meanwhile, the fluctuating pressure correlativety of different frequency component between two circumferential points at the same temporality in the windware is less than that in the wake region. The same conclusion has been obtained by Abu-Sitta [4] in wind-tunnel model tests. $C_R(\theta, \Delta\theta)$ is called the cross correlation coefficient of fluctuating pressure between two points with zero stagnation time. It characterizes the relation with each other of fluctuating pressure between two points at the same time. From Fig. 10 it will be seen that a wind gust has only a limited spatial spread and consequently, the maximum load is not able to appear everywhere on the shell surface at the same time. The distribution of the correlation coefficients depends on the reference angle Θ as well as on the direction of $\Delta\Theta$. If $\Delta\Theta$ runs upstream the zones of pressure and suction has a strong negative correlation, on the contrary they are only slightly correlated if $\Delta\Theta$ runs downstream. This conclusion agrees quite well with the results obtained by Pröpper [9] in the wind-tunnel model tests.

(2). <u>Measurement of Displacement Response on Full-Scale Tower</u>

In the reference [6] the normal displacement response of a fixed point on the shell surface at 20 m height level has been measured by means of laser. The curve of time history is shown in Fig. 11. The confidence intervals of mean value $\bar{W}$ and RMS σ_W of gust response of normal displacement with respect to 99% confidence probability is given in Table 4 which are based on 10 min. time history of gust response recorded from 810827-1-2 storm. In this table, the results in theory are obtained by means of the procedures given in references [1] and [2]. Two kinds of critical damping 0.015 and 0.04 as well as aerodynamic damping are taken into account in analysis. It will be seen that the theoretical values of gust response agree quite well with the results measured on full-scale cooling tower. The maximum relative error of gust response factor G_W is about 12% and the maximum relative error of maximum normal displacement W_{max} is about 8.3%.

Terms / Procedure		Confidence interval			Confidence interval	
		$\bar{W}$ (mm)	σ_W (mm)	$\sigma_W/\bar{W}$	W_{max} (mm)	G_W
Full-scale measurement 810827-1-2		0.47-0.51	0.16-0.19	0.32-0.40	1.10-1.20	2.20-2.50
Theoretical values	ζ=0.015	0.56	0.20	0.36	1.30	2.40
	ζ=0.04	0.56	0.18	0.31	1.20	2.20

Table 4. The Comparison of Normal Displacement Between Theory and Full-Scale Measurement

4. Reliability of Hyperbolic Cooling Tower

The function $p_z(z)$ is defined as the probability density of the difference between strength y and load x, i.e. $z = y-x$. The risk of structural failure p is defined as the probability under the condition $z < 0$, that is

$$p_f = \int_0^{\infty} p_z(z)\,dz = \int_0^{\infty} R_L(y)\,f_s(y)\,dy \tag{13}$$

in which we assume that x and y are independent random variables. If the effect of imperfection on gust response is taken into account, it is necessary to have an overstress by a factor $\alpha^* = 0.2 - 0.3$ in the meridional reinforcement of perfect shell to ensure the tower safe for design [11]. The relations between the probability of failure and the relative dispersion parameter of wind field are given in Fig. 12. For perfect shell the probability of failure is $p_f = 0.012 - 0.096$, but for imperfect shell $p_f = 0.024 - 0.16$. The reliability with respect to the design age limit is equal to $6.0 \cdot 10^{-5} - 12.0 \cdot 10^{-4}$. It will be seen that the parameter of wind field is large, the probability of failure is more lower.

Conclusions

1. The gust response factor G_w given by reference [2] agrees quite well with the measured value on full-scale tower. In $\zeta = 0.04$ which is normally specified for cooling tower design, the maximum relative error of W_{max} is 8.3%.
2. The distribution of mean wind pressure $C_p(50, \Theta)$ suggested by the present paper is valid for "smooth" surface cooling towers.
3. The reliability with respect to the design age limit is equal to $6.0 \cdot 10^{-5} - 12.0 \cdot 10^{-4}$ for the limited state design procedure of cooling towers in China.

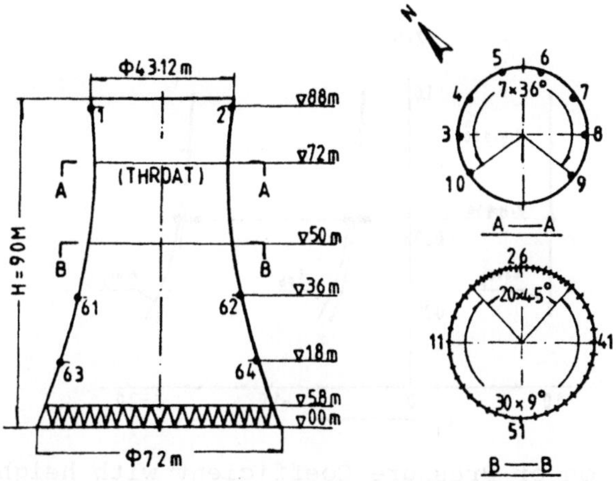

Fig. 1. Location of Pressure Measurement Points for Maoming Tower

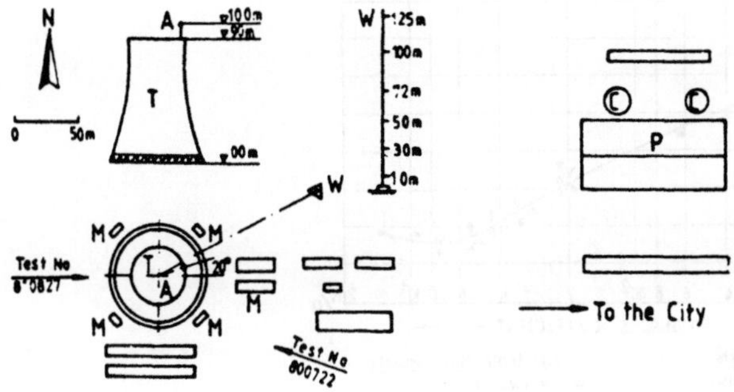

T Natural Draft Cooling Tower
A Anemometer Mast

M Measurement Station

W Wind Measuring Tower
C Chimneys

P Main Building of Power Plant

Fig. 2. Ground Plan of Maoming Power Station

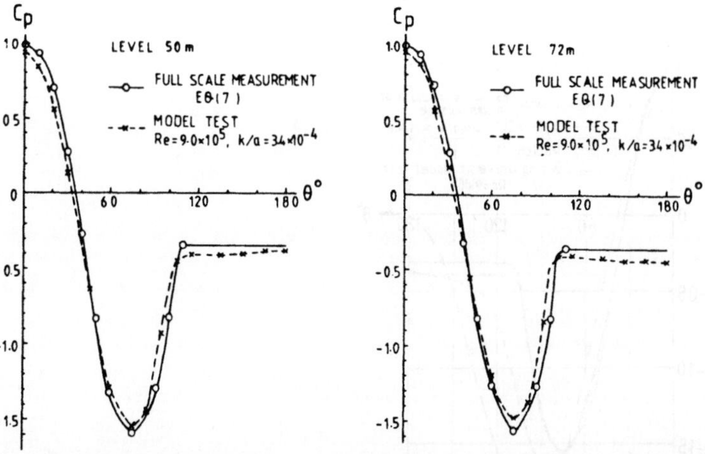

Fig. 3. Comparison of C_p of Full-Scale Measurement and Model Test

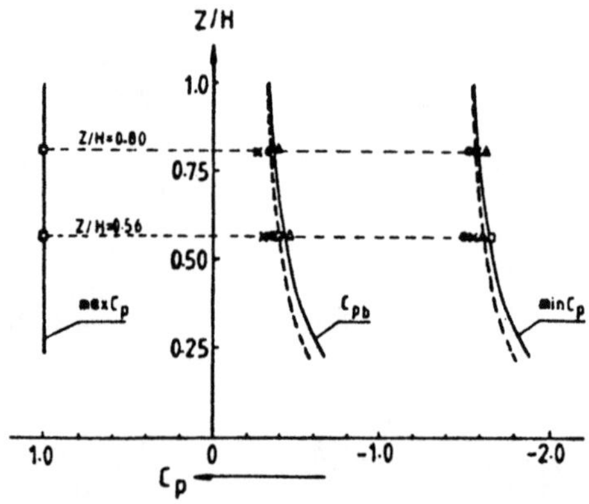

Fig. 4. Variation of Pressure Coefficient with height in Atmosheric Turbulence

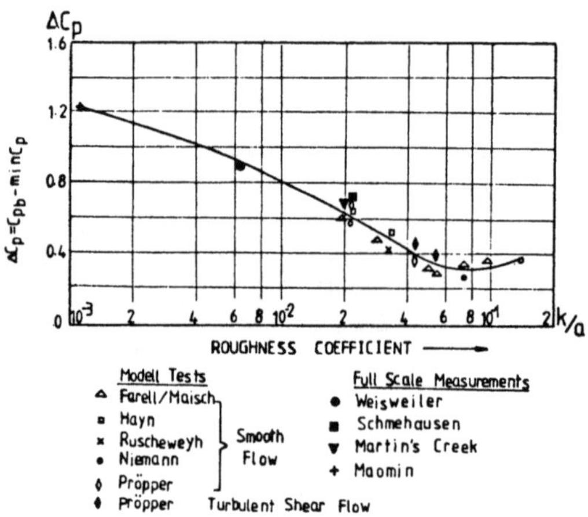

Fig 5. Influence of Surface Roughness on Pressure-Rise Coefficient

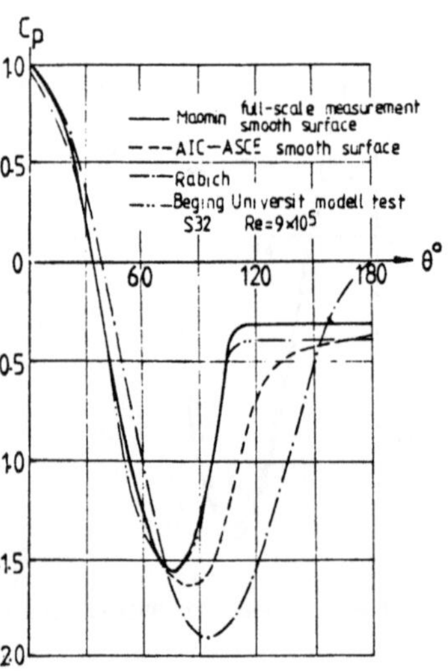

Fig. 6. Comparison of C_p Different Typical Distribution

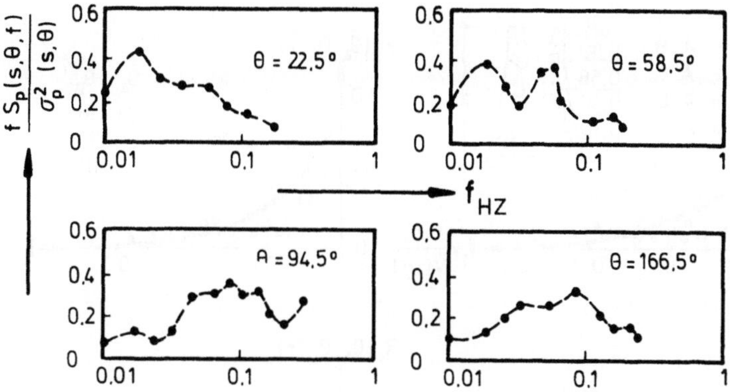

Fig. 7. Power Spectral Densities Measured on Maoming Tower

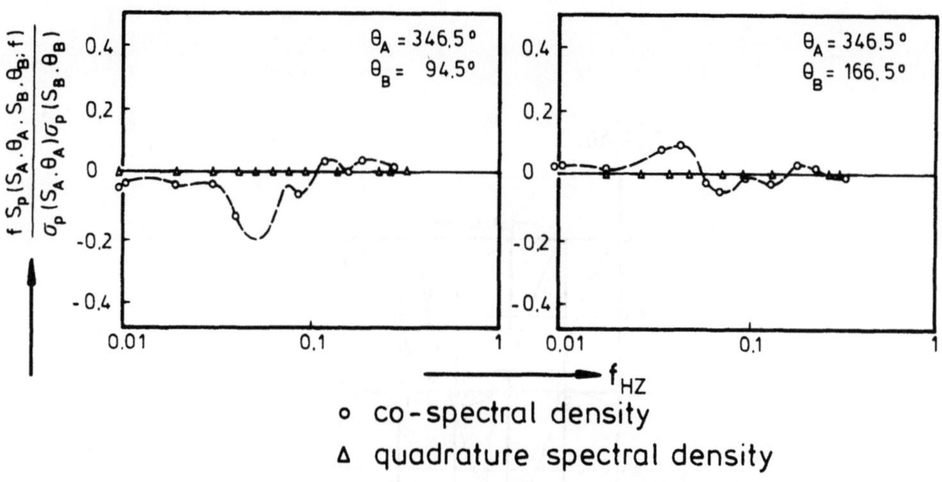

Fig. 8. Cross-spectral Densities on Maoming Tower

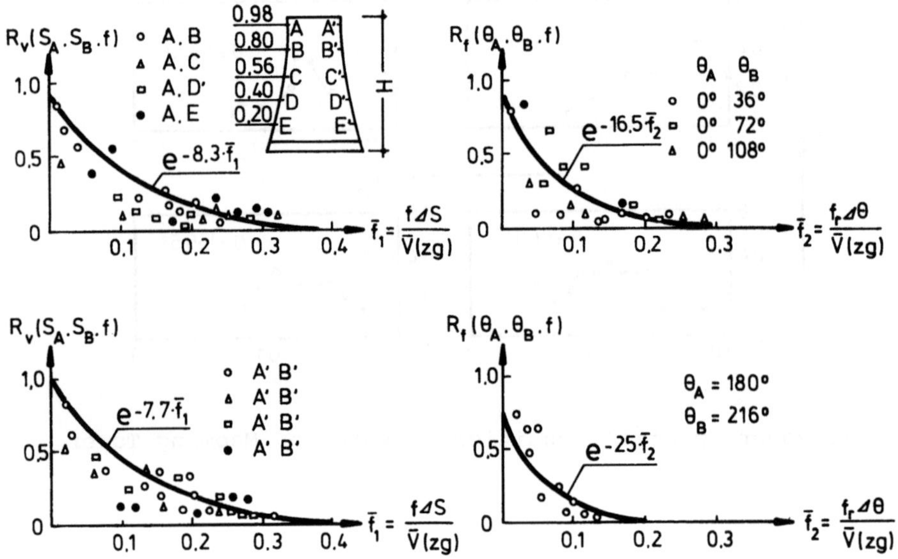

Fig. 9. Coherence Function Along Meridian and Circumference Measured on Maoming Tower

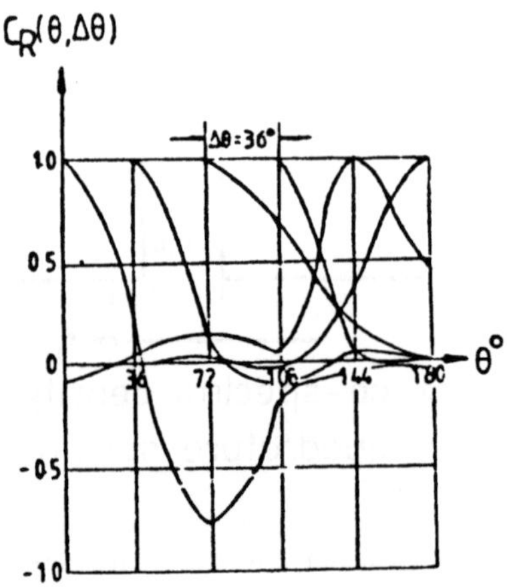

Fig. 10. Cross Correlation Coefficient of Fluctuating Pressure

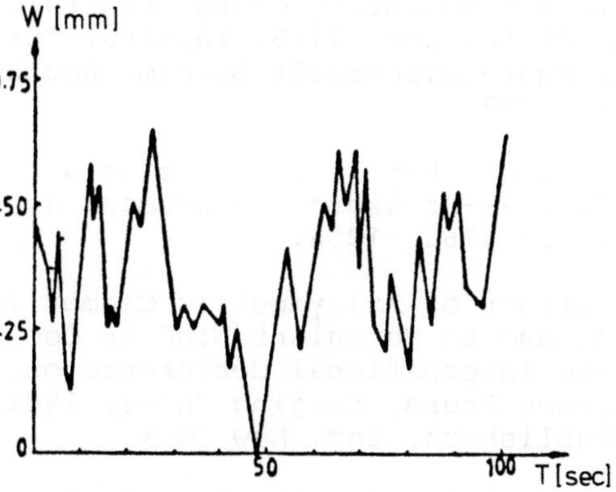

Fig. 11. Time History of Gust response of Normal Displacement

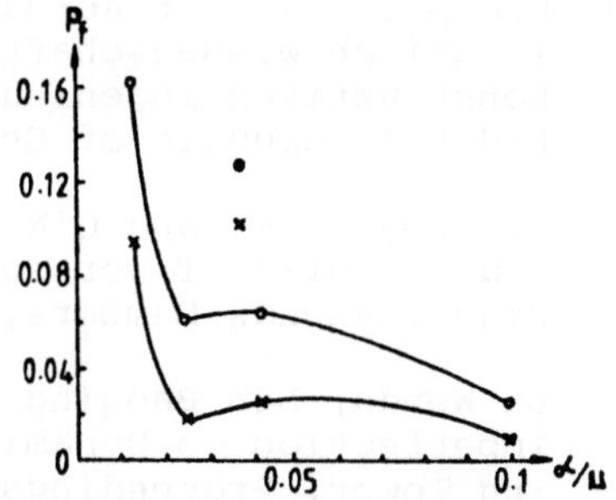

Fig. 12. Relationship between Probability of Failure and Relative Dispersion Parameter of Wind Field

References

1. Lu Wenda and Peng Jingxin, Finite Element Solution of Hyperbolic Cooling Tower Supported by Column System, Research Report of Xian Institute of Thermal Engineering, June, 1976, China.

2. Lu Wenda, Wan Younchi, Peng Jingxin and Lin Baoching, Dynamic Response of Hyperbolic Cooling Tower to Turbulent Wind, Journal of Electrical Engineering CSEE, No. 1, 1982 (in Chinese).

3. Fang Deming and Jin Xuepei, Wind Pressure Measurement System of Hyperbolic Cooling Tower, Research Report of Xian Institute of Thermal Engineering, December, 1978, China.

4. Wind-tunnel Study of the Wind Pressure Distribution Around Hyperbolic Cooling Tower, Report of Department of Mechanics, Peking University, October, 1977 (in Chinese).

5. Lu Wenda, Peng Jingxin and Wan Younchi, FFT Computation of Spectrum and Correlative Function Based on Overlapping of Seperate Section, Proceedings of the Conference on Mathematics, Xian, October, 1979 (in Chinese).

6. Jin Minzhong et al, Measurement Study of Gust Response on Full-Scale Hyperbolic Cooling Tower, Journal of Xian Institute of Metallurgy and Construction Engineering, July, 1982 (in Chinese).

7. Cesar Farell, Oktag Guven and Federico Maisch, Mean Wind Loading on Rough-Walled Cooling Tower, Journal of the Engineering Mechanics Division, ASCE, Vol. 102, No. EM6, 1976.

8. H.-J. Niemann, Wind Effects on Cooling Tower Shells, Journal of the Structural Division, ASCE, Vol. 106, No. ST3, March, 1980, PP. 643-661.

9. Pröpper, H., Zur Aerodynamischen Belastung Großer Kühltürme, Technisch-wissenschaftliche Mitteilung, 77-3, Institut für Konstruktiven Ingenieurbau, Ruhr-Universität Bochum, Bochum, Federal Republic of Germany, 1977.

10. Cowdrey, C.F. and O'Neill, P.G.G., Report of Tests on a Model Cooling Tower for C.E.A.----Pressure Measurement at High Reynolds Numbers, NPL-Aero-316a, 1956.

11. Lu Wenda, Lin Baoqing, The Effect of Axisymmetric Geometric Imperfection on Dynamic Response to Turbulent Wind in Cooling Tower, Proceedings of the International Conference on Finite Element Methods, Science Press, Beijing China, 1982, Gordon and Breach, Sciene Publishers, Inc. New York.

12. Lu Wenda, "Wind Loading on Hyperbolic Cooling Tower", Proceedings of the Fifth Conference on Wind Engineering of Tall Structures, Guanzhou, Guangdong Province, December, 1982 (in Chinese).

Wind Effects on High Cooling Towers

M. Robert

Electricite de France,
Direction des Etudes et Recherches,
Department T.T.A.

Abstract

In 1974, in parallel with the spreading of cooling towers and their increasing height, E.D.F. initiated a wide programme of experimental and numerical studies about large towers behavior.

The objectives of this programme were to provide E.D.F. with design criteria adapted to the new generation of cooling towers, after the validity of the assumption used for small size structures or structures with different shapes (chimneys) has been systematically questioned.

This programme includes four trends:

- The determination of loadings
- The static response of the tower
- The dynamic response of the tower
- The influence of imperfections and cracks, and the way they could be taken into account.

Many results are presented, some of them somehow unexpected.

The consequences for the design of the current E.D.F. cooling towers are discussed.

A Discussion of the Treatment of Wind Loading in Natural Draught Cooling Towers with Special Reference to Reliability and Simplified Dynamic Factors

A.G. Davenport, N. Isyumov

The University of Western Ontario,
Faculty of Engineering Science,
London, Ontario, Canada

Abstract

At the time of the introduction of natural-draught cooling towers to the U.S. and the Ferry Bridge Cooling Tower collapse, the Boundary Layer Wind Tunnel Laboratory undertook the first aeroelastic studies of tall hyperbolic cooling towers in natural turbulent wind. These studies were later extended to include earthquake excitation. Since then a considerable amount of wind tunnel and full-scale information has been obtained on cooling towers. The purpose of this paper is to reflect on this information and make use of simplified approaches developed for the treatment of complex dynamic wind forces and discuss the overall reliability of the response to wind.

1984
IASS-RUB

Dynamics

Seismic Analysis of Cooling Towers

JOHN P. WOLF

Electrowatt Engineering Services Ltd. CH-8022 Zürich

1. Introduction

Hyperbolic, natural-draught cooling towers to be erected in seismically active zones are routinely analysed for earthquake excitation. The axisymmetric shell is discretised with higher-order finite-element frusta with an isoparametric expansion in the meridional and a Fourier expansion in the circumferential direction ([1, 2]). The stiffening-ring beam, the foundation-ring beam, if present, and the supporting columns are synthesised into a dynamic-stiffness matrix compatible with the axisymmetric shell element [3]. The same applies also for the model of the soil which is represented as a spring and possibly as a damper with frequency-independent coefficients. For uniform earthquake excitation arising from vertically propagating waves, only the first and zeroth symmetric harmonics are excited in a linearly elastic analysis for the horizontal and vertical components, respectively, of the earthquake. Under the assumption that classical modes exist, only the first mode or the first two modes have to be considered. The calculation is performed directly in the time domain.

For all investigations carried out in this paper, the nuclear-power plant's 144-m-high reinforced-concrete cooling tower with a base radius $a = 58.54$ m in Leibstadt, Switzerland is used. Its main dimensions are specified in Fig. 1 [4]. The hyperbolic shell of variable thickness is stiffened at its base and exhibits an eccentric stiffening-ring beam at the top. The shell is supported by a total of $n = 36$ pairs of columns. Each pair of columns forms a V with an in-plane angle of inclination $\alpha = \pm 11.1^0$, and rests on a separate foundation. The selected angle corresponds to an equidistant support of the lower edge of the shell by the columns. The material pro-

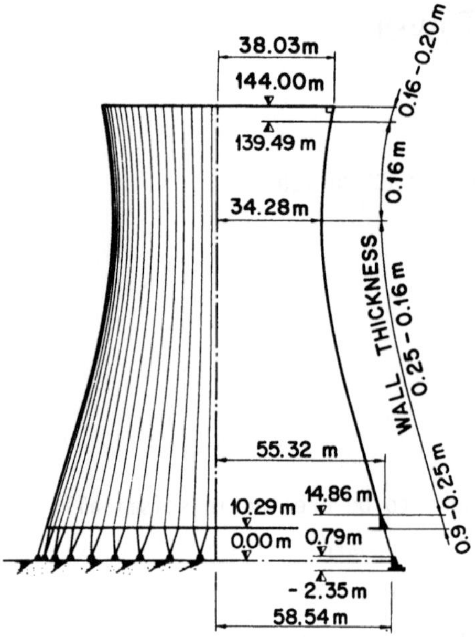

Fig. 1: Natural-draught cooling tower Leibstadt

perties of the shell are as follows: modulus of elasticity = 30 GPa, density = 2.5 Mg/m^3, Poisson's ratio = 0.2. The shear modulus G, the density ρ, Poisson's ratio ν and the material damping ratio ζ_g of the soil, which is regarded as an elastic halfspace, are equal to 0.12 GPa, 2.4 Mg/m^3, 0.33 and 0.05, respectively. The corresponding shear-wave velocity c_s ($= \sqrt{G/\rho}$) equals 223 m/s. The radial, circumferential and vertical components of the displacements of the first mode of the first harmonic U_1^s, V_1^s, W_1^s with a frequency of 1.96 Hz are plotted along the height of the tower in Fig. 2. Clearly visible is the large relative displacement between the

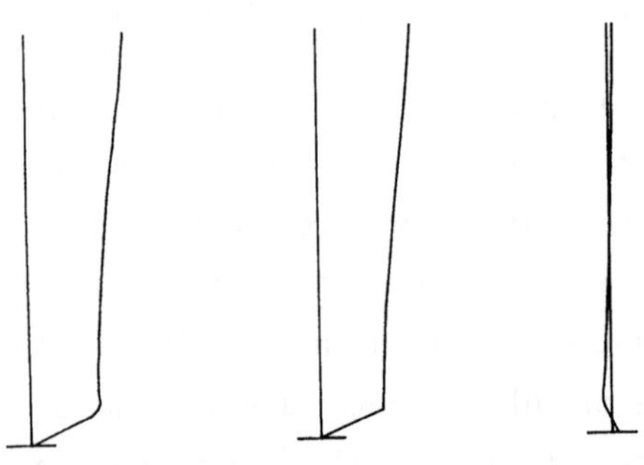

U_1^s V_1^s W_1^s

Fig. 2: Shape of first mode of first symmetric harmonic

upper and lower ends of the columns. This leads to large forces in the foundation, in the columns and in the lower edge beam of the shell. Hence, for increasing seismic loads, these parts of the tower are the first whose design is governed by the loading case of earthquake instead of by the wind. If the modeshape were equal to a horizontal rigid-body translation, no other modes would contribute to the global base-shear force.

2. Simplified Seismic Analysis

The characteristic seismic-elastic response for vertically propagating waves can also be captured using simpler models. For instance, the shell can be represented with beam elements with shear distortion and rotary inertia [5]. For parametric studies, a further simplification is possible: The shell, although its wall thickness is extremely small compared to a characteristic length, can be modelled as a rigid body. This is possible because the supporting columns are extremely flexible in the horizontal direction. The shape of the first mode of the first symmetric harmonic (which dominates the response for a uniform horizontal earthquake excitation) shown in Fig. 2 demonstrates this fact. Thus the model shown in Fig. 3a can be used [6].

a)

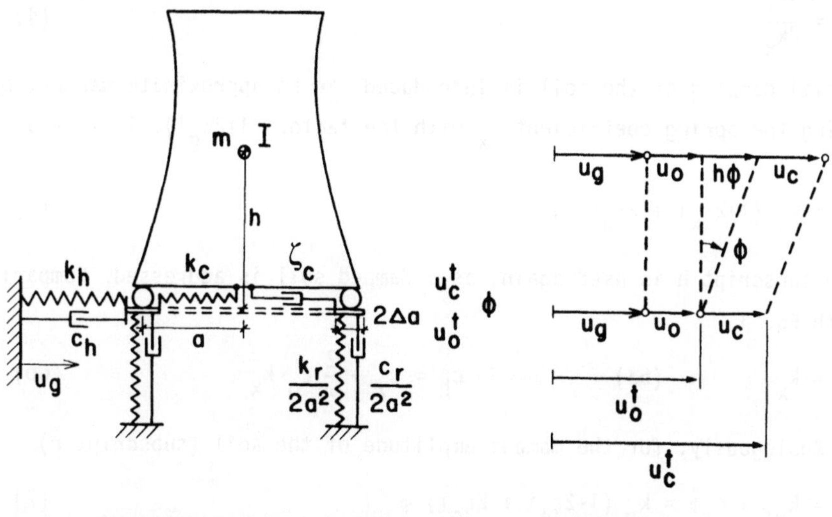

Fig. 3: Simplified model consisting of rigid shell on flexible columns founded on soil

a) Dynamic model

In the following, the influence of the flexibility of the soil on the dynamic response is examined, i.e. the effect of soil-structure interaction is evaluated. The results can only be summarised here. For a detailed discussion, see Ref. [6]. Harmonic motion with the excitation frequency ω is addressed. The horizontal-force amplitude P_h of the damped soil (subscript h) is expressed approximately as

$$P_h = k_h u_o + c_h \dot{u}_o \tag{1}$$

where k_h and c_h are the spring and damping coefficients and u_o denotes the horizontal-displacement amplitude of the base relative to the ground excitation u_g (free field). For a soil without material damping, the corresponding equation is written as

$$P_x = k_x u_o + c_x \dot{u}_o \tag{2}$$

where the subscript x denotes the horizontal direction for a purely elastic soil. Substituting $\dot{u}_o = i\omega u_o$ in Eq. 2 leads to

$$P_x = k_x \left(1+i\omega \frac{c_x}{k_x}\right) u_o = k_x (1+2\zeta_x i) u_o \tag{3}$$

where ζ_x represents the ratio of the viscous radiation damping in the horizontal direction

$$\zeta_x = \frac{\omega c_x}{2k_x} \tag{4}$$

The material damping of the soil is introduced, in an approximate manner, by multiplying the spring coefficient k_x with the factor $(1+2\zeta_g i)$. This results in

$$P_h = k_x (1+2\zeta_x i + 2\zeta_g i) u_o \tag{5}$$

where the subscript h is used again, as a damped soil is addressed. Comparing Eq. 5 with Eq. 1

$$k_h = k_x \quad \text{(6a)} \qquad c_h = c_x + \frac{2}{\omega} \zeta_g k_x \quad \text{(6b)}$$

follows. Analogously, for the moment amplitude of the soil (subscript r)

$$M_r = k_r \phi + c_r \dot{\phi} = k_r (1+2\zeta_\phi i + 2\zeta_g i) \phi \tag{7}$$

applies, with

$$\zeta_\phi = \frac{\omega c_\phi}{2k_\phi} \qquad (8)$$

$$k_r = k_\phi \qquad (9a) \qquad\qquad c_r = c_\phi + \frac{2}{\omega}\zeta_g\, k_\phi \qquad (9b)$$

ϕ denotes the rocking amplitude. The subscript ϕ denotes the rocking direction in the undamped soil.

The columns are modelled in the horizontal direction as a spring with a constant k_c and a hysteretic-damping ratio ζ_c. The amplitude of the relative displacement between the upper and lower ends of the columns is denoted as u_c. In the vertical direction the columns are assumed to be rigid. The shell is represented as a rigid body with mass m and moment of inertia I. Its centre of mass is located at a height h above the foundation.

The equations of motion are formulated as

$$\begin{bmatrix} \frac{\omega_c^2}{\omega^2}(1+2\zeta_c i)-1 & -1 & -1 \\ -1 & \frac{\omega_h^2}{\omega^2}(1+2\zeta_x i+2\zeta_g i)-1 & -1 \\ -1 & -1 & [\frac{\omega_r^2}{\omega^2}(1+2\zeta_\phi i+2\zeta_g i)-1](1+\frac{I}{h^2m}) \end{bmatrix} \begin{Bmatrix} u_c \\ u_o \\ h\phi \end{Bmatrix} = \begin{Bmatrix} 1 \\ 1 \\ 1 \end{Bmatrix} u_g \qquad (10)$$

where the following natural frequencies with simple physical interpretations are introduced.

$$\omega_c^2 = \frac{k_c}{m} \qquad (11a) \qquad\qquad \omega_h^2 = \frac{k_h}{m} \qquad (11b) \qquad\qquad \omega_r^2 = \frac{k_r}{h^2m+I} \qquad (11c)$$

The coefficient matrix is a function of ω_c, ω_h, ω_r, ω, I/h^2m, ζ_c, ζ_x, ζ_ϕ, and ζ_g.

When using this simple model for a transient excitation $u_g(t)$, it is appropriate to work directly in the time domain. For instance, the second equation of Eq. 10 is equal to

$$m(\ddot{u}_c(t)+\ddot{u}_o(t)+h\ddot{\phi}(t)) + c_h\dot{u}_o(t) + k_h u_o(t) = -m\ddot{u}_g(t) \qquad (12)$$

As the separate foundations of equivalent radius Δa are quite far apart, they can be regarded as independent of one another when calculating the dynamic-stiffness coefficients of the total foundation. The following crude approximations are used

$$k_x = n\,\frac{8G\Delta a}{2-\nu} \tag{13a}$$

$$c_x = n\,\frac{4.6}{2-\nu}\,\rho\, c_s\, \Delta a^2 \tag{13b}$$

$$k_\phi = \frac{4G\Delta a}{1-\nu}\sum_1^n x_i^2 = n\,\frac{4G\Delta a}{1-\nu}\,\frac{a^2}{2} \tag{13c}$$

$$c_\phi = \frac{4}{1-\nu}\,\rho\, c_s\, \Delta a^2 \sum_1^n x_i^2 = n\,\frac{4}{1-\nu}\,\rho\, c_s\, \Delta a^2\,\frac{a^2}{2} \tag{13d}$$

x_i denotes the distance from the centre line.

The following dimensionless parameters are introduced: ratio of stiffness of columns to that of soil $\bar{s} = \omega_c a/c_s$, $\bar{a} = \Delta a/a$, n, slenderness ratio $\bar{h}$ = h/a, mass ratio $\bar{m} = m/(\rho a^3)$, $I/(h^2 m)$, ν, ζ_c, and ζ_g. The equations of motion (Eq. 10) can then be expressed as a function of these parameters and of ω/ω_c.

$$\frac{\omega_h^2}{\omega^2}\,(1+2\zeta_x i+2\zeta_g i) - 1 = \frac{\omega_c^2}{\omega^2}\,\frac{8n\bar{a}}{(2-\nu)\bar{s}^2\bar{m}}(1+0.58\,\bar{s}\bar{a}\,\frac{\omega}{\omega_c}\, i+2\zeta_g i) - 1 \tag{14a}$$

$$[\frac{\omega_r^2}{\omega^2}\,(1+2\zeta_\phi i+2\zeta_g i)-1](1+\frac{I}{h^2 m})$$
$$= [\frac{\omega_c^2}{\omega^2}\,\frac{2n\,\bar{a}}{(1-\nu)\bar{s}^2\bar{m}\,\bar{h}^2(1+\frac{I}{h^2 m})}(1+\bar{s}\bar{a}\,\frac{\omega}{\omega_c}\, i+2\zeta_g i)-1](1+\frac{I}{h^2 m}) \tag{14b}$$

The tower shown in Fig. 1 has the following properties, in addition to those specified in Section 1: h = 65.64 m, Δa = 2.5 m, m = 18.55 Gg, I = 46500 Ggm^2, k_c = 5.05 GN/m, ζ_c = 0.025. In Figure 3b, the significant relative displacement between the upper and lower ends of the columns $|u_c|/|u_g|$ determined at resonance (i.e. for that ω/ω_c for which the largest value results whereby Eqs. 10 and 14 are used) is plotted versus $\bar{s}$ which represents the flexibility of the soil. In Figure 3c, the corresponding total displacement at the base $|u_o + u_g|/|u_g|$ and the rotation $|h\phi|/|u_g|$ versus $\bar{s}$ are shown. Even for the quite firm soil with c_s = 223 m/s corresponding to $\bar{s} \sim 4.3$, soil-structure interaction is important, which is characterised by a significant rocking response ϕ and by a non-negligible translation at the base relative to the ground u_o. The column forces (which are a function of u_c) are reduced in comparison to those for the same tower on rock ($\bar{s}$ = 0). The influence of the flexibility of the soil on other results is discussed in Ref. [2].

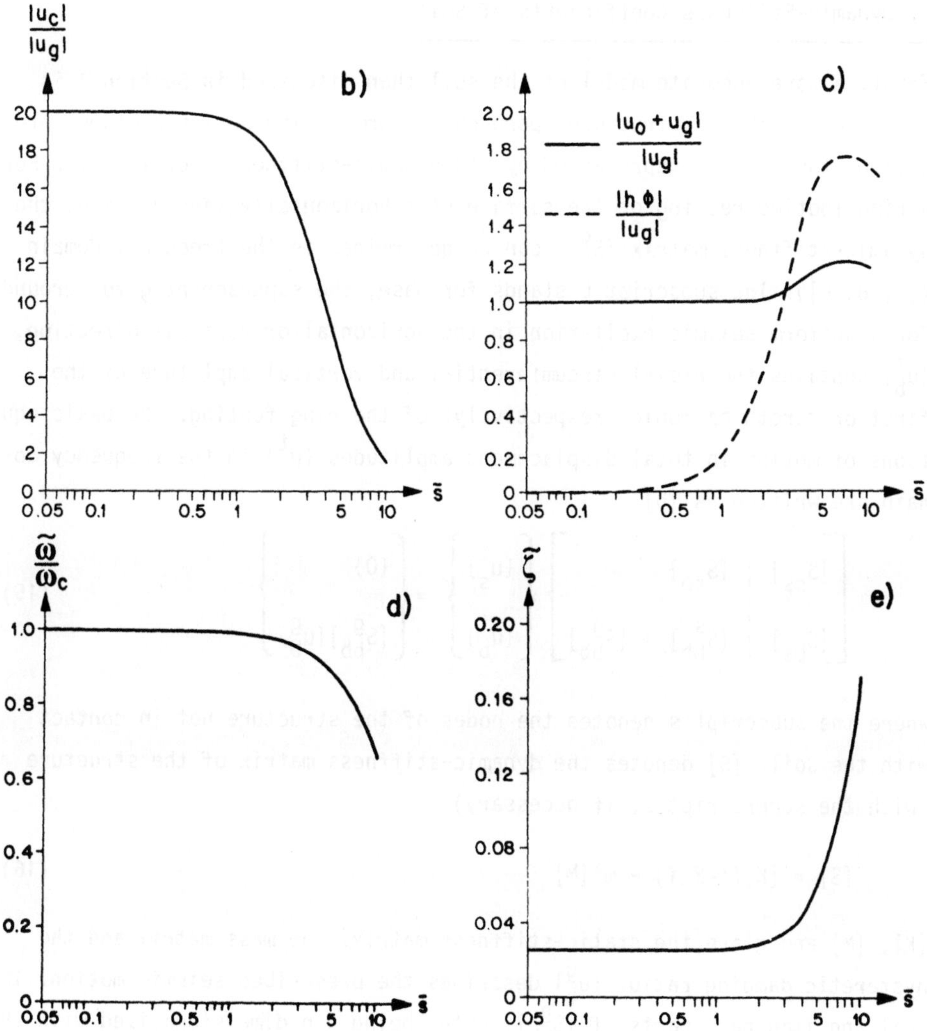

Fig. 3: Simplified model consisting of rigid shell on flexible columns founded on soil

b) Relative displacement of columns

c) Total displacement at base and rocking

d) Equivalent natural frequency

e) Equivalent damping ratio

To gain further physical insight, an equivalent one-degree-of-freedom system is introduced. Its properties (natural frequency $\tilde{\omega}$, ratio of hysteretic damping $\tilde{\zeta}$) are selected such that, at resonance, the same response u_c results as for the coupled system described by Eq. 10. $\tilde{\omega}/\omega_c$ and $\tilde{\zeta}$ are plotted versus $\bar{s}$ in Figs. 3d and e. For very soft soil ($\bar{s}$ large) the activated radiation damping leads to large values of $\tilde{\zeta}$.

3. Dynamic-Stiffness Coefficients of Soil

First, a more accurate model of the soil than discussed in Section 1 is examined. In the substructure approach of formulating soil-structure interaction, the soil is represented by its dynamic-stiffness coefficients. For a ring footing resting on the surface of a horizontally layered site, the dynamic-stiffness matrix $[S^g_{bb}]$ can be determined in the frequency domain ([7, 8, 6]). The subscript b stands for base, the superscript g for ground. For a uniform seismic excitation in the horizontal or vertical direction, $\{u^t_b\}$ contains the radial, circumferential and vertical amplitude of the first or zeroth harmonic, respectively, of the ring footing. The basic equations of motion in total-displacement amplitudes $\{u^t\}$ in the frequency domain are written as [6]

$$\begin{bmatrix} [S_{ss}] & [S_{sb}] \\ [S_{bs}] & [S^s_{bb}] + [S^g_{bb}] \end{bmatrix} \begin{Bmatrix} \{u^t_s\} \\ \{u^t_b\} \end{Bmatrix} = \begin{Bmatrix} \{0\} \\ [S^g_{bb}]\{u^g_b\} \end{Bmatrix} \tag{15}$$

where the subscript s denotes the nodes of the structure not in contact with the soil. [S] denotes the dynamic-stiffness matrix of the structure (with the superscript s, if necessary)

$$[S] = [K](1+2\zeta i) - \omega^2[M] \tag{16}$$

[K], [M] and ζ are the static-stiffness matrix, the mass matrix and the hysteretic damping ratio. $\{u^g_b\}$ describes the prescribed seismic motion. The real and imaginary parts of $[S^g_{bb}]$, after being non-dimensionalised with the static values $[K^g_{bb}]$, are decomposed into spring coefficients [k] and damping coefficients [c] as

$$[S^g_{bb}] = [K^g_{bb}]([k] + i\, a_o[c]) \tag{17}$$

with the dimensionless frequency

$$a_o = \frac{\omega\, a}{c_s} \tag{18}$$

a is a characteristic length, e.g. the average radius of the annular ring beam. [k] and [c] depend strongly on the frequency. This is illustrated in Fig. 4, where the corresponding vertical coefficients k_z and c_z for a rig-

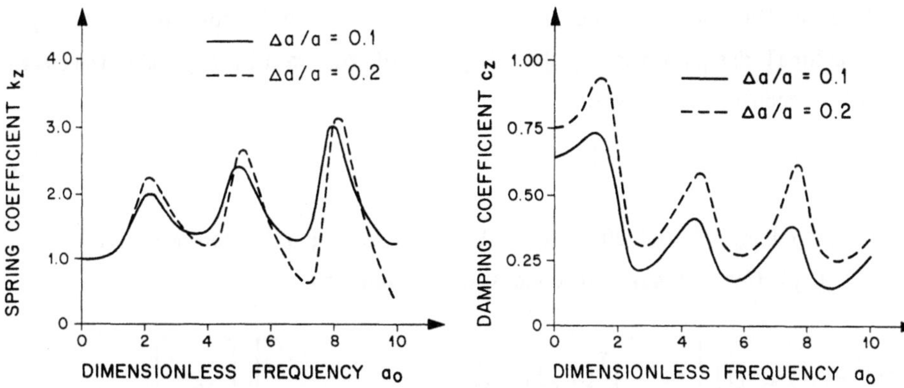

Fig. 4: Dynamic-stiffness coefficient in the vertical direction, annular ring on surface of undamped halfspace, Poisson's ratio = 0.33

id annular ring of width Δa resting on the surface of an undamped halfspace are plotted versus a_0 [6]. Thus, in general, an analysis in the frequency domain has to be performed.

On the basis of Eq. 15 the most general soil-structure interaction analysis can be performed, as long as it is elastic. Various transformations of Eq. 15 are possible [6], which lead to valuable physical insight and for a specific type of structure and excitation can result in a more appropriate form of the same equations of motion. One alternate formulation is summarised in this paper. First, the base response motion in total displacements $\{u_b^t\}$ is split up into the known seismic excitation motion $\{u_b^g\}$ and the relative motion $\{u_b\}$. Second, the latter, which is unknown, is quasi-statically transmitted into the structure with the remaining unknown dynamic part denoted as $\{u_s^d\}$.

$$\{u_b^t\} = \{u_b^g\} + \{u_b\} \tag{19a}$$

$$\{u_s^t\} = [T_{sb}]\{u_b\} + \{u_s^d\} \tag{19b}$$

Each column of the transformation matrix can be visualised as the static displacements of the unloaded nodes s of the structure when a unit displacement is imposed at a specific node b, and zero in all others. Third, $\{u_s^d\}$ is expressed as a linear combination of the first few modeshapes of the structure built in at its base

$$\{u_s^d\} = [\Phi]\ \{z\} \tag{20}$$

$[\Phi]$ contains the (scaled) modeshapes and $\{z\}$ the amplitudes. With the square of the natural frequencies on the diagonal of the matrix $[\Omega]$, the following eigenvalue problem is solved.

$$[K_{ss}]\ [\Phi] = [M_{ss}][\Phi][\Omega] \tag{21}$$

Substituting Eqs. 19 and 20 in Eq. 15 and assuming a lumped-mass matrix ($[M_{sb}] = [0]$) for the sake of conciseness leads to

$$\begin{bmatrix} [\Omega](1+2\zeta i)-\omega^2[I] & -\omega^2[\Phi]^T[M_{ss}][T_{sb}] \\ -\omega^2[T_{sb}]^T[M_{ss}][\Phi] & \begin{array}{l}([K^s_{bb}]-[K_{bs}][K_{ss}]^{-1}[K_{sb}])(1+2\zeta i) \\ -\omega^2([M^s_{bb}]+[T_{sb}]^T[M_{ss}][T_{sb}])+[S^g_{bb}]\end{array} \end{bmatrix} \begin{Bmatrix} \{z\} \\ \{u_b\} \end{Bmatrix} =$$

$$\omega^2 \begin{bmatrix} [\Phi]^T[M_{ss}][T_{sb}] \\ [M^s_{bb}]+[T_{sb}]^T[M_{ss}][T_{sb}] \end{bmatrix} \{u^g_b\} + \begin{bmatrix} [0] \\ (-[K^s_{bb}]+[K_{bs}][K_{ss}]^{-1}[K_{sb}])(1+2\zeta i) \end{bmatrix} \{u^g_b\} \tag{22}$$

For a uniform earthquake excitation $\{u^g_b\}$, the second term on the right-hand side vanishes, leading to a load vector of the form "mass times earthquake acceleration". This formulation takes advantage of the fact that, although classical modes do not exist for the total system, uncoupling of the modes does occur for the structure built in at its base.

Alternate approximate formulations exist. Using the first few modeshapes of the total system as generalised displacements, the coupling of the transformed equations of reduced order is restricted to the damping terms. The dynamic-stiffness matrix of the soil $[S^g_{bb}]$ is approximated as

$$[S^g_{bb}(\omega)] = [K^g_{bb}] + i\omega[C^g_{bb}] \tag{23}$$

where the "static"-stiffness matrix $[K^g_{bb}]$ and the damping matrix $[C^g_{bb}]$ are both independent of the frequency. The frequency at which they are evaluated, i.e. the natural frequency, is determined by iteration. As another possibility, complex eigenvectors can be used to decouple the equations of motion of the total system. Equation 23 still has to be used iteratively.

Embedded foundations can also be analysed, whereby the corresponding dynamic-stiffness matrix of the soil can be calculated using the boundary-element method. Slight modifications occur in the equations of motion [6]. In particular, it is necessary to distinguish between the so-called scattered motion and that of the free field, which is denoted as $\{u_b^g\}$ in this article. This theory, however, is not discussed in this paper.

4. Horizontally Propagating Waves

Second, as a step towards a more realistic wave pattern, the body waves are assumed to propagate at an angle and in addition, surface waves are introduced. These waves propagate horizontally across the base of the structure with an apparent velocity c_a. They lead to reduced effective horizontal and vertical translational seismic input motions, as compared to those resulting from assuming the same waves as vertically incident. However, a rotational input is also created. The dimensions of the tower at its base are so large that even for quite large apparent velocities, travelling effects are significant. Even in an elastic analysis,the separate foundations, which can be connected by a ring beam, will, for horizontally propagating waves, produce local motions which are associated with higher-order harmonics of the displacements of the axisymmetric shell.

Equation 15 (or Eq. 22) can be used to analyse soil-structure interaction for horizontally propagating waves as well. The term $\{u_b^g\}$ in the load vector is affected.

Often for a cooling tower, which is not a high-safety-class structure, for economical reasons, only approximate dynamic-soil investigations are performed. This does not allow the wave pattern, which can occur at the site, to be calculated rigorously [6]. Certain "reasonable" wave patterns are assumed to exist. The response spectrum of the 10-sec artificial time history for the horizontal direction closely follows that of the U.S. NRC Regulatory Guide 1.60. Two horizontally propagating wave patterns are examined. The vertical component is determined from the horizontal one by making different assumptions, which, in general, violate the laws of elastodynamics. The absolute values of the horizontal and vertical amplitudes for all frequencies are selected as equal. Retrograde and prograde motions are selected and denoted as such. For both types of waves, the peak vertical acceleration is approximately equal to that in the horizontal direction. The apparent velocity c_a = 500 m/s is used in the calculations, which is reasonable when considering the dispersion curves for surface waves of this site. For comparison, the horizontal and the vertical components of the motion are assumed to originate from vertically incident waves ($c_a = \infty$), leading to a uniform earthquake excitation. For the vertical component the retrograde wave is used. In addition, the horizontal component alone is processed, with c_a = 500 m/s, and for $c_a = \infty$.

Evaluating the results, it is convenient to compare the following maximum-dynamic values: The total base shear force Q, which characterises the overall lateral response; the maximum normal column force N_{max} (tension), taken along the circumference without dead weight (the force that governs the design of the columns); and finally the total horizontal acceleration $\ddot{u}^t$ at the top of the shell (this acceleration acts as an indicator of local shell motions). In Table 1, these three values, normalised to the corresponding ones for the horizontal and vertical motions with vertical incidence, are presented [9].

Type of analysis	Wave pattern	Total shear force Q	Max column force N_{max}	Acceleration $\ddot{u}^t$
Linear	Hor/vert $c_a = \infty$	1	1	1
	Hor $c_a = \infty$	1.00	0.98	0.88
	Hor $c_a = 500$	0.59	0.76	0.59
	Retrograde	1.74	1.78	1.85
	Prograde	0.56	0.81	2.24
Non-linear	Hor/vert $c_a = \infty$	0.67	0.58	1.16
	Hor $c_a = \infty$	0.72	0.55	1.14
	Hor $c_a = 500$	0.53	0.48	0.74
	Retrograde	0.81	0.78	2.50
	Prograde	0.44	0.47	2.47

Table 1: Ratios of maximum response

The horizontal component only is examined at the beginning. Deleting the vertical earthquake for the case $c_a = \infty$ reduces N_{max} and, to a somewhat larger degree, $\ddot{u}^t$. When this horizontal component propagates, the resulting self-cancelling effect of the translational input motion reduces all three response values. Since the reduction for $\ddot{u}^t$ is the same (0.59) as that for Q, the horizontal component of the propagating wave does not seem to excite

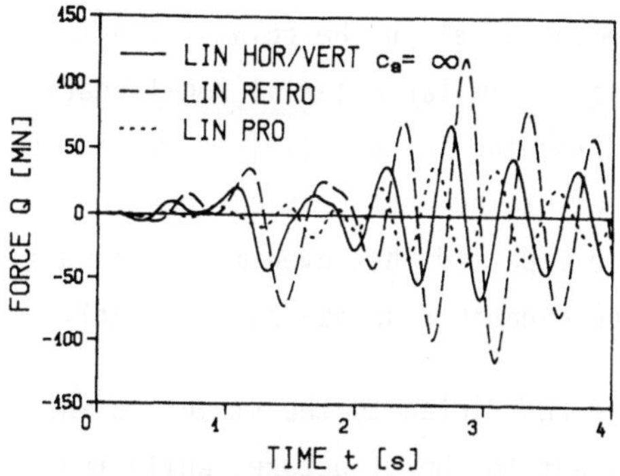

Fig. 5: Time history of total shear force at base, travelling-wave effect

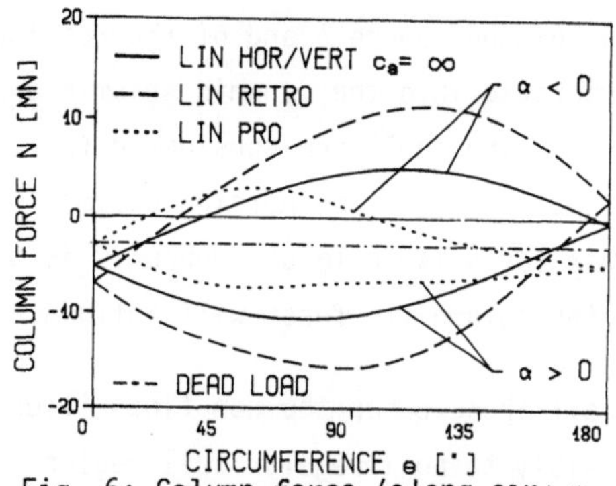

Fig. 6: Column force (along circumference) corresponding to maximum-total shear force (tavelling-wave effect,linear)

local shell modes. It is well known that the horizontal seismic loading resulting from the horizontal input component is in phase with that arising from the rocking input component for a retrograde wave. Introducing the propagating vertical component, the retrograde waves lead, as expected, to larger Q's and N's, while the prograde waves lead to smaller ones. This is also apparent in Figs. 5 and 6, in which the first four seconds of the time history of Q and of the normal force N in the columns with positive and negative angles of inclination α around the circumference are plotted for that time when the maximum value of Q occurs (maximum ground acceleration 0.12 g). The dominance of the motion in the fundamental frequency = 1.96 Hz for all three cases is evident in Fig. 5. However, the rocking component of the effective seismic input excites the local shell modes for both the retrograde and the prograde waves, as is apparent from $\ddot{u}^t$ in Table 1.

5. Material Non-linearities

Third, the analysis including material non-linearities is addressed. For a specific design, various kinds of material non-linearities, which also influence the stiffness, will arise for increasing earthquake excitation. In this paper, those associated with the columns and their foundation are examined [4, 9]. The non-linear behaviour can consist of partial uplift and slipping of the foundation. These non-linearities in the area of contact of the foundation and the soil reduce the overall response significantly, as compared to a linear-elastic analysis. In addition, the non-linear material laws of the concrete and of the reinforcement steel in the columns are incorporated into the dynamic seismic analysis. Similar material non-linearities in the shell are considered for wind loading in Ref. [10]. As mentioned in Section 1, the supporting columns represent the most critical part of the tower for seismic loads. Hence it is to be expected that even minor changes in the column stiffness will influence the dynamic response significantly.

In all these cases, the non-linearities are restricted to the structure and possibly to an irregular soil region adjacent to the structure, while the far field of the unbounded soil is assumed to remain linearly elastic with material damping. Considering the irregular soil region, if present, as part of the structure, the line joining the nodes with subscript b separates the non-linear from the linear region. To analyse such a case, a procedure which works directly in the time domain has to be used. The basic equations of motion in the frequency domain (Eq. 15) transformed to the time domain lead to expressions involving convolution integrals [11].

$$\begin{bmatrix}[M_{ss}] & [M_{sb}] \\ [M_{bs}] & [M^s_{bb}]\end{bmatrix}\begin{Bmatrix}\{\ddot{u}^t_s(t)\} \\ \{\ddot{u}^t_b(t)\}\end{Bmatrix} + \begin{Bmatrix}\{P_s(t)\} \\ \{P^s_b(t)\}\end{Bmatrix} + \begin{Bmatrix}\{0\} \\ \int_0^t [S^g_{bb}(t-\tau)]\{u^t_b(\tau)\}d\tau\end{Bmatrix} = \begin{Bmatrix}\{0\} \\ \int_0^t [S^g_{bb}(t-\tau)]\{u^g_b(\tau)\}d\tau\end{Bmatrix} \quad (24)$$

$\{P(t)\}$ represents the non-linear internal forces of the structure. The dynamic-stiffness matrix of the soil in the time domain $[S^g_{bb}(t)]$ contains the forces required to produce unit-impulse displacements. $[S^g_{bb}(t)]$ and $[S^g_{bb}(\omega)]$ form a Fourier-transform pair.

$$[S^g_{bb}(t)] = \frac{1}{2\pi}\int_{-\infty}^{+\infty}[S^g_{bb}(\omega)]\, \exp(i\omega t)\, d\omega \quad (25)$$

Equation 24 can be solved directly in the time domain, taking the non-linearities into account, using, e.g., an explicit integration scheme.

Various simplifications are possible. For instance, introducing frequency-independent matrices $[K^g_{bb}]$ and $[C^g_{bb}]$ as defined in Eq. 23, the convolution integrals in Eq. 24 are removed. For the term on the left-hand side, this leads to

$$\int_0^t [S^g_{bb}(t-\tau)]\{u^t_b(\tau)\}\, d\tau = [K^g_{bb}]\{u^t_b(t)\} + [C^g_{bb}]\{\dot{u}^t_b(t)\} \tag{26}$$

For a linear structure, the equations of motion in the time domain corresponding to Eq. 22 are straightforwardly formulated. Using the "constant stiffness-initial stress" approach, non-linearities in the structure lead to equivalent loads [4].

For the non-linear analysis of the tower, the relationship between the normal force N and the axial strain ε of the column of a diameter 0.85 m with 15 rebars of diameter 35 mm is based on the constitutive laws of concrete and steel (Fig. 7). Details are specified in Ref. [4]. The vertical reaction force R_w of an individual foundation is limited by the ultimate bearing capacity of the soil = -1 MPa, which results in R^y_w = 20 MN. The corresponding vertical reaction-displacement relationship (R_w-w) is plotted in Fig. 8. The radial and tangential reactions are limited by Coulomb's friction law with a coefficient = 0.58. The shell is assumed to remain linearly elastic.

The same earthquake records as defined in Section 4 are used (maximum acceleration = 0.12 g). Introducing non-linearities reduces the corresponding Q and N_{max} for all earthquake excitations (Table 1). This is also demonstra-

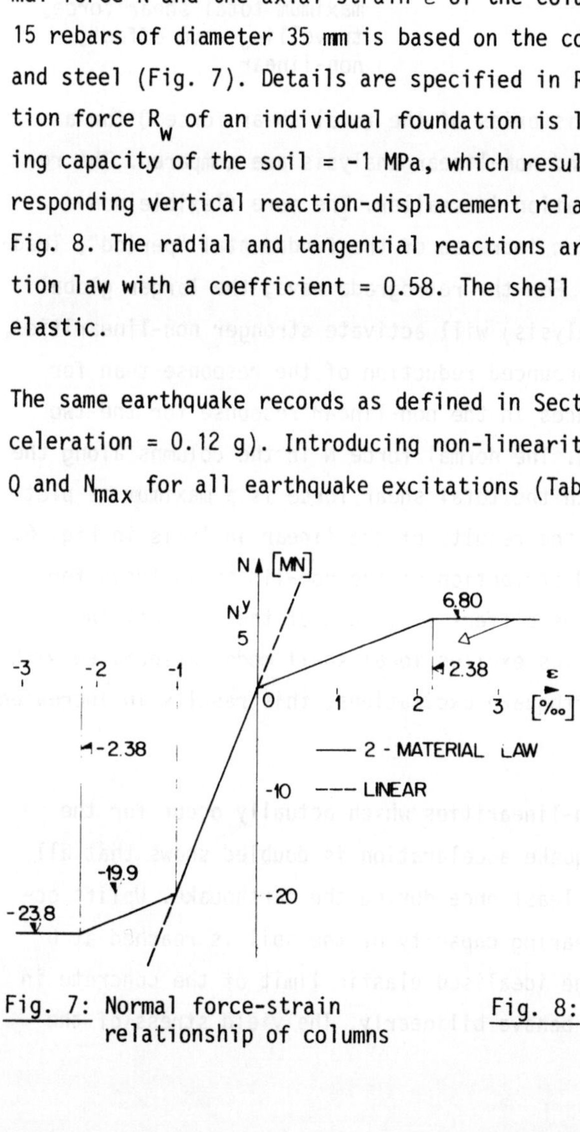

Fig. 7: Normal force-strain relationship of columns

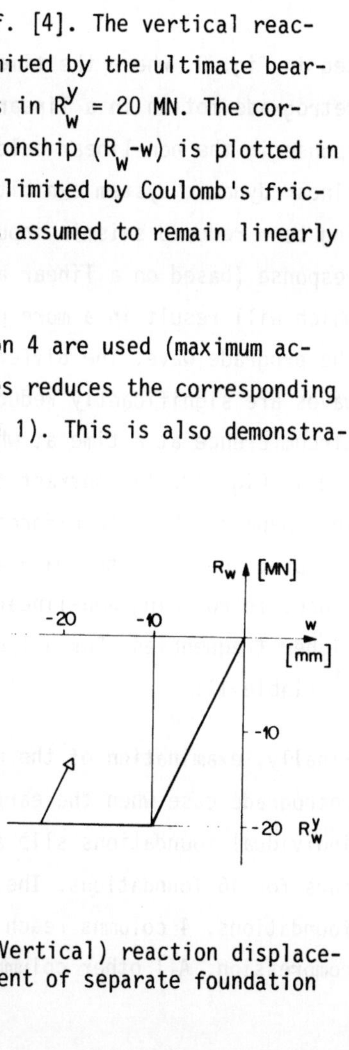

Fig. 8: (Vertical) reaction displacement of separate foundation

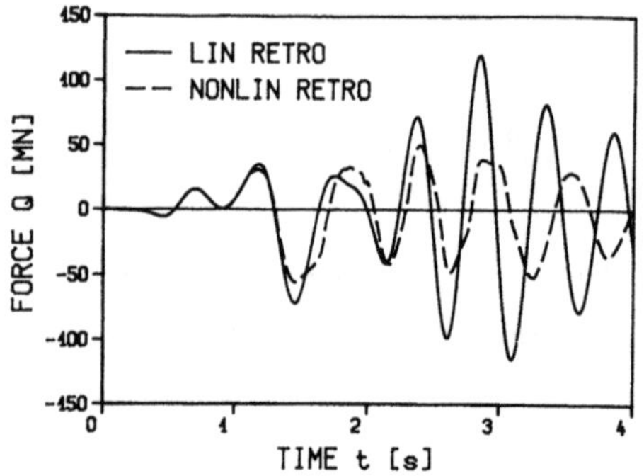

Fig. 9: Time history of total shear force at base, non-linear effect

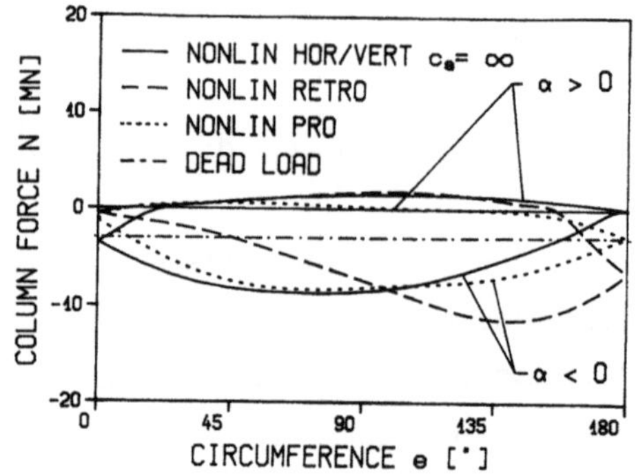

Fig. 10: Column force (along circumference) corresponding to maximum total shear force, travelling-wave effect, non-linear

ted in Fig. 9, where the time histories of the total shear force Q for a retrograde motion in a linear and non-linear analysis are compared. The response in the non-linear calculation is smaller. The more flexible non-linear dynamic system exhibits an increase of the fundamental "period", leading to a reduced seismic input. For the retrograde wave, the larger global response (based on a linear analysis) will activate stronger non-linearities, which will result in a more pronounced reduction of the response than for the prograde wave. The differences in the non-linear response for the two waves are significantly reduced. The normal force N in the columns along the circumference at a time at which the total shear force is a maximum is plotted in Fig. 10. In contrast to the results of the linear analysis in Fig. 6, the shape of the column-force distribution of the non-linear analyses for the prograde and retrograde waves differs from that of the vertical incidence. Introducing non-linearities excites local shell modes associated with higher frequencies. For all earthquake excitations, this results in increased $\ddot{u}^t$ (Table 1).

Finally, examination of the non-linearities which actually occur for the retrograde case when the earthquake acceleration is doubled shows that all individual foundations slip at least once during the earthquake. Uplift occurs for 16 foundations. The bearing capacity of the soil is reached at 6 foundations. 4 columns reach the idealised elastic limit of the concrete in compression. All other columns behave bilinearly. The yield stress of the

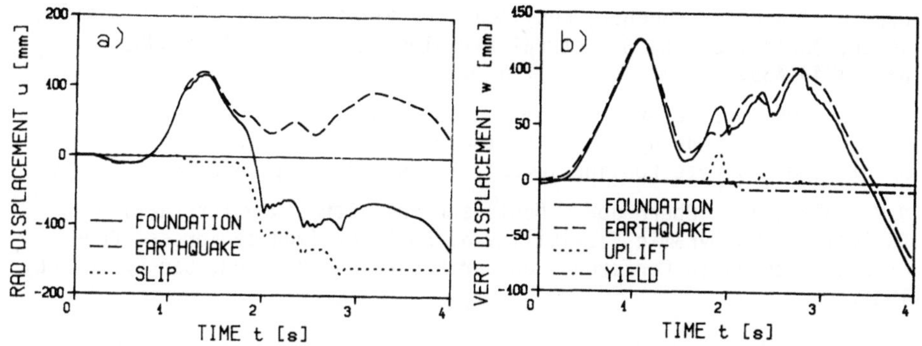

Fig. 11: (Total) displacement time history of foundation at $\Theta = 175^0$, non-linear retrograde, 0.24 g

a) Radial direction b) Vertical direction

reinforcement steel in the columns is never reached. As an illustration, the total-displacement time histories in the radial and vertical directions (u and w) are plotted for the foundation at 175^0 measured from the direction of the earthquake motion. Slip and uplift are clearly visible (Figs. 11a and b).

References

[1] A.S.L. Chan and A. Firmin, The Analysis of Cooling Towers by the Matrix Finite-Element Method, Aeronautical Journal, Vol. 74, 1970, pp. 826-835.

[2] O.M. El-Shafee and P.L. Gould, Cooling Tower Shells Founded on Soil, Journal of the Structural Division, ASCE, Vol. 108, 1982, pp. 800-813.

[3] A.S.L. Chan and J.P. Wolf, Cooling tower supporting columns and reinforcing rings in small and large displacement analyses, Computer Methods in Applied Mechanics and Engineering, Vol. 13, 1978, pp. 1-26.

[4] J.P. Wolf and P.E. Skrikerud, Influence of Geometry and of the Constitutive Law of the Supporting Columns on the Seismic Response of a Hyperbolic Cooling Tower, Earthquake Engineering and Structural Dynamics, Vol. 8, 1980, pp. 415-437.

[5] W.B. Krätzig and K. Meskouris, Simplified Earthquake Analysis of Natural Draught Cooling Towers, Institut für konstruktiven Ingenieurbau, Ruhr-Universität Bochum, Report 78-5, 1978.

[6] J.P. Wolf, Dynamic Soil-Structure Interaction, Prentice-Hall, Englewood Cliffs, N.J., 1985.

[7] O.M. El-Shafee and P.L. Gould, Dynamic Axisymmetric Soil Model for a Flexible Ring Footing, Earthquake Engineering and Structural Dynamics, Vol. 8, 1980, pp. 479-498.

[8] E. Kausel and J.M. Roesset, Stiffness Matrices for Layered Soils, Bulletin of the Seismological Society of America, Vol. 71, 1981, pp. 1743-1767.

[9] J.P. Wolf and K.M. Bucher, Nonlinear Traveling-Wave Effects on Cooling Tower, Journal of the Structural Division, ASCE, Vol. 108, 1982, pp. 1424-1439.

[10] J. Form and H.L. Peters, Economical aspects of design and execution of large natural-draught cooling towers, IASS Symposium Very Tall Reinforced Concrete Cooling Towers, Paris, 1978, pp. 342-388.

[11] J.P. Wolf and P. Obernhuber, Nonlinear Soil-Structure Interaction Analysis using Dynamic Flexibility of Soil for Impulse Forces, Proceedings 8th World Conference on Earthquake Engineering, San Francisco, 1984.

Seismic Response of Cooling Towers on Interactive Shallow or Deep Foundations

B. J. LEE

Research Associate
Petroleum and Energy Research Institute
University of Tulsa
Tulsa, Oklahoma, USA

P. L. GOULD

Harold D. Jolley Professor and Chairman
Dept. of Civil Engineering
Washington University
St. Louis, Missouri, USA

INTRODUCTION

An analytical method has been developed to determine the seismic response of rotational shell structures, such as cooling towers, supported either by shallow foundations, in the form of a ring footing, or by deep foundations, in the form of long piles. The soil-structure interaction effects for the former case, and the soil-pile-structure interaction effects for the latter case are included through the development of a boundary system at the contact area between the superstructure and the substructure. The soil has been modelled as a layered viscoelastic medium with constant hysteretic damping. The corresponding finite element model utilizes efficient energy transmitting boundaries. Both time and frequency domain analyses were performed on the system. The details of the formulation are given in Refs. 1 and 2 along with a comprehensive bibliography of pertinent literature. In this paper some numerical results of practical significance are presented and discussed.

DESCRIPTION OF SHELL-PILE-SOIL SYSTEM

The structure under consideration is the reinforced concrete cooling tower shell shown in Fig. 1. The tower is assumed to have a deep foundation in the form of vertical piles. The shell meridian consists of three curves with slope continuity at the junction points (nodal points #4 and #7) as described by the equations of the shell meridian given in Table 1.

To examine the effect of base flexibility on the seismic response, a complex response (frequency domain) analysis is

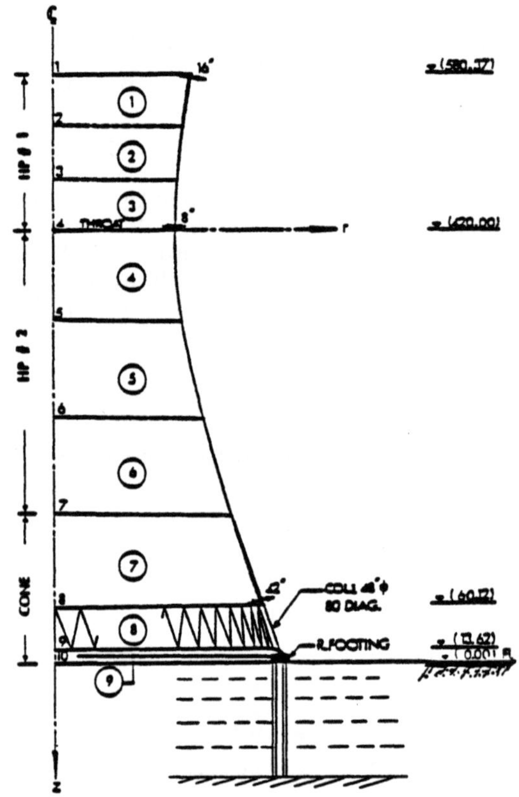

Fig. 1. Cooling tower on vertical piles

Table 1. Shell meridian of the structure under study

Shell Type	Nodes From	Nodes To	Curve Equation
HP #1	1	4	$z^2 - 123.68377\ r^2 + 27587.5165\ r - 1536846.5 = 0$
HP #2	4	7	$z^2 - 9.40153\ r^2 + 1302.5923\ r - 25462.9 = 0$
CONE	7	10	$0.3z - r + 87.2112 = 0$

conducted. Four different base conditions are considered, i.e., rock founded (fixed base), 334 piles, 167 piles and a flexible base. These cases are selected to provide a wide variation of site conditions and also to facilitate the detection of trends in the structural response. The properties of the five soil layers are specified in the table of Fig. 2. The shear moduli are compatible with the strains derived from earthquake records [3,4]. Gap development between the soil and the piles in the upper portion can be accounted for in the present formulation by considering the gap region to be void, but this is not considered here. The data related to the piles is also given in Fig. 2 and hysteretic damping of all soil layers is taken as 10%.

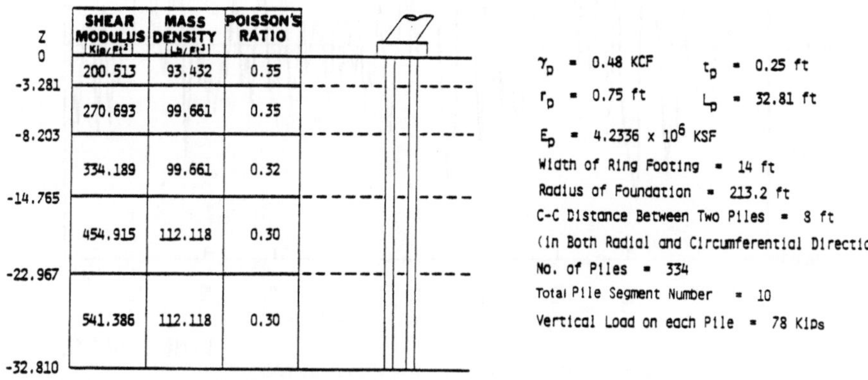

Fig. 2. Foundation detail for pile-supported cooling tower

The tower of Fig. 1 with four different base conditions is analized under the El-Centro earthquake (5/18/40) EW Comp. with input in the form of Fig. 3. For the complex response analysis the time step is taken as 0.02 second, which gives a total of 2^9 = 512 points for the duration of 10.22 seconds. The damping coefficients for the structure itself (α = 0.715779 and β = 0.003356) are obtained based on 5% damping ratio for the first two modes of vibration (the modes of vibration are obtained from a free vibration analysis with the base fixed).

RESULTS OF COMPLEX RESPONSE ANALYSIS

The variation of discrete values of normal displacement amplitude $|w|$ at the first resonant frequency (i.e. the first vibration mode) along the structure for the different base conditions is shown in Fig. 4. The flexible base case is representative of a cooling tower structure supported on a ring footing assuming that the soil-pile impedance functions approach zero, while the fixed base case means that displacements u, v, w, β_ϕ and β_θ are constrained at the base node. The foundation flexibilities of 334 piles and 167 piles lie between those of the fixed base and the flexible base.

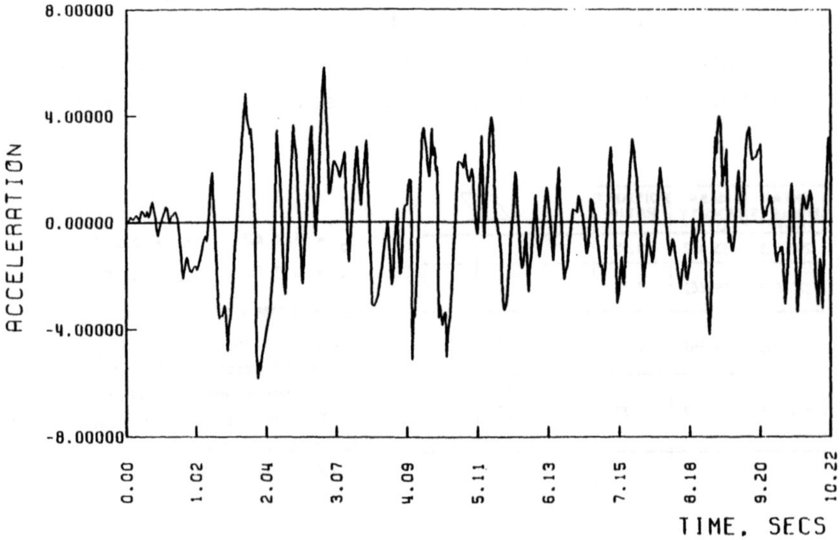

Fig. 3. Input earthquake motion

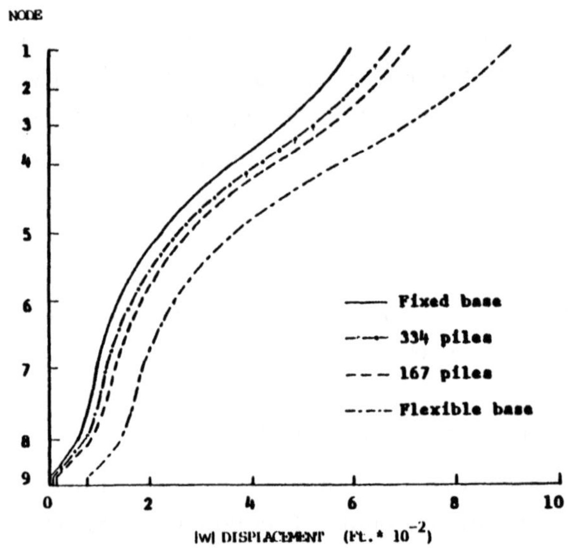

Fig. 4. Frequency history plot for harmonic number = 1
Component no. 3 at node no. 1

As indicated in this figure, the structure tends to move as a whole except for the fixed base case, i.e., rigid body motions occur if the base is not assumed to be fixed. Thus, in calculating the actual deformation, rigid body motions should be taken into consideration [1,2]. It is seen that the base flexibility can have a significant effect on the results. Generally speaking, the flexible base will reduce the natural frequency and increase the relative displacements. The fixed and flexible base cases are shown as a basis of comparison.

After elimination of rigid body effects, the normal displacement at the top of the shell $|w_1(\omega)|$ for three foundation conditions (fixed base, 334 piles, 167 piles) is plotted on Fig. 5. The amplitudes of the resonant peaks of $|w_1(\omega)|$ near the first fundamental frequency, $\bar{\omega}_1$, decrease with increased foundation flexibility. This suggests that the damping of the system increases as the foundation becomes more flexible because of the energy which is dissipated by radiation into the supporting system. Note also that the amplitude $|w_1(\omega)|$ converges toward the value for a fixed base as the pile number increases. Increasing the number of piles effectively increases the stiffness of the foundation, primarily reducing rocking but also affecting translation of the base. As a result of these effects, the natural frequency of the pile-supported cooling tower approaches that corresponding to a fixed base assumption. The reduction in the

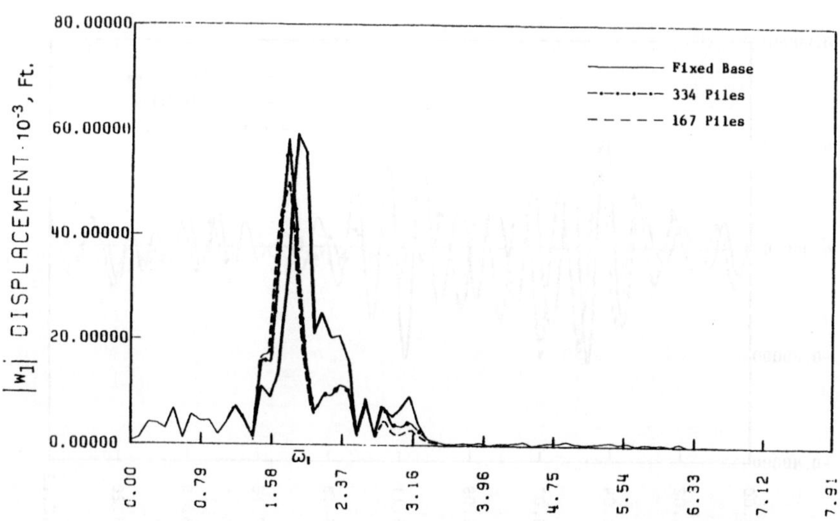

Fig. 5. Frequency history plot for harmonic number = 1
Component no. 3 at node no. 1

first resonant frequency due to a softer interacting system produced by including foundation flexibility is found to be in the range of 6% from the fixed base frequency, while the resonant frequencies for 334 piles and 167 piles are not much different from each other. On the other hand, the drastic decrease in the amplitude of the second resonant peak, as shown in Fig. 5, may be attributed to the increase in damping provided by the soil-pile systems at the higher frequency [1].

RESULTS OF TIME DOMAIN ANALYSIS

The normal displacement, $w_1(t)$, in the time domain at node 1 is shown in Fig. 6, for which the inverse Fourier transform was performed on the corresponding discrete values of response in the frequency domain. It can be seen in this figure that the fixed base produces deformations which envelop all pile cases, except at a few points. Thus, the maximum response, which is of main concern in the present analysis, is reduced by interaction.

To assess the importance of soil-pile-structure interaction on the stress resultants and stress couples in the shell (Fig. 7), the resultant forces for some selected nodes at $\theta = 0°$ are shown in Figs. 8 through 10. It may be observed that the fixed base

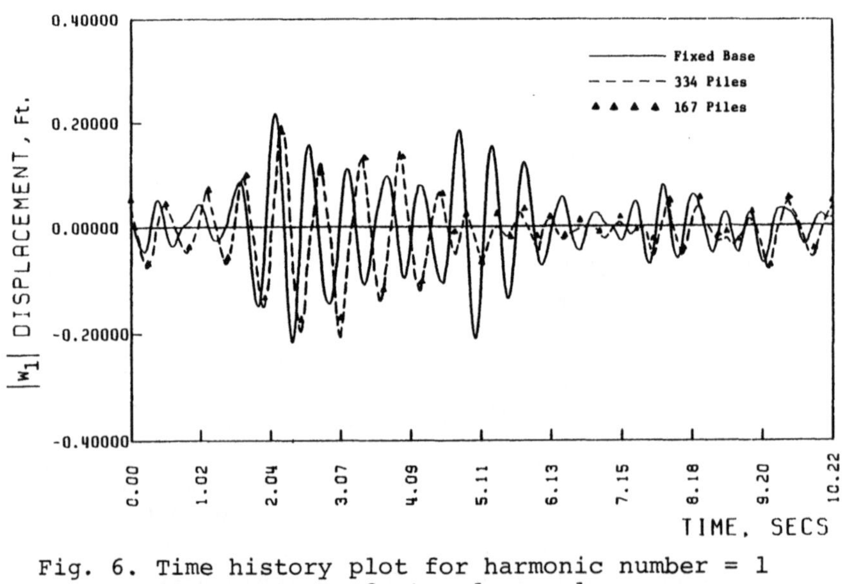

Fig. 6. Time history plot for harmonic number = 1
Component no. 3 at node no. 1

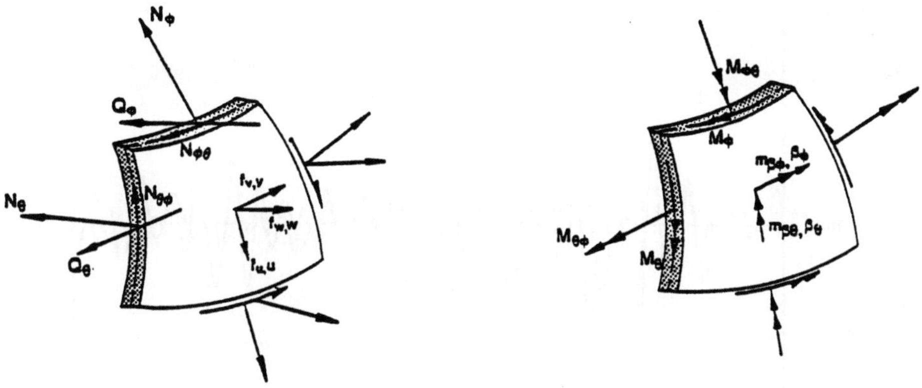

Stress resultants   Stress-couples

Fig. 7. Stress resultants and couples

case produces maximum resultant forces larger than for both pile cases. Since the response curves for the 167 pile case are close to those for the 334 pile case, only the peak values for the former case are recorded with triangular symbols. The maximum responses are listed in Table 2. A comparison of the maximum response values shows that the interaction effect indeed reduces the stresses in the shell when a strong motion earthquake loading is applied. The percentage of reduction of absolute maximum resultant forces is given in Table 3 in comparison with the fixed base case. Obviously, the difference between

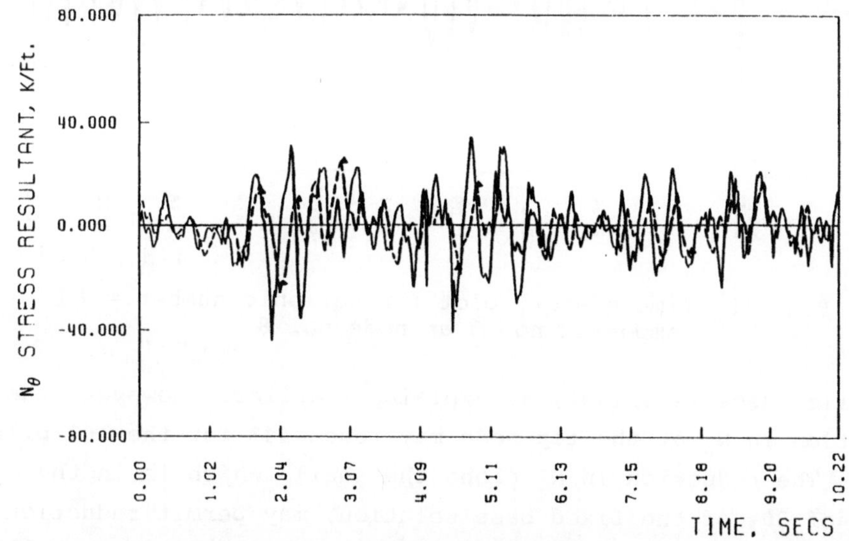

Fig. 8. Time history plot for harmonic number = 1
Component no. 2 at node no. 1

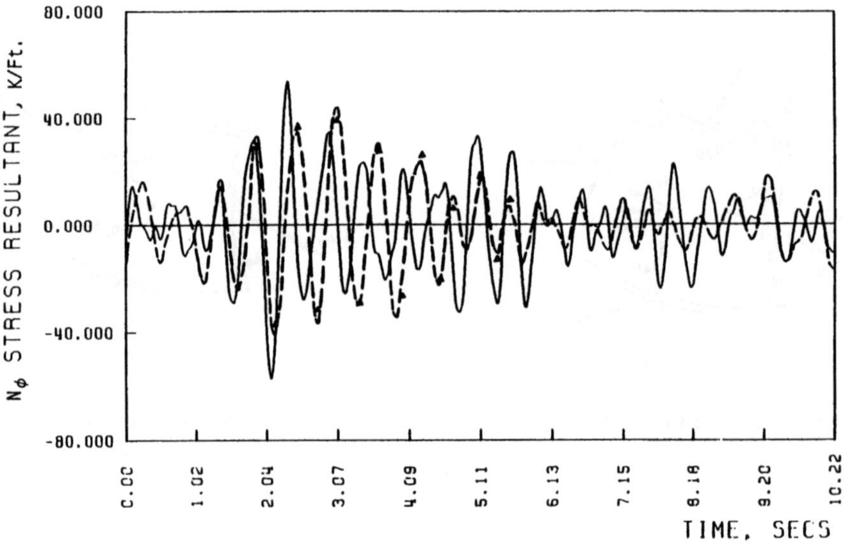

Fig. 9. Time history plot for harmonic number = 1
Component no. 1 at node no. 5

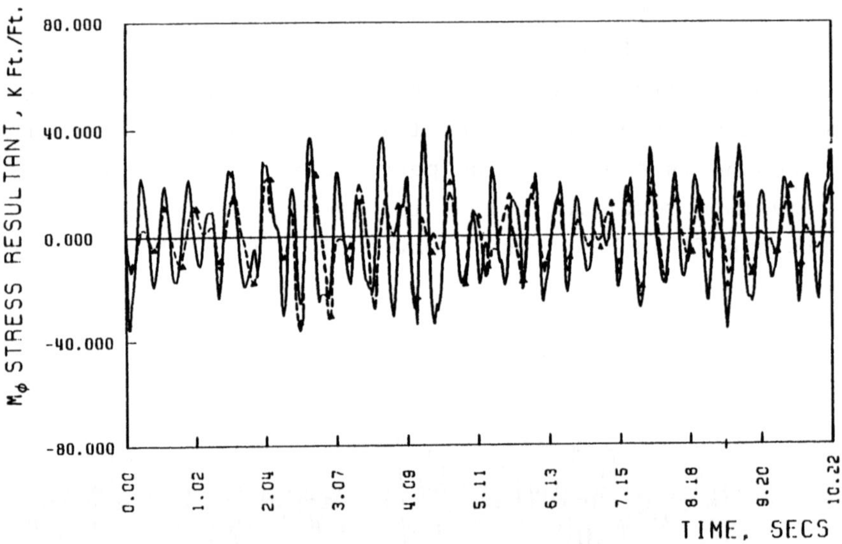

Fig. 10. Time history plot for harmonic number = 1
Component no. 3 at node no. 8

two pile cases is slight, as explained earlier. However, the reduction in N_θ at the top node may reach 44% for the 167 pile case. The reduction in N_ϕ along the shell, which is in the range of 30% of the fixed base solution, may permit reduction of the shell cross section and the meridional steel in the shell, resulting in a considerable cost savings.

Table 2. Maximum response (at θ=0°) in the shell

Max. Resultant Forces & Displ.*	N_θ	N_ϕ	N_ϕ	N_ϕ	N_ϕ	M_ϕ	M_θ	w
Case \ Node	1	4	5	7	8	8	8	1
Fixed	+34.44 (4.82)**	+79.00 (5.00)	+54.79 (2.38)	+41.49 (2.38)	+44.62 (1.82)	+40.50 (4.64)	+8.47 (4.64)	+0.2176 (2.16)
	-43.83 (1.90)	-75.93 (4.76)	-57.14 (2.06)	-48.53 (2.06)	-53.33 (2.06	-36.01 (8.70)	-7.04 (4.44)	-0.2188 (2.36)
334 Piles	+25.04 (2.96)	+61.06 (2.56)	+43.83 (3.08)	+33.66 (3.08)	+44.71 (3.08)	+28.96 (2.78)	+5.32 (2.78)	+0.2060 (2.16)
	-25.39 (2.04)	-64.31 (2.20)	-39.06 (2.20)	-32.88 (2.16)	-42.27 (8.06)	-35.92 (2.56)	-7.00 (2.56)	-0.1900 (3.08)
167 Piles	+24.31 (2.96)	+61.06 (2.56)	+39.21 (3.08)	+29.35 (3.08)	+41.57 (3.08)	+22.70 (2.78)	+4.38 (2.78)	+0.1960 (2.16)
	-22.35 (2.04)	-61.06 (2.20)	-36.79 (2.20)	-28.18 (2.16)	-39.80 (8.06)	-33.57 (2.56)	-6.58 (2.56)	-0.1738 (3.08)

* UNIT: Stress Resultants K/Ft Displacement Ft.
Stress Couples K·Ft/Ft

** The values in parentheses refer to time.

Table 3. Reduction percentage of stress resultants and stress couples compared to the fixed base case

	Reduction Percentage (%)						
Resultant Forces	N_θ	N_ϕ	N_ϕ	N_ϕ	N_ϕ	M_ϕ	M_θ
Case \ Node	1	4	5	7	8	8	8
334 piles	42	18.6	23.3	30.6	16.1	11.3	17.3
167 piles	44	22.7	31.4	39.5	22.0	17.1	22.3

The axial force, bending moment and shear force in the columns are calculated at $\theta = 0°$ and the results are shown in Table 4. It may be noted that there is a sharp decrease in the shear force while the axial force and bending moment are also decreased.

Table 4. Maximum column forces at θ=0° (horizontal ground motion)

Case \ Force	Axial Force (Kip)	Bending Moment (Kip·ft)	Shear Force (Kip)
Fixed	862.117	654.711	746.112
334 Piles	722.768	580.687	291.792
167 Piles	671.975	542.650	291.467

CONCLUSION

The complex response analysis of a cooling tower on a pile foundation shows that with fewer piles, the overall flexibility of the shell increases as the foundation stiffness decreases. Consequently, the inertial force on the shell is reduced. The study also reveals a drastic reduction in the displacement amplitude corresponding to the second resonant frequency, attributable to the increase in damping capacity of the soil-pile system at higher frequencies. As expected, the natural frequencies of pile-supported cooling towers approach those corresponding to a fixed base case due to the reduction of rocking and translation as the number of piles increases. The remaining slight reduction in the resonant frequency is due to the softer interacting system resulting from foundation flexibility. The comparison of the maximum response values for stress resultants and stress couples shows that the interaction effect indeed reduces the response of the shell structure when a strong motion earthquake loading is applied. The decrease in N_ϕ along the shell is in the range of 30% of the fixed-base solution and may reach 50% for some components. This may permit reduction of the shell cross section and the meridional steel in the shell, resulting in considerable cost saving. Perhaps the parts of the shell structure most affected by the soil-pile-structure interaction are the column supports, as may be seen from Table 4. It is noted that there is a sharp decrease in the shear force while the axial force and bending moment are also decreased.

ACKNOWLEDGEMENTS

The partial support of the early phases of this work by the National Science Foundation, through Grant No. PFR-7900012, is gratefully acknowledged. However, any opinions, findings, conclusions or recommendations expressed herein are those of the writers and do not necessarily reflect the views of the National Science Foundation.

REFERENCES

1. Lee, B.J.: Seismic analysis of shells of revolution on interactive shallow or deep foundations. Dissertation; Sever Institute of Technology, Washington University, St. Louis, MO (Aug. 1983).

2. Lee, B.J.; Gould, P.L.: Complex response analysis of shells of revolution including uniform base translation and rock-

ing. Earthquake Engineering and Structural Dynamics. In press.

3. Wolf, J.P.; Von Arx, G.A.: Horizontally travelling waves in a group of piles taking pile-soil-pile interaction into account. Earthquake Engineering and Structural Dynamics; Vol. 10 (1982) 225-237.

4. Seed, H. Bolton; et al.: Soil moduli and damping factors for dynamic response analyses. Report No. EERC 70-10, Earthquake Engineering Research Center, University of California, Berkeley, CA (Dec. 1970).

Analysis of Cooling Towers due to Horizontal Incident Earthquake Motions Propagating Vertically from Base Rock

S. Kato and M. Hirota

Toyohashi University of Technology, Toyohashi and Research Institute of Nippon Kokan, Kawasaki, Japan.

P. L. Gould

Washington University, St.Louis, Mo., USA.

Summary

An analysis of soil-structure interaction response of cooling towers was performed based on the frequency domain response. The method makes use of the finite element scheme of a rotational shell and solid elements. The effects of radiation damping into the far field as well as the internal damping of soil layers are revealed as important, and reduce the response of towers.

INTRODUCTION

Interest in the design methods and behavior of cooling towers has increased particularly after the collapse of the Ferrybridge and Ardeer towers. The focus has been mainly on the collapse mechanism due to dynamic wind, buckling and imperfections.

Dynamic analysis for earthquake motions has been alternatively tried by Gould et al(1), Cedolin et al(2) and Wolf et al(3) to look into the effects of supporting columns or foundations. However the research on the dynamics of large cooling towers including surrounding soil media has not been performed extensively.

Thus the present work aimed first developing a computer program to estimate the soil-structure interaction for shells of revolution including soil media and second at investigating the effects of the interaction on the shell response.

METHOD OF ANALYSIS OF SOIL STRUCTURE INTERACTION

The method to analyze structure in the present work is based on a rotational shell element for shells and a solid rotational element for soils. The response analysis makes use of the Fast Fourier Transform(FFT). Earthquake motions and the responses to them are decomposed into a Fourier expansion of limitted terms.

Once the results in the frequency domain are found, the inverse Fast Fourier Transform(IFFT) of them gives the corresponding time-domain solutions.

The present computer program concerns with two structural systems in Figs.1 and 2 for idealizing soils and structures subject to incident horizontal earthquake motions $u_{gI}(t)$ propagating vertically from the base rock. The system, Fig.1, is based on the far field concept proposed by Waas et al(4), however, with a modification of dash-pots at the bottom in the far field. The second one, Fig.2, has the Lysmer's dash-pots at both the side and the bottom of the core region. The dash-pots at the bottom are active only for horizontal displacement, but those at the side boundary are effective for both the horizontal and vertical displacements. Then the systems can approximate structures on a elastic half space. In both systems, a natural soil column is included to incorporate the effect of radiation damping due to the difference between the displacements of the core region and natural soil column and is assumed to receive no influence from the response of the core region except for the properties itself.

Preliminary analyses of a disc and a ring footing on a soil layer due to stationary oscillating forces were performed to find which system is more practical. Since a good agreement between the two systems, Fig.2. The latter require neither any eigenvalue problems nor Hankel functions with complex and large argument.

The resulting equation, expressed in the frequency domain, of the structural system subject to a horizontal incident earthquake motion $u_{gIk} \cdot e^{i\omega k}$ takes the form:

$$([K]-\omega_k^2[M])\{y_k\}= 2u_{gIk} \cdot \omega_k^2[M]\{\beta\}+[K_B]\{y_k-y_k^*\} \qquad (1)$$

where $[K]$,$[M]$ and $[K_B]$ are the mass and stiffness matrices of the core region and the boundary stiffness matrix to represent the radiation damping; $\{\beta\}$ shows a factor vector to represent the effect of the earthquake motion on the mass $[M]$; ω_k is the circular frequency of a component, $u_{gIk}\ e^{i\omega k}$, of the incident wave $u_{gI}(t)$ at the base rock; $\{y_k\}$ and $\{y_k^*\}$ are the displacements of the core and the natural soil column, relative to $2u_{gIk}\ e^{i\omega k}$.

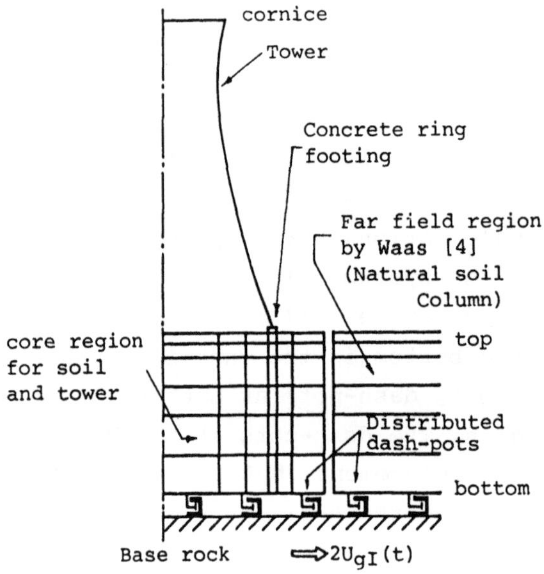

Fig.1 Modified Waas's model for incident earthquake motion $u_{gI}(t)$.

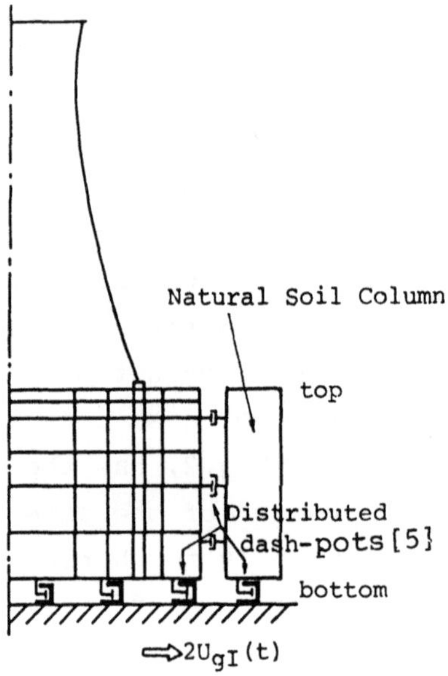

Fig.2 Lysmer's model modified to incorporate a incident motion.

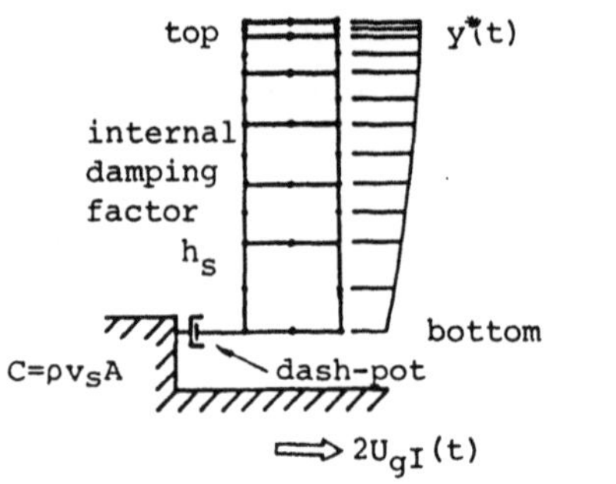

Fig.3 System for a natural soil column.

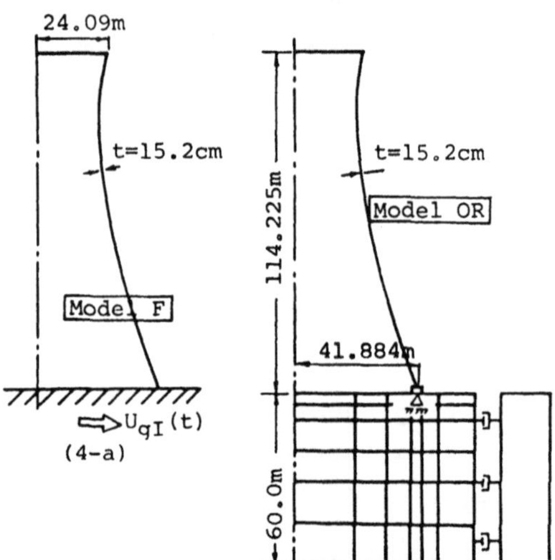

Fig.4 Models for checking the necessary $\Delta\omega$ to apply FFT to time series response.

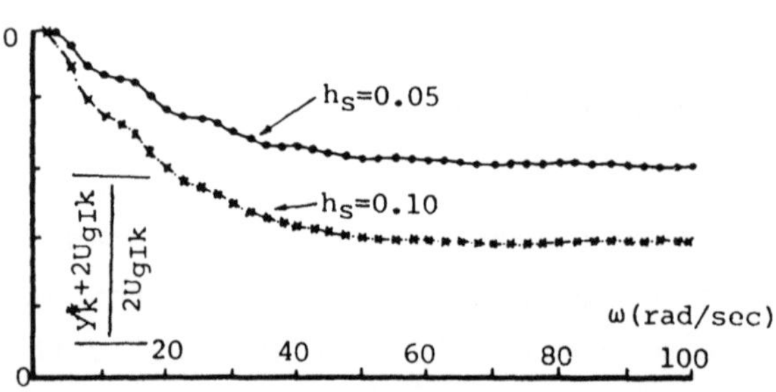

Fig.5 Transfer function of a natural soil column , Fig.3 and Table 1.

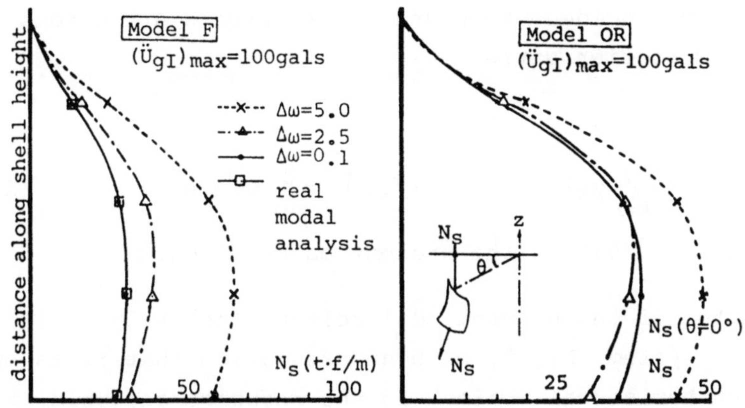

Fig.6 Variation of the maximum meridional stress N_s(t·f/m) on the different frequency increment Δω.

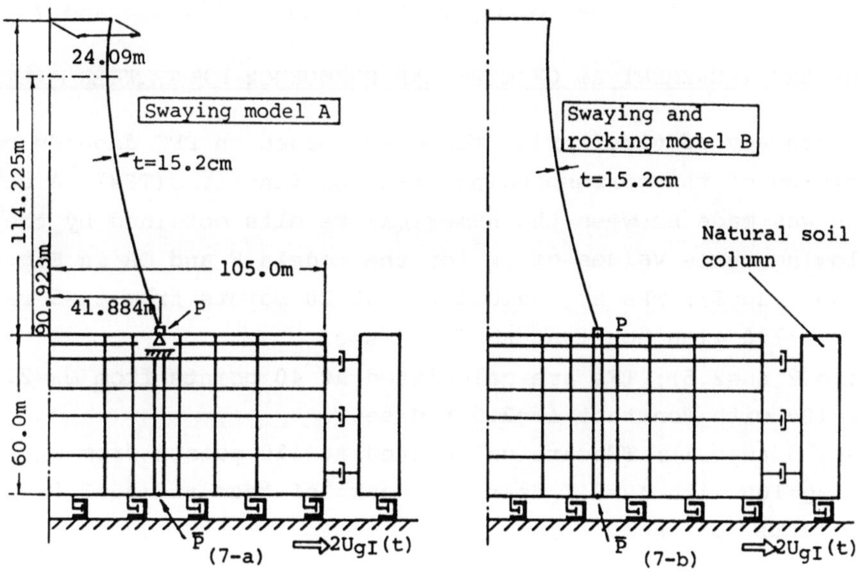

Fig.7 Analyzed models A and B due to El-Centro NS(u_{gImax}= 100gals) at the base rock.

Table 1 Material constants of the tower and the homogeneous soil layer for the models F, OR, A and B (Figs.4,7).

Constants \ Material	Cooling tower and Ring footing	Natural soil column and Soil column
Youngs modulus or shear modulus	$E_C=2.1\times10^6$ t·f/m^2	$G_s=10000$ t·f/m^2
Poisson's ratio	$\nu_C=0.16$	$\nu_s=0.25$
Weight per unit volume	$\gamma_C=2.4$ t·f/m^3	$\gamma_s=1.8$ t·f/m^3
Internal damping factor	$h_C=0.02$	$h_s=0.05$ or 0.10

In Eq.(1), the incident wave and displacements are decomposed into complex Fourier series.

$$u_{gI}(t)=\sum_{k=1}^{N} u_{gIk}\ e^{i\omega k}$$

$$\{y(t)\}=\sum_{k=1}^{N}\{y_k\}\ e^{i\omega k}\ ,\quad \{y^{*}(t)\}=\sum_{k=1}^{N}\{y_k^{*}\}\ e^{i\omega k} \tag{2}$$

where N was set 2048 in the present calculation.

In the matter of the natural soil column, the analysis is ased on a system, Fig.3, with an assumption that it is subject to only the horizontal motion. In the response analysis, the acceleration record of El-Centro NS was used throughout as the incident earthquake motion and the material constants were assumed as Table 1 for the models F, OR, A and B(Figs.3,4 and 7).

NECESSARY INCREMENT $\Delta\omega$ OF CIRCULAR FREQUENCY FOR TRANSFER FUNCTIONS

The accuracy of time series responses based on FFT depends on the precision of the corresponding transfer functions(TFs). A comparison was made between the numerical results obtained by the following three values of $\Delta\omega$ for the models F and OR in Fig.4.

(Case 1 $\Delta\omega=5$); TFs are calculated at 20 points from $\omega_0=5$ rad/sec to $\omega_E=100$ with constant $\Delta\omega=5$ rad/sec.

(Case 2 $\Delta\omega=2.5$); TFs are calculated at 40 points from $\omega_0=2.5$ to $\omega_E=100$ with constant $\Delta\omega=2.5$ rad/sec.

(Case 3 $\Delta\omega=0.1$); TFs are calculated at 120 points from $\omega_0=1.0$ to $\omega_E=100$ with $\Delta\omega=0.1$ near the peaks of them and $\Delta\omega=2.5$ at the plateaus of them.

The maximum distribution of the meridional stress $N_s(t\cdot f/m)$ for the models F and OR due to El-Centro NS(u_{gImax}=100gals) is shown in Fig.6. From the present check, a decision was made to use the increment $\Delta\omega=0.1$ near the peaks of TFs in the following analysis.

SOIL STRUCTURE INTERACTIVE RESPONSE OF TOWERS ON A HOMOGENEOUS LAYER

(1) Solutions in the frequency domain analysis; A brief description is outlined for the time series responses based on the frequency domain analysis.

(Step A) Calculation of the continuous TFs by the third order

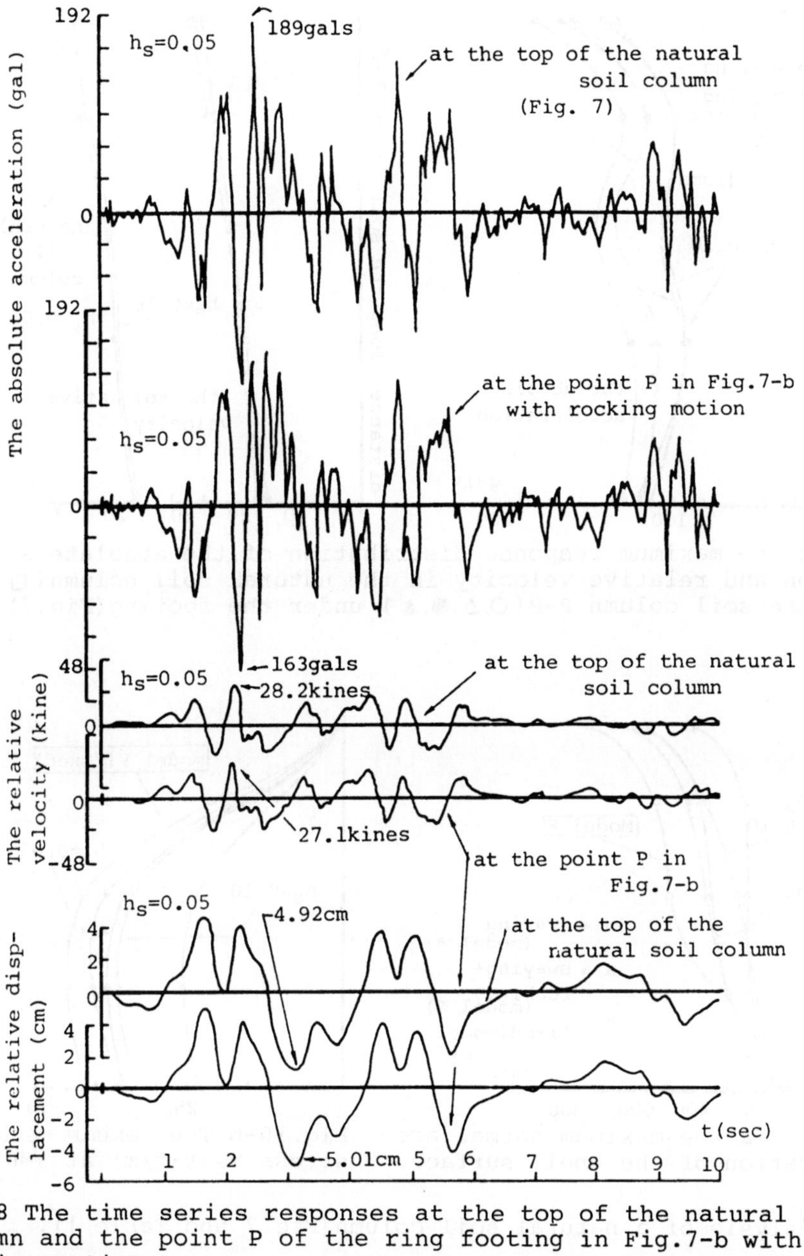

Fig.8 The time series responses at the top of the natural soil column and the point P of the ring footing in Fig.7-b with rocking motion.

interpolation of the TFs obtained at the discrete point ω_j.

(Step B) Calculation of the Fourier coefficients of the selected earthquake motion by FFT.

(Step C) Calculation of the time series response from the above products by IFFT.

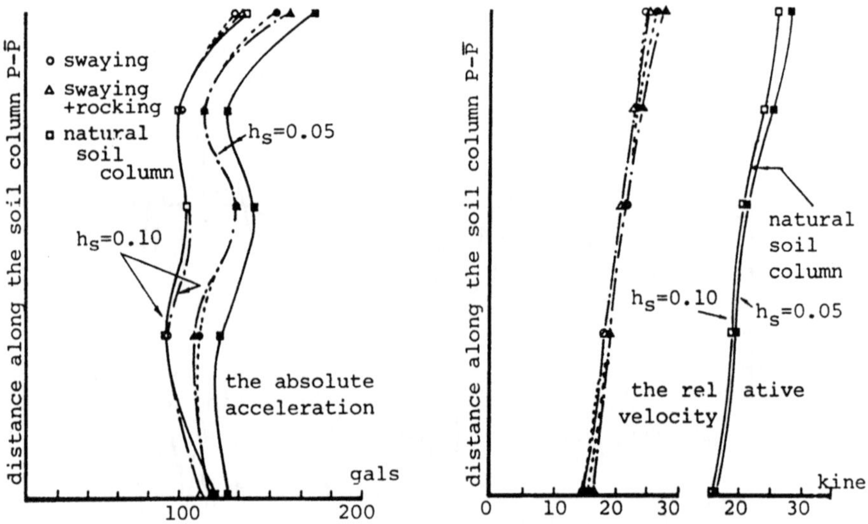

Fig.9 The maximum response distribution of the absolute acceleration and relative velocity in the natural soil column(□ , ■) and the soil column P-P (○,△,●,▲) under the footing(Fig.7).

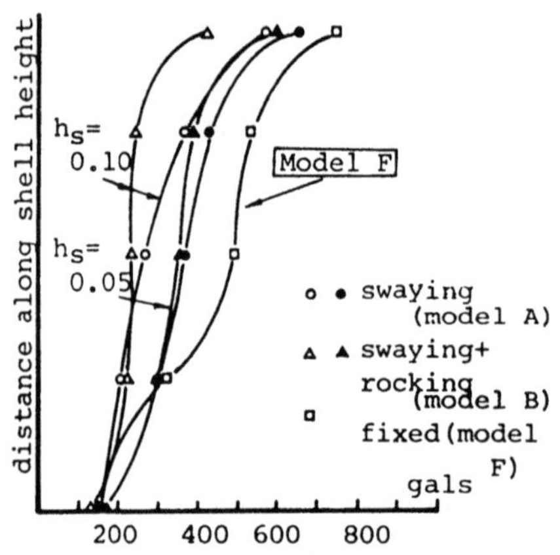

Fig.10-a the maximum normal acceleration of the shell surface.

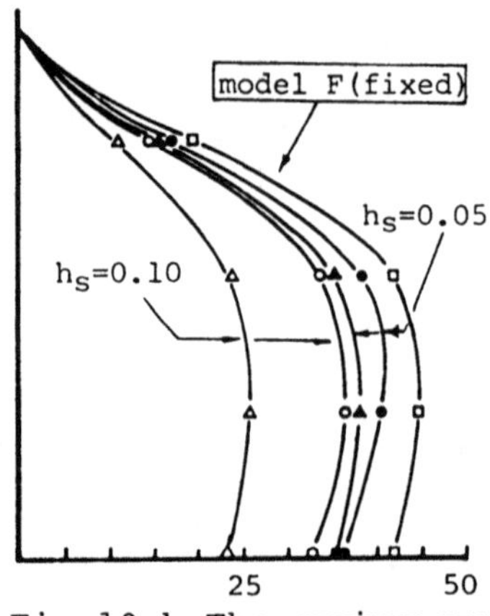

Fig.10-b The maximum meridional stress N_s(t·f/m) at θ=0°.

(2) Analysis of a natural soil column(Fig.7 and Table 1).
The TF of the absolute horizontal displacement at the top of the natural soil column is shown in Fig.5 with the time series response included in Fig.8 and the maximum distribution included in Fig.9.

(3) Soil-structure response of cooling towers on a homogeneous layer; The effect of soil-structure interaction was examined

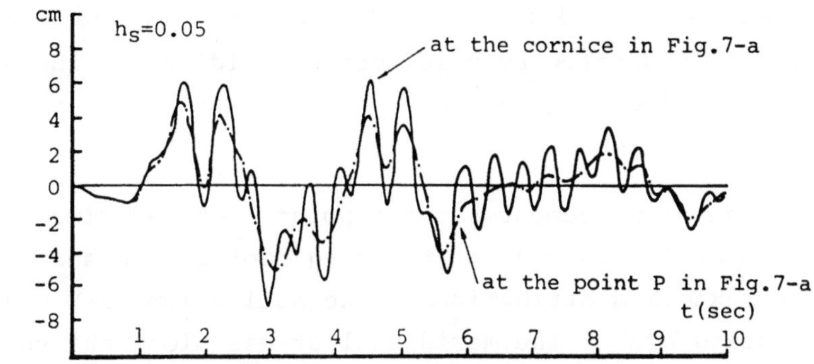

Fig.11 The time series responses of the normal displacements at the cornice and the point P in Fig.7-a without rocking motion.

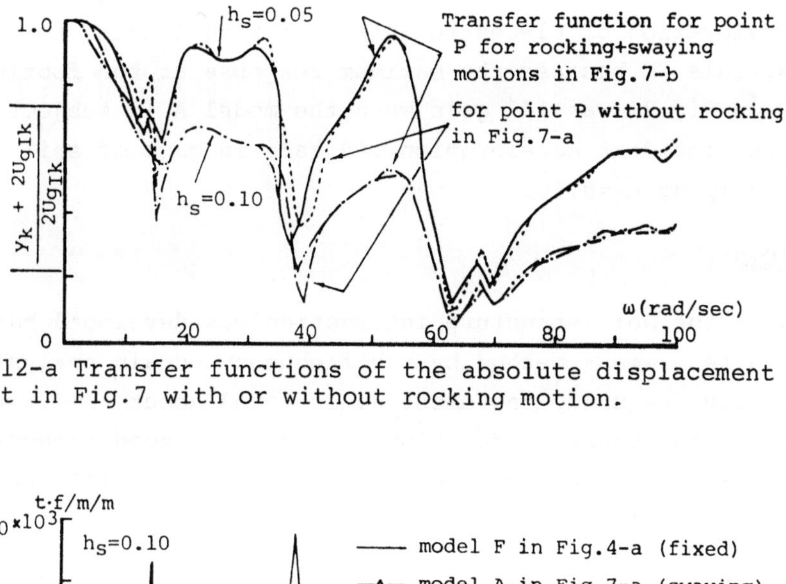

Fig.12-a Transfer functions of the absolute displacement at the point in Fig.7 with or without rocking motion.

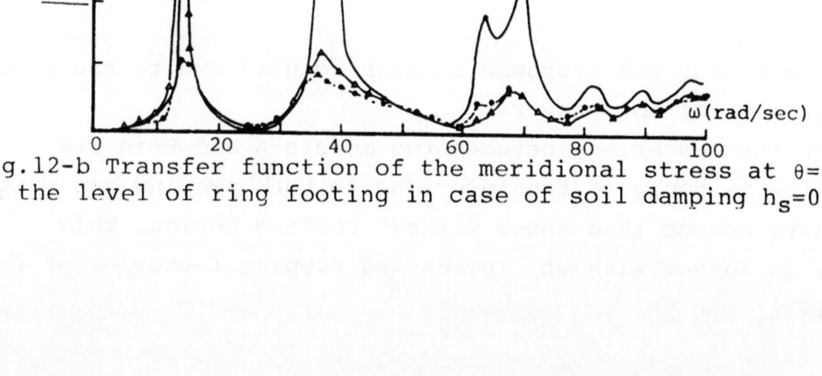

Fig.12-b Transfer function of the meridional stress at θ=0°, at the level of ring footing in case of soil damping h_s=0.10.

using two models A and B in Fig.7. In the model A, the concrete ring footing is supported by piles being rigid in tension and compression free from bending so that the tower is due to swaying motion without rocking one.

The TFs for the displacements at the point P of the ring footing and the meridional stress N_s at the ring footing are shown in Fig.12. The maximum distributionsof the soil column P-P under the ring footing and of the meridional stress along the shell height are given in Figs.9 and 10 with a time series response of the normal displacements of the tower, model A, given in Fig.11. Fig.10 includes the result for the model F, being fixed at the footing, the response of which is obtained by setting the maximum input acceleration of El-Centro NS at the ring footing to be 141 gals. This is because the maximum response at the footing (point P in Fig.7) was 141 gals when the model A is subject to the maximum incident acceleration 100 gals in case of soil internal damping h_s=0.10.

CONCLUSIONS

The program for soil-structure interaction was developed based on the finite element method by the frequency domain analysis. The necessary frequency increment $\Delta\omega$ is found about $\Delta\omega$=0.1 around the peaks of the transfer functions and gives a good agreement with the result by the real modal analysis for a cooling tower fixed at the footing.

The examination of the results for soil-structure response of towers with or without rocking motion leads the following;

(A) The response of cooling towers including soil-structure is much smaller than those without interaction as shown in Figs.9 and 10.

(B) And moreover the response is much reduced due to the internal damping h_s of soil layers.

(C) From the comparison between two models A and B in Fig.7, smaller the response of the tower due to both swaying and rocking motions become than those without rocking motion. This tendency is larger with an increasing damping factor h_s of the soil layer.

The present work is only in the initial phases and the authors have a plan to investigate the effects of other factors such as multi-layer soil, pile rigidity, and other earthquake motions, such as Taft, Hachinohe and Tokachi-oki.

ACKNOWLEDGMENT

The authors are grateful to Dr. B.J. Lee at Washington University for collaboration at the outset of the present work and Mr. A. Harikane at Toyohashi University of Technology for collaborating with the first author and providing numerical data with a good drawing of figures. The computations for the study were performed at the data processing centers at Toyohashi University and Nagoya University.

REFERENCE

1. Gould,P.L. et al.: Dynamic analysis of column-supported hyperboloidal shells. Earthquake Eng. Struct. Dyn. 1974,2,269.

2. Cedolin,L. et al.: Dynamic analysis of cooling towers using a frontal based frequency solver, Eng. Struct.,1980,2,157.

3. Wolf,J.P. et al.: Nonlinear response of hyperbolic cooling tower to horizontally propagating seismic waves. Proceedings of ASCE(Preprint), 1981.

4. Waas,G. et al.: Dynamic analysis of footings on layered media. ASCE, Vol.101, No.EM5, Oct., 1975, pp.679-693.

5. Lysmer,B,J. et al.: Finite dynamic model for infinite media. ASCE, Vol.95, No.EM4, Aug., 1969, pp359-877.

6. M. Hirota.: Dynamic interaction analysis between shells and soil foundation. The dissertation for master's degree, Toyohashi University of Technology, 1983(in Japanese)

7. Analysis for soil-structure interaction effects for nuclear power plants. ASCE, 1979.

The Earthquake Design of Reinforced Concrete Natural Draught Cooling Towers

H. Tottenham

Hume Tottenham & Bennet,
Broadwater House,
Romsey, Hants. U.K.

Abstract

The application of the membrane theory of shells to the problem of finding the natural frequencies and normal modes of vibration of hyperboloid shells of revolution is outlined. A comparison of the results so obtained with published results based upon the full bending theory shows that a sufficient accuracy is obtained for them to be used for design purposes.

Using conventional random vibration theory and a Housner-Tajimi type specification of the ground acceleration spectrum it is possible to determine the response spectrum of the tower. For this it is necessary to evaluate not only the natural frequencies and modes of vibration but also the corresponding generalized forces and generalized masses. The necessary expressions are derived.

To assist in design tables of these quantities are given in dimensionless form for the range of geometrical parameters used in cooling tower design.

The Free Vibration Analysis of the Thin, Elastic Hyperboloidal Shell by the Analytical-Numerical Methods

Z. JAMICKI

P.B.CH."CHLODNIE KOMINOWE"-Gliwice,Poland

Summary

Free vibrations in the hyperbolic shell of revolution are considered in this report.Differential equations of motion are formulated according to the linear theory of thin,elastic shells,given by V.Z.Vlasov.The solutions are obtained by the method of separation of variables and using the approximation of functional coefficients by means of Hermite polynomial series.The algorithm constructed makes it possible to express the expansion coefficients as a function of integration constants (of coordinates of eigenvector).Solving the boundary problem for eigenvalues,the frequencies of free vibrations and normalized coordinates of eigenvector are defined numerically.Finally,the results of numerical example are included.

Introduction

The subjekt of these analytical considerations is a problem of free vibrations of the hyperbolic cooling tower shell,and especially the presentation of analytical-numerical method permitting determination of the oscillating motion parameters. In the professional literature,there are known solutions obtained by means of numerical methods,for example,using a finite-element method [1] or finite-difference method [2].In the face of scarce experimental investigations and many doubts raised,the development of analytical-numerical methods,particularly for more and more complex analytical models,can contribute to a better likelihood of theoretical calculations. The problem of dynamics should not be ignored in design computations for the construction of cooling tower shell with a height of about 150 m and above.

Formulation of the Problem

The analytical model used for determining the free vibrations of cooling tower shell is a thin,elastic,istropic mid-surface having the shape of hyperboloid of revolution,defined by the formula

$$\frac{x^2}{a^2} + \frac{y^2}{a^2} - \frac{z^2}{b^2} = 1 , \qquad (1)$$

or using the curvilinear coordinates:

$$x=a.\cos\alpha 2.\mathrm{ch}\alpha 1 \;,\; y= a.\sin\alpha 2.\mathrm{ch}\alpha 1 \;,\; z=b.\mathrm{sh}\alpha 1 \;, \qquad (2)$$

where

x,y,z-coordinates in cartesian system,

$\alpha 1$,$\alpha 2$-curvilinear coordinates (right-handed system),

a,b -hyperbola parameters.

We assume that the shell displacements are infinitesimal and the Kirchhoff-Love hypotheses are hold.We shall use of the geometrical relations of the strain and displacement states and internal force equations,as given by V.Z.Vlasov [3].We accept the notation of forces and displacements for the elementary panel of shell as shown in Fig.1.

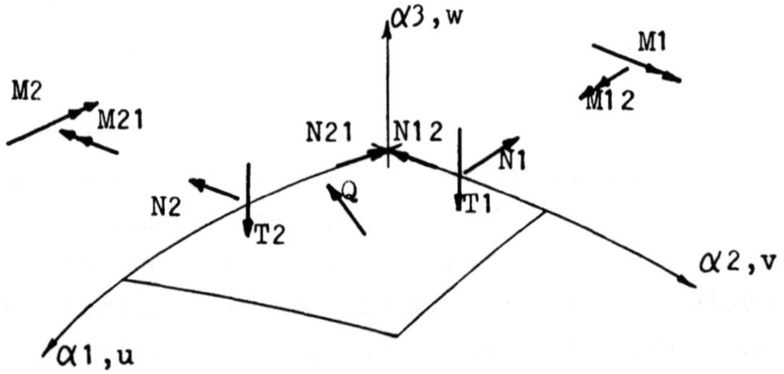

Fig.1.The notation of forces,moments and displacements.

The Lame's coefficients of the first quadratic form A1,A2 and the main radii of curvatures R1,R2 are defined by formulae:

$$A1=[(a.sh\alpha 1)^2+(b.ch\alpha 1)^2]^{1/2}\ ,\quad A2=a.ch\alpha 1\ ,$$

$$R1=-\frac{1}{a.b}A1^3\ ,\quad R2=\frac{a}{b}A1\ . \qquad (3)$$

The equations for the equilibrium of forces and the moments of elementary panel of hyperboloidal shell will be written in the form:

$$\frac{1}{A1.A2}\Big[\frac{\partial(A2.N1)}{\partial\alpha 1}+\frac{\partial(A1.N21)}{\partial\alpha 2}-\frac{\partial A2}{\partial\alpha 1}N2\Big]+\frac{T1}{R1}+Q1=0\ ,$$

$$\frac{1}{A1.A2}\Big[\frac{\partial(A2.N12)}{\partial\alpha 1}+\frac{\partial(A1.N2)}{\partial\alpha 2}+\frac{\partial A2}{\partial\alpha 1}N21\Big]+\frac{T2}{R2}+Q2=0\ ,$$

$$\frac{1}{A1.A2}\Big[\frac{\partial(A2.T1)}{\partial\alpha 1}+\frac{\partial(A1.T2)}{\partial\alpha 2}\Big]-\frac{N1}{R1}-\frac{N2}{R2}+Q3=0\ , \qquad (4)$$

$$\frac{1}{A1.A2}\Big[-\frac{\partial(A2.M1)}{\partial\alpha 1}+\frac{\partial(A1.M21)}{\partial\alpha 2}+\frac{\partial A2}{\partial\alpha 1}M2\Big]-T1=0\ ,$$

$$\frac{1}{A1.A2}\Big[\frac{\partial(A2.M12)}{\partial\alpha 1}-\frac{\partial(A1.M2)}{\partial\alpha 2}+\frac{\partial A2}{\partial\alpha 1}M21\Big]-T2=0\ ,$$

$$N12-N21+\frac{M12}{R1}-\frac{M21}{R2}=0\ .$$

The strains of shell fibres,which are located in a distance from the mid-surface,will be developed after V.Z.Vlasov,with respect to the variable z:

$$e1=\varepsilon 1+\varkappa 1.z+\varphi 1.z^2,\quad e2=\varepsilon 2+\varkappa 2.z+\varphi 2.z^2,\quad e12=\varepsilon 12+\varkappa 12.z+\varphi 12.z^2, \qquad (5)$$

where:

$$\varepsilon 1=\frac{1}{A1}\frac{\partial u}{\partial\alpha 1}+\frac{1}{R1}w\ ,\quad \varepsilon 2=\frac{1}{A2}\frac{\partial v}{\partial\alpha 2}+\frac{A2'}{A1.A2}u+\frac{1}{R2}w\ ,$$

$$\varepsilon 12=\frac{1}{A1}\frac{\partial v}{\partial\alpha 1}+\frac{1}{A2}\frac{\partial u}{\partial\alpha 2}-\frac{A2'}{A1.A2}v\ ,$$

$$\varkappa 1=\frac{1}{A1}\Big(\frac{1}{R1}\Big)'u+\frac{A1'}{A1}\frac{\partial w}{\partial\alpha 1}-\frac{1}{A1^2}\frac{\partial^2 w}{\partial\alpha 1^2}-\frac{1}{R1^2}w\ ,$$

$$\varkappa 2=\frac{1}{A2}\Big(\frac{1}{R2}\Big)'u-\frac{1}{A2^2}\frac{\partial^2 w}{\partial\alpha 2^2}-\frac{A2'}{A1.A2}\frac{\partial w}{\partial\alpha 1}-\frac{1}{R2^2}w\ , \qquad (6)$$

$$\varkappa 12=\left(\frac{1}{R1}-\frac{1}{R2}\right)\left[\frac{1}{A2}\frac{\partial u}{\partial \alpha 2}-\frac{1}{A1}\frac{\partial v}{\partial \alpha 1}+\frac{A2'}{A1.A2}v\right]-$$

$$-2\frac{1}{A1 \circ A2}\left[\frac{\partial^2 w}{\partial \alpha 1 \partial \alpha 2}-\frac{A2'}{A2}\frac{\partial w}{\partial \alpha 2}\right], \qquad (6)$$

$$\varphi 1=-\frac{1}{R1}\varkappa 1\ ,\ \varphi 2=-\frac{1}{R2}\varkappa 2\ ,\ \varphi 12=\frac{1}{2}\left(\frac{1}{R1}-\frac{1}{R2}\right)^2\varepsilon 12-\frac{1}{2}\left(\frac{1}{R1}+\frac{1}{R2}\right)\varkappa 12,$$

( )' -denotes the differential with respect to $\alpha 1$.
The forces and internal moments in the shell are defined by expressions:

$$N1=F(\varepsilon 1+\nu.\varepsilon 2)-D\left(\frac{1}{R1}-\frac{1}{R2}\right)\varkappa 1\ ,\ N2=F(\varepsilon 2+\nu.\varepsilon 1)+D\left(\frac{1}{R1}-\frac{1}{R2}\right)\varkappa 2\ ,$$

$$N12=\frac{1}{2}(1-\nu)F.\varepsilon 12-\frac{1}{4}(1-\nu)D\left(\frac{1}{R1}-\frac{1}{R2}\right)\left[\varkappa 12-\left(\frac{1}{R1}-\frac{1}{R2}\right)\varepsilon 12\right],$$

$$N21=\frac{1}{2}(1-\nu)F.\varepsilon 12+\frac{1}{4}(1-\nu)D\left(\frac{1}{R1}-\frac{1}{R2}\right)\left[\varkappa 12+\left(\frac{1}{R1}-\frac{1}{R2}\right)\varepsilon 12\right], \qquad (7)$$

$$M1=-D\left[\varkappa 1+\nu.\varkappa 2+\frac{1}{R2}(\varepsilon 1+\nu \circ \varepsilon 2)\right],\ M2=-D\left[\varkappa 2+\nu.\varkappa 1+\frac{1}{R1}(\varepsilon 2+\nu.\varepsilon 1)\right],$$

$$M12=\frac{1}{2}(1-\nu)D\left[\varkappa 12+\frac{1}{R2}\varepsilon 12\right],\qquad M21=\frac{1}{2}(1-\nu)D\left[\varkappa 12+\frac{1}{R1}\varepsilon 12\right],$$

where

$$F=\frac{E.h}{(1-\nu^2)}\ ,\quad D=\frac{E.h^3}{12(1-\nu^2)}\ , \qquad (8)$$

E, ν -Young's modulus of elasticity, Poisson's ratio,
h= h($\alpha 1$) -shell thickness.

By eliminating the forces T1,T2 from the equilibrium equations, we obtain the sought differential equations motion:

$$L1\frac{\partial^2 u}{\partial \alpha 1^2}+L2\frac{\partial^2 u}{\partial \alpha 2^2}+L3\frac{\partial u}{\partial \alpha 1}+L4.u+L5\frac{\partial^2 v}{\partial \alpha 1 \partial \alpha 2}+L6\frac{\partial v}{\partial \alpha 2}+L7\frac{\partial^3 w}{\partial \alpha 1^3}+$$

$$L8\frac{\partial^3 w}{\partial \alpha 1 \partial \alpha 2^2}+L9\frac{\partial^2 w}{\partial \alpha 1^2}+L10\frac{\partial^2 w}{\partial \alpha 2^2}+L11\frac{\partial w}{\partial \alpha 1}+L12.w=-\frac{1}{F}Q1\ ,$$

$$L13\frac{\partial^2 u}{\partial \alpha 1 \partial \alpha 2}+L14\frac{\partial u}{\partial \alpha 2}+L15\frac{\partial^2 v}{\partial \alpha 1^2}+L16\frac{\partial^2 v}{\partial \alpha 2^2}+L17\frac{\partial v}{\partial \alpha 1}+L18.v+L19\frac{\partial^3 w}{\partial \alpha 2^3}+$$

$$L20\frac{\partial^3 w}{\partial \alpha 1^2 \partial \alpha 2}+L21\frac{\partial^2 w}{\partial \alpha 1 \partial \alpha 2}+L22\frac{\partial w}{\partial \alpha 2}=-\frac{1}{F}Q2\ , \qquad (9)$$

$$L23\frac{\partial^3 u}{\partial\alpha1^3} + L24\frac{\partial^3 u}{\partial\alpha1\partial\alpha2^2} + L25\frac{\partial^2 u}{\partial\alpha1^2} + L26\frac{\partial^2 u}{\partial\alpha2^2} + L27\frac{\partial u}{\partial\alpha1} + L28.u +$$

$$L29\frac{\partial^3 v}{\partial\alpha1^2\partial\alpha2} + L30\frac{\partial^3 v}{\partial\alpha2^3} + L31\frac{\partial^2 v}{\partial\alpha1\partial\alpha2} + L32\frac{\partial v}{\partial\alpha2} + L33\frac{\partial^4 w}{\partial\alpha1^4} + L34\frac{\partial^4 w}{\partial\alpha2^4} +$$

$$L35\frac{\partial^4 w}{\partial\alpha1^2\partial\alpha2^2} + L36\frac{\partial^3 w}{\partial\alpha1^3} + L37\frac{\partial^3 w}{\partial\alpha1\partial\alpha2^2} + L38\frac{\partial^2 w}{\partial\alpha1^2} + L39\frac{\partial^2 w}{\partial\alpha2^2} + L40\frac{\partial w}{\partial\alpha1} +$$

$$L41.w = \frac{1}{F} Q3 , \qquad (9)$$

where

LP=LP(α1) -functional coefficients.

A Solution to the Homogeneous System of Differential Equations of Motion

We seek the solution of the differential equations of motion as the sum of particular integrals of seperated variables:

$$u(\alpha1,\alpha2,t)=\sum_{N=0}^{NK} UN(\alpha1).\cos(N.\alpha2).T(t) ,$$

$$v(\alpha1,\alpha2,t)=\sum_{N=0}^{NK} VN(\alpha1).\sin(N.\alpha2).T(t) , \qquad (10)$$

$$w(\alpha1,\alpha2,t)=\sum_{N=0}^{NK} WN(\alpha1).\cos(N.\alpha2).T(t) .$$

We require that the function u,v,w satisfies the equations of motion, boundary conditions for each N-th harmonic form, and be differentiable to an appropriate order.
By introducing a relation

$$\frac{\partial^2 T(t)}{\partial t^2} / T(t) = -\omega N^2 , \qquad (11)$$

into the equatons of motion the sought solution for the function T(t) is found

$$T(t)=T0.\cos(\omega N.t+\varphi 0) , \qquad (12)$$

where

T0,φ0 -integration constants derived from initial conditions,

ωN-frequency of free vibrations.
Let us introduce the eigenvalue into the notation of equations of motion

$$\Omega N=-\frac{1}{F}\varrho.h.\omega N^2=-\frac{(1-\nu^2)}{E}\varrho.\omega N^2=-\frac{(1-\nu^2)\gamma}{E.g}\omega N^2 \quad , \qquad (13)$$

where

ϱ, γ -density,specific weight.

The next step of solution of the equations of motion assumes the expansion of all variables with respect to $\alpha 1$,i.e. h(α1), LP(α1),UN(α1),VN(α1),WN(α1) into the appropriate series of Hermite polynomials,and then we compare the expressions for equations of motion adjacent to the similar $\alpha 1^s$.On the basis of 3(s+1) algebraic equations obtained,we shall define the linearly dependent coefficients of series of function UN,VN,WN. Having taken into consideration the function u,v,w in the form of separated variables,and eliminating the motion UN from the equations,we shall get the system of differential equations of the 8-th order (2-nd order with respect to UN(α1),VN(α1) and 4-th order with respect to WN(α1)):

$$L101.UN''+L103.UN'+(L104-N^2.L102-\Omega N)UN+N.L105.VN'+N.L106.VN+$$

$$L107.WN'''+L109.WN''+(L111-N^2.L108)WN'+(L112-N^2.L110)WN=0 \ ,$$

$$-N.L113.UN'-N.L114.UN+L115.VN''+L117.VN'+(L118-N^2.L116-\Omega N)VN+$$

$$-N.L120.WN''-N.L121.WN'+N(N^2.L119-L122)WN=0 \ , \qquad (14)$$

$$L123.UN''+(L124+N^2.L125+L100.\Omega N)UN'+(L126+N^2.L127)UN+N.L128.VN''+$$

$$L129.VN'+N(L131-N^2.L130)VN+L132.WN^{IV}+L133.WN'''+(L135+N^2.L136)WN''+$$

$$(L137+N^2.L138)WN'+(L139+N^2.L140+N^4.L134+\Omega N)WN=0 \ ,$$

where

LP(α1) -functional coefficients.

In case of variable shell thickness $h(\alpha 1)$, we approximate the discretely preset values of $h(\alpha 1, i)$ using the Hermite polynomical series

$$\chi(\alpha 1)=\sum_{n=0}^{nk} \chi n.Hn(\alpha 1) \ , \qquad (15)$$

where

χn -series coefficients defined by the least square method,

$$Hn=\sum_{k=0}^{n}(-1)^{k}.2^{(n-k)}.Tk.\alpha 1^{r}, \ r=n-2.k \ , \text{for } r<0 \leftarrow r=0 \ ,$$

$$T0=1 \ , \quad Tk=T0 \ \frac{(n-2.k+2)(n-2.k+1)}{2.k} \ . \qquad (16)$$

Similarly, for the discretely computed values of $LP(\alpha 1, i)$, we calculate the coefficients $lP(s)$ from the relation

$$LP=\sum_{s=0}^{sk} lP(s).Hs(\alpha 1) \ . \qquad (17)$$

Considering the order of differentials of function UN,VN,WN, that occur in the differential equations of motion, we shall make use of their expansion into the Hermite polynomial series to the order of (sk+2),(sk+2),(sk+4):

$$UN(\alpha 1)=\sum_{n=0}^{sk+2} Un.Hn \ , \ VN(\alpha 1)=\sum_{n=0}^{sk+2} Vn.Hn \ , \ WN(\alpha 1)=\sum_{n=0}^{sk+4} Wn.Hn \ , \qquad (18)$$

where

sk -the degres of approximation of functional coefficients LP.

The linearly independent solution of function UN,VN,WN with respect to $\alpha 1$ will be the sum of particular integrals:

$$\overline{UN}(\alpha 1)=U1.H1+U2.H2+V1.0+V2.0+W1.0+W2.0+W3.0+W4.0 \ ,$$

$$\overline{VN}(\alpha 1)=U1.0+U2.0+V1.H1+V2.H2+W1.0+W2.0+W3.0+W4.0 \ , \qquad (19)$$

$$\overline{WN}(\alpha 1)=U1.0+U2.0+V1.0+V2.0+W1.H1+W2.H2+W3.H3+W4.H4 \ .$$

It can be proved that solution written in such a manner is differentiable to the 2-nd and 4-th order, respectively, and the integrals are truly linearly independent, since the Vronskian

$$\widetilde{W}(\overline{UN},\overline{VN},\overline{WN},\alpha 1) = 0 \ . \tag{20}$$

The linearly dependent coefficients Un, Vn, Wn will be defined from the system of algebraic equations, obtained from differential equations of motion

$$[a_{\alpha,\beta}]\{x_\beta\} = [b_{\alpha,\gamma}]\{y_\gamma\} \ , \tag{21}$$

from which we get

$$\{x_\beta\} = [\varphi_{\beta,\gamma}]\{y_\gamma\} \ , \tag{22}$$

where

$$[\varphi_{\beta,\gamma}] = [a_{\alpha,\beta}]^{-1}[b_{\alpha,\gamma}] \ ,$$

$$\{y_\gamma\} = \{U1, U2, V1, V2, W1, W2, W3, W4\} \ , \tag{23}$$

$$\{x_\beta\} = \{U0, U3, U4, \ldots, Usk+2, V0, V3, V4, \ldots, Vsk+2,$$

$$W0, W5, W6, \ldots, Wsk+4\} \ .$$

The matrix elements $a_{\alpha,\beta}$ and $b_{\alpha,\gamma}$ are the combination of expressions 1P(s-k), N, ΩN. We shall calculate their final values in the iterative procedure with respect to the eigenvalue ΩN, which reduces to zero the characteristic determinant of boundary conditions. Therefore, we can write the displacement functions UN(α1), VN(α1), WN(α1) in the form:

$$UN(\alpha 1) = \{y_\gamma\}[\varphi_{1,\gamma}.H0+z1.H1+z2.H2+\sum_{\beta=2}^{sk+1}\varphi_{\beta,\gamma}.H(\beta+1)] \ , \tag{24}$$

$$VN(\alpha 1) = \{y_\gamma\}[\varphi_{sk+2,\gamma}.H0+z3.H1+z4.H2+\sum_{\beta=sk+3}^{2.sk+2}\varphi_{\beta,\gamma}.H(\beta-sk)] \ ,$$

$$WN(\alpha 1) = \{y_\gamma\}[\varphi_{2.sk+3,\gamma}.H0+z5.H1+z6.H2+z7.H3+z8.H4+$$

$$+\sum_{\beta=2.sk+4}^{3.sk+3} \varphi_{\beta,\gamma}.H(\beta-2.sk+1)] \quad , \tag{24}$$

where

$$zl=\begin{cases}1 & \text{for } l=\gamma \\ 0 & \phantom{\text{for }} l\neq\gamma\end{cases} , \ l\in[1,\gamma],\ \gamma\in[1,8],\ \beta\in[1,3(sk+1)]. \tag{25}$$

The Boundary Problem Relating to Eigenvalues

In the analysis of the boundary problem we assume continuous, homogeneous conditions,and we require that they are satisfied for any arbitrary coordinate $\alpha 2$ and the time t,separately for every N-th parallel harmonic form.We shall define the eigenvalues from the condition of reducing to zero the characteristic determinant for 8 boundary equations,however,the eigenvector coordinates (integration constants) shall be defined with accuracy to a constant parameter,whose value shall result from the normalization condition.

We shall write the conditions for lower boundary ($\alpha 1=\alpha 1,d$) in the alternative form:

- for the hinged support of shell:

$$u(\alpha 1,\alpha 2,t)=0 \ , \ v(\alpha 1,\alpha 2,t)=0 \ , \ w(\alpha 1,\alpha 2,t)=0 \ , \ M1(\alpha 1,\alpha 2,t)=0, \tag{26}$$

- for fixed lower boundary the shell:

$$u(\alpha 1,\alpha 2,t)=0 \ , \ v(\alpha 1,\alpha 2,t)=0 \ , \ w(\alpha 1,\alpha 2,t)=0 \ , \ \alpha(\alpha 1,\alpha 2,t)=0, \tag{27}$$

where

$\alpha(\alpha 1,\alpha 2,t)$ -the angle of rotation of shell along the coordinate $\alpha 1$.

- for the elastic support of shell [1]:

$$\widehat{N1}(\alpha 1,\alpha 2,t)=N1(\alpha 1,\alpha 2,t)-\overline{N1}(\alpha 1,\alpha 2,t)=0 \ ,$$

$$\widehat{M1}(\alpha 1,\alpha 2,t)=M1(\alpha 1,\alpha 2,t)-\overline{M1}(\alpha 1,\alpha 2,t)=0 \ ,$$

$$\widehat{N12}^{*}(\alpha 1,\alpha 2,t)=N12^{*}(\alpha 1,\alpha 2,t)-\overline{N12}^{*}(\alpha 1,\alpha 2,t)=0, \tag{28}$$

$$\widehat{T1}^*(\alpha1,\alpha2,t)=T1^*(\alpha1,\alpha2,t)-\overline{T1}^*(\alpha1,\alpha2,t)=0\ , \tag{28}$$

where

$\overline{N1},\overline{M1},\overline{N12}^*,\overline{T1}^*$ -the forces resulting from elastic support of shell.

We shall write the conditions for the upper free boundary ($\alpha1=\alpha1,g$) in the form:

$$N1(\alpha1,\alpha2,t)=0\ ,\quad M1(\alpha1,\alpha2,t)=0\ ,$$

$$N12^*(\alpha1,\alpha2,t)=N12(\alpha1,\alpha2,t)+\frac{M12(\alpha1,\alpha2,t)}{R2}=0\ , \tag{29}$$

$$T1^*(\alpha1,\alpha2,t)=T1(\alpha1,\alpha2,t)+\frac{1}{A2}\frac{\partial M12(\alpha1,\alpha2,t)}{\partial\alpha2}=0.$$

In order that 8 boundary conditions (of 4 for every shell boundary) could be satisfied for any arbitrary values of y ,

$$[R_{\mu,\gamma}(\Omega N)]\ \{y_\gamma\}=0\ , \tag{30}$$

such an eigenvalue ΩN^*, should be defined as to reduce to zero the characteristic determinant

$$\det[R_{\mu,\gamma}(\Omega N^*)]=0\ . \tag{31}$$

We shall define the sought eigenvalue ΩN^* by iterative procedure, and then the frequency of free vibrations can be calculated from the formula

$$fN=\sqrt{\frac{-\Omega N^*.E}{4.\pi(1-\nu^2)\varrho}}\ . \tag{32}$$

The matrix $[R_{\mu,\gamma}(\Omega N^*)]$ is of 7-th order, therefore the coordinates of eigenvector $\{y_\gamma\}$ are interrelated as follows:

$$y1:y2:y3:y4:y5:y6:y7:y8=$$

$$=A_{\mu,1}:A_{\mu,2}:A_{\mu,3}:A_{\mu,4}:A_{\mu,5}:A_{\mu,6}:A_{\mu,7}:A_{\mu,8}\ , \tag{33}$$

where

$A_{\mu,\gamma}$ -the algebraic complement of element μ,γ in the matrix $[R_{\mu,\gamma}]$.

The eigenvector coordinates are defined with accuracy to the constant parameter CN:

$$\{y_\gamma\} = A_{\mu,\gamma} \cdot CN \ . \tag{34}$$

We shall select its value from the displacement normalization condition WN(α1):

$$CN = \frac{1}{WN^{(max)}(\alpha 1)} \ , \tag{35}$$

and then we shall calculate the standard values of displacement:

$$\|UN(\alpha 1, CN)\| \ , \ \|VN(\alpha 1, CN)\| \ , \ \|WN(\alpha 1, CN)\| \ . \tag{36}$$

In the iterative procedure leading to the determination of eigenvalue,numerical calculations are performed in a loop starting from the coefficients $a_{\alpha,\beta}$ and $b_{\alpha,\gamma}$,to the checking for reducing to zero the characteristic determinant of boundary conditions.

Numerical Example

For a cooling tower,whose technical data has been given in the studies [1,2] the obtained frequency of free vibrations of shell are presented in Table 1.

Table 1.Natural Frequencies for a Hyperboloidal Shell, Fixed at the Base.

N	Natural frequencies (Hz),Mode 1			
	FEM(Ref.1)	Fin.Diff.(Ref.2)	IASS(Ref.4)	Pres.Work
1	3,29	3,33	-	-
2	1,76	1,78	1,84	1,76
3	-	1,39	1,39	-
4	1,18	1,20	1,21	1,03
5	1,03	1,04	1,12	1,03
6	-	1,15	1,30	-

References

1. Sen,S.K.;Gould,Ph.L.;Suryotomo,H.:High precision static and dynamic finite element of shells of revolution with applications to hyperboloidal shells on discrete supports. Washington Univ.Res.Rep.No.23.Struct.Div.November 1973.

2. Hashish,M.G.;Abu-Sitta,S.H.:Free vibration of hyperbolic cooling towers.Journ.of the Mech.Div.ASCE.EM2.April 1971.

3. Vlasov,V.Z.:Obscaja teoria obolocek.GITTL.Moskva 1949.Leningrad.

4. Recommendations for the design of hyperbolic or other similarly shaped cooling towers.IASS.Work.Group No.3.Brussels.1977.

The Finite Element Solution and Model Tests of Free Vibration of Hyperbolic Cooling Tower

Lu Wenda[1], Cheng Yuzhuo[2], Peng Jingxin[3], Wang Yangqi[4]

1. Institute of Applied Mathematics and Mechanics, Shanghai University of Technology, Shanghai, China
2. Xian Institute of Metallurgy and Construction Engineering, Xian, China
3. Xian Institute of Thermal Engineering, Xian, China
4. Xian Institute of Thermal Engineering, Xian, China

Abstract

In recent years, many a paper on free-vibration of hyperbolic cooling towers has been published. However, the effect of dead loading has not been taken into consideration and the bottom of tower shell has been assumed as fixed boundary condition. In the present paper, by using the finite element solution of hyperbolic cooling tower supported by column system based on curved axisymmetric finite elements we obtain the additional work produced by the dead loading of tower shell during free-vibration. Then the geometric stiffness matrix is given. Besides, the effect of stiffness and the mass of column system are also considered.

In the present paper, the free-vibration from model tests on cooling tower has been investigated. This model is made by electroplating which is geometrically similar to the new designed hyperbolic cooling tower with the height of 150 m. The lower end of the column system of the model is embedded in a base frame. The model is oscillated by means of a crystal stuck to its surface. The natural frequencies are readed off from a frequency detector during resonance. Meanwhile, the mode shapes are recorded on holographic films by means of laser holometry. The computed frequencies and mode shapes by means of finite element method agree quite well with the results from model tests.

1. Introduction

In November 1965, three cooling towers with a height of 113 m at Ferrybridge Power Station suddenly collapsed under a storm. It was not long before, the hyperbolic cooling tower with a height of 91 m at the Ardeer Nylon Works of Imperial Chemical Industries Ltd in the United Kingdom had a serious meridional curvature imperfection in the shell construction. So that, in September 1973, it was suddenly collapsed under a storm. People were awakened by a series of collapses of towers, and led to a series of investigations about the dynamic characteristics and the dynamic responses of cooling tower subjected to turbulence wind. Recently a number of papers [1, 2, 3, 4] have considered these problems. None of the investigations, however, has accounted for the effect of dead weight loading on the natural frequencies and gust wind response of hyperbolic cooling towers. In 1973, V.I. Weingarten has investigated the effect of dead weight loading on the natural frequencies and earthquake response, but the bottom of tower shell has been assumed as a fixed boundary condition. In the present paper, the natural frequencies and mode shapes of a hyperbolic cooling tower supported by a column system subjected to dead weight loading are investigated analytically and experimentally. By using the finite element solution of hyperbolic cooling tower supported by column system based on curved axisymmetric finite elements we obtain the additional work produced by the dead weight loading of tower shell during free-vibration. Then the geometric stiffness matrix can be given. Besides, the effect of stiffness and the mass of column system are also considered. The model of cooling tower is made by electroplating which is geometrically similar to the recent designed hyperbolic cooling tower with the height of 150 m. The lower end of the column system of the model is embedded in base frame. Holographic interferometry is used to determine the natural frequencies and mode shapes. The experimental dates agree quite well with the finite element solutions.

2. Fundamental Equations

The finite element method based on the Novozhilov shell theory is used to determine the frequencies and mode shapes of hyperbolic cooling towers. The displacements are expanded into Fourier series in the circumferential direction. The higher order Hermite's interpolation functions are used in the meridional direction. In the axisymmetric rotational shell element, it is defined that the Fourier series coefficients of displacement vector (u, v, w), their derivatives of first order with respect to the arc length s, $(\frac{\partial u}{\partial s}, \frac{\partial v}{\partial s}, \frac{\partial w}{\partial s})$, and their derivatives of second order $\frac{\partial^2 u}{\partial s^2}, \frac{\partial^2 u}{\partial s^2}$ are the generalized nodal circular displacements. In view of the linear property of this problem, each Fourier harmonic term can be considered separately. Using Hamilton's principle for the cooling tower, the matrix equation of free-vibration in terms of generalized coordinates can be obtained as follows:

$$\left[\frac{d}{dt}\left(\frac{\partial T}{\partial \dot{\alpha}_k}\right)\right] + \left[\frac{\partial U}{\partial \alpha_k}\right] - \left[\frac{\partial W}{\partial \alpha_k}\right] = 0 \tag{1}$$

in which the effect of damp is neglected, T and U are the kinetic and strain energies, respectively; the term W_g denotes the work done by the dead loading of the tower shell. The displacement functions of reference [6] are adopted. For the ith element of shell we have:

$$\underset{\sim}{V}^* = \underset{\sim}{T} \sum_{k=0}^{M} \underset{\sim}{X}_{ck} \underset{\sim}{A} \underset{\sim}{\alpha}_k \tag{2}$$

in which $\underset{\sim}{V}^* = (u^*, v^*, w^*)$, $\underset{\sim}{X}_{ck}$ = meridional, circumferential interpolation function matrix; $\underset{\sim}{T}$ = transformation matrix between the displacement vector $\underset{\sim}{V}^*$ in the local system and the displacement vector $\underset{\sim}{V}$ in the datum system. Thus for any element of shell we have the equation of free-vibration as follows:

$$\underset{\sim}{M}_k \ddot{\underset{\sim}{\rho}}_k + (\underset{\sim}{K}_k + \underset{\sim}{K}_{gk}) \underset{\sim}{\rho}_k = 0 \tag{3}$$

in which $\ddot{\underset{\sim}{\rho}}_k = \frac{d^2}{dt^2} \underset{\sim}{\rho}_k$, $\underset{\sim}{M}_k$, $\underset{\sim}{K}_k$, $\underset{\sim}{K}_{gk}$ = mass, stiffness and geometric stiffness matrices of shell structure in the datum system, respectively. All the matrices are assembled from element submatrices in the standard manner with submatrices $\underset{\sim}{M}^e$, $\underset{\sim}{K}^e$, $\underset{\sim}{K}_g^e$ given by following relations:

$$\underset{\sim}{M}^e = \pi l_i \int_{-1}^{1} \underset{\sim}{A}^T \underset{\sim}{X}^T \underset{\sim}{X} \underset{\sim}{A} \rho h r \, dx, \quad \text{element mass matrix;}$$

$$\underset{\sim}{K}^e = \pi l_i \int_{-1}^{1} \underset{\sim}{A}^T \underset{\sim}{\beta}_k^T \underset{\sim}{D} \underset{\sim}{\beta}_k \underset{\sim}{A} r \, dx, \quad \text{element stiffness matrix;}$$

$$\underset{\sim}{K}_g^e = \pi l_i \int_{-1}^{1} \underset{\sim}{A}^T \underset{\sim}{G}^T \underset{\sim}{N} \underset{\sim}{G} \underset{\sim}{A} r \, dx, \quad \text{element geometric stiffness matrix}$$

In the following, we discuss how to determine the geometric stiffness matrix of the shell element. When the shell has been set into free-vibration, W_g is the "pre-stress" work done in the membrane and bending deformations of free-vibration by the membrane stress resultants (N_φ, N_θ, $N_{\theta\varphi}$) and the transverse shear stress resultants (Q_φ, Q_θ) due to the dead weight loading. Because the bending moment and the twisting moment have no any change during free-vibration of shell, therefore in the calculation of W_g their effects can not be considered. Meanwhile, we do not taken account of the effects of bending deformations on the membrane deformations. According to the virtual work principle, W_g of a typical element can be easily obtained as follows:

$$W_g^e = -\frac{l_i}{2}\int_0^{2\pi}\int_{-1}^{1} \underset{\sim}{\omega}^T \underset{\sim}{N}\, \underset{\sim}{\omega}\, r\, d\theta\, dx \tag{4}$$

in which

$$\underset{\sim}{N} = \begin{bmatrix} N_\varphi & N_{\theta\varphi} & Q_\varphi & 0 \\ N_{\theta\varphi} & N_\theta & 0 & Q_\theta \\ Q_\varphi & 0 & 0 & 0 \\ 0 & Q_\theta & 0 & 0 \end{bmatrix}, \quad \underset{\sim}{\omega} = \begin{bmatrix} \beta_\varphi \\ \beta_\theta \\ \varepsilon_\varphi \\ \varepsilon_\theta \end{bmatrix} = \underset{\sim}{\bar{L}}\, \underset{\sim}{V}^*, \quad \underset{\sim}{\bar{L}} = \begin{bmatrix} \frac{1}{r_\varphi} & 0 & -\frac{1}{r_\varphi}\frac{\partial}{\partial\varphi} \\ 0 & \frac{\sin\varphi}{r} & -\frac{1}{r}\frac{\partial}{\partial\theta} \\ \frac{1}{l_i}\frac{\partial}{\partial x} & 0 & \frac{1}{r_\varphi} \\ \frac{\cos\varphi}{r} & \frac{1}{r}\frac{\partial}{\partial\theta} & \frac{\sin\varphi}{r} \end{bmatrix}$$

ε_φ, ε_θ = meridional, circumferential membrane strains;

β_φ, β_θ = meridional, circumferential rotations;

r_φ = meridional radius of curvature;

r = radius of parallel circle normal to the axis of revolution;

$x=\frac{s-s_i}{l_i}-1$ = dimensionless variable; s = arc-lenght; si = arc-lenght of the ith element; li = half-lenght of arc of the ith element.

From

$$\frac{\partial W_g^e}{\partial \underset{\sim}{\alpha}_k} = -\underset{\sim}{K}_g^e\, \underset{\sim}{\alpha}_k \tag{5}$$

the $\underset{\sim}{K}_g^e$ can be obtained, in which

$$\underset{\sim}{G} = \begin{bmatrix} -\frac{\cos\varphi}{l_i}\lfloor x' \rfloor_2 & 0 & -\frac{\sin\varphi}{l_i}\lfloor x' \rfloor_2 \\ k\frac{\cos\varphi}{r}\lfloor x \rfloor_2 & \frac{\sin\varphi}{r}\lfloor x \rfloor_1 & \frac{k}{r}\sin\varphi\lfloor x \rfloor_2 \\ -\frac{\sin\varphi}{l_i}\lfloor x' \rfloor_2 & 0 & \frac{\cos\varphi}{l_i}\lfloor x' \rfloor_2 \\ 0 & \frac{k}{r}\lfloor x \rfloor_1 & \frac{1}{r}\lfloor x \rfloor_2 \end{bmatrix}$$

where angle φ = measured between the axis of revolution and the normal to a meridian of shell surface; $\lfloor x \rfloor_1 = \lfloor 1\ x\ x^2\ x^3 \rfloor$, $\lfloor x \rfloor_2 = \lfloor 1\ x\ x^2\ x^3\ x^4\ x^5 \rfloor$, $\lfloor x' \rfloor_2 = \lfloor 0\ 1\ 2x\ 3x^2\ 4x^3\ 5x^4 \rfloor$.

3. Mass matrix of V-typical Column system

The beam element of V-typical column system will be assumed to be a straight bar of uniform cross section, then the mass matrix of the beam element can be given as follows:

$$\underset{\sim}{M} = \int_V \rho\ \underset{\sim}{a}^T\ \underset{\sim}{a}\ dv \tag{6}$$

in which we only take the constant term out of expansion in natural frequency powers of the shape function matrix $\underset{\sim}{a}$, that is, the mass matrix is a identity mass matrix. Since the bending planes xy and xz will be assumed to coincide with the principal axes of the cross section of beam element and the ox axis will coincide with the centroidal axis of the beam. The mass matrix of ij-beam element in the beam coordinate system is given as:

$$\underset{\sim}{M}_{ij} = l\int_0^1\int_A \rho\ \underset{\sim}{a}^T\ \underset{\sim}{a}\ d\xi\ dA = \rho\ A\ l \left[\begin{array}{c|c} m_{ii} & m_{ij} \\ \hline m_{ji} & m_{jj} \end{array}\right] \tag{7}$$

in which

$$\underset{\sim}{a} = \begin{bmatrix} 1-\xi & 0 & 0 \\ 6(\xi-\xi^2)\zeta & 0 & 1-3\xi^2+2\xi^3 \\ (1-4\xi+3\xi^2)l\zeta & 0 & (-\xi+2\xi^2-\xi^3)l \\ 0 & -(1-\xi)l\zeta & (1-\xi)l\eta \\ \xi & 0 & 0 \\ 6(-\xi+\xi^2)\zeta & 0 & 3\xi^2-2\xi^3 \\ (-2\xi+3\xi^2)l\zeta & 0 & (\xi^2-\xi^3)l \\ 0 & -l\xi\zeta & l\xi\eta \end{bmatrix} \tag{8}$$

$\xi = \frac{x}{l}$, $\zeta = \frac{y}{l}$, $\eta = \frac{z}{l}$ are denoted the non-dimensional variables; l = length of beam.

We transform the mass matrix of ij-beam element to the V-typical column coordinate system, then

$$\bar{\underset{\sim}{M}}_{ij} = \left[\begin{array}{c|c} T_{jj}^T \, m_{jj} \, T_{jj} & T_{jj}^T \, m_{ji} \, T_{ii} \\ \hline T_{ii}^T \, m_{ij} \, T_{jj} & T_{ii}^T \, m_{ii} \, T_{ii} \end{array}\right] = \left[\begin{array}{c|c} \bar{m}_{jj} & \bar{m}_{ji} \\ \hline \bar{m}_{ij} & \bar{m}_{ii} \end{array}\right] \tag{9}$$

$$\underset{\sim}{T}_{ii} = \begin{bmatrix} -\sin\alpha & \cos\alpha & 0 & 0 & 0 \\ 0 & 0 & 1 & 0 & 0 \\ 0 & 0 & 0 & \sin\alpha & -\cos\alpha \\ 0 & 0 & 0 & -\cos\alpha & -\sin\alpha \end{bmatrix}$$

$$\underset{\sim}{T}_{jj} = \begin{bmatrix} -\sin\alpha & \cos\alpha & 0 & 0 & 0 \\ 0 & 0 & 1 & 0 & 0 \\ 0 & 0 & 0 & -\sin\alpha & \cos\alpha \\ 0 & 0 & 0 & \cos\alpha & \sin\alpha \end{bmatrix}$$

A pair of V-typical columns is constructed by ij-beam and kj-beam. $\bar{\underset{\sim}{M}}_{kj}$ may be obtained using $\pi - \alpha$ instead of α in $\bar{\underset{\sim}{M}}_{ij}$, meanwhile $\sin(\pi-\alpha) = \sin\alpha$, $\cos(\pi-\alpha) = -\cos\alpha$, and $\sin(\pi-\alpha)\cos(\pi-\alpha) = -\sin\alpha\cos\alpha$, are taken into account, then the mass matrix of the V-typical column in the V-typical column coordinate system can be given as:

$$\underset{\sim}{\bar{M}}^{*} = \left[\begin{array}{c|cc} \bar{M}_{jj} + \bar{M}'_{jj} & \bar{M}_{ji} & \bar{M}_{jk} \\ \hline \bar{M}_{ij} & \bar{M}_{ii} & 0 \\ \bar{M}_{kj} & 0 & \bar{M}_{kk} \end{array}\right] \tag{10}$$

If the lower ends of beam element are assumed as fixed edge, then we have

$$\underset{\sim}{\bar{M}}^{*} = \underset{\sim}{\bar{M}}_{jj} + \underset{\sim}{\bar{M}}'_{jj} \tag{11}$$

$$= \rho\, Al \begin{bmatrix} \bar{m}_{11} & 0 & 0 & 0 & 0 \\ 0 & \bar{m}_{22} & 0 & 0 & 0 \\ 0 & 0 & \bar{m}_{33} & \bar{m}_{34} & 0 \\ 0 & 0 & \bar{m}_{43} & \bar{m}_{44} & 0 \\ 0 & 0 & 0 & 0 & \bar{m}_{55} \end{bmatrix}$$

in which $\bar{m}_{11} = 2/3 \sin^2\alpha$; $\bar{m}_{22} = 2/3 \cos^2\alpha$; $\bar{m}_{33} = 2(13/35 + 6Iy/5Al^2)$;

$\bar{m}_{34} = \bar{m}_{43} = -2(11l/210 + Iy/10Al) \sin\alpha$;

$\bar{m}_{44} = 2(l^2/105 + 2Iy/15A) \sin^2\alpha + Ix \cos^2\alpha/3A)$;

$\bar{m}_{55} = 2(l^2/105 + 2Iy/15A) \cos^2\alpha + Ix \sin^2\alpha/3A)$;

$Ix = \int_A (y^2+z^2)\, dA$ = polar moment of inertia of the cross section about x-axis;

$Iy = \int_A z^2\, dA$ = moment of inertia of the cross section about y-axis;

A = cross section; ρ = the density of the shell material

Inasmuch as the V-typical column coordinate system keeps with the local coordinate system of the shell ($\vec{\tau}$, $\vec{\theta}$, $\vec{n}$), $\underset{\sim}{\bar{M}}^{*}$ is the mass matrix of a pair of V-typical column in the local coordinate system of the shell. $\vec{\tau}$ = the unit vector tangent to the surface in the meridian; $\vec{\theta}$ = the unit vector tangent to the surface in the parallel circle; $\vec{n}$ = the unit vector normal to the surface.

Supposing that the number of V-typical columns is equal to C , whereas the interval $(o,2\pi)$ is divided into C equal portions, and the left end of the first portion is set on $\theta = 0^{o}$, then the circumferential angle of the V-typical column is

$$\theta_{\lambda} = 2\pi\ (\lambda-1)\ /\ C\ , \qquad (\lambda = 1,\ 2,\ ----,\ C)\ .$$

If only the kth harmonic term is taken into account, meanwhile takes note of the transformation matrix relationship between the generalized displacements $\bar{u}^*$, $\bar{w}^*$, $\bar{\beta}^*_{\varphi}$ $\bar{\beta}^*_{\theta}$ and the generalized displacements of the shell:

$$\begin{bmatrix} \bar{u}^* \\ \bar{v}^* \\ \bar{w}^* \\ \bar{\beta}^*_{\varphi} \\ \bar{\beta}^*_{\theta} \end{bmatrix} = \begin{bmatrix} 1 & 0 & 0 & 0 & 0 & 0 & 0 & 0 \\ 0 & 1 & 0 & 0 & 0 & 0 & 0 & 0 \\ 0 & 0 & 1 & 0 & 0 & 0 & 0 & 0 \\ \frac{1}{r_{\varphi}} & 0 & 0 & 0 & 0 & -\frac{1}{l_i} & 0 & 0 \\ 0 & \frac{1}{r_{\theta}} & \frac{k}{r} & 0 & 0 & 0 & 0 & 0 \end{bmatrix} \begin{bmatrix} u^* \\ v^* \\ w^* \\ u^{*\prime} \\ v^{*\prime} \\ w^{*\prime} \\ u^{*\prime\prime} \\ w^{*\prime\prime} \end{bmatrix} \qquad (12)$$

then we have

$$\underset{\sim}{\hat{M}}^* = C/2\ \underset{\sim}{\bar{A}}^T\ \underset{\sim}{M}^*\ \underset{\sim}{\bar{A}} \qquad (13)$$

The transformation matrix between the generalized displacements in the local system and the generalized displacements in the datum system is obtained as follows:

$$\underset{\sim}{\bar{\bar{A}}} = \begin{bmatrix} -\sin\varphi & 0 & \cos\varphi & 0 & 0 & 0 & 0 & 0 \\ 0 & 1 & 0 & 0 & 0 & 0 & 0 & 0 \\ \cos\varphi & 0 & \sin\varphi & 0 & 0 & 0 & 0 & 0 \\ l_i\ \cos\varphi/r_{\varphi} & 0 & -l_i\ \sin\varphi/r_{\varphi} & -l_i\ \sin\varphi & 0 & l_i\ \cos\varphi & 0 & 0 \\ 0 & 0 & 0 & 0 & l_i & 0 & 0 & 0 \\ -l_i\ \sin\varphi/r_{\varphi} & 0 & l_i\ \cos\varphi/r_{\varphi} & l_i\ \cos\varphi & 0 & l_i\ \sin\varphi & 0 & 0 \\ 0 & 0 & 0 & 0 & 0 & 0 & 0 & 0 \\ 0 & 0 & 0 & 0 & 0 & 0 & 0 & 0 \end{bmatrix}$$

(14)

then the mass matrix of V-typical column system in the datum coordinate system can be obtained as follows:

$$\hat{\underset{\sim}{M}} = \bar{\bar{\underset{\sim}{A}}}^T \, \hat{\underset{\sim}{M}}^* \, \bar{\bar{\underset{\sim}{A}}} \tag{15}$$

4. <u>Equation of Free-Vibration</u>

By means of the transformation relationship between the uncoupled generalized displacements and the assembled generalized displacements, i.e.

$$\underset{\sim}{\alpha}_k = \bar{\underset{\sim}{a}} \, \underset{\sim}{\rho}_k \tag{16}$$

in which

$$\bar{\underset{\sim}{a}} = \left[\begin{array}{cccccc} I_8 & 0 & 0 \cdots 0 & 0 \cdots 0 & 0 \\ 0 & I_8 & 0 \cdots 0 & 0 \cdots 0 & 0 \\ \vdots & & & & \\ 0 & I_8 & 0 \cdots 0 & 0 \cdots 0 & 0 \\ 0 & 0 & I_8 \cdots 0 & 0 \cdots 0 & 0 \\ \vdots & \vdots & \vdots \quad \vdots & \vdots \quad \vdots & \vdots \\ 0 & 0 & 0 \cdots I_8 & 0 \cdots 0 & 0 \\ 0 & 0 & 0 \cdots 0 & I_8 \cdots 0 & 0 \\ \vdots & \vdots & \vdots \quad \vdots & \vdots \quad \vdots & \\ 0 & 0 & 0 \cdots 0 & 0 \cdots I_8 & 0 \\ 0 & 0 & 0 \cdots 0 & 0 \cdots 0 & I_8 \end{array} \right] \tag{17}$$

I_8 = (8·8) identity matrix .

Since the resulting virtual work (a scale quantity) must obviously be independent of the coordinate system, so that the free-vibration equation of hyperbolic cooling tower supported by column system can be obtained as:

$$\underset{\sim}{M}_k \, \ddot{\underset{\sim}{\rho}}_k + \underset{\sim}{K}^*_k \, \underset{\sim}{\rho}_k = 0 \tag{18}$$

in which

$\underset{\sim}{M}_k = \bar{\underset{\sim}{a}}^T \underset{\sim}{M}_O \bar{\underset{\sim}{a}}$ system mass matrix;

$\underset{\sim}{K}_k^* = \bar{\underset{\sim}{a}}^T \underset{\sim}{K}_O^* \bar{\underset{\sim}{a}}$ system stiffness matrix;

$\underset{\sim}{M}_O =$ uncoupled total element mass matrices;

$\underset{\sim}{K}_O^* = \underset{\sim}{K}_O + \underset{\sim}{K}_{gO}$ uncoupled total element stiffness matrices and geometric stiffness matrices.

Let $\underset{\sim}{\delta}_k = \underset{\sim}{q}_k \sin(\omega t+\varphi)$, then the characteristic equation of the kth harmonic is given as follows:

$$(\underset{\sim}{K}_k + \underset{\sim}{K}_{gk} - \omega^2 \underset{\sim}{M}_k)\, \underset{\sim}{q}_k = 0 \qquad (19)$$

in which the mth element is a V-typical column system which system stiffness matrix is given by reference [6]. This equation is solved by means of subspace iteration procedure.

5. Procedure of Model Tests

The model of a hyperbolic cooling tower supported by sixty pairs of V-typical columns as shown in Fig. 2. was manufactured by electroplating which was geometrically similar to the recent designed hyperbolic cooling tower with 150 m height. The geometry of the cooling tower is shown in Fig.3. which is 1/450 scale model with variable thickness. The lower end of the column system of this model was embedded in base frame. The material properties of the shell were $E = 0.78 \cdot 10^6$ kg/cm^2; $\nu = 0.37$; and $\rho = 8.51 \cdot 10^{-6}$ kg$\cdot$s^2/cm^4. The photograph of the experimental set-up is shown in Fig. 4.. A frequency signal source which was connected to a crystal struck to the shell surface and power-amplifier, was used to excite the model (see Fig. 5.). A frequency detector was used to read off the natural frequencies during resonance. Laser holographic interferometry was used to record the tower shell mode shapes at resonant frequencies. The arrangement plan of laser holographic interferometry used for obtained time average holograms of mode shapes are sketched in Fig. 6..

6. Results

Since the digital frequency detector was used to determine the natural frequencies of tower shell which minimum read is equal to 0.1 Hz, and the order of magnitude of measured displacement used by holographic interferometry is about to the order of wave length, therefore the experimental procedure is very accurate. In Fig. 7. and table 1., the experimental data measured on model tests and the solutions by means of finite element method are given. It will be seen that they agree quite well with one another. The maximum relative error of individual frequency value will not excess of 10%. The time average holograms of several typical mode shapes are given in Fig. 8.. From Fig. 9., it will be seen that the mode shapes plotted from experimental datum are in good agreement with the finite element solutions.

Conclusions

1. The finite element solutions and the experimental procedure of holographic interferometry suggested by the present paper can be used to determine the dynamic characteristics of hyperbolic cooling towers.

2. The top lintel has a large effect on the natural frequencies, in particular, on the natural frequencies of higher order meridional mode shapes.

This investigation was supported by the Ministry of Water Resources and Electric Power.

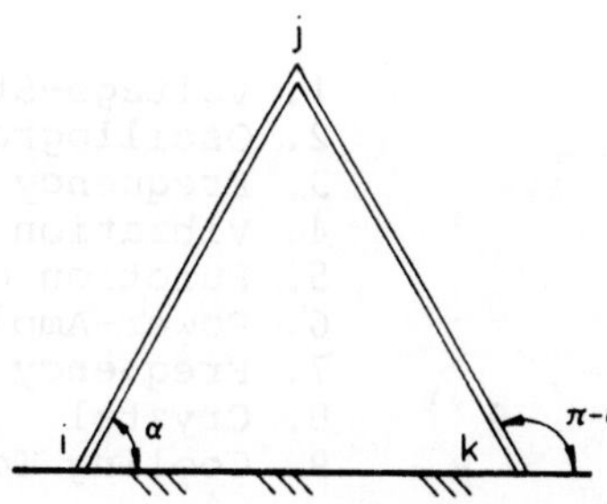

Fig. 1. V-typical Column

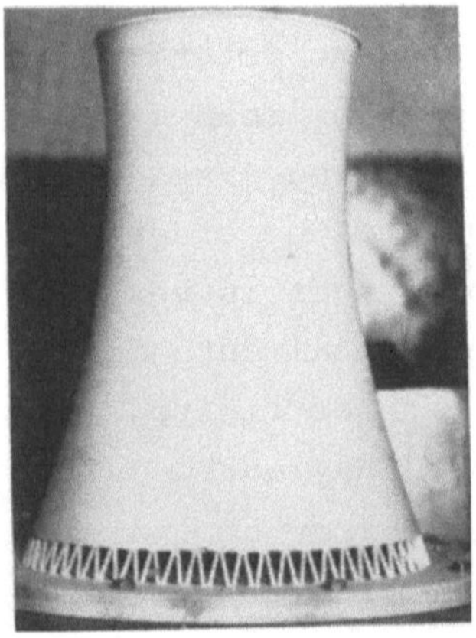

Fig. 2. Model of Hyperbolic Cooling Tower

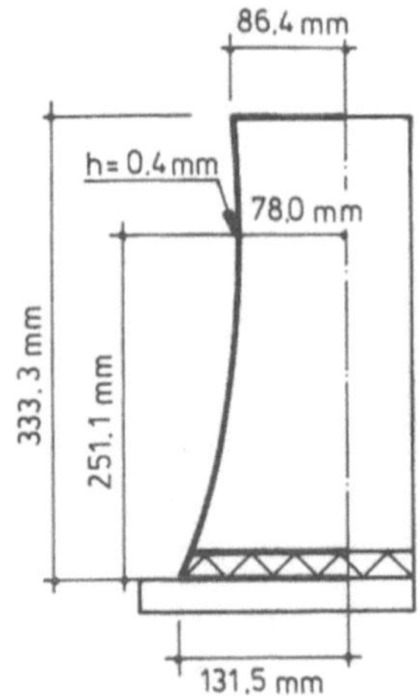

Fig. 3. Geometry of Cooling Tower Model

Fig. 4. Experimental Set-Up of Cooling Tower

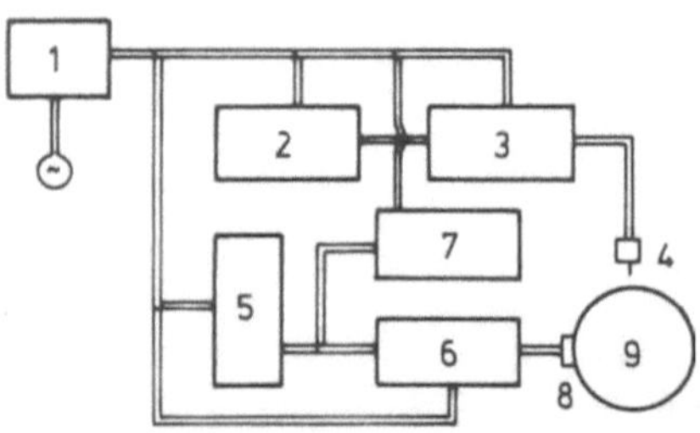

1. Voltage-Stabilized Source
2. Oscillograph
3. Frequency Detector
4. Vibration Pickup
5. Function Generator
6. Power-Amplifier
7. Frequency Meter
8. Cryztal
9. Cooling Tower

Fig. 5. Experimental Electric Set-up

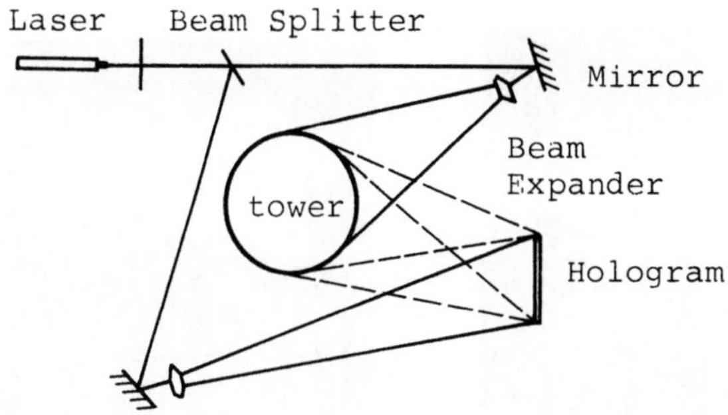

Fig. 6. Set-Up for Laser Holographic Interferometry Method

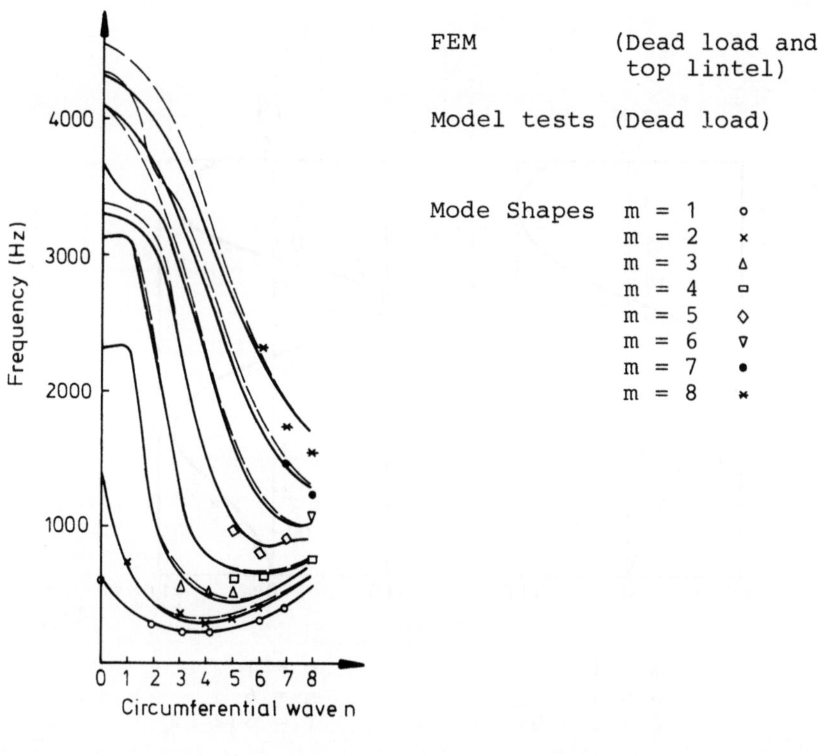

Fig. 7. Frequency Versus Circumferential Wave Number

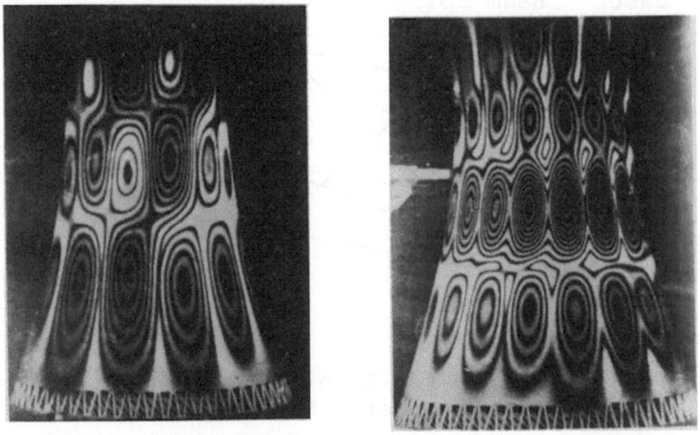

Fig. 8. Holographic Films of Vibration Modes of Hyperbolic Cooling Tower

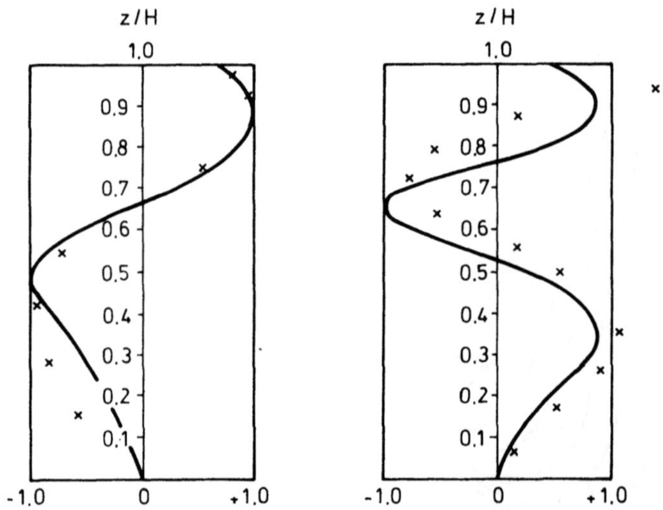

m = 1
n = 4
f = 248 Hz

m = 1
n = 6
f = 332 Hz

Fig. 9. Mode Shapes of Hyperbolic Cooling Tower

m		Natural Frequencies in Cycles Per Seconcl								
		n= Circumferential Wave Number								
		0	1	2	3	4	5	6	7	8
1	a	627	----	323	247	233	285	364	475	----
	b	593	----	318	255	248	295	332	430	----
2	a	1414	743	----	349	316	345	446	----	----
	b	1431	740	----	363	292	317	421	----	----
3	a	----	----	----	623	585	515	----	----	----
	b	----	----	----	580	532	547	----	----	----
4	a	----	----	----	----	772	680	691	692	----
	b	----	----	----	----	804	633	679	728	----
5	a	----	----	----	----	----	1031	860	901	932
	b	----	----	----	----	----	1004	861	946	967
6	a	----	----	----	----	----	----	----	1016	1043
	b	----	----	----	----	----	----	----	1050	1105
7	a	----	----	----	----	----	----	----	1451	1285
	b	----	----	----	----	----	----	----	1467	1236
8	a	----	----	----	----	----	----	2240	1914	1703
	b	----	----	----	----	----	----	2353	1759	1549

m= Meridional Mode Shape Number.

a= Finite Element Solution.

b= Experimental Solution.

Table 1. Experimental and Theoretical Results on the Natural Frequencies of Hyperbolic Cooling Tower.

References

1. Hashish, M.G. and Abu-Sitta, S.H.: Free Vibration of Hyperbolic Cooling Tower. Journal of Engineering Mechanics Division Proceeding of the ASCE, 1971. 4, EM2.

2. Sen, S.K. and Gould, P.L.: Free Vibration of Shell of Revolution Using FEM. Journal of the Engineering Mechanics Division, ASCE, Vol. 100, No. EM2, April 1974.

3. Abu-Sitta, S.H. and Hashish, M.G.: Dynamic Wind Stress in Hyperbolic Cooling Tower. Journal of Structural Division ASCE, Vol. 99, No. ST9, September 1973.

4. Singh, M.P. and Gupta, A.K.: Gust factors for Hyperbolic Cooling Tower. Journal of the Structural Division, ASCE, Vol. 102, No. ST2, Proc 11894, February 1976.

5. Weingarten, V.I.; Masri, S.F.; Lashkari, M. and Kahyal: Effect of Gravity Loading on the Earthquake Response of Cooling Tower. Fifth World Conference on Earthquake Engineering, Rome, 1973.

6. Wenda, L. and Jingxin, P.: Finite Element Solution of the Hyperbolic Cooling Tower Supported by Column System, Research Report of Xian Institute of Thermal Engineering, June 1976, China.

7. Wenda, L.; Jingxin, P. and Yangqi, W.: Effect of Dead Load on the Dynamic Property of Hyperbolic Cooling Tower, Thermal Electrical Power, No. 5, February 1982, China.

8. Jiayin, H. and Yuzhuo, Ch.: Application of the laser Holographic Interferometry Method in the Investigation of Dynamic Characteristics of Structures. Journal of Xian Institute of Metallurgy and Construction Engineering, Vol. 1-2, 1979, Xian, China.

1984
IASS-RUB

Stability and Nonlinear Behavior

The Problem of Kinetic Instabilities in The Natural Wind-Theoretical and Numerical Investigations

S. SPIERIG and E. STEIN

Universität Hannover, Institut für Baumechanik und Numerische Mechanik

Summary

Natural draught cooling towers are very thin rotational shells with negative Gaussian curvature. Their static stability may decrease by vibrations. This effect should not be neglected in practical stability analysis. In general both effects are coupled and cannot be handled separately. The coupled problem is formulated and an approximate solution is described making use of known solutions for the loss of static stability by geometric imperfections.

1. Introduction

Contemporary natural draught cooling towers are very thin shell structures which possess relatively little resistance against bending. In contrast to the deadweight of the structure, which causes membrane stresses, the wind loading causes bending stresses within the structure which vary in time as does the wind. Thus bending vibrations may be excited which have to be taken into account when the stability of the structure is investigated. Spectacular collapses of cooling towers in Ferrybridge (1965) and Bouchain (1979) [1,2] stress the necessity of a dynamic stability analysis. Moreover local damages have been observed at several other cooling towers that cannot be explained when considerations are limited to statics.

If the smallest natural frequency is significantly higher than the frequency of relevant excitation (e.g. Karman vortex excitation), then the usual way of separate investigations for the static stability and the vibration problem is sufficient. This is true for cooling towers of contemporary size with a sufficiently stiff upper edge beam. In view of taller towers in future (with lower natural frequencies) and possible alterations of structural details it is desirable to study the coupled kinetic stability problem. It is possible that the relevant stiffnes of the system decreases in the presence of vibrations in a way that the safety against buckling becomes too small or even buckling takes place [3].

In this paper a formulation for the coupled kinetic stability problem is given. Then an approximate decoupled solution is proposed using known informations about the loss of stability due to geometric imperfections of the shell.

2. Structure under Consideration

The present investigation will be performed for the cooling tower structure shown in Fig. 1 a. For comparison purposes, in addition a comparable cylindrical shell structure (Fig. 1 b) is investigated.

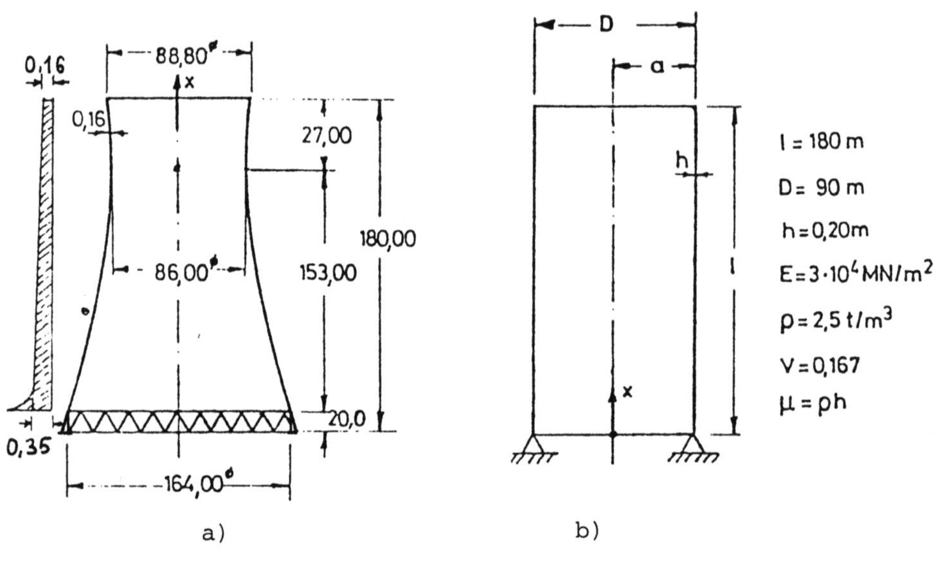

Fig.1. Structures under consideration
a) Cooling tower (all dimensions in meter)
b) Circular cylindrical shell

3. Loading Conditions

The structure is loaded by its deadweight and by aerodynamic forces. The latter consists mainly of the pressure forces acting upon the outer surface of the structure due to the natural wind. Moreover there are suction forces acting as a negative pressure upon the inner surface of the shell, caused by the flow inside the cooling tower.

The distribution of pressure caused by the wind may be given by the formula

$$q(z,\varphi,t) = q(z)\cdot c(\varphi)\cdot f(t). \qquad (1)$$

The quantity q(z) denotes the distribution of the dynamic pressure at the stagnation meridian over the height z. Appropriate values are given in Fig. 2a. The coefficient $c(\varphi)$ describes the circumferential distribution. The circumferential coordinate $\varphi = 0$ designates the stagnation point (Fig. 2b). The different

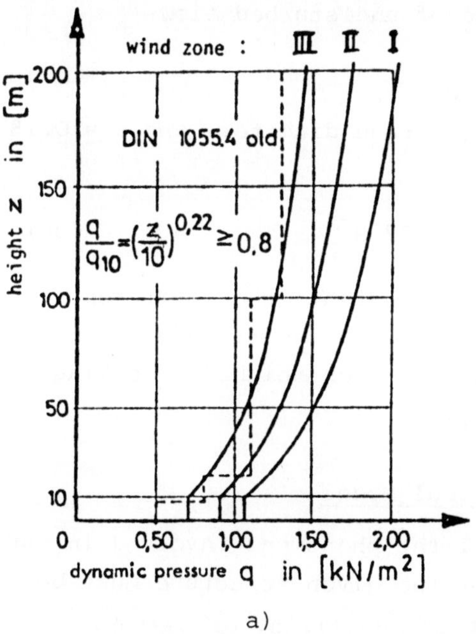

a)

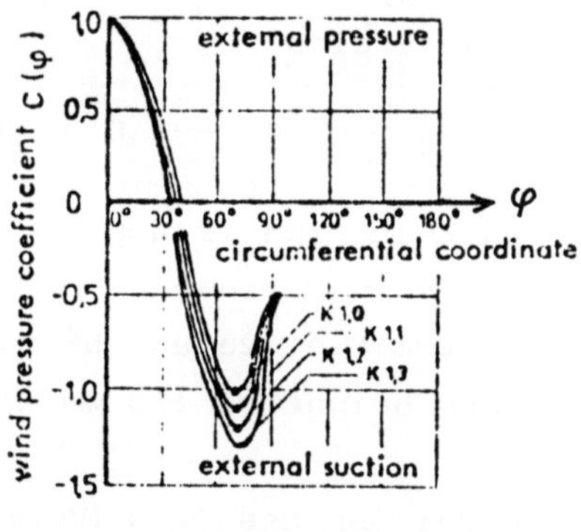

b)

Fig.2. Distribution of dynamic pressure due to natural wind
a) Distribution over height
b) Circumferential distribution

curves correspond to different magnitudes of surface roughnes [4]. The factor f(t) accounts for the time-dependent character of of the pressure distribution. The true wind is a stochastic process. For sake of simplicity however, f(t) is taken to be composed of a constant and a harmonic part [3]:

$$f(t) = 0{,}7 - 0{,}3 \sin \Omega t \tag{2}$$

The harmonic part may be caused by natural air turbulence as well as by a Karman vortex process happening either at the cooling tower itself or at perhaps neighbouring towers or other obstacles upstream.

The frequency of the Karman vortex process at the cooling tower can be calculated by Strouhal's law

$$f = \frac{u_\infty S}{D} , \tag{3a}$$

with

u_∞	average wind speed of undisturbed flow
S	Strouhal number
D	diameter of the tower.

Assuming $u_\infty = 50$ m/s, D = 100 m (average diameter) and S = 0.18 [5], (3a) yields

$$f = 0.09 \text{ Hz or } \Omega = 2\pi f = 0.57 \text{ s}^{-1} . \tag{3b}$$

The inner suction is approximately constant in circumferential as well as in meridional direction and may be taken as $-q(h^*)$ with q from Fig. 2a and h^* average between height of cooling tower and height of its waist [2].

4. General Aspects of a Mathematical Model

For a complete description of all the phenomena involved in our problem the mathematical model of the given structure must be capable to reproduce correctly the stabilizing or destabilizing influence respectively of membrane stresses on the bending deformations of the shell. The state of membrane stresses consists of three parts:

(1) Membrane stresses belonging to the basic state of the structure (caused by loadings constant in time such as deadweight, inner suction and constant part of wind pressure).

(2) Additional membrane stresses connected with the deformation of the structure from the basic state into some neighbouring state (due to vibration or buckling).

(3) Additional membrane stresses caused by the time-varying part of the wind forces.

The latter part (3) might give rise to a vibration of the parameter excitation type. This type of excitation endangers zones with membrane pressure, exspecially those with pressure in both directions. Such zones exist in the flanks of the cooling tower below the waist [6]. For towers of contemporary size and structure with an edge beam at the upper edge however, no such instability is to be expected [3]. Hence part (3) may be neglected in practical stability analysis.

Part (2) of the membrane stresses yields nonlinearities when taken into account completely. If only limiting values of stability are desired but no postbuckling behaviour, then linearization is admissible. Such a linearized model is suited as well for determining static stability as for determining the vibration process of the shell structure caused for instance by Karman excitation.

After discretization, for instance by the finite element method, we arrive at

$$\underline{M}\,\ddot{\underline{V}} + \underline{C}\,\dot{\underline{V}} + \underline{K}\,\underline{V} = \underline{P}(t) \qquad (4)$$

with

- $\underline{M}$ mass matrix
- $\underline{C}$ damping matrix
- $\underline{K}$ stiffnes matrix
- $\underline{V}$ vector of nodal displacements
- P vector of external nodal loads.

The stiffnes matrix can be splitted up [2] :

$$\underline{K} = \underline{K}_e + \lambda \underline{K}_g \tag{5}$$

with

- $\underline{K}_e$ elastic stiffnes matrix
- $\underline{K}_g$ geometric stiffnes matrix (due to stresses associated with the basic state of stress)
- λ intensity factor for basic state loading.

5. Concept for the Investigation of the Kinetic Stability

The linearized vibration problem (4, 5) will be solved using modal analysis [7]. Introducing the vector $\underline{q}$ of generalized displacements instead of $\underline{V}$ by the transformation

$$\underline{V} = \underline{\Phi}\, \underline{q} \quad , \tag{6}$$

with

$$\underline{\Phi} = [\underline{\Phi}_1, \underline{\Phi}_2, \ldots, \underline{\Phi}_M] \tag{7}$$

being the modal matrix of the associated conservative system ($\underline{C}$ = 0), we arrive at the set (8) of uncoupled differential equations, provided the damping matrix $\underline{C}$ is of a suitable form such that $\underline{C}$ can be diagonalized simultaneously with $\underline{M}$ and $\underline{K}$

$$\ddot{q}_i + 2D_i \omega_{io} \dot{q}_i + \omega_{io}^2 q_i = \frac{\Omega_i(t)}{m_i} \quad . \tag{8}$$

The meaning of the quantities in (8) is:

ω_{io}	undamped circular frequency of i-th natural mode
D_i	Lehr's damping value for i-th nat. mode
$\underline{\Omega}_i(t) = \underline{\Phi}_i^T \underline{P}(t)$	generalized external load
$m_i = \underline{\Phi}_i^T \underline{M}\, \underline{\Phi}_i$	modal mass for i-th natural mode.

If we assume $\underline{P}(t)$ to result from the time varying part of equations (1, 2), the right-hand side of (8) is of the form

$$\frac{Q_i(t)}{m_i} = b_i \sin \Omega t \ . \tag{9}$$

Then the steady state solution of (8) is given by (see [8] e.g.)

$$q_i(t) = \hat{q}_i \sin (\Omega t - \varepsilon_i) \tag{1o}$$

with

$$\hat{q}_i = \frac{b_i}{\omega_{io}^2} \left\{ \left[1 - (\frac{\Omega}{\omega_{io}})^2 \right]^2 + \left[2D_i \frac{\Omega}{\omega_{io}} \right]^2 \right\}^{-\frac{1}{2}} \tag{11}$$

$$\varepsilon_i = \text{arc tan} \frac{2D_i \Omega / \omega_{io}}{1-(\Omega/\omega_{io})^2} \ . \tag{12}$$

If we insert the solution of the vibration problem given by (1o, 11, 12) into (6) we arrive at the amplitudes of displacements

$$\hat{\underline{V}}_t = \sum_{i=1}^{M} \Phi_i \hat{q}_i(t) \ . \tag{13}$$

The subscript t means that the displacements $\underline{V}_t$ are time dependent due to the time varying part of the wind loading. The vector $\underline{V}_t$ is added to the displacements $\underline{V}_s$ caused by static loading.

It can easily be seen that the system is kinetically instable if static instability occures. Static instability is associated with a non-regularity of the resultant stiffness matrix (5) which follows from (4, 5) with $\ddot{\underline{V}} = \underline{O}$, $\dot{\underline{V}} = \underline{O}$ [2]. In this case the natural frequency condition

$$(-\omega^2 \underline{M} + \underline{K}_e + \lambda \underline{K}_g)\underline{V} = \underline{O} \tag{14}$$

resulting form (4, 5) with $\underline{C} = \underline{O}$ yields zero undamped natural frequencies $\omega_{io} = 0$ and thus $\underline{V}_t$ according to (13, 1o, 11) will tend to infinity.

Kinetic instability must be expected, however, even if static loading is fairly below the point of static instability but with a superimposed vibration of a sufficiently large amplitude (kinetic snap through). A complete analysis of this kinetic instability problem is impossible with the mathematical model (4) since nonlinearities are neglected.

For practical kinetic stability analysis of cooling towers the following procedure seems to be a sufficiently reliable approximation of the problem: Calculate the pattern of vibration amplitudes using the simple linearized model as shown above and then investigate static stability as usual (with static loads including the constant part of windloading in equ. (2)) but assuming fictitious geometric imperfections identical with the vibration pattern calculated before.

6. Derivation of Matrices

Since a suitable conical ring element for axial symmetric strain and stress distribution was not available here at the time, the FE-discretization employed the plane quadrilateral 4-node element with 4 DOF's per node which is implemented in SAP IV. The elements are arranged in 3o sectors in circumferential direction and 15 rings in direction of the height of the structure with a resulting size of an element of roughly 1o m x 1o m. The generated consistent $\underline{K}_c$ and $\underline{M}$ matrices - generated with SAP IV - are then condensed using a Fourier-distribution in circumferential direction. This results in different matrices $\underline{M}^{(n)}$ and $\underline{K}_c^{(n)}$ for any wave - number n (n = 1, 2,...,N). Compared with the original matrices $\underline{K}_c$ and $\underline{M}$ their order is less by the factor 1/p with p = number of sectors in the FE-discretization.

The damping matrix $\underline{C}$ or $\underline{C}^{(n)}$ respectively in the Fourier-condensed form need not to be constructed explicitely when modal analysis is applied. All to do is to choose appropriate damping values D_i to be inserted in (8). There are no mathematical limitations so far. If desired (e.g. when a step by step integration of (4) is wished rather than modal analysis), construction of $\underline{C}$ is easily possible by the relation [12]

$$\underline{C} = 2\,\underline{M}\,\underline{\Phi}\,\underline{\delta}\,\underline{\Phi}^T\,\underline{M} \qquad (16)$$

with

$$\underline{\delta} = \lceil D_i\,\omega_i/M_i \rfloor\,. \qquad (17)$$

(16) is an improvement of the well known Caughey - polynomial

$$\underline{C} = \underline{M}\left[\sum_{j=0}^{\nu-1} \gamma_j\,(\underline{M}^{-1}\underline{K})^j\right] \qquad (18)$$

with ν = order of matrices (= number of DOF's) [13]. The Caughey formula allows to represent any modal damping value by appropriate choice of the coefficients γ_i such as the Witfeld-formula does. In case of multiple eigenvalues however, (18) does not decouple completely in modal analysis while (17) does.

7. Numerical Results

Numerical computations have not been completed when this manuscript had to go to press. The results will be given in the lecture on this paper.

References

1. The Institution of Civil Engineers: Natural Draught Cooling Towers - Ferrybridge and After. Proceedings, Conference London, 1967.

2. Krätzig, B., Şanal, Z. and Wittek, U.: Zum Einfluß oberer Randaussteifungen auf das dynamische Verhalten von Kühlturmschalen. Beton- Stahlbetonbau 79 (1984) 74-80.

3. Nguyen, B., Rosemeier, G. E., Spierig, S. and Stein, E.: Dynamisches Verhalten von Kühlturmschalen infolge von Windkräften unter besonderer Berücksichtigung kinetischer Instabilitäten. Beton- und Stahlbetonbau 76 (1981) 169-174.

4. Niemann, H.-J.: Ein Windlastkonzept für Naturzugkühltürme. Konstruktiver Ingenieurbau Berichte, Heft 29/30, Vulkan-Verlag Essen, 1977.

5. Ruschewey, H.: Dynamische Windwirkungen durch Interferenzen an benachbarten schlanken Bauwerken. Konstruktiver Ingenieurbau Berichte, Heft 35/36, Vulkan-Verlag Essen, 1981.

6. Rosemeier, G.E., Spierig, S. und Stein, E.: Dynamisches Verhalten von Kühlturmschalen infolge von Windkräften unter besonderer Berücksichtigung kinetischer Instabilitäten. 4. Berichtskolloquium des DFG-Schwerpunktprogramms "Bauwerke unter aerodynamischer Belastung".Göttingen 8.Okt. 1979.

7. Pestel, E. C. and Leckie, F. A.: Matrix Methods in Elastomechanics. Mc Graw Hill 1963.

8. Klotter, K.: Technische Schwingungslehre, Bd I A, Springer Berlin, Heidelberg, New York, 1978.

9. Eckstein, U., Krätzig, W. B. und Wittek, U.: Wirklichkeitsnahe Grenzbeullasten von Rotationsschalen mit unterschiedlicher Gauss'scher Krümmung. Techn.-wissensch. Mittlg., Inst. f. Konstr. Ingenieurbau, Ruhr-Universität Bochum, Nr. 79-5, 1979.

10. Wittek, U.: Beitrag zum Verhalten der Strukturen bei endlichen Verformungen unter besonderer Beachtung des Nachbeulmechanismus dünner Flächentragwerke. Techn.-wissensch. Mittlg. Inst. f. Konsruktiver Ingenieurbau, Ruhr-Universität Bochum, Nr. 80-1, 1980.

11. Wunderlich. W., Rensch, H.-J. and Obrecht, H.: Analysis of Elastic-plastic Buckling and Imperfection-Sensivity of Shells of Revolution. In: Buckling of Shells. Proceed. State-of-the-Art Colloquium, University Stuttgart, May 1982. E. Ramm (Ed.), Springer, Berlin ,Heidelberg, New York, 1982.

12. Witfeld, H.: Modale Dämpfung in linearen Schwingungssystemen. Lecture GAMM-Tagung Regensburg 1984. To be published in ZAMM.

13. Caughey, T. K.: Classical Normal Modes in Damped Linear Dynamic Systems. Journ. Appl. Mech. 27 (1960).

Buckling of Reinforced Concrete Cooling Tower Shells

Dr. E. DULÁCSKA

Institute of Building Types Design
Budapest, HUNGARY

Summary

The results of the nonlinear shell buckling theory has been extended to apply to the reinforced concrete cooling tower shells. A simple relation is derived for the snap-through critical load of the shell. The solution contains the special properties of reinforced concrete, i.e., cracks, plasticity, creep and quantity of reinforcement.

The critical loads of the cooling tower shells

Among the hyperbolic shells of revolution the simplest one is the hyperboloid of revolution. This surface can be generated by rotating a pair of hyperbolas around their axis of symmetry which does not intersect them. However, hyperbolic surfaces of revolution may also be generated by rotating, e.g., a parabola or any other curve around the axis.

Hyperbolic shells are mostly used as cooling towers and are made, as a rule, of reinforced concrete. The dimensions of the cooling towers increased steadily increasing for years without their stability problem being thoroughly investigated. After the collapse of three cooling towers at Ferry-bridge in 1965, the research in to the stability of hyperbolic shells of revolution began.

In recent years rather accurate computer calculations have been developed for the determination of the critical load. Lehmkämper calculated the critical loads of cooling tower shells with various dimensions and stiffening ribs for several loading cases. Veronda and Weingarten performed computer calculations on shells previously tested experimentally, taking the actual boundary conditions into account, and they found a rather close agreement. These investigations included several loading cases and different boundary conditions. For the ratio of the experimental to the calculated critical load the following mean values and variational coefficients /i.e. standard deviations divided by the mean values/ were obtained: - vertical load: o.96/o.1o/, - lateral pressure: 1.o4/o.12/. Unfortunately, not every engineer has access to the afore-mentioned computer calculations. Moreover, it will take a long time before, on the basis of numerical examples calculated by the exact computer programmes, practical recommendations can be set up. Hence, it seems necessary to present some less accurate but easy-to-use results and methods obtained by more simple means.

There are two dominant kinds of loads acting on hyperbolic cooling towers: their own weight and wind pressure. Both loads cause both meridian-directed and hoop stresses, as contrasted to cylinders with vertical axis where their own weight does not cause any hoop stresses. Hence, the influence of both stresses on buckling should be investigated simultaneously. Since, however, it is much simpler to deal separately with the two kinds of stresses, following Krätzig [10] we shall deal separately with the two basic loading cases. In case /a/ no stress arises in the meridian direction, while in case /b/ there are no hoop stresses. We shall denote the critical hoop force pertaining to case /a/ with $n^o_{y,cr}$, while the critical meridian force of case /b/ will be called $n^o_{x,cr}$. The superscript "zero" refers in each case to the zero value of the other normal force.

The theoretical and experimental investigations [12], definitely showed that in the case of simultaneous action of both normal forces the condition of stability can be expressed /slightly to the benefit of safety/ by the Dunkerley theorem:

$$n_x/n^o_{x,cr}+n_y/n^o_{y,cr} = 1. \qquad /1/$$

The boundary conditions at the upper and lower edges of the hyperbolic shell, to be used in stability investigations, depend on the structural solution of the cooling tower, and are as follows: At the upper edge mostly a stiffening ring is applied, due to which the following boundary conditions apply for the buckling of the shell: $w=\partial^2 w/\partial x^2=v=n_x=0$. During erection /and also in the case of smaller towers/ there is no stiffening ring, so that the boundary conditions become: $\partial^2 w/\partial x^2 = \partial^3 w/\partial x^3 = n_x = n_{xy} = 0$.

At the lower edge, in most cases, also a stiffening ring is applied, sometimes realized by the thickening of the shell wall. The ring or the lower edge is mostly supported by a truss resting on the soil, through which the cooling air can stream into the tower.

Hyperbolic shells have four characteristic buckling modes. In the case of a free upper edge, the specific distribution of the wind pressure may cause a buckling pattern shown in Fig. 1/d, which we shall call free-edge buckling.

The meridian-directed forces may cause local buckling in a diamond-shaped pattern shown in Fig. 1/b. This phenomenon is similar to the buckling of cylinders under compression. The boundary conditions /even comparatively weak edge supports/do not influence this buckling mode, as a rule, since in the vicinity of the upper edge ring the meridian-directed forces are small so that they cause no local buckling here, while near the lower edge the usual thickening of the shell wall prevents local buckling.

The meredian-directed forces may also cause an axisymmetric buckling shown in Fig. 1/a, which is again similar to the case of the axially compressed cylinder.

Finally, a reticulated buckling pattern, extending to the entire surface, has to be mentioned, which we call overall buckling /Fig. 1/c/. This buckling mode is markedly influenced by the boundary conditions.

In the following we shall treat, in turn, the critical loads pertaining to each buckling mode. The notations are explained in Fig. 2.

The b u c k l i n g o f t h e f r e e e d g e is caused by the lateral pressure of the wind. It may be onerous during erection when the strength of the concrete is still low and the upper stiffening ring is not yet built.

Most authors set up the following formula for the critical wind pressure /i.e. the maximum pressure value of the circumferential distribution of the wind load/: $p_{cr} = CE[t/R_o]^{7/3}$ /2/
where R_o is the throat radius.

For the factor C they suggest values on the basis of model tests, depending on the geometric ratios /principal curvatures and height to throat radius/ of the shell:

Der and Fidler [2]: C=o.o6o~o.o77;

Walther and Wölfel [15]:

- taking extreme cases into account: C=o.o48~o.3o9;
- considering practical cases only: C=o.o72~o.22o.

The discrepancies between these values are caused by the various support conditions of the lower edge and by the different geometric ratios of the shell.

Cole, Abel and Billington, and also Walther and Wölfel, investigated by more exact computer calculations the influence of various factors and compared their results with those of several other authors.

Their findings are:

/a/ If the shell has a free upper edge without any stiffening ring, wind pressure causes buckling on the upper section of the shell. Besides the weakness of the free edge, there is the additional reason that in the upper section of the tower an increased local internal wind suction develops, due to the vortices caused by the free edge. A stiffening ring applied on the upper edge eliminates the free-edge buckling, and the wind pressure becomes onerous for the overall buckling. In such cases the wind-induced buckling bulges in the middle part or in the lower half of the tower. The upper stiffening ring thus increases the critical wind load of the shell with constant wall thickness by about 1o~2o %. If the wall thickness of a shell with upper stiffening ring is increased towards the bottom along the lower half of the shell by up to 3 or 4 times, as is usual in cooling tower practice, the critical wind pressure augments by about 5o~1oo % as compared to the shell with constant wall thickness.

/b/ As compared with the uniform lateral pressure, the actual wind pressure distribution causes an increase of 2o~3o % in the critical load /referred to the peak value of the wind pressure/.

/c/ Constant wind profile along the height reduces the critical pressure by about 2o % as compared with the maximum intensity of the actual wind profile, increasing upwards.

/d/ The interaction of the horizontal wind pressure p and of the vertical own weight q can be closely approximated by the straight line of Dunkerley:

$$p/p_{cr}+g/q_{cr}=1. \qquad /3/$$

/e/ The factor C in Eq. /2/ depends on the ratio of the two radiuses of curvature of the shell. In the case of a rigid lower support /u=0/, the value of C may be computed from the relation:

$$C=0.33\sqrt{R_y/R_x} \qquad /4/$$

which was obtained by approximating the results of numerical examples in the range $0.05<R_y/R_x<0.25$. Here R_x denotes the radius of curvature of the meridian line at the halfway height of the shell according to Eq. /9/, while R_y is the horizontal radius of the shell at the same height /see Fig.2/.

/f/ Hyperbolic surfaces are much less sensitive to initial imperfections than elliptic ones. This follows from the fact that the lower critical load of the geometrically perfect shell /computed by the nonlinear theory/ is only slightly less than the linear critical load.

The l o c a l b u c k l i n g caused by the own weight and the wind load together.

The compressive forces acting in meridian direction also cause hoop forces in a hyperbolic shell of revolution, so that its critical load is less than that of a substituting cylinder with the radius R equal to R_y of the hyperbolic shell /see Fig.2/

According to the investigations of Rosemeier [13], the critical loads computed on the replacement cylinder can be used for the hyperbolic shell of revolution provided they are multiplied by the reduction factor: $1/[1+R_y/R_x]$. /5/

Thus, we arrive at the following results:

$$n_{x,cr}^{lin}=\left(0.6Et^2/R_y\right)/\left(1+R_y/R_x\right). \qquad /6/$$

In practical cases, the local buckling analysis has to be carried out at various heights, since the wall thickness as well as the stresses vary along the height.

A x i s y m m e t r i c b u c k l i n g was investigated by Kohli [7]. He derived for the critical value of the meridian force the expression:

$$n_{x,cr} = \lambda E t^2 / R_y . \qquad /7/$$

For R_y sec Fig.2. The factor λ has the value o.96~o.93 for shells with an upper stiffening ring and o.46~o.43 for unstiffened shells. The higher λ-values were obtained for the linear critical load while the lower values give the lower critical load /of the geometrically perfect shell/ obtained by the nonlinear theory .

Since these values exceed those of the other buckling modes, axisymmetric buckling, as a rule, need not be investigated.

The linear critical load of the o v e r a l l b u c k l i n g , which extends to the entire surface, is usually described by the following formula, similar to Eq. /7/:

$$n_{x,cr} = \lambda E t^2 / R_y . \qquad /8/$$

For the factor λ, various researchers obtained different results from the evaluation of model tests:

Krätzig: λ=o.o79; Rosemeier: λ=o.o7~o.lo; Der and Fidler: λ=o.18; Wianeczki: λ=o.12; Mateja: λ=o.24~o.27.
The lowest value among the above results is only one fourth of the highest one.

As possible reasons for these discrepancies we may mention the different geometric ratios and boundary conditions occurring in the model tests and computations of the various authors.

Krätzig [10] calculated numerical tables, from which we can establish the following expression for λ:

$$\lambda = /1.13 \sim 1.85/ \sqrt[3]{t/[4R_o]} .$$

His tables make it possible to compute the values of $n_{x,cr}^{lin,o}$ and $n_{y,cr}^{lin,o}$ separately, as functions of the ratio throat to lower radii R_o/R_1 and of the dimensionless height factor ξ_1, sec Fig.2. Unfortunately, these tables do not allow the influence of the dimensions of the upper part of the tower /above the throat/ and of the ratio R_y/t to be properly taken into account. The critical loads computed by these tables give results that are 3o~5o % higher than those obtained by exact computations confirmed by model tests. Due to all these circumstances, we now present, on the basis of [5] the results of an approximate method which takes into consideration the influence of the geometric parameters. The main point of the approximation is that the investigation is carried out on a hyperbolic paraboloid shell by which we replace the hyperbolic shell of revolution in question. Since the hypar is a translational surface and, therefore, has no twist, it is less rigid than hyperbolic shell of revolution, and has a somewhat lower critical load, which is, however, much easier to determine. Its radius of curvature in the y direction R_y is assumed to be equal to the horizontal radius R_y of the shell of revolution at its half way height, i.e. at the mid-distance between the two edges /see Fig.2/. For the other radius of the hypar R_x we have chosen, at the half-way height of the tower, the radius of curvature of the flat parabola arc which lies in common plane with the axis of revolution and passes through the end points of R_u, R_y, and R_ℓ /see Fig.2/ i.e.:

$$R_x = \frac{o{,}25\ H^2}{R_u + R_\ell - 2R_y} . \qquad /9/$$

By using these radiuses, the curvatures of the replacement hypar closely approximate the average curvatures of the hyperbolic shell of revolution in question.

We can write the equation of the replacement hypar surface as follows:

$$z = y^2/2R_y - x^2/2R_x . \qquad /1o/$$

The boundary conditions are:-along the upper edge /stiffening ring/: $w=\partial^2 w/\partial x^2 = v = n_x = 0$, -along the lower edge /rigid support/: $w=\partial^2 w/\partial x^2 = v = u = 0$.

We performed this computation and calculated the critical load in [5] according to the rules of the energy method. The following expressions for the critical meridian and hoop normal forces were obtained:

$$n_{x,cr}^{lin,o} = \lambda_x^o\ Et^2/R_y ; \quad n_{y,cr}^{lin,o} = \lambda_y^o Et^2/R_y . \qquad /11/\ /12/$$

It was found by several nonlinear investigations on the overall buckling that the lower critical load /of the perfect shell/ is only a few per cent /maximum 5%/ lower than the linear critical one. Hence, knowledge of the latter seems to be sufficient to perform the stability investigation of hyperbolic shells of revolution.

The numerical values of the factors λ_x^o and λ_y^o are compiled in Tables 1 and 2, Since by linear interpolation we would commit an error to the detriment of safety, we recommend a parabolic interpolation between the values of the tables.

Due to the variation of the meridian directed force, its value valid at the halfway height of the tower should be compared with that of the force. The interaction of the meridian and hoop forces can be accounted by the Dunkerley formula.

We compared the critical forces $n_{x,cr}^{lin}$ given by the exact calculations and verified by model tests with those computed with the λ-values of Tables 1 and 2, and found that the λ-values of the tables yield results about 10% lower than the more exact ones.

Values of the factor λ_x^o Table 1.

$n_{x,cr}^{o,lin}=\lambda_x^o Et^2/R_y$.

$R_y/t=$	100			200			300		
$H/R_y=$	2	4	6	2	4	6	2	4	6
$R_x/R_y=2$	.224	.137	----	.180	.113	----	.157	.160	----
4	.289	.188	.183	.240	.159	.155	.209	.133	.174
6	.344	.255	.192	.276	.193	.170	.249	.172	.146
8	.392	.264	.238	.333	.228	.170	.286	.192	.161
10	.424	.294	.246	.364	.257	.208	.329	.234	.183

Values of the factor λ_y^o Table 2.

$n_{y,cr}^{o,lin}=\lambda_y^o Et^2/R_y$.

$R_y/t=$	100			200			300		
$H/R_y=$	2	4	6	2	4	6	2	4	6
$R_x/R_y=2$	.115	.070	----	.087	.067	----	.080	.081	----
4	.092	.055	.040	.060	.040	.034	.056	.034	.038
6	.062	.040	.036	.057	.034	.026	.046	.028	.022
8	.057	.039	.029	.044	.029	.022	.045	.026	.024
10	.057	.035	.025	.040	.026	.022	.035	.024	.018

The snap - through behaviour of homogeneous, elastic shells

When considering a shell of perfect geometry, the differential equations of equilibrium and compatibility valid for small deflections lead to an eigenvalue problem. The lowest eigenvalue obtained is the so-called linear /"classical"/ critical force of the shell /N_{cr}^{lin}/. The surface load causing this force is the linear critical load of the shell /p_{cr}^{lin}/.

In most cases the carrying capacity of the perfect shell decrease until the deflection due to buckling reaches a certain value w_{cr} . The force and load belonging to w_{cr} are called the lower critical force N_{cr}^{lower} and the lower critical load p_{cr}^{lower} respectively /this notation was used by KÁRMÁN/. However, not even this value of p_{cr}^{lower} was confirmed by experiments. Failure loads appeared between the linear critical load and the lower critical one. Further investigations found that the reason why the shells buckled at a load smaller than the linear critical one was their initial imperfection.

The maximum amplitude of imperfection will be denoted by w_o. Investigations on shells with initial imperfections were carried out as well. The essence of the results was that the initial deformation increases with the load intensity. The load intensity reaches a maximum value, and, after that, it decreases with increasing deformation /"snap-through phenomenon"/. This maximum load is denoted by N_{cr}, or p_{cr} /upper critical load/. Actually, the shells buckle at the upper critical load /see Fig.3./. The ratio of the upper to the linear critical load of homogeneous shells,

$$\varrho = p_{cr,el}/p_{cr}^{lin} , \qquad /13/$$

for cooling tower shells can be represented by the curves of Fig.4, plotted against the ratio w_o/t.

The stability condition of reinforced concrete shells

The s a f e t y of the r.c. shell a g a i n s t b u c k l i n g is sufficient, if $p < p_{cr}/\gamma$.

Here p is the actual load acting on the shell, p_{cr} is the critical load of the shell, and γ is the factor of safety from [8] /see Table 4./. The behaviour of the r.c. shell is influenced by the geometric data of the shell and of the cross sections, and by the material properties of both the concrete and the reinforcement. On the whole, reinforced concrete differs from the elastic homogeneous material in the following points:

- the compressed concrete zone creeps,
- the tension zone of the concrete cracks, thus the stiffness of the cross section drops, and the position, quantity and quality of the reinforcement play an important role,
- the materials of the shell behave elasto-plastically.

The method to be outlined in the following takes this effects into account.

The critical load of the r.c. shell can be written in the following form [8] : $p_{cr}= \zeta \varrho_{rc}\, p^{lin}_{cr(\varphi)}$. /14/

Here ζ represents the effects of plasticity and ϱ_{rc} represents the effect of snap through caracter, craks and reinforcement respectively, and $p^{lin}_{cr(\varphi)}$ is the classical critical load of the shell made of elastic material, which depends from the φ of the creep factor.

The e f f e c t o f c r e e p may be estimated by reducing the value of the modulus of elasticity, E_{co}, of concrete according to the formula $E_c=E_{co}/(1+\varphi_c)$. /15/

Here E_c is the modulus of elasticity of the concrete reduced due to creep, $E_{co}=675o\sqrt{f_c'}$ /N/mm^2/ /16/

is the initial modulus of elasticity /without creep/, f_c' is the cylinder strength of the concrete, and $\varphi_c=4-2\log_{1o}(f_c')$ is the final value of the creep factor. /17/

If only a part of the load p_o is acting from beginning, while another part p_t begins to act only at a later date t, we may reduce the creep factor φ_c accordingly. By so doing we arrive at the following approximate formula for the reduced creep factor:

$$\varphi_c'= \varphi_c\left(1+o.75k\ p_t/p_o\right) / \left(1+p_t/p_o\right). \qquad /18/$$

The coefficient k represents the decreasing propensity of the concrete to creep with increasing time. It has the value 1.8 if the shell is loaded at the age of one week, and 1.o if it is loaded at the age of one month, while it decreases to k=o.5 for concretes loaded when they are older then one year.

Finally we can write: $p_{cr}^{lin}(\varphi_c) = p_{cr}^{lin} / (1+\varphi_c)$. /19/

If some of the loads act only for a short term, such as wind load or the loads due to seismic activity, the modulus of elasticity should be only partially reduced in proportion to the ratio of short-term load to all loads.

The initial imperfection consists of two parts. One is the calculable imperfection to be computed by the bending theory of shells. Its amplitude will be denoted by $w_{o,calc}$. The other part is the accidental imperfecion due to inaccuracies of erection, the amplitude of which will be denoted by $w_{o,accid}$. The following empirical formula can be proposed for accidental imperfection:

$$w_{o,accid} = 0.05\ t + \frac{R}{2000}\left(\frac{a}{R/1000t + 1000t/R}\right), \qquad /20/$$

where t and R are the thickness and the radius of curvature. The factor "a" represents the influence of the accuracy of the erection method. For example we can assume a=1 for r.c. shells with rigid formwork, while for sliding shuttering we may take a=6.

The coincidence of the maximum values of both imperfections is rather improbable. Thus, according to the rule of probability theory, we may take the following value for the design imperfection w_o:

$$w_o = \sqrt{w_{o,calc}^2 + 1.4\ w_{o,calc}\ w_{o,acc} + w_{o,acc.}^2} \qquad /21/$$

/It should be remarked, that w_o is not the maximum imperfection amplitude, but the mean value of the absolute values of the maximum amplitudes, i.e. practically half the maximum measurable amplitude. The deviation from w_o is covered by the safety factor, since the scatter of the critical load is mostly caused by the deviation of the imperfection amplitude from w_o/.

The relation between imperfection and eccentricity may be to analyse as follows. If we impose a small deformation w onto a shell with a given geometry and state of stresses, we can determine the pertaining bending moment and the change in the membrane forces at any point and in any direction with the aid of the classical bending theory. Dividing the bending moment by the modified value of the membrane force we arrive at the magnitude of the eccentricity e_o, which is not necessarily equal to the imperfection amplitude w_o. Performing this investigation for several shell surfaces and for various states of stresses in hyperbolic shells we obtained the value for the ratio e_o/w_o=o.5o.

The effect of reinforcement and cracking may be taken into account as follows.
We calculate the quotient $n\mu = E_sA_s/E_cA_c$ /22/,

where the area of steel A_s is taken in one direction only. From Table 3 we obtain the value of coefficient ψ corresponding to the ratio n of the moduli of elasticity of steel and concrete, to the percentage μ of steel related to the concrete cross section area, and to the arrangement of reinforcement.

Values of factor ψ Table 3.

$n\mu$	o	o.o5	o.1o	o.15	o.2o	o.3	o.4	o.5
η=o.2	o	o.178	o.285	o.373	o.453	o.579	o.73o	o.85o
η=o.5	o	o.139	o.212	o.269	o.316	o.393	o.457	o.51o

In the case of a single layer reinforcement /placed in the middle of the thickness/ η=o.5, while for a double layer one /placed on the two sides of the cross section/ we may take $\eta\approx$o.2. Here η is the concrete covering of the reinforcement related to the wall thickness.

Then, for a given value of ψ and of the imperfection to thickness ratio w_o/t, the reduced value of the upper critical load $p_{cr,el,rc}$ for the r.c. shell can be obtained from the relation

$$p_{cr,el,rc} = \varrho_{rc}\, p_{cr}^{lin}/(1+\varphi_c) \qquad /23/$$

where values of ϱ_{rc} are compiled from [8] in Table 4. In the Table 4. the ϱ(o.5) is the value of ϱ for w_o/t=o.5/sec Fig.5./

If the reinforcement, i.e. the rigidities, are not equal in the x and y directions, the shell can nevertheless be considered as approximately isotropic for practical purposes using the average factor $\varrho_{rc,s} = \left(\varrho_{rc,x}\ \varrho_{rc,y}\right)^{1/2}$. /24/

Values of ϱ_{rc} and the safecty factor γ Table 4.

Buckling form	ψ	$\frac{w_o}{t}$=o	o.1	o.2	o.4	o.6	1.o
Local buckling	1.o	1.oo	o.63	o.45	o.29	o.22	o.14
	o.6	1.oo	o.62	o.42	o.22	o.14	o.o9
ϱ(o.5) $\approx$ o.25	o.4	1.oo	o.62	o.4o	o.18	o.1o	o.o6
	o.2	1.oo	o.62	o.38	o.14	o.o6	o.o3
γ=3.o	o.o	1.oo	o.62	o.37	o.1o	o.o2	o
Axisymmetric,	1.o	1.oo	o.98	o.96	o.92	o.88	o.82
free edge, and	o.6	1.oo	o.84	o.72	o.59	o.54	o.49
overall buckling	o.4	1.oo	o.76	o.6o	o.43	o.36	o.33
ϱ(o.5) $\approx$ o.9o	o.2	1.oo	o.69	o.48	o.26	o.19	o.16
γ=2.6	o.o	1.oo	o.62	o.37	o.1o	o.o2	o

Plastic properties of materials may be taken into account according to [8] with the aid of the "quadratic Dunkerley interaction formula", $\zeta = p_{cr}/p_{cr,el,rc} = \left\{1/\left[1+\left(P_{cr,el,rc}/P_{pl}\right)^2\right]\right\}^{1/2}$. /25/

References

1. Abel, J.F. - Billington, D.F.: Buckling of Cooling Towers. Journ. Struct. Div. ASCE. ST1o /1982/ 2162-2174.

2. Der, T.I. - Fidler, R.: A Model Study of the Buckling Behaviour of Hyperbolic Shells. Proc. of the Institute of Civil Engineers London, 41. /1968/ Szept. 1o5-118.

3. Dulácska, E. : Buckling of Reinforced Concrete Shells. Journ. Struct. Dic. ASCE Dec. /1981/ 2831-24o1.

4. Dulácska, E. : Explanation of the chapter on stability of the "Recommendations for Reinforced Concrete Shells and Folded Plates, 1979". and a proposal to its improvement. IASS Bulletin No.77 /1981/

5. Dulácska, E.-Nagy, J.-Bódi, I.: Overall Buckling of Hyperbolic Shells of Revolution with Unmovable Lower Edges. Acta Technica Sci.Hung. 92 /1981/ 167-187.

6. Hayman, B. -Chilver, A.R.: The Effect of Structural Degeneracy on the Stability of Cooling Towers. Proc. IASS Colloquium on Cooling Towers. Brussels, May 1971.

7. Kohli, J.: Beitrag zum axialsymmetrischen Ausbeulen einer einschaliger Hyperboloidschale. Dissertation, Karlsruhe, 1968.

8. Kollár, L.-Dulácska,E.: Buckling of Shells for Engineers. John Wiley and Sons, Chichester /1984/.

9. Krätzig, W.B.: Statische und dynamische Stabilität der Kühlturmschhale Naturzug - Kühltürme. Festigkeitsberechung und Konstruction Tagung von 19.Ápril. 1968. Vulkan-Verlag. Dr.W. Classen, Essen.

1o. Krätzig, W.B.: Grosse Naturzug Kühltürme aus Stahlbeton VBG Kraftwerkerstechnik 55. /1975/ H.3. 191-197.

11. Mateja, O.: O badaniach nod statecznescia Hyperboloidalnych chlodni wiezonych. Inzynieriai Budownictwo. 23.No.11 /1966/ 428-431.

12. Mungan, I.: Buckling stress states of Hyperboloidal shells. Journ. Struct. Div. ASCE.1o2 ST 1o, 2oo5-2o2o.

13. Rosemeier, G.: ZurStabilität von Hyper und Hyperboloidschalen. Bauingenieur, 48. /1973/ 437-444.

14. Veronda, D.R.-Weingarten, V.I.: Stability of Hyperboloidal Shells. Journ. struct.Div. ASCE ST7 /1975/ 1585-16o2.

15. Walter, H.J.-Wölfel, R.: Stabilitätsverhalfen hyperbolischer Köhlturmschalen und Windbelastung.Bauplanung-Bautechnik 32 /1978/ 51o-512.

16. Wianecki, I.: Stabilité d'une coque en forme d'hyperboloide de revolution sous des charges uniformément reparties sur les bords et symetriques rapport a son axe. Cahiers de la recherche, nr.19.Eyrolles, Paris /1963/

17. Wittek, U.: Überlick un theoretische Einführung in das Stabilitätsverhalten von Kühlturmschalen. Konstruktiver Ingenierbau Berichte, Bochum. 1977. Heft 29/3o /63-7o/

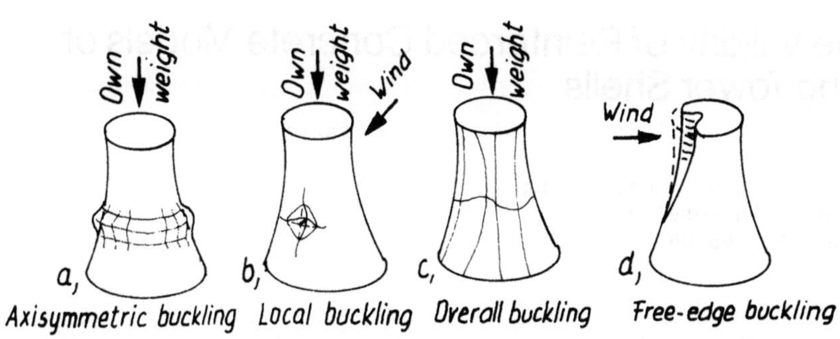

FIG. 1. The buckling modes of the hyperbolic shells of revolution

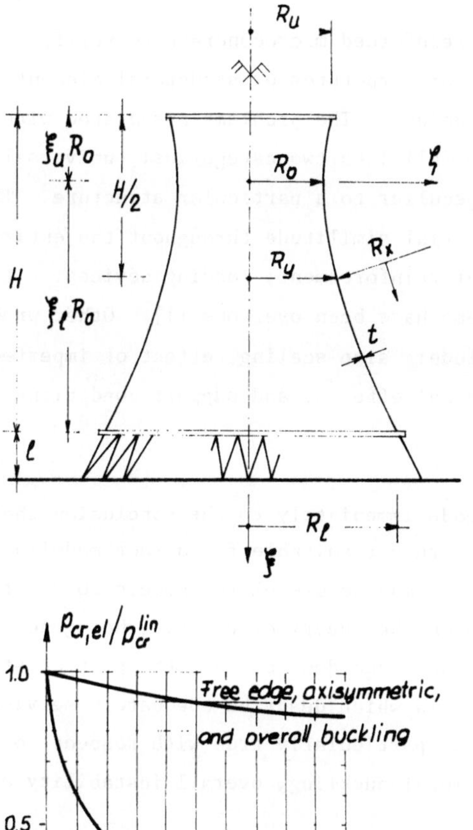

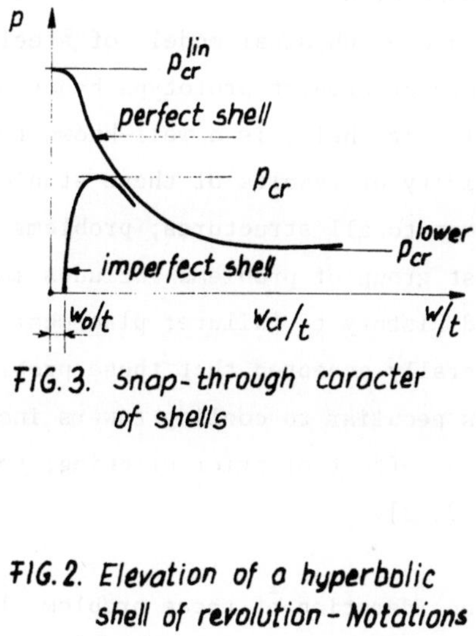

FIG. 3. Snap-through caracter of shells

FIG. 2. Elevation of a hyperbolic shell of revolution - Notations

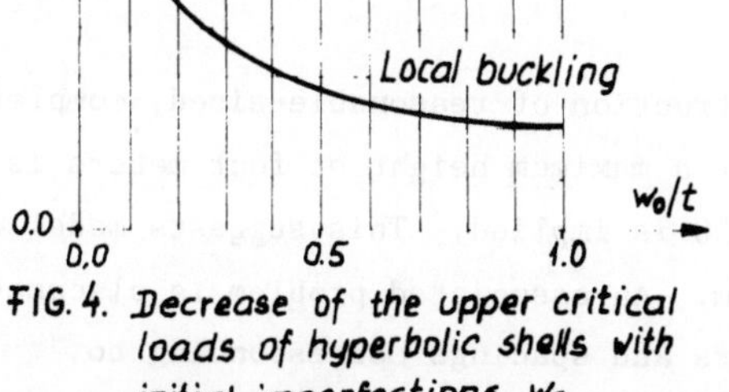

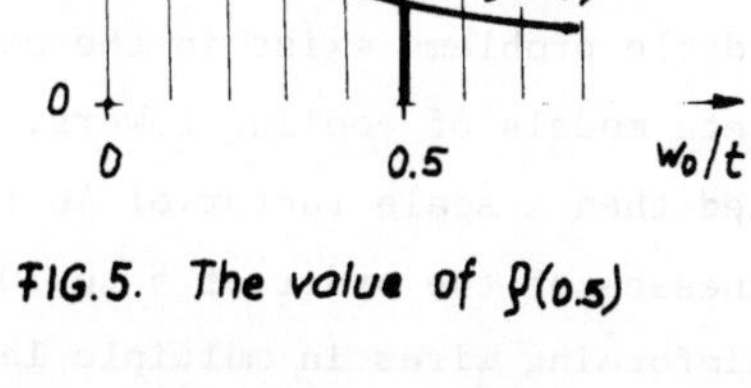

FIG. 4. Decrease of the upper critical loads of hyperbolic shells with initial imperfections w_0

FIG. 5. The value of $\rho(0.5)$

On the Validity of Reinforced Concrete Models of Cooling Tower Shells

Stuart E. Swartz, Ali Nikaeen, Hossein Mozaffarian

Department of Civil Engineering
Kansas State University
Manhattan, KS 66506

INTRODUCTION

The use of physical models of steel-reinforced microconcrete to verify theory or predict prototype behavior of structures or structural elements including shells is a well-known technique. The problems associated with the validity of results of these studies fall into two categories: universal or common to all structures; problems peculiar to a particular structure. The first group of problems includes material similitude throughout the entire load history to failure; placement of reinforcement; loading devices. It is generally accepted that these problems have been overcome [1]. Other problems peculiar to cooling towers include: size scaling, effect of imperfections, effect of prior cracking; thermal effects; and support conditions [1, 2, 3].

A consideration of these problems leads immediately to the conclusion that materials such as metals or plastics are not suitable for direct modeling of concrete cooling towers. However, they may be useful for models to verify formulas or predictive methods based on the theory of elastic, homogeneous and isotropic bodies. In fact, this has been done but now the present state-of-the-art demands more realistic models which describe adequately service level response and collapse. This is particularly true with respect to determining the mode of failure, eg., local buckling, overall instability or some other mode.

Formidable problems exist in the construction of reasonable-sized, complete concrete models of cooling towers. If a maximum height of four meters is adopted then a scale factor of 40 to 50 is implied. This suggests model wall thicknesses of the order of 5 to 10 mm. An associated problem is placement of reinforcing wires in multiple layers and spacings corresponding to recommendations of various practices.

In the writers' experience the major problem to be addressed with respect to modeling service level response and failure in these shells is that of achieving accuracy in overall geometry and wall thickness. This paper discusses some aspects of these.

ELASTIC MODELS

Following the failures of some cooling towers in the 1960's and 70's considerable interest developed in the prediction of the performance of these shells. A summary of these efforts is included in References 2 and 4.

With respect to possible buckling failure modes the extensive experimental studies by Der and Fiddler [5] may be mentioned. These models were constructed of electroformed copper and also PVC with a scale factor of 1/275. The copper models were of very uniform thickness while the PVC models had a thickness variation of about 15%. Failure occurred during wind tunnel tests in a snap-through mode with very large deformations. The mode of failure was clearly influenced by supported and free edge boundary conditions.

Several elastic shell models of cold epoxy resin and hardener to a scale of about 1/125 were constructed and tested by Mungan [6]. These were not true hyperboloids but the generating curve was approximated by a circular arc. The models were machined using a special lathe. Reported thicknesses are given in Reference 6 and relative thickness variations for the various models are given in Table 1. It is seen that the maximum thickness deviations were +26, -22% (for Model 5). These models were tested with combinations of lateral pressure and axial load. Unlike the Der and Fidler models, these models buckled in a relatively localized region. In part as a result of these tests, a localized buckling theory - the BSS approach - has been proposed [7].

One other model tested - which has not been reported in the literature - is one built by the Marley Co. of glass fiber reinforced polyester [8]. The scale was 1/50 with a height of about 10 ft. and thickness of 1/8 in. + 1/32 in. (ie deviation in thickness of 25%). This model was tested with a vacuum and experienced localized buckling.

CONCRETE MODELS

The construction of a concrete model of practicable proportions implies the prototype material cannot be used because of the very thin walls. Accordingly, a "micro-concrete" of specially-graded sand, cement and water is used.

The resulting material will generally satisfy strength, cracking and ductility requirements - eg compressive and tensile stress - strain response identical to the prototype - provided careful mix designs are done [1]. All of the models reported here satisfy this important requirement.

Possibly the first realistic model of a relatively large scale was constructed and tested at the Cement and Concrete Association in England by Best, et. al. [9]. This model had a scale of 1/25 compared to prototypes at that time with a height of 13 ft. and a nominal wall thickness of 0.2 in.

More recently, a large scale model was constructed by Mostafa [10] and coworkers using precast elements connected together by post-tensioned, steel cables. The model height was 8 ft. and a nominal thickness of 0.5 in. was used. Thus, the model thickness was not scaled in the same proportion as the height. Testing to simulate wind load effects was conducted during different stages of construction. The measured strains agreed well with theoretical predictions but measured displacements were considerably greater than those predicted. The author felt that this was due to the segmental construction technique which resulted in a structure that was not monolithic. The author also stated the precasting method resulted in very accurate geometric properties - including thickness - but data were not given.

The last model to be discussed is that constructed at Kansas State University. This is shown in Fig. 1. The model was cast in three lifts using a double-walled form [11]. Form separation during placement of the mix in the bottom lift resulted in very large thickness errors as shown in Table 2 but the other two lifts have much more uniform thickness. Deviations from average thickness for lifts 2 and 3 are given in Table 3. Thickness contours for these lifts are given in Fig. 2. The maximum thickness deviations from average are +82% and -60%.

The maximum deviation in horizontal radius was ±3% and the maximum deviation in a vertical distance along a meridion was ±2.5% based on slope.

A complete deflection map was obtained for different load levels using dial gages and a technique described in Reference 12. The applied loading was uniform lateral pressure achieved by suction. Deflections obtained showed an asymmetrical pattern at all loads even though loading was perfectly symmetrical. This behavior has not been reported by others - actually no deflection maps have been given by other investigators - and may be attributed directly to the imperfections present in the model.

When failure occurred it was by local buckling followed immediately by an implosion [12]. This was clearly related to local imperfections and lends credence to the BSS approach [7]. However, it should be noted that clearly it is impossible to model imperfections that may be present in the prototype and therefore it is not possible to extrapolate these results. It is possible to apply methods of analysis to the model, treating it as a miniature prototype and including measured geometry and material properties. Sufficient data were obtained to allow this.

CONCLUSIONS

1. The use of microconcrete models is a feasible means to investigate service level response and failure of hyperboloidal cooling tower shells.
2. Overall control of the shell geometry can be obtained with the same level of accuracy as the prototype. However, percent variations in thickness probably will exceed industry standards for prototypes - perhaps by a considerable amount.
3. Extensive data on geometry, thickness, displacements and strains can be taken with a model more economically than can be done on the prototype.
4. Because of imperfection sensitivity especially in thickness these models cannot be used directly to predict local buckling failures in prototypes. However, buckling modes can be determined and correlations with existing and proposed analytical procedures made.

ACKNOWLEDGEMENTS

The research reported herein was supported in part by the National Science Foundation, Grant No. ENG 78-18415.

REFERENCES

1. Sabnis, Gajanan M., Harris, Harry G., White, Richard, N. and Mirza, M. Saeed., Structural Modeling and Experimental Techniques, Prentice-Hall, Inc., Englewood Cliffs, N.J., 1983.

2. Billington, David P. and Harris, Harry G., "Test Methods for Concrete Shell Buckling," Concrete Shell Buckling, ACI Publication SP-67, American Concrete Institute, Detroit, MI, 1981.

3. "Recommendations for Reinforced Concrete Shells and Folded Plates," International Association for Shell and Spatial Structures, Madrid, 1979.

4. Abel, J. F. and Gould, P. L., "Buckling of Concrete Cooling Tower Shells," Concrete Shell Buckling, ACI Publication SP-67, American Concrete Institute, Detroit, MI, 1981.

5. Der, T. J. and Fidler, R., "A Model Study of the Buckling Behavior of Hyperbolic Shells," Proceedings, Institution of Civil Engineers, Vol. 41, Sept. 1968.

6. Mungan, Ihsan, "Buckling Stress States of Hyperboloidal Shells," Journal of the Structural Division, American Society of Civil Engineers, Vol. 102, No. ST 10, Oct. 1976.

7. Mungan, Ihsan, "Buckling of Reinforced Concrete Cooling Tower Shells: BSS Approach," Journal of the American Concrete Institute, Proceedings, Vol. 79, No. 5, Sept.-Oct. 1982.

8. Ting, B.Y., "Buckling Test of a Hyperbolic Shell Subjected to Uniform Pressure," The Marley Co., Mission, Kansas (private communications).

9. Best, B. C., Evans, D. J. and Long. J. E., "The Construction and Testing of a 1/25 Scale Model of Micro Concrete Cooling Towers," Cement and Concrete Association, Great Britain, 1971.

10. Mostafa, M. T., "An Investigation of Segmentally Constructed Hyperbolic Cooling Towers," Report to the National Science Foundation from the Dept. of Civil Engineering, North Carolina State Univ., Raleigh, N.C., Nov. 1980.

11. Gates, T. E., Hu, K. K., McDonald, C. R. and Swartz, S. E., "Construction and Testing of Micro-Concrete Models of Cooling Tower Shells," International Developments in Experimental Mechanics, Society for Experimental Stress Analysis, 1982.

12. Mozaffarian, Hossein, "Testing of a Repaired Micro-Concrete Model of a Cooling Tower Shell," M.S. Thesis, Dept. of Civil Engineering, Kansas State University, Manhattan, KS 1983.

Table 1 Relative Thickness Variations in Models Tested by Mungan[6]

MODELS	1			2			3			4			5		
$\frac{Z}{H_{SM}}$	Average Thickness (mm)	+%	-%	Average Thickness (mm)	+%	-%	Average Thickness (mm)	+%	-%	Average Thickness (mm)	+%	-%	Average Thickness (mm)	+%	-%
8/9	2.18	7	9	2.23	8	6	2.06	9	8	--	--	--	1.80	12	14
7/9	2.19	7	10	2.13	10	6	2.04	12	10	--	--	--	1.74	17	14
6/9	2.18	9	9	2.03	7	7	2.00	15	12	1.51	6	7	1.70	19	18
5/9	2.09	10	8	1.87	4	6	1.85	16	16	1.53	7	8	1.56	22	17
½ (throat)	2.06	12	10	1.76	5	3	1.81	15	16	1.55	2	4	1.41	26	22
4/9	2.14	9	8	1.76	8	6	1.84	11	11	1.59	6	9	1.53	20	15
3/9	2.20	9	9	1.88	6	4	1.97	11	13	1.61	4	6	1.69	12	12
2/9	2.22	9	9	1.93	8	4	2.00	13	11	1.58	4	5	1.81	7	6
1/9	2.24	7	6	1.97	15	7	2.04	15	13	1.57	7	8	1.92	9	6

(1 mm = 0.0394 in)

Table 2. Shell Thicknesses, Shell No. 1

Thickness, inches

Merid. Location	A	B	C	D	E	F	G	H	I	J	K	L	Dist. From Throat Z, in.
1	0.41	1.02	0.37	0.56	0.71	0.48	0.45	0.58	0.72	0.51	0.25	0.68	- 33
2	0.25	0.75	0.61	0.32	0.61	0.43	0.31	0.41	0.74	0.48	0.29	0.18	- 24
3	0.61	0.54	0.42	0.75	0.97	0.58	0.79	0.78	0.85	0.58	0.58	0.73	- 12
4	0.60	0.52	0.33	0.76	0.41	0.64	0.46	0.49	0.65	0.61	0.57	0.59	0
5	0.72	0.47	0.38	0.65	0.54	0.65	0.61	0.32	0.59	0.42	0.49	0.69	+ 12
6	0.41	0.40	0.31	0.96	0.75	0.89	0.67	0.61	0.62	0.51	0.32	0.78	+ 24
7	0.33	0.61	0.42	0.72	0.86	0.81	0.42	0.62	0.65	0.36	0.51	0.73	+ 36
8	1.45	1.11	1.15	1.07	1.24	0.85	0.78	1.44	1.45	0.58	1.61	1.28	+ 48
9	0.83	1.27	1.27	0.87	1.05	1.10	0.73	1.45	0.57	0.68	0.95	0.60	+ 60
10	0.75	2.21	2.57	0.44	0.97	1.07	1.35	1.48	0.79	0.65	0.86	0.81	+ 72
11	1.07	3.30	2.70	0.91	0.72	1.28	1.31	1.81	0.78	0.52	0.68	1.01	+ 84

1 in. = 25.4 mm

Table 3. Percent Deviation From Mean Thickness for

Lifts 2 and 3 of Shell No. 1

Merid. Location	Avg. Thickness, inches	A	B	C	D	E	F	G	H	I	J	K	L
1	0.56	-27	82	-34	0	26	-15	-20	3	28	- 9	-55	21
2	0.45	-44	67	36	-29	36	- 4	-31	-9	65	7	-35	-60
3	0.68	-11	-21	-38	10	42	-15	16	14	25	-15	-15	7
4	0.55	8	- 6	-40	38	-26	16	-17	-11	18	10	3	7
5	0.54	32	-14	-30	19	- 1	19	12	-41	8	-23	-10	27
6	0.60	-32	-34	-49	59	24	48	11	1	3	-15	-47	29
7	0.59	-44	4	-28	23	47	38	-28	6	11	-39	-13	24

1 in. = 25.4 mm

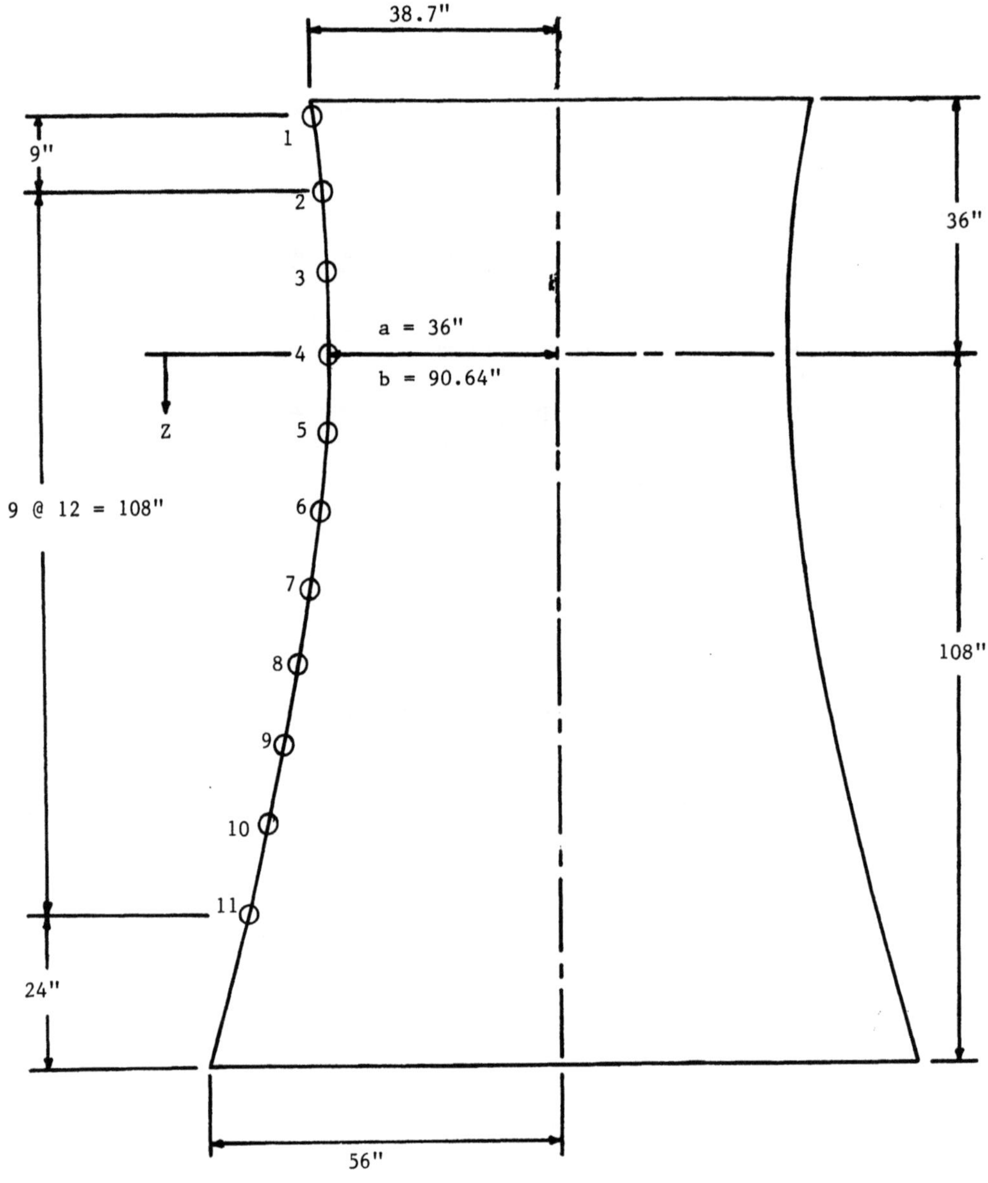

⊙ Locations for thickness, geometry, displacement data

1 in. = 0.0254m

Fig. 1. Shell Geometry of KSU Model

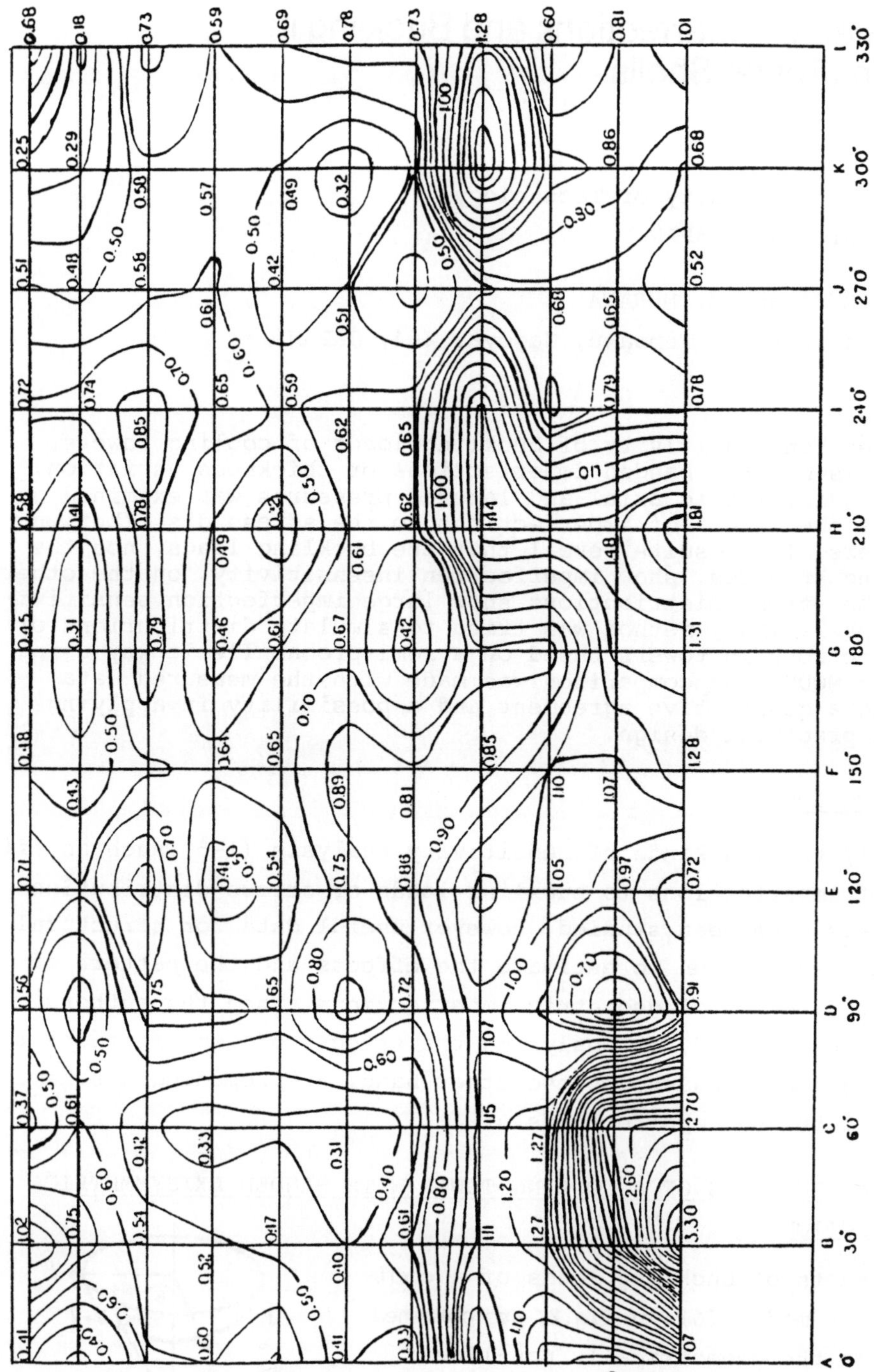

Fig. 2. Thickness Contours

Geometric Imperfections and Buckling in Cooling Tower Shells

S. KATO

Toyohashi University of Technology

Toyohashi, 440, JAPAN

A. MIYAMURA and M. MURATA

Meijo University, Tempaku, Nagoya, 464, JAPAN

Summary

Imperfection sensitivity of buckling loads of cooling towers, with axisymmetric banded imperfections or thickness variation, due to axisymmetric axial and lateral pressures was examined as well as the effect of imperfections to the stress distributions in towers. The results reveal that the buckling loads, not the buckling stresses, show imperfection insensitivity, on the other hand the stress distributions show large imperfection sensitivity. A preliminary method was tried to simulate distributions of imperfections in towers based on a semi-probabilistic way using ENTROPY MODEL. An comparison was made with the measured data, showing a qualitative agreement and a possibility in applying in the practical design.

INTRODUCTION

Recently various kinds of sensitivity analysis (1-5) such as effects of imperfections to buckling loads or stress distributions in towers have been studied, however useful data for structural design seem scarce to estimate the effects and the pattern of the distribution of geometric imperfections. Then the authors aimed, on one hand, at analysing the effects of imperfections to buckling and stresses, on the other hand, at proposing a scheme to simulate the imperfections.

BUCKLING ANALYSIS OF A COOLING TOWER WITH BANDED AXISYMMETRIC IMPERFECTIONS

The results of buckling loads of a cooling tower model (6), in Fig.2, are obtained based on the geometrically non-linear finite element analysis (7) on an assumption that the behavior was expressed by a combination of one axisymmetric and one asymmetric displacements chosen from

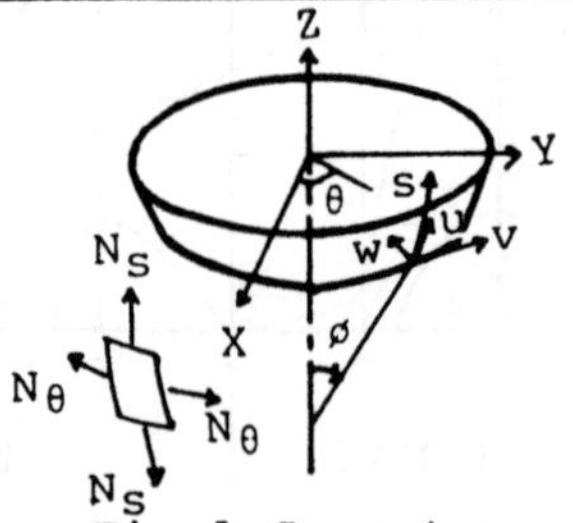

Fig.1 Geometry of shells

expanded terms in Fourier series.

$$\begin{Bmatrix} u(s,\theta) \\ v(s,\theta) \\ w(s,\theta) \end{Bmatrix} = \sum_{j=1}^{N} \begin{Bmatrix} u_j(s)\cos N_j\theta \\ v_j(s)\cos N_j\theta \\ w_j(s)\cos N_j\theta \end{Bmatrix}, \begin{Bmatrix} u_I \\ v_I \\ w_I \end{Bmatrix} = \sum_{i=0}^{M} \begin{Bmatrix} u_{0i}(s)\cos n_i\theta \\ v_{0i}(s)\cos n_i\theta \\ w_{0i}(s)\cos n_i\theta \end{Bmatrix} \quad (1)$$

where u,v,w in Fig.1, are the displacements decomposed into Fourier terms with harmonic numbers N_j and u_I, v_I, w_I are the corresponding geometric initial imperfections, decomposed into terms with harmonic numbers n_i. The assumed boundary conditions at both ends and the imperfections are listed in Table 1.

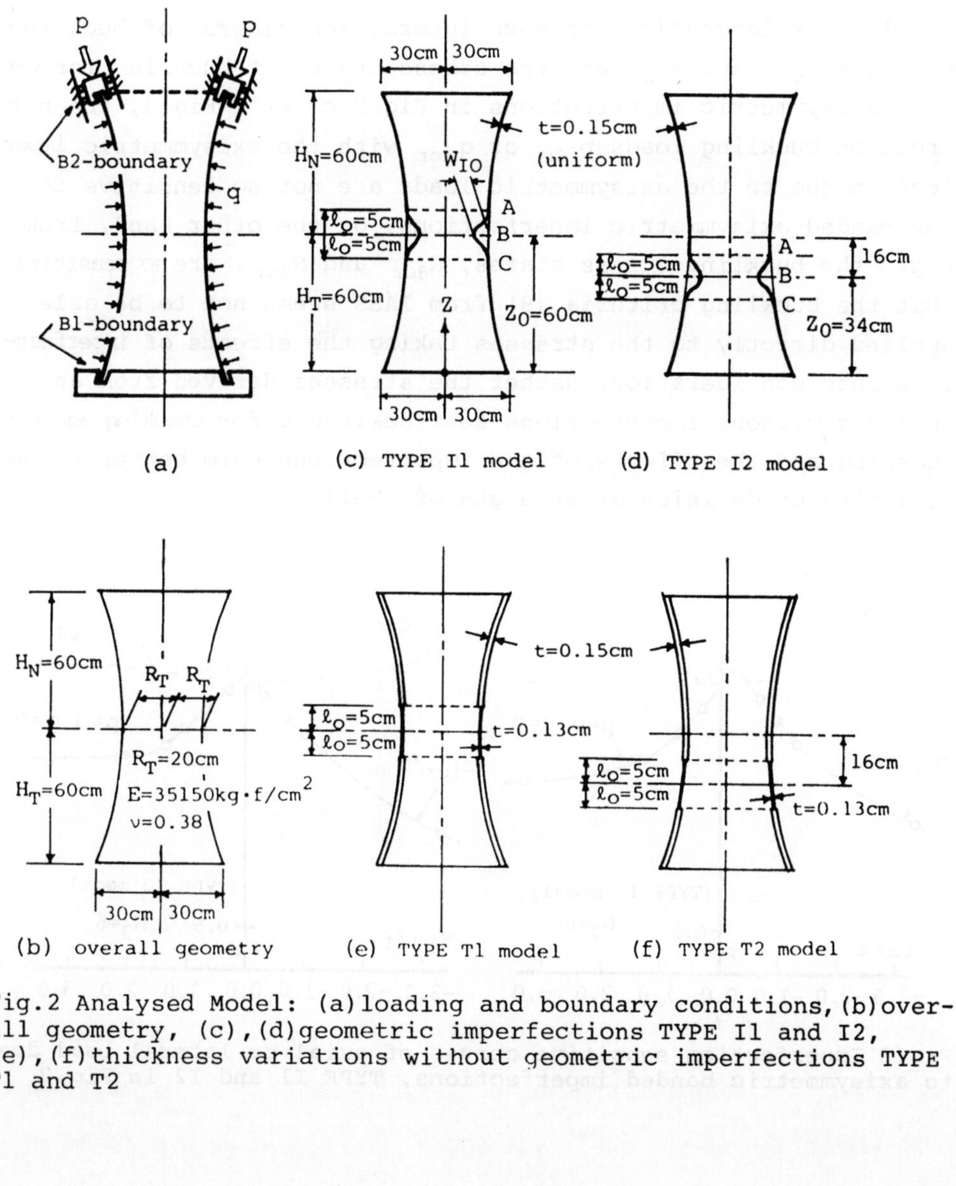

Fig.2 Analysed Model: (a) loading and boundary conditions, (b) overall geometry, (c), (d) geometric imperfections TYPE I1 and I2, (e), (f) thickness variations without geometric imperfections TYPE T1 and T2

Table 1 Boundary conditions and assumed axisymmetric imperfections

(1-a)

	B1-boundary	B2-boundary
axisymmetric deformation $N_j=0$	$u=0$ $v=0$ $w=0$ $\phi_s=0$	$u\neq0$ $v=0$ $w=0$ $\phi_s=0$
asymmetric deformation $N_j=0$	$u=0$ $v=0$ $w=0$ $\phi_s=0$	$u=0$ $v=0$ $w=0$ $\phi_s=0$

(1-b)

Axisymmetric imperfection
between AB in Fig.2 $w_{00}(z)$ $=w_{I0}[1-3(z-z_0)^2/\ell_0^2+2(z-z_0)^3/\ell_0^3]$
between BC in Fig.2 $w_{00}(z)$ $=w_{I0}[3(z-z_0+\ell_0)^2/\ell_0^2-2(z-z_0+\ell_0)^3/\ell_0^3]$

Fig.3 to 5 show the imperfection sensitivity due to the uniform axial(p) or lateral(q) loads,an interaction diagram of buckling stresses $N_{\theta cr}$ and N_{scr} and the stress states at buckling for various axisymmetric imperfections in Fig.2-c. From Fig.3, the bifurcation buckling loads p_{icr} or q_{icr} with the axisymmetric imperfection due to the axisymmetric loads are not so sensitive to the banded axisymmetric imperfections, on the other hand, from Fig.5 the buckling stress states, $N_{\theta cr}$ and N_{scr}, are so sensitive that the buckling criteria (9) from IASS seems not to be able applied directly to the stresses taking the effects of imperfections into consideration. Rather the stresses derived from an analysis without imperfections seem desirable for checking against buckling and the effects of the imperfections seem better to be reflected on decision of strength of shell.

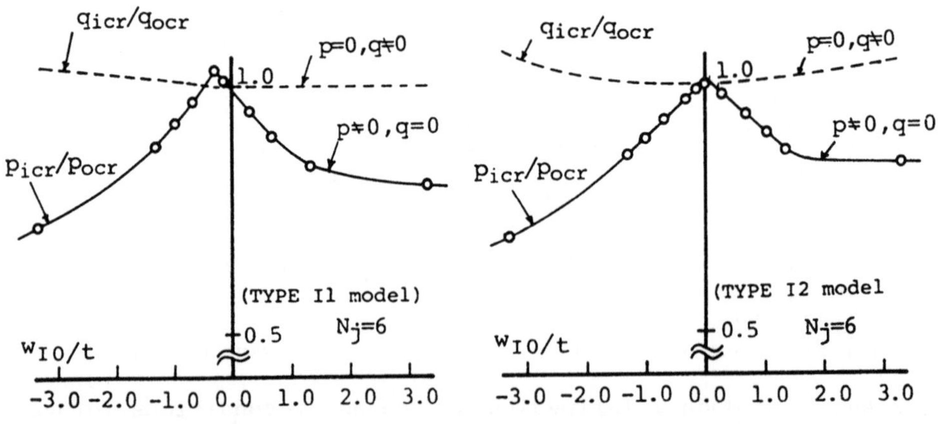

Fig.3 Imperfection sensitive curves of axial or lateral load due to axisymmetric banded imperfections, TYPE I1 and I2 in Fig.2

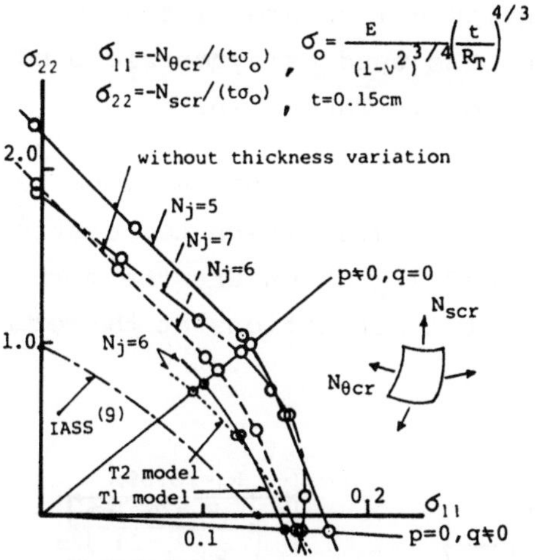

Fig.4 Interaction of buckling stresses of shells with thickness variation(T1, T2 models in Fig.2) without geometric initial displacements

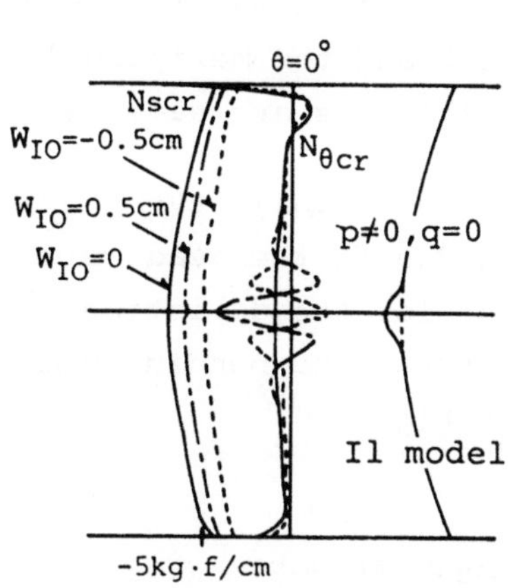

Fig.5 Stress states of buckling of shells with axisymmetric imperfection due to axial force p≠0, q=0

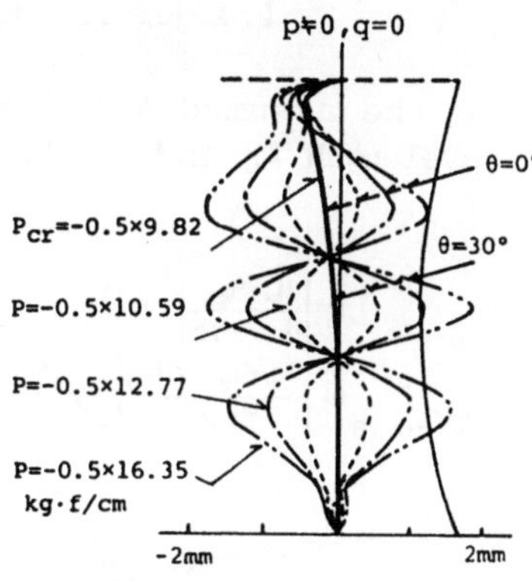

Fig.6-a Buckling deformation w(s,θ) of a shell in Fig.2-b without thickness variation and geometric imperfection (t=1.5 mm)

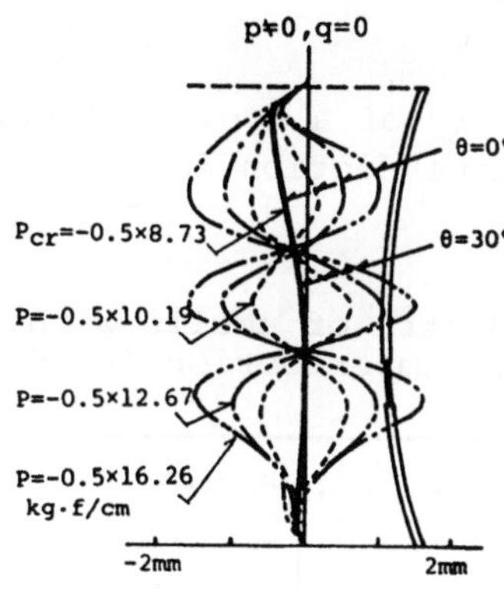

Fig.6-b Buckling deformation of a shell in Fig.2-f with thickness variation (TYPE T2 model)

EFFECTS OF GEOMETRIC IMPERFECTIONS TO STRESS DISTRIBUTIONS SUBJECT TO DEAD LOAD

The model in Fig.7 was analysed to get some information about the effects of axisymmetric geometric imperfections to stress distributions, especially to hoop stress N_θ and the meridional bending moment M_s. The assumed configurations are listed in Table 2 and the meridional shape of the used axisymmetric banded imperfections are same as the equation in Table 1-b.

Different twenty seven towers with axisymmetric banded imperfections due to the dead weight result in Fig.8, where two distinct curves for the deviations from $N_{\theta 0}$ and $M_{\theta 0}$ which are the meridional stress and bending moment without imperfections at the section BB in Fig.7.

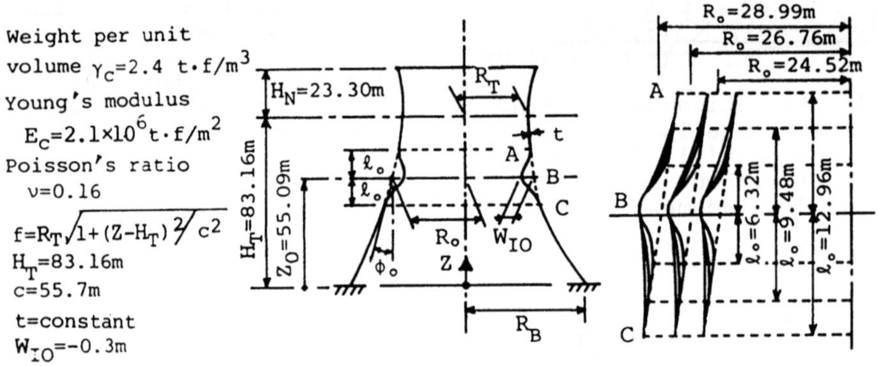

Fig.7 Geometry of a cooling tower and the assumed shape for axisymmetric imperfection for various wavelength ℓ_0 and shell radia R_0

Table 2 Geometry of shells with axisymmetric imperfections (n_i=0)

shells	ℓ_0 (m)	t (cm) (three cases)	R_T, R_0, R_ϕ, ϕ_0
A-1	6.32	15.23	R_T=21.9 m
A-2	9.48	22.85	R_0=24.52 m R_ϕ=208.3 m
A-3	12.64	30.46	ϕ_0=12.03°
B-1	6.32	15.23	R_T=23.9 m
B-2	9.48	22.85	R_0=26.76 m R_ϕ=227.3 m
B-3	12.64	30.46	ϕ_0=10.92°
C-1	6.32	15.23	R_T=25.9 m
C-2	9.48	22.85	R_0=28.99 m R_ϕ=246.3 m
C-3	12.64	30.46	ϕ_0=11.81°

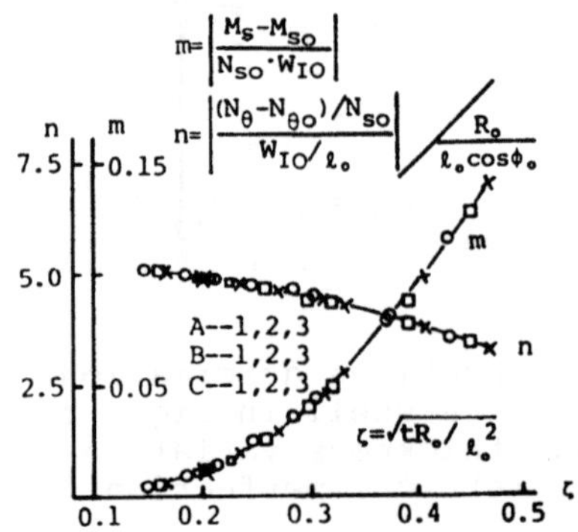

Fig.8 Deviations of stresses due to axisymmetric imperfection at BB section under the dead load

Then the non-dimensional terms n and m with ζ in Fig.8 seem useful to estimate the effects of the axisymmetric banded imperfections.

On the other hand, Fig.11 shows the effects of the asymmetric geometric imperfections to N_θ and M_s. The analysed towers are shown in Fig.9 and Table 3 where only one term, in Eq.(1), was used as the asymmetric imperfection with a harmonic number n_i and one additional parameter ℓ/ℓ_0 The meridional configuration of the imperfection is the same one as the equation in Table 1-b except n_i. Fig.10 shows the distributions of stresses due to the dead load of the concrete shell, expressing that the effect to the circumferential bending moment M_θ is also so large that the meridional cracks along the meridian may occur with a high possibility together due to a high hoop tension. Fig.11 shows also that the effect of asymmetric imperfections is especially prominant in the meridional bending moment M_s. Together with Fig.10, Fig.11 shows the necessity of a research into a qualitative effect of local imperfections with high harmonic numbers.

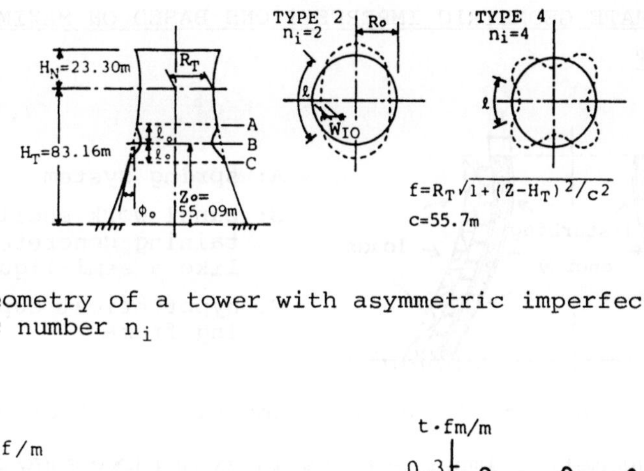

Fig.9 Geometry of a tower with asymmetric imperfection of a harmonic number n_i

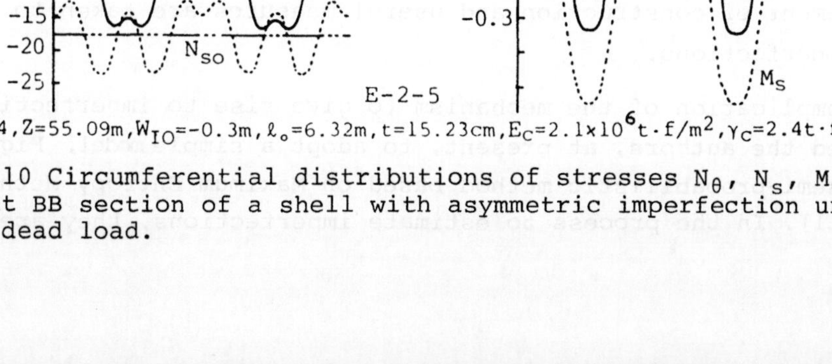

$n_i=4, Z=55.09m, W_{IO}=-0.3m, \ell_o=6.32m, t=15.23cm, E_C=2.1\times10^6 t\cdot f/m^2, \gamma_C=2.4t\cdot f/m^3$

Fig.10 Circumferential distributions of stresses N_θ, N_s, M_θ and M_s at BB section of a shell with asymmetric imperfection under the dead load.

Table 3 Geometry of shells with asymmetric imperfections
R_0=24.52m, ϕ_0=10.03, R_T=21.9m
w_{IO}=-0.3m

shells		t(cm)	ℓo(m)	N	ζ	ℓ/ℓo
Type-2 (nj=2)	D-1	7.88	6.32	5	0.220	6.09
	D-2	15.23		5	0.306	6.09
	D-3	32.99		5	0.450	6.09
	D-4	31.54	12.64	5	0.220	3.05
Type-4 (nj=4)	E-1	7.88	6.32	6	0.220	3.05
	E-2-5	15.23		5	0.306	3.05
	E-2-6			6	0.306	3.05
	E-2-7			7	0.306	3.05
	E-3	32.99		6	0.450	3.05
	E-4	8.25	3.16	7	0.450	6.09

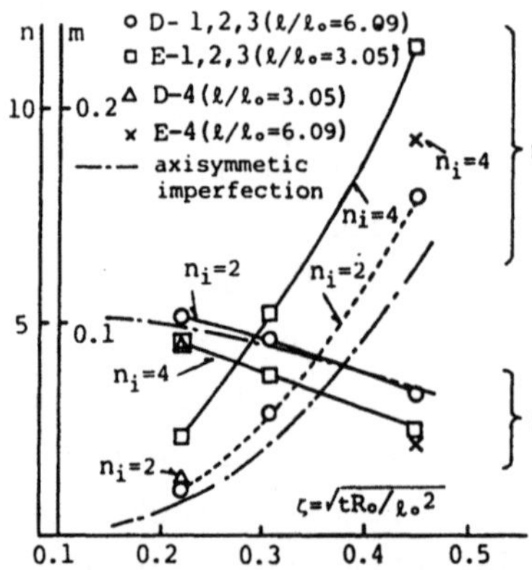

$$m=\left|\frac{M_s-M_{so}}{N_{so}\cdot W_{IO}}\right|,\quad n=\frac{|(N_\theta-N_{\theta o})/N_{so}|}{|W_{IO}/\ell_o|}\Big/\left(\frac{R_o}{\ell_o\cos\phi_o}\right)$$

Fig.11 Deviations of stresses due to asymmetric imperfection at BB section under the dead weight of concrete

METHOD TO ESTIMATE GEOMETRIC IMPERFECTIONS BASED ON MAXIMUM ENTROPY CONCEPT

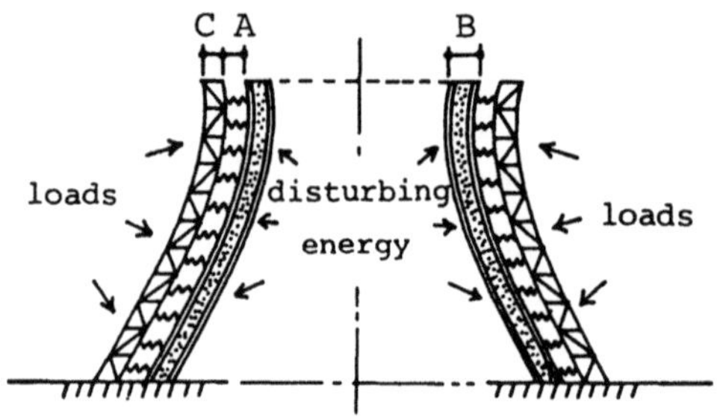

A: spring system

B: panel work shell containing concrete still like a semi-liquid

C: hypothetical supporting frame

Fig.12 Assumed mechanism of causing geometric imperfections

The causes of geometric imperfections in cooling towers may include various factors such as deviations in construction surveying for towers, rigidities of supporting frames and panel works to cast concrete, temporary and dead loads during construction and others being so complicated. On the other hand, the control of quality, improvement of construction and useful measures are taken to reduce imperfections.

Deep complication of the mechanism to give rise to imperfections enforced the authors, at present, to adopt a simple model, Fig.12, and a semi-probabilistic method based on Maximum Entropy Method (MEM)(11). In the process to estimate imperfections, they are assumed to be devided into two parts independent from each other.

One of the imperfections is the average or overall imperfection, the cause of which can be assumed to be deterministic and be the temporary and dead loads to the supporting frame before the concrete-hardening. During construction the rigidity, being much lower than the hardened concrete tower, of the supporting frame keeps the tower in shape. Then, the average imperfection is calculated on an assumption that the supporting frame, having the same geometry as the tower but a reduced rigidity than the concrete tower, is subjected to the dead load including the supporting frame and the concrete for shell.

The other of the imperfection is the local one with randomness. The mechanism of random imperfection is assumed to occur from the displacement of the panel work, containing concrete being still soft, subjected to a random disturbance. Also the panel work is assumed to be a shell composed of panels and soft concrete shell and to be supported by a distributed spring over the surface of the supporting frame. The system is illustrated in Fig.12. Even if a disturbance is so large to cause a random imperfection, the resulting imperfection can be reduced when a controlling effort is invested. The authors simply represent the effort by the magnitude of the constant of the attached spring between the supporting frame and panel work shell. Then the random imperfection comes up from a conflict between the controlling effort and the random disturbance, and the conflict is to be expressed by a kind of energy balance as follows.

$$V_I = U_p - V_c \qquad (2)$$

where U_p is the disturbing energy and V_c is the controlling effort to be accounted by energy. The difference $U_p - V_c$ is the resulting energy to actually cause the random imperfections.

Since the disturbing energy U_p is also considered random, Eq.2 is desirably converted by the following ensamble mean.

$$E[V_I] = E[U_p] - E[V_c] \qquad (3)$$

To formulate mathematically Eq.3 the following assumption are introduced.

(a) about imperfection : The imperfection is assumed to occur as the initial displacement only in the normal direction of towers and to be devided into the average and random imperfections.

$$w_I(s,\theta)=\sum_{i=0}^{m} a_i(s)\cos i\theta + \sum_{j=1}^{n} \xi_j w_j(s)\cos N_j\theta \qquad (4)$$

where the first term is the average and the second is the random imperfections. ξ_j and $w_j(s)$ are the random variables with zero mean and the mode shape, being vibrational modes in the present calculation, in the meridional direction.

(b) about U_p : The disturbing energy U_p is uniformly distributed in the structural system of the spring and the panel work shell, expressed as

$$E[U_p]=\gamma^2 A \qquad (5)$$

where A is the total area of the surface of the panel work shell, and γ^2 is the average disturbance.

(c) about V_c : The controlling energy V_c can be expressed by a spring system as

$$E[V_c]=\int_A E[\frac{1}{2}\rho_w\{\sum_{i=1}^{n}\xi_i w_i(s)\cos N_i\theta\}^2]dA \qquad (6)$$

where ρ_w is the spring constant, being high when a high controlling energy is introduced.

(c) about V_I : The energy inducing imperfections V_I is assumed to change into a strain energy, stored in the panel work shell by resulting in the imperfection, which is calculated as if the imperfection was the elastic displacement. The assumption leads the expression as

$$V_I=\int_A\frac{1}{2}[\mathbb{L}_b(\sum_{i=1}^{n}\xi_i w_i(s)\cos N_i\theta)]^T\cdot H\cdot$$
$$[\mathbb{L}_b(\sum_{j=1}^{n}\xi_j w_j(s)\cos N_j\theta)]dA \qquad (7)$$

where $\mathbb{L}_b$ is the differential operator (10),(11) and H is the elastic constant matrix, being assumed as follws in the present calculation.

$$H=\frac{E_F\cdot t_0}{1-\nu_F{}^2}\begin{pmatrix}1 & \nu_F & 0\\ \nu_F & 1 & 0\\ 0 & 0 & 2(1-\nu_F)\end{pmatrix} \qquad (8)$$

E_F, ν_F and t_0 are the equivalent Young's modulus , Poisson's ratio and the thickness respectively, based on an assumption that the panel work can be replaced into an equivalent isotropic shell.

Determination of the covariance function C_{ij} for ξ_k: The entropy

D with the restriction of Eq.3 can be written as

$$D=\log\{(2\pi e)^{\frac{n}{2}}|\mathbb{C}|^{\frac{1}{2}}\}-\lambda\{E[V_I]+E[V_c]-E[U_p]\} \tag{9}$$

where an assumption that the random variables ξ_i, i=1,2,··, n, have a Gaussian probability density function $p(\boldsymbol{\xi})$.

$$p(\boldsymbol{\xi})=\frac{1}{(2\pi e)^{n/2}\;|\mathbb{C}|^{1/2}}\exp[-\frac{1}{2}\{\boldsymbol{\xi}\}^T[\mathbb{C}]^{-1}\{\boldsymbol{\xi}\}] \tag{10}$$

$$\{\boldsymbol{\xi}\}^T=\{\xi_1,\xi_2,\cdots,\xi_n\},\quad [\mathbb{C}]=\text{ matrix from } C_{ij}$$

The maximum condition of entropy, which shows the most natural state, with C_{ij} and the Lagrange multiplier λ determine C_{ij} and λ.

$$C_{ij}^{-1}=2\lambda\{B_{ij}+\beta_{ij}\}\quad,\quad \lambda=\frac{n}{2A\gamma^2}$$

$$B_{ij}=\frac{1}{2}\int_A \mathbb{L}_b^T\ (w_i\cos N_i\theta)\cdot H\cdot \mathbb{L}_b(w_j\cos N_j\theta)\,dA$$

$$\beta_{ij}=\frac{1}{2}\int_A w_i\cos N_i\theta\cdot\rho_w\cdot w_j\cos N_j\theta\ dA \tag{11}$$

C_{ij}^{-1} = the elements of the inverse matrix of $[\mathbb{C}]$

$$C_{ij}=E[(\xi_i-\bar{\xi}_i)(\xi_j-\bar{\xi}_j)]\quad,\quad \bar{\xi}_i=E[\xi_i]$$

Numerical examples ; according to the present theory based on MEM, a few calculations were performed for two types of the distribution of ρ_w. The magnitude and the distribution of ρ_w, the spring constant in Table 5, may be regarded as the degree of the control to imperfections. In the analysis, the material properties for the supporting frame in Fig.12 were assumed as Table 4.

Table 4 Assumed properties of the supporting frame in Fig.12

axisymmetric dead load	twice the dead load of the completed concrete tower in Fig.15
the weight of the tower	$\gamma_c=2.4\ t\cdot f/m^3$
Young's modulus of concrete	$E_c=2.1\times10^6\ t\cdot f/m^2$
equivalent Young's modulus of the supporting frame	$E_F=E_c/100$
equivalent Poisson's ratio	$\nu_F=0.16$
equivalent thickness	the same as the thickness of the concrete tower in Fig.15

The results for the random imperfections $\xi_i w_i(s)\cos N_i\theta$, in Eq.4, are shown in Fig.13 where the normalized distributing patterns for the harmonic number k of the Fourier series are shown. In Fig.13 an approximation for

$$E[\{\sum_{i=1}^{n}\xi_i w_i \cos N_i\theta\}^2]=\sum_{i=1}^{n}\sum_{j=1}^{n} C_{ij}w_i w_j \cos N_i\theta \cos N_j\theta \qquad (12)$$

was used as Eq.13 since the cross terms C_{ij} for $i \neq j$ are small enough to neglect.

$$E[\{\sum_{i=1}^{n}\xi_i w_i \cos N_i\theta\}^2]=\sum_{i=1}^{n} C_{ii}w_i{}^2\cos^2 N_i\theta$$

$$=\sum_{k=0}^{m}(\sigma_{\xi_w}^{k})^2\cos^2 k\theta \qquad (13)$$

where the terms with the same harmonic number k are collected and eight modes were used along the meridian, then, totally the number , n, of the used modes is 80 and m=9.

Table 5 Distribution of the spring constant ρ_w along the meridian

INDEX OF CASES		$\bar{\rho}_w$	$\bar{\rho}_w = {}_N\rho_w/(\pi \cdot E_F \cdot t_0/H \cdot R_T)$
Type A	A-1	0.02	
	A-2	0.06	
	A-3	0.10	
Type B	B-1	0.02	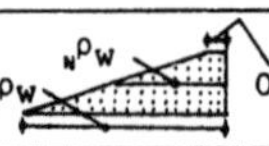
	B-2	0.06	
	B-3	0.10	

The results in Fig.13 show that, if ρ_w increases, $\sigma_{\xi_w}^{k}$ approaches to a flat form for k and a large ρ_w at a lower part decreases the imperfection for higher number of k. The distributions of the coefficient $a_k \pm\sigma_{\xi_w}^{k}$ in Eq.4 are shown in Fig.14 for several points along the shell. The results are obtained under the following condition as;

$$\frac{1}{A}\int_A \sqrt{E[\{\sum_{i=1}^{n}\xi_i w_i \cos N_i\theta\}^2]}\,dA=1.5t_0 \qquad (14)$$

$$t_0=0.127 \text{ m}$$

where the mean of ξ_i was assumed zero for the random imperfections and $a_k\,(k=0)$ of the average or overall imperfection was obtained for the supporting frame subjected to the deterministic load shown in Table 4.

To validate the present theory requires a comparison with actually measured imperfections, however, up to the present, the authors have obtained only one case published by Ellinas et al(3). According to the data which were made by smoothing the original data by Ellinas around cracks, the following errors for the radii at three points, z=20, 40, 60m above the soffit level.

$$R_I = \sum_{i=0}^{9} b_i \cos i\theta \qquad (15)$$

The results are shown in Fig.16. The immediate comparison of b_i in Fig.16 with $a_k \pm \sigma_{\xi w}^{k}$ in Fig.14 is impossible, because the former is a data of a sample and the latter is a statistical quantity. But the two quantities are considered similar in a rough tendency. However this does not mean that the present theory will be directly useful to estimate imperfections. One of the configuration of simulated imperfections for towers is shown in Fig.17, which will be expected to be useful in design in future.

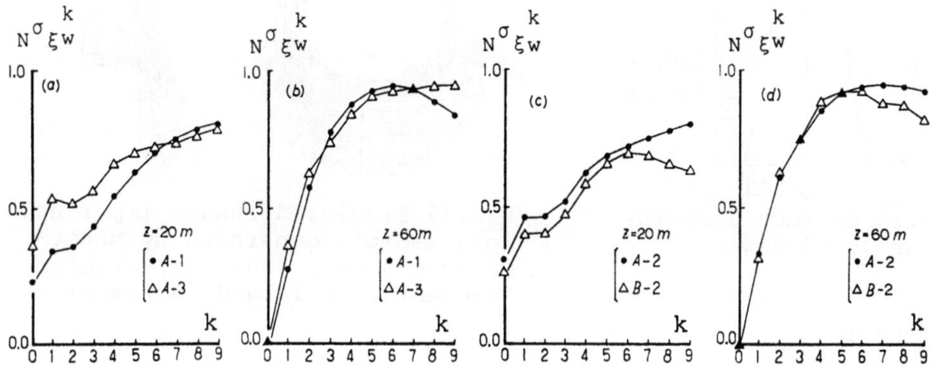

Fig.13 Estimated statistical quantity of the random imperfection corresponding to each ρ_w distributions in Table 5

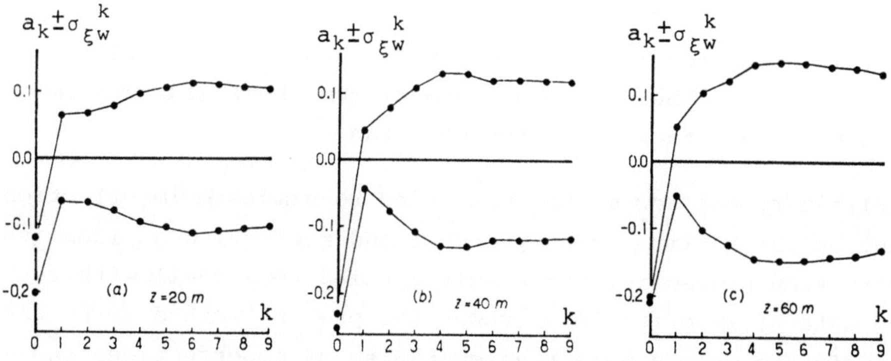

Fig.14 Characteristics of estimated imperfections with the assumptions shown in Table 4 (CASE B-2)

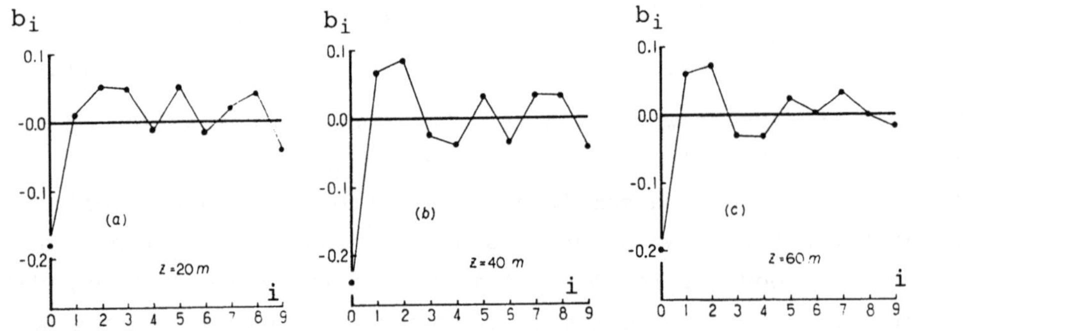

Fig.16 Distributions of the calculated coefficient b_i in Eq.15 (measured data (3))

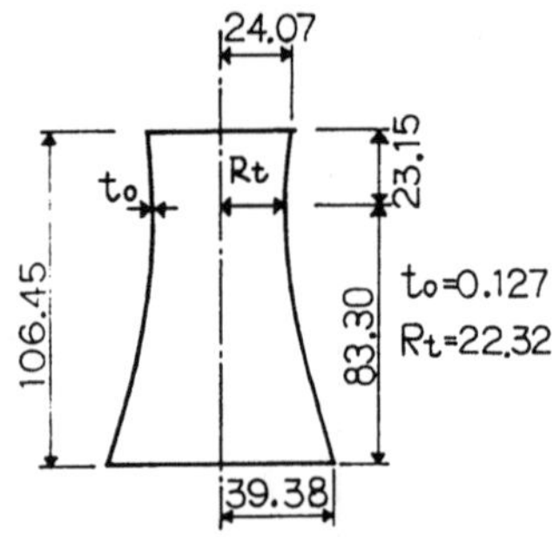

Fig.15 Perfect Geometry of analysed tower

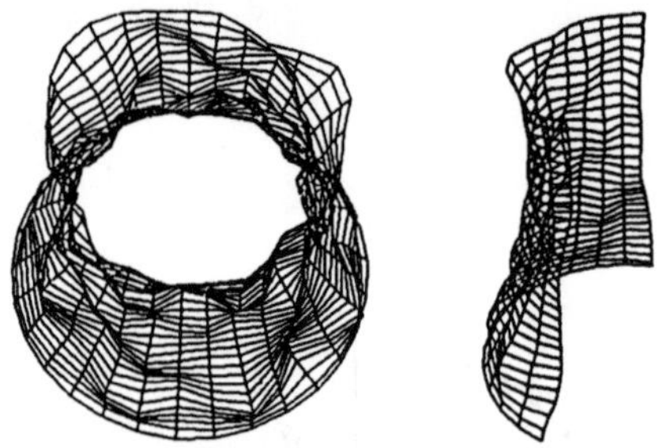

Fig.17 Simulated random imperfections (sample generated by Monte-Carlo method), (the initial displacement is enlarged 20 times)

CONCLUSION

Imperfection sensitivity of buckling loads was obtained and the analysis reveals that the buckling loads are not sensitive to a banded axisymmetric imperfection. On the other hand, the effects of imperfections to stresses are revealed great, and especially the effect of imperfections with high harmonic numbers, in case of initial imperfection being expanded into Fourier series, are so large in both the meridional and circumferential bending moments that shells may crack along the meridian.

A preliminary but new method was tried to simulate imperfections basing on the Maximum Entropy Method and reflecting an idealized construction procedure. The result applied to a shell with recorded data of imperfections, shows the present method to be useful in future when a more real mechanism of imperfections and a real construction procedure would be formulated into the theory.

ACKNOWLEDGEMENT

The study was partially supported by the scientific research funds from the Minister of Education of JAPAN, and the first author express a sincere gratitude to Professor P.L.Gould for giving an opportunity to study the problem in his office. Also much thanks are to Mr. A. Muto for his collaboration and numerical study for simulation of imperfections.

REFERENCES

1. Calladine, C.R.: Structural Consequences of Small Imperfections in Elastic Thin Shells of Revolution. Int. J. Solids Structures, Vol.8, 1972, pp.679-697

2. Kemp, K.O., Croll, J.G.A.: The role of geometric imperfections in the collapse of a cooling tower. Engineering Structures Vol.54, No.1, Jan., 1976, pp.33-36

3. Ellinas, C.P., Croll, J.G.A., Kemp, K.O.: Cooling towers with Circumferential Imperfections. Journal of the Structural Devision, A.S.C.E., No. ST12, Dec., 1980, pp.2405-2423

4. Al-Dabbagh, A., Gupta, A.K.: Meridional Imperfection in Cooling Tower Design. Journal of the Structural Division, A.S. C.E., No.ST6, June, 1979, pp.1089-1102

5. Zintilis, G.M., Croll, J.G.A.: Pressure Buckling of End Supported Shells of Revolution. Engineering Structures, Vol. 4, Oct., 1982, pp.222-232

6. Mungan, I.: Buckling Stress States of Hyperboloidal Shells. Journal of the Structural Devision, A.S.C.E., No.ST10, Oct. 1976, pp.2005-2020

7. Kato, S., Yokoo, Y.: Effects of Geometric Imperfections on Stress Distributions in Cooling Towers. Engineering Structures, Vol.2, July, 1980, pp.150-156

8. Mungan, I.: Buckling Stresses of Stiffened Hyperboloidal Shells. Journal of the Structural Division, A.S.C.E., No. ST8, Aug., 1979, pp.1589-1604

9. I.A.S.S. Working Group Nr.3; Recommendations for the Design of Hyperbolic or Other Similarly Shaped Cooling Towers. BRUSSELS, 1977

10. Kato, S., Murata, M., Chiba, Y., Matsuoka, O.: Elastic Buckling of Cooling Towers Due to Axisymmetric Loads. Trans. of A.I.J., No.327, 1983, pp.40-49 (In Japanese)

11. Murata, M., Miyamura, A., Tagawa, K., Kato, S.: Statistical Analysis of Initial Imperfections by Entropy Model. Trans. of A.I.J., No.330, 1983, pp.39-47 (In Japanese)

Physically Linear Buckling Analysis of Reinforced Concrete Cooling Towers – Design Necessity or Academic Excercise?

H. A. MANG* AND F. TRAPPEL**

Institute for Structural Analysis and Strength of Materials
Technical University of Vienna
Karlsplatz 13, A-1040 Wien, Austria

Summary

The theoretical part of the paper contains the fundamentals of physically and geometrically nonlinear analysis of thin shells by the finite element method including determination of instability points on the load-displacement path. The numerical investigation consists of a comparison of crack and ultimate loads of two different reinforced concrete cooling towers, subjected to dead load and quasistatic wind load, with buckling loads resulting from three different modes of physically linear buckling analysis. With regards to the first cooling tower, three different idealizations of the crown of the shell are taken into account. It is shown that for both shells the ultimate load is considerably smaller than the smallest of the buckling loads obtained from the three modes of physically linear buckling analysis. It is concluded that the latter, although important in its own right as an academic exercise, is not a design necessity. Moreover, it is demonstrated that the crack load, representing a lower bound to the ultimate load, can be estimated through linear extrapolation from linear elastic strength analysis.

Introduction

Proof of a sufficient degree of safety of wind-loaded reinforced concrete (RC) cooling tower shells against physically linear buckling is generally considered to be an important item in the design process of these shells. However, there are differences of opinion concerning the proper mode of physically linear buckling analysis. For example, Zerna and other authors of his school [1] , [2] , [3] advocate the intuitively appealing concept of buckling as a locally induced phenomenon in the sense that the process of loss of stability is initiated at this point of the shell at which a "critical" state of biaxial compression is reached first. With regards to the more conventional concept of global buckling, several modes of physically linear buckling analysis were proposed, ranging from geometrically linear, axisymmetric analysis, on one end of the spectrum [4] , [5] , to geometrically nonlinear, nonaxisymmetric analysis, on the other end [6] , [7] , [8] . Axisymmetric buckling analy-

* Professor, ** Research Assistant

sis may be based, for example, on the concept of equivalent axisymmetric pressure. This fictitious axisymmetric load is obtained as the pressure acting on the luff meridian. The main advantage of this mode of buckling analysis is its computational economy. A disadvantage is that the obtained buckling pressure may be on the unsafe side of the corresponding result from the "actual" wind load [9].

Results from physically linear buckling analysis strongly depend on the selected mode of this category of buckling analyses. For example, for the special case of geometrically linear, nonaxisymmetric buckling analysis the result of an "initial linear buckling analysis" may differ considerably from the one of a "classical linear buckling analysis" [9].

The aim of the present paper is to answer the question whether physically linear buckling analysis of RC cooling towers is at all relevant to the design of these shells. In order to answer this question buckling results will be compared with corresponding results from ultimate load analysis considering nonlinearity and fracture of concrete and yielding of steel. Zerna, Mungan, and Steffen [10] have reported on such an analysis of a cooling tower shell. Since their investigation is restricted to axisymmetry, fracture of concrete in biaxial compression is found to be the dominant failure mode. This result is fundamentally different from the one obtained by the authors of Ref. [9] from nonaxisymmetric ultimate load analysis of a wind-loaded cooling tower-shell, characterized by predominantly tensile failure of concrete followed by yielding of steel.

In order to answer the title question, crack and ultimate loads of two different RC cooling towers will be compared with buckling loads resulting from three different modes of physically linear buckling analysis.

Buckling and ultimate analysis of thin shells by the finite element method

The starting point for the formulation of the finite element method (FEM) employed by the writers for buckling and ultimate load analysis of cooling tower shells is the following hybrid virtual work principle:

$$-\int_{V^{(0)}} \sigma_{ij}\,\delta\varepsilon_{ij}\,dV + \int_{V^{(0)}} X_i^{(0)}\delta u_i\,dV + \int_{S_\sigma^{(0)}} \hat{T}_i^{(0)}(p_{Sj},u_j)\,\delta u_i\,dS^{(0)} + $$

$$+ \int_{\tilde{S}_u^{(0)}\in S_u^{(0)}} [\,T_i^{(0)}\delta u_i + (u_i - \bar{u}_i)\delta T_i^{(0)}]\,dS^{(0)} = 0. \qquad (1)$$

Eq. (1) represents the total Lagrangian formulation of a weak form of the equilibrium equations for the considered body [Fig. 1(a)] at time t. $V^{(0)}$ is the volume of the body at time $t^{(0)}$, that is, at the initial state which is characterized by the symbol "$^{(0)}$". $S_{\sigma}^{(0)}$ denotes the part of the surface $S^{(0)}$ on which tractions $\hat{T}_i^{(0)}$ depending on the given pressure function p_{Si} and on the unknown displacements u_i are acting. $S_u^{(0)}$ is the part of $S^{(0)}$ on which displacements $\bar{u}_i$ are specified. σ_{ij} is the Second Piola-Kirchhoff stress tensor, ε_{ij} is the Green-Lagrange strain tensor, $X_i^{(0)}$ is the vector of volume forces and $T_i^{(0)}$ are tractions acting on the part $\tilde{S}_u^{(0)}$ of $S_u^{(0)}$ on which kinematic boundary conditions are relaxed. Usually, computational economy is the reason for partial relaxation of such boundary conditions. The tractions $T_i^{(0)}$ are independent unknowns in the sense of Lagrange multipliers. The virtual Green-Lagrange strains are obtained as

$$\delta\varepsilon_{ij} = \frac{1}{2}\,(\delta u_{i,j} + \delta u_{j,i} + u_{k,i}\delta u_{k,j} + u_{k,j}\delta u_{k,i}) \tag{2}$$

where $u_{i,j} = \partial u_i/\partial a_j$ with $a_i = x_i$ according to Fig. 1(a).

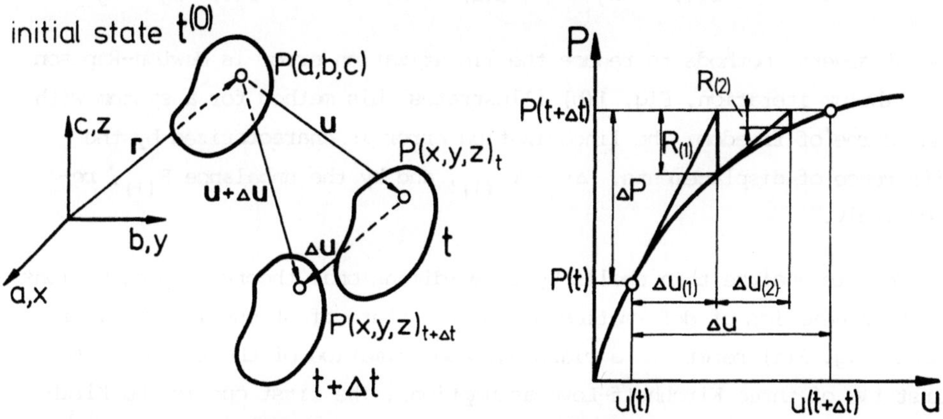

(a) state of displacements at times t and t+Δt, respectively (b) Newton - Raphson equilibrium iteration

Fig. 1. Incremental-iterative procedure for physically and/or geometrically nonlinear analysis of deformable solids

Consideration of physical and/or geometric nonlinearity suggests an incremental-iterative analysis procedure. For this purpose Eq. (1) is formulated at time $t + \Delta t$ where Δt represents an increment of time. Subtraction of Eq. (1) from the so-obtained relationship and linearization of the result yields an incremental virtual work equation. It is given as

$$-\int_{V^{(0)}} (\Delta\sigma_{ij}\,\delta\varepsilon_{ij} + \sigma_{ij}\,\Delta\delta\varepsilon_{ij})\,dV + \int_{S_\sigma^{(0)}} \Delta\hat{T}_i^{(0)}\delta u_i\,dS^{(0)} +$$

$$+\int_{\tilde{S}_u^{(0)}\in S_u^{(0)}} (\Delta T_i^{(0)}\delta u_i + \Delta u_i\,\delta T_i^{(0)})\,dS^{(0)} = 0 \qquad (3)$$

where

$$\Delta\sigma_{ij} = D_{ijkl}(\varepsilon_{mn})\,\Delta\varepsilon_{kl} \qquad (4)$$

is an increment of the Second Piola-Kirchhoff stress tensor, assumed to depend linearly on $\Delta\varepsilon_{ij}$ representing an increment of the Green-Lagrange strain tensor. The constitutive tensor D_{ijkl} is assumed to depend only on ε_{ij}. $\Delta\delta\varepsilon_{ij}$ is obtained as

$$\Delta\delta\varepsilon_{ij} = \frac{1}{2}\,(\Delta u_{k,i}\delta u_{k,j} + \Delta u_{k,j}\delta u_{k,i}) \qquad (5)$$

where $\Delta u_{k,i}$ is an increment of $u_{k,i}$. $\Delta\hat{T}_i^{(0)}$ is approximated as

$$\Delta\hat{T}_i^{(0)} \doteq \Delta\hat{T}_{i(1)}^{(0)}(\Delta p_{Sj}) + \Delta\hat{T}_{i(2)}^{(0)}(\Delta p_{Sj}, u_j) + \Delta\hat{T}_{i(3)}^{(0)}(p_{Sj}, \Delta u_j). \qquad (6)$$

One of several methods to reduce the linearization error is Newton-Raphson equilibrium iteration. Fig. 1(b) illustrates this method for a system with one degree of freedom. The linearization error is characterized by the difference of displacements, $\Delta u - \Delta u_{(1)}$, and by the unbalance $R_{(1)}$, respectively.

In order to analyze thin shells by a two-dimensional theory, approximations of the kinematics of deformation and of the state of stress must be introduced. Fig. 2(a) refers to a classical approximation of the former -- the first two of three Kirchhoff-Love assumptions. The first one is the Kirchhoff normal-hypothesis. It can be formulated mathematically as

$$\gamma_{\alpha\gamma} = 0, \quad \gamma_{\beta\gamma} = 0 \qquad (7)$$

where $\gamma_{\alpha\gamma}$ and $\gamma_{\beta\gamma}$ are transverse engineering shear strains. The curvilinear coordinate system α, β, γ is assumed to be orthogonal. The second Kirchhoff-Love assumption is

$$w(\gamma) \doteq w(\gamma = 0). \qquad (8)$$

The last Kirchhoff-Love assumption is concerned with neglecting the influence of the transverse normal stress σ_ν in the constitutive equations.

One of the most popular shell theories based on the Kirchhoff-Love assumptions is the one by Koiter-Sanders [11], [12], used by the writers. Consideration of geometric nonlinearity is restricted to taking account of moderately large rotations. This is sufficient for ultimate load analysis of RC cooling tower shells [9].

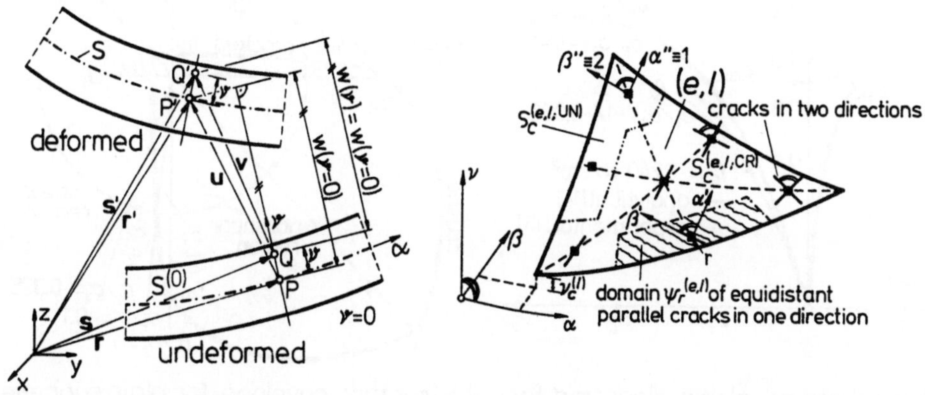

(a) Kirchhoff - Love assumptions (b) middle surface of partially cracked concrete-layer l of finite element e

Fig. 2. Thin-shell analysis by the finite element method

A concept for ultimate load analysis of RC shells by the FEM, developed at Technical University of Vienna, is described in detail in the doctoral dissertation of Floegl [13]. Curved, triangular, C^1-conforming, thin-shell elements [19], subdivided into a sufficiently large number of thin layers such that approximately a plane state of stress may be assumed to exist in each layer serve as the analysis tool. Fig. 2(b) shows the middle surface of a layer of a finite element (layer l of element e). The layer is assumed to be partially cracked. $S_C^{(e,l;UN)}$ denotes the uncracked and $S_C^{(e,l;CR)}$ the cracked part of the layer. The state of stress is determined at seven integration points of the layer.

The local coordinate system α'', β'' in Fig. 2(b) refers to the principal directions of stress in the uncracked concrete. The reason for this choice of local coordinates is that the constitutive equations for the intact concrete are referred to principal directions. Fig. 3(a) shows a comparison of typical stress-strain diagrams for plain concrete on the basis of the employed constitutive model, developed by Liu, Nilson and Slate [14], with diagrams from experiments conducted by Kupfer [15]. In this figure, ε_1 and ε_2 are the principal strains and σ_1 and σ_2 are corresponding principal stresses; ε_{1u} and ε_{2u} are the fracture strains; f_{cu} is the prism strength of concrete.

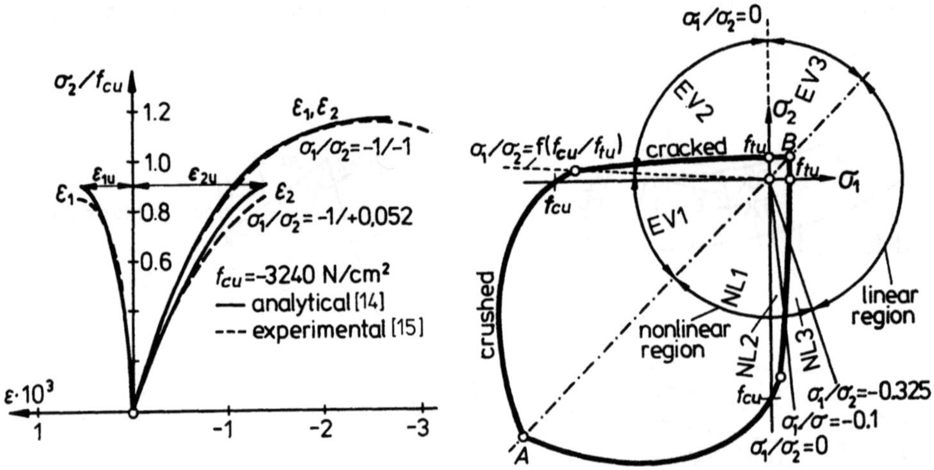

(a) typical stress-strain diagrams for plain concrete subjected to bilinear states of stress

(b) fracture envelope for plain concrete subjected to biaxial states of stress

Fig. 3. Constitutive modelling of plain concrete

The local coordinate system α', β' in Fig. 2(b) refers to the normal and the parallel, respectively, to a band of equidistant parallel cracks distributed over the tributary domain of integration point r of the considered concrete layer. Thus, fracture of concrete is taken into account within the framework of the concept of "smeared" cracks. Cracks in one or two directions are considered. The second direction is restricted to be orthogonal to the first one. Although this simplifying assumption is inconsistent with the state of strain, it is considered to be admissible [9]. Fig. 3(b) illustrates the fracture envelope of concrete, representing the geometric locus of biaxial states of fracture stresses σ_{1u}, σ_{2u}. Each one of the three symbols EV1, EV2 and EV3 denotes a certain part of the half of the fracture envelope above the symmetry line through points A and B. The three parts are described by different analytical expressions [15]. For $-\infty < \sigma_1/\sigma_2 = \alpha_2 \leq f(f_{cu}/f_{tu}) = \alpha_{2L}$, where f_{tu} is the uniaxial tensile strength of concrete, fracture occurs by cracking normal to the direction of ε_1. For $\alpha_{2L} < \alpha_2 \leq 1$, concrete fails by crushing. In this case, concrete loses all of its strength. When concrete cracks, it retains its capacity to carry internal forces in the direction parallel to the cracks so long as $f_{cu} < \sigma_2 < f_{tu}$. Because of tension stiffening, concrete also retains part of its strength normal to the cracks. In the aforementioned concept for ultimate load analysis of RC shells by the FEM, tension stiffening is taken into account [16].

Each one of the three symbols NL1, NL2 and NL3 in Fig. 3(b) refers to a range of values α_2 for which a certain form of nonlinear constitutive equations for the intact concrete holds [13]. Thus, NL1, NL2 and NL3 characterize the nonlinear region of the constitutive equations for the intact concrete.

The stress-strain diagram of the reinforcement steel is assumed to be bilinear and identical in tension and compression.

On the basis of the Love-Kirchhoff assumptions virtual work equations for physically and geometrically nonlinear analysis of thin shells can be formulated analogous to the equations (1) and (3) for nonlinear analysis of deformable solids. Discretization of the former within the framework of the FEM enables derivation of a system of linear algebraic equations for incremental-iterative FE analysis of thin shells. These equations are given as

$$\left[\begin{array}{c:c} \mathbf{K}_T(\mathbf{q}^2_{(r-1)}) - \hat{\mathbf{K}}_T(p_S + \Delta p_S, \mathbf{q}_{(r-1)}) & \mathbf{G}^T \\ \hdashline \mathbf{G} & \mathbf{0} \end{array}\right] \begin{Bmatrix} \Delta \mathbf{q}_{(r)} \\ \Delta \boldsymbol{\lambda}_{(r)} \end{Bmatrix} = \begin{Bmatrix} \mathbf{R}_{(r-1)} \\ \mathbf{0} \end{Bmatrix} \tag{9}$$

where $\mathbf{K}_T$ is the tangent stiffness matrix, $\hat{\mathbf{K}}_T$ is the tangent pressure stiffness matrix [17], $\mathbf{G}$ is a constraint matrix required for enforcing C^1-interelement continuity, $\Delta \mathbf{q}_{(r)}$ is the vector of increments of node displacements, $\Delta \boldsymbol{\lambda}_{(r)}$ is a vector of increments of discrete Lagrange multipliers and $\mathbf{R}_{(r-1)}$ is the vector of out-of-balance forces after r-1, r = 1,2,..., steps of an equilibrium iteration for the considered load increment. (For ultimate load analysis of cooling tower shells subjected to dead load and quasistatic wind load, the rate at which the deformation takes place is of little interest. Therefore, instead of considering increments of time, increments of the pressure function, $\Delta p_S \equiv \Delta p_{S\nu}$ are considered.) $\mathbf{K}_T$ depends quadratically on the total node displacements $\mathbf{q}_{r-1}$ before step r of the equilibrium iteration. $\hat{\mathbf{K}}_T$ depends linearly on $p_S + \Delta p_S$ and on $\mathbf{q}_{r-1}$.

The tangent stiffness matrix can be written as [9]

$$\mathbf{K}_T(\mathbf{q}^2) = \mathbf{K} + \mathbf{K}_\sigma(\sigma(\mathbf{q}^2)) + \mathbf{K}_L(\mathbf{q}^2) = \mathbf{K} + \mathbf{N}_1(\mathbf{q}) + \mathbf{N}_2(\mathbf{q}^2) \tag{10}$$

where $\mathbf{K}$ is the small-displacement stiffness matrix, $\mathbf{K}_\sigma$ is the initial-stress stiffness matrix, $\mathbf{K}_L$ is the initial-displacement stiffness matrix, $\mathbf{N}_1$ is the first geometric stiffness matrix and $\mathbf{N}_2$ is the second geometric stiffness matrix. The coefficients of $\mathbf{N}_1$ ($\mathbf{N}_2$) depend linearly (quadratically) on the node displacements. The tangent pressure stiffness matrix can be written as [17]

$$\hat{\mathbf{K}}_T(p_S,\mathbf{q}) = \hat{\mathbf{K}}(p_S) + \hat{\mathbf{N}}(p_S,\mathbf{q}) \tag{11}$$

where $\hat{\mathbf{K}}$ is the small-displacement pressure stiffness matrix and $\hat{\mathbf{N}}$ is the geometric pressure stiffness matrix.

The condition for an instability point on the load-displacement path is obtained as [9]:

$$\text{Det}\begin{bmatrix} \mathbf{K}_T - \hat{\mathbf{K}}_T & \mathbf{G}_T \\ \mathbf{G} & \mathbf{0} \end{bmatrix} = 0. \tag{12}$$

The instability point may either be a bifurcation point or a relative maximum on the load-displacement path, indicating snap-through of the shell, or the absolute maximum on this path signalling the ultimate load.

In case of restriction to a physically linear buckling analysis, only the first two of the three listed modes of loss of stability are possible. In case of restriction to a physically and geometrically linear buckling analysis, only bifurcation buckling is possible. One of two frequently used modes of physically and geometrically linear buckling analysis is based on linearization of $\mathbf{K}_T - \hat{\mathbf{K}}_T$, characterized by disregarding $\mathbf{N}_2$ and $\hat{\mathbf{N}}$ and setting

$$\mathbf{N}_1(\mathbf{q}) = \mathbf{N}_1(\mathbf{q}^*) + \chi\mathbf{N}_1(\bar{\mathbf{q}}) \tag{13}$$

$$\hat{\mathbf{K}}(p_S) = \hat{\mathbf{K}}(\bar{p}_S) \tag{14}$$

where $\mathbf{q}^*$ and $\bar{\mathbf{q}}$ are vectors of node displacements resulting from dead load and from a reference pressure function $\bar{p}_S \equiv \bar{p}_{S\gamma}$, respectively; χ is the load intensity factor [9]. The condition for a bifurcation point on the linearized load-displacement path is given as

$$\text{Det}\left[\begin{array}{c|c} \mathbf{K} + \mathbf{N}_1(\mathbf{q}^*) + \chi[\mathbf{N}_1(\bar{\mathbf{q}}) - \hat{\mathbf{K}}_1(\bar{p}_S)] & \mathbf{G}^T \\ \hline \mathbf{G} & \mathbf{0} \end{array}\right] = 0. \qquad (15)$$

Eq. (15) yields the so-called "initial linear buckling pressure" [9] which has no physical significance. The second mode of physically and geometrically linear buckling analysis is characterized by disregarding $\mathbf{K}_L$ and linearizing $\mathbf{K}_\sigma$ and $\hat{\mathbf{K}}_T$ resulting in

$$\mathbf{K}_\sigma(\sigma(\mathbf{q}^2)) = \mathbf{K}_\sigma(\sigma(\mathbf{q}^*)) + \chi\mathbf{K}_\sigma(\sigma(\bar{\mathbf{q}})) \qquad (16)$$

and in Eq. (14), respectively. The condition for a bifurcation point is

$$\text{Det}\left[\begin{array}{c|c} \mathbf{K} + \mathbf{K}_\sigma(\sigma(\mathbf{q}^*)) + \chi[\mathbf{K}_\sigma(\sigma(\bar{\mathbf{q}})) - \hat{\mathbf{K}}_1(\bar{p}_S)] & \mathbf{G}^T \\ \hline \mathbf{G} & \mathbf{0} \end{array}\right] = 0. \qquad (17)$$

Eq. (17) yields the so-called "classical linear buckling pressure" [9]. Providing the rotations of the tangents to the parameter lines α, β vanish before loss of stability in the linear range of the constitutive equations, in the limit of finite element representations Eq. (17) gives the exact solution.

Numerical investigation

Cooling tower at Voitsberg, Styria, Austria

Fig. 4 shows characteristic dimensions, the thickness profile and the finite element mesh of a hyperbolic cooling tower of approximately 93 m height erected in Voitsberg, Styria, Austria, designated as cooling tower Voitsberg III. Details concerning the geometric form of the generatrix of the shell, the arrangement of the reinforcement and the material properties of concrete and steel are given in Ref. [18]. Because of symmetry of the quasistatic wind load with respect to the luff and the lee meridian, only one half of the shell needs to be considered for the analysis. Fig. 5 shows the distribution of the wind-load function according to Ref. [2]. Fig. 6 illustrates three different idealizations of the crown of the shell shown in Fig. 4(b). Idealization A is characterized by a radial shift of the interior web of the $\cup$-shaped cornice ring towards the exterior web leading to the union of the two webs. Idealization B of the crown of the shell is characterized by a gradual thickening of the shell such that the area of the cross-section is equal to the one of the cornice ring. Idealization C pertains to an unstiffened crown of the cooling tower shell.

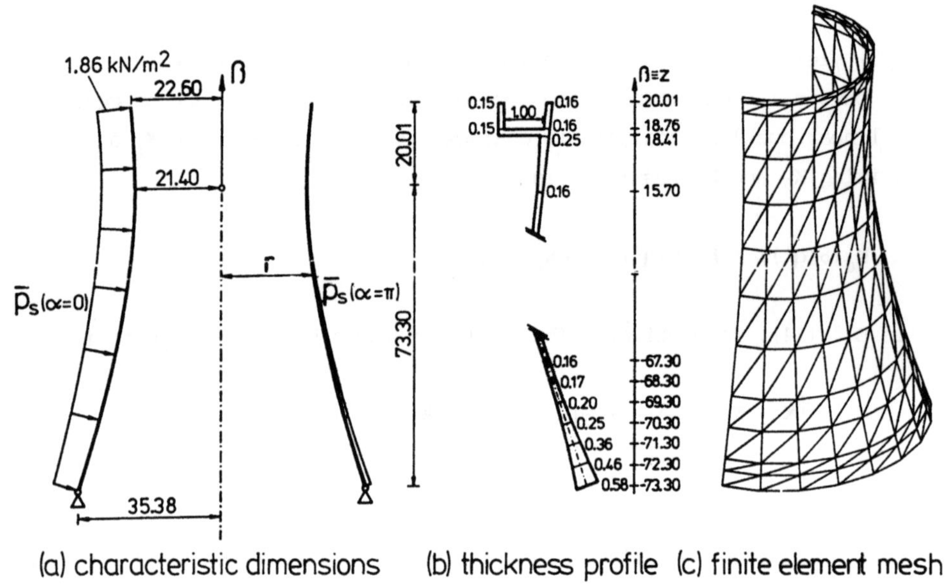

Fig. 4. Cooling tower Voitsberg III, Styria, Austria

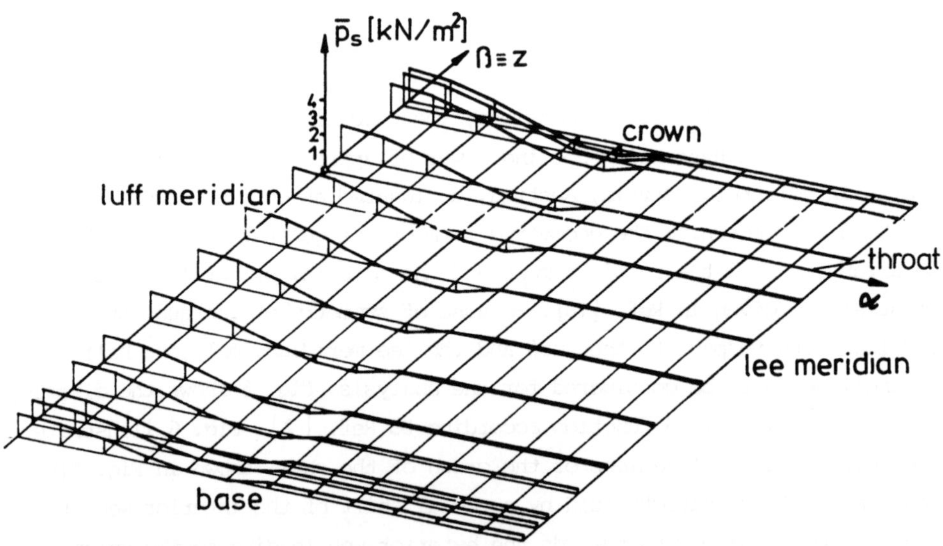

Fig. 5. Distribution of wind-load function

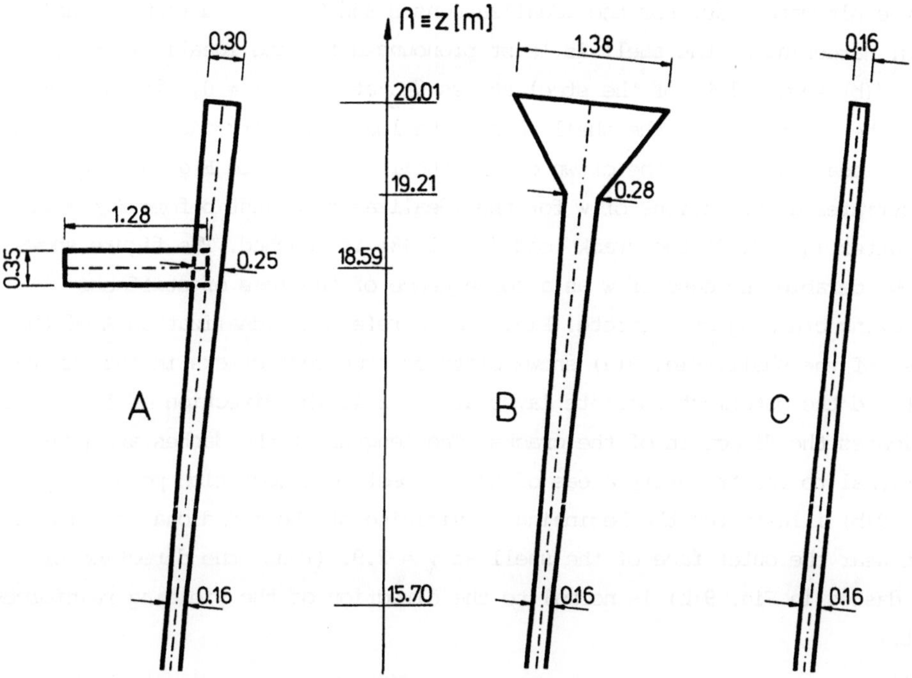

Fig. 6. Three different idealizations of crown of shell

Fig. 7(a) shows plots of the transverse displacement w at point $\alpha = 0$, $\beta = -18.8$ m versus the load intensity factor χ for the three idealizations of the crown of the shell. Point P_A divides the load-displacement path referring to the idealization A into two parts characterized by intact concrete (state I) and by fractured concrete (state II), respectively. The second part consists of three sections, termed "crack plateau", "hardening" and "yield plateau". The crack plateau is characterized by a considerable increase of the displacements at practically no increase of the load. This is the consequence of a sudden loss of stiffness resulting from concrete cracking. The hardening of the shell is the result of an increased activation of the load-carrying capacity of the reinforcement after the beginning of concrete cracking. The yield plateau is the consequence of the loss of stiffness resulting from yielding of the reinforcement. The value of the steel stress, on one hand, and failure of the equilibrium iteration to converge, on the other hand, indicate that the yield plateau has been reached. From the standpoint of the analysis the respective load represents the ultimate load. With regards to the idealization A of the crown of the shell, the aforementioned point P_A determines the crack load which is a lower bound to the ultimate load. For the idealizations B and C the situation is analogous. Hardening of the shell is least pronounced for the idealization C. Fig. 7(b) shows plots of the steel stress $\sigma_{S\beta}$ at point $\alpha = 0$, $\beta = -45.6$ m near the outer face of the shell versus the load intensity factor χ for the three idealizations of the crown of the shell according to Fig. 6. Fig. 8 illustrates distributions of w for the idealization A just before ($\chi = 2.16$) and after ($\chi = 2.20$) the crack load ($\chi = 2.18$) is reached. The figure shows a considerable increase of w as a consequence of the loss of stiffness following cracking of concrete. Fig. 9 also refers to idealization A of the crown of the shell. Fig. 9(a) shows plots of "smeared" cracks in the innermost and the outermost concrete layer at $\chi = 2.4$. The direction of the dashes indicates the direction of the cracks. The lengths of the dashes are proportional to the tributary areas of the respective integration points. Fig. 9(b) illustrates the beginning of yielding of the meridional reinforcement near the outer face of the shell at $\chi = 2.9$. (Thus, the direction of the dashes in Fig. 9(b) is normal to the direction of the yielding reinforcement.)

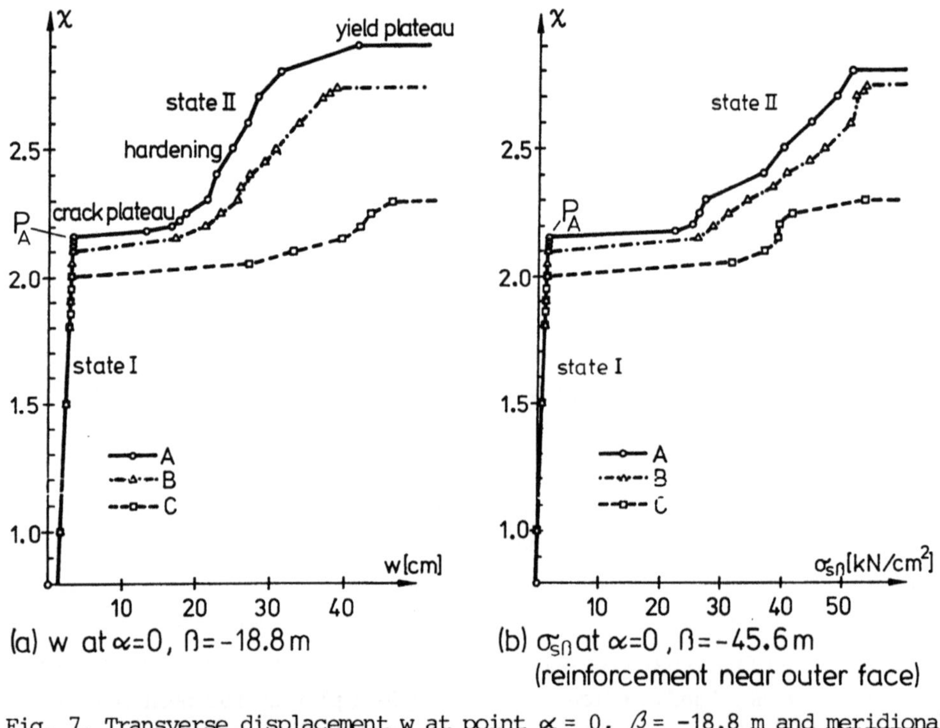

Fig. 7. Transverse displacement w at point $\alpha = 0$, $\beta = -18.8$ m and meridional steel stress $\sigma_{S\beta}$ at point $\alpha = 0$, $\beta = -45.6$ m, respectively, versus load intensity factor χ for three different idealizations of the crown of the shell

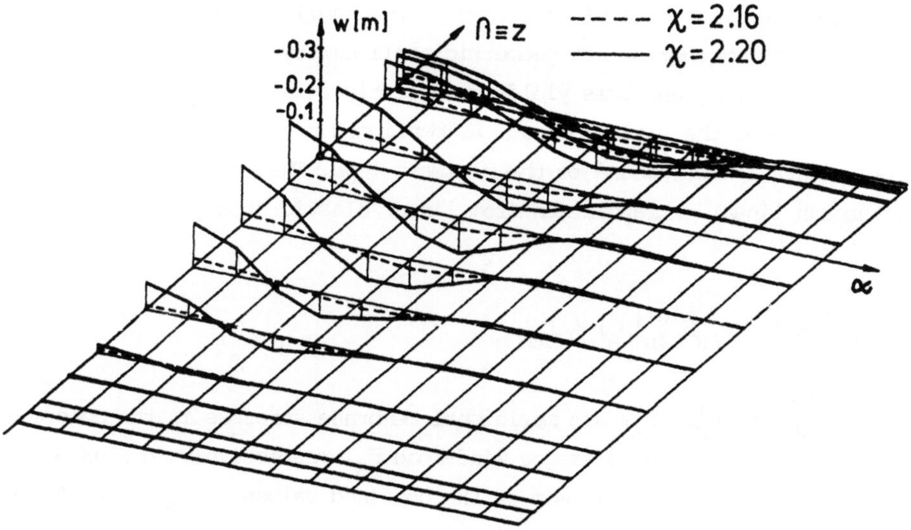

Fig. 8. Distribution of transverse displacement w at χ = 2.16 and 2.20, respectively, for idealization A of crown of shell

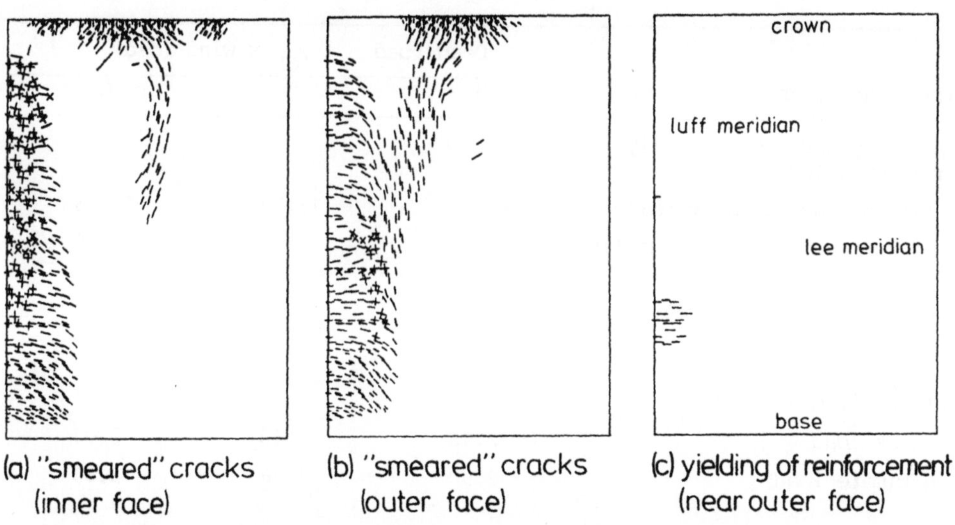

Fig. 9. "Smeared" cracks of concrete in two different concrete layers at χ = 2.4 and yielding of meridional reinforcement near outer face at χ = 2.9 for idealization A of crown of shell

Table 1 contains a comparison of crack and ultimate loads with buckling loads resulting from three different modes of physically linear buckling analysis for the three idealizations of the crown of the shell according to Fig. 6. It is seen that, whether or not geometric nonlinearity is considered, physically linear buckling analysis yields unrealistic results which are far on the unsafe side of the ultimate load. As mentioned previously, the crack load represents a lower bound to the ultimate load. The former can be estimated through linear extrapolation from linear elastic strength analysis as [9]

$$\sigma^{(0)}_{C\phi L}(\bar{p}_V) + \chi_{CR}\sigma^{(0)}_{C\phi L}(\bar{p}_S) \doteq f_{tu} \tag{18}$$

where $\sigma^{(0)}_{C\phi L}(\bar{p}_V)$ and $\sigma^{(0)}_{C\phi L}(\bar{p}_S)$ are meridional membrane stresses in the concrete from dead load $\bar{p}_V$ and the reference wind-load $\bar{p}_S$, respectively, evaluated for the luff meridian. The minimum of the obtained values for χ is the "critical" load intensity factor χ_{CR} indicating initial cracking. Table 1 shows that the described estimate is acceptable.

Table 1. Cooling tower at Voitsberg, Styria, Austria: Comparison of crack and ultimate loads with buckling loads resulting from physically linear buckling analyses for three different idealizations of crown of shell

Mode of analysis	Dead load + (χ_{CR} x wind load)		
	A	B	C
Physically and geometrically linear buckling			
(a) initial linear buckling	13.82	13.78	12.16
(b) classical linear buckling	43.80	---	41.90
Physically linear, geometrically nonlinear buckling	---	59*	48*
Ultimate load analysis			
crack load	2.18	2.15	2.05
ultimate load	2.90	2.76	2.30
Estimate of crack load based on linear elastic analysis	2.27	2.25	2.14

*$\Delta\chi$ = 3

Cooling tower at Port Gibson, Miss., USA

Fig. 10 shows characteristic dimensions, the thickness profile and the finite element mesh of a cooling tower of approximately 150 m height erected in Port Gibson, Miss., USA. The figure also contains information on the wind-load function. Details concerning the geometric form of the generatrix of the shell, the arrangement of the reinforcement, the material properties of concrete and steel and the wind-load function are given in Ref. [9].

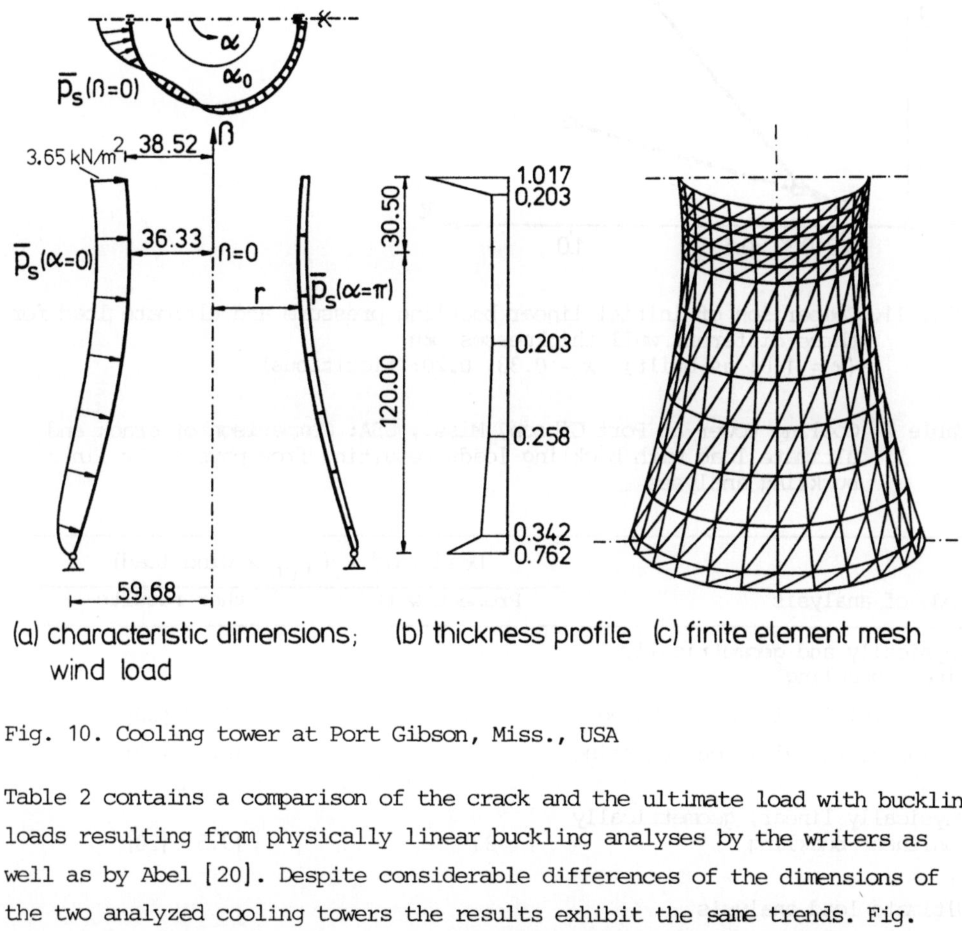

Fig. 10. Cooling tower at Port Gibson, Miss., USA

Table 2 contains a comparison of the crack and the ultimate load with buckling loads resulting from physically linear buckling analyses by the writers as well as by Abel [20]. Despite considerable differences of the dimensions of the two analyzed cooling towers the results exhibit the same trends. Fig. 11 shows a comparison of the initial linear buckling pressure, representing the minimum of the buckling loads resulting from the three considered modes of physically linear buckling analysis, with the ultimate load for three different wall thicknesses $\varkappa h$ where $\varkappa$ is a proportionality factor. $\varkappa = 1.0$ refers to the built shell; $\varkappa = 0.33$ and 0.20, respectively, pertain to a fictitious, proportional reduction of the wall thickness. It is seen that for the considered shell initial linear buckling would only be relevant for $\varkappa < 0.2$ for which, however, χ_{CR} would be much smaller than 1.

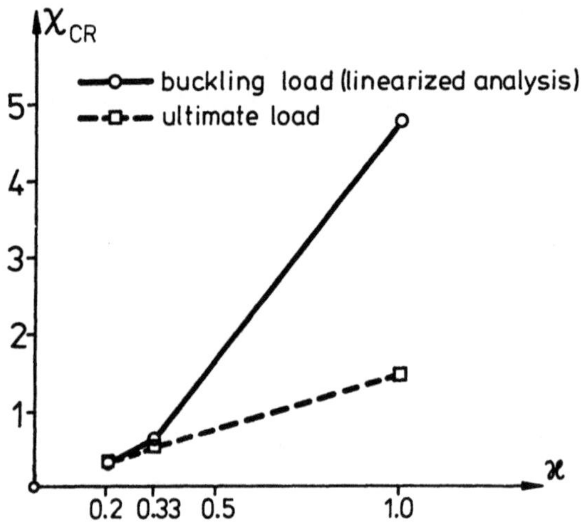

Fig. 11. Comparison of initial linear buckling pressure and ultimate load for three different wall thicknesses $\varkappa h$
($\varkappa$ = 1.0: as built; $\varkappa$ = 0.33, 0.20: fictitious)

Table 2. Cooling tower at Port Gibson, Miss., USA: Comparison of crack and ultimate load with buckling loads resulting from physically linear buckling analyses

	Dead load + (χ_{CR} x wind load)	
Mode of analysis	Present work	Other results
Physically and geometrically linear buckling		
(a) initial linear buckling	4.80	4.97 [20]
(b) classical linear buckling	---	18.7 [20]
Physically linear, geometrically nonlinear buckling	20*	19.5 [20]
Ultimate load analysis		
crack load	1.34	---
ultimate load	1.49	---
Estimate of crack load based on linear elastic analysis	1.28	---

$*\Delta\chi = 2$

Conclusions

On the basis of the presented numerical investigation of two different cooling towers with strongly different dimensions the following conclusions can be drawn:

1. The ultimate load is considerably smaller than the corresponding initial linear buckling load representing the minimum of the buckling loads resulting from three different modes of physically linear buckling analysis.

2. The above conclusion also holds for different idealizations of the crown of the cooling tower Voitsberg III.

3. With regards to the cooling tower shell at Port Gibson, initial linear buckling was found to become relevant only for an unrealistically small thickness of the shell.

4. The crack load indicating initial cracking in the domain of the luff meridian is a lower bound to the ultimate load.

5. Linear extrapolation from linear elastic strength analysis permits a good estimate of the crack load.

Ultimate load analysis of the two cooling tower shells showed that the process of degradation of the structural integrity of these shells is induced locally through cracking of concrete. Therefore, from the standpoint of the design of RC cooling tower shells for dead load and quasistatic wind load proof of a sufficient degree of safety against initial cracking is indispensable. Physically linear buckling analysis, however, although important in its own right as a challenging academic exercise, is not a design necessity.

The present paper was restricted to perfect RC cooling tower shells. However, it is extremely unlikely that consideration of imperfections would result in a reversal of the observed trend. The reason for this statement is that the buckling pressure obtained from physically linear, geometrically nonlinear analysis, representing the geometrically relevant mode of physically linear buckling analyses of wind-loaded shells, is by one order of magnitude larger than the ultimate load.

Acknowledgements

The help of P. Torzicky concerning computational aspects is gratefully acknowledged. The excellent co-operation with the Computing Center of Technical University of Vienna, Austria, was much appreciated. Financial support from the Foundation for Promotion of Scientific Research of Austria is also acknowledged.

References

1. Mungan, I.: Experimentelle Beuluntersuchungen und Stabilitätsnachweise für Kühlturmschalen. Kühlturm Symp. 1977 (Zerna W., ed.), Vol. 29/30 of Berichte aus dem Institut für Konstruktiven Ingenieurbau der Ruhr-Universität Bochum. Essen: Vulkan 1977

2. Krätzig, W.P.; Peters, H.L.; Zerna, W.: Naturzugkühltürme aus Stahlbeton - Derzeitiger Stand und Entwicklungsmöglichkeiten. Beton- und Stahlbetonbau 73 (1978) 37-42

3. Zerna, W.; Almannai, A.; Basar, Y.; Mungan, I.: Randbedingungen und Beulverhalten von Kühlturmschalen. Beton- und Stahlbetonbau 74 (1979) 200-204

4. Cole, P.P.; Abel, J.F.; Billington, D.P.: Buckling of cooling-tower-shells: Bifurcation results. Proc. ASCE, J. Struct. Div. 101 (1975) 1205-1222

5. Abel, J.F.; Billington, D.P.; Nagy, O.A.; Wiita-Dworkin, C.: Buckling of cooling towers. Proc. ASCE, J. Struct. Div. 108 (1982) 2162-2174

6. Yeh, C.H.; Shieh, W.J.: Stability and dynamic analysis of cooling towers. Proc. ASCE, J. Power Div. 99 (1973) 339-347

7. Chang, S.C.: An integrated finite element nonlinear shell analysis system with interactive computer graphics. Ph.D. thesis, Cornell University, 1981

8. Mehl, M.: Über das Tragverhalten von Naturzugkühltürmen aus Stahlbeton. Dr.techn. thesis, Technische Universität Wien, 1982

9. Mang, H.A.; Floegl, H.; Trappel, F.; Walter, H.: Wind-loaded reinforced concrete cooling towers: Buckling or ultimate load? Engrg. Struct. 5 (1983) 163-180

10. Zerna, W.; Mungan, I.; Steffen, W.: Bestimmung der Beulsicherheit von Schalen aus Stahlbeton unter Berücksichtigung der physikalisch-nichtlinearen Materialeigenschaften. Deutscher Ausschuß für Stahlbeton, Vol. 315, Berlin: Wilhelm Ernst 1980

11. Koiter, W.T.: A consistent first approximation in the theory of thin elastic shells. Proc. IUTAM Symp. on Theory of Thin Shells (Koiter, W.T., ed.), Amsterdam: North-Holland 1960

12. Budiansky, B.; Sanders, L.J.: On the "best" first order linear shell theory. Progress in solid mechanics: The Prager Anniversary Volume. New York: Macmillan 1963

13. Floegl, H.: Traglastermittlung dünner Stahlbetonschalen mittels der Methode der Finiten Elemente unter Berücksichtigung wirklichkeitsnahen Werkstoffverhaltens sowie geometrischer Nichtlinearität. Dr.techn. thesis, Technische Universität Wien, 1982

14. Liu, T.C.Y.; Nilson, A.H.; Slate, F.O.: Biaxial stress-strain relations for concrete. Proc. ASCE, J. Struct. Div. 98 (1972) 1025-1034

15. Kupfer, H.: Das Verhalten des Betons unter mehrachsiger Kurzzeitbelastung unter besonderer Berücksichtigung der zweiachsigen Beanspruchung. Deutscher Ausschuß für Stahlbeton, Vol. 229, Berlin: Wilhelm Ernst 1973

16. Floegl, H.; Mang, H.A.: On tension stiffening in cracked reinforced concrete slabs and shells considering geometric and physical nonlinearity. Ing.-Archiv 51 (1981) 215-242

17. Mang, H.A.; Gallagher, R.H.: On the unsymmetric eigenproblem for the buckling of shells under pressure loading. J. Appl. Mech. 105 (1983) 95-100

18. Trappel, F.: Vergleichende Beul- und Traglastuntersuchungen windbeanspruchter Kühlturmschalen aus Stahlbeton. Dipl.-Ing. thesis, Technische Universität Wien, in progress

19. Thomas, G.R., Gallagher, R.H.: A triangular thin shell finite element: Linear analysis. NASA Constractor Rep. 2482, Washington, D.C.: NASA 1975

20. Abel, J.F.: Private communication, 1981

Nonlinear Behavior of Cooling Tower Shells

I. Mungan, J. Ruhwedel and M. Winter

Institut für Konstruktiven Ingenieurbau,
Ruhr-Universität Bochum,
Bochum, West Germany

Summary

Being the most slender shell erected in reinforced concrete the veil of a cooling tower exhibits a pronounced nonlinear behavior, both geometrical and physical, combined with instabilities of bifurcation or limit point type. The study of the different aspects of this intricate problem is the objective of some recent research projects. One of the points needing clarification is the extension of the buckling concept based on the local critical stress states, as obtained under axisymmetric loading conditions, to the wind loaded case. Buckling tests on aeroelastic cooling tower models carried out in wind tunnel provide considerable information on the initiation and propagation of the buckling process in terms of measured versus calculated strains and deflections, for cases with and without stiffening rings. The effects of reinforced concrete nonlinearities and formation of cracks is studied numerically. The analysis is based on layered quadrilateral Semi-Loof Shell Elements with 32 degrees of freedom. Keeping the dead load constant and increasing the wind load incrementally the nonlinear behavior is tracked until failure occurs. In this way it is possible to assess some optimal design criteria with respect of the resistance against both instability and material failure. A productive discussion on a reasonable safety margin for cooling tower shells is only possible after extensive investigations of this kind.

Introduction

Nonlinear behavior of cooling tower shells is one of the main research subjects at the Ruhr-Universität Bochum. At the beginning the attention was focused on the bifurcation buckling of hyperboloidal shells subjected to axisymmetric loads. The observation done in tests was that nonlinearity originates locally at the weakest region of the shell. This fact led to the idea of basing the buckling design of a cooling tower shell on the worst stress state with respect of buckling along the

Circumference. This behavior was even more pronounced in tests on hyperboloidal models stiffened by means of rings. After a lot of numerical parametric investigations it was possible to develop a design procedure applicable both for unstiffened and stiffened cooling tower shells. In a further research work the assumption of linear elastic behavior of the shell material was abandoned. Introducing a biaxial material law with a failure criteria for concrete and a bilinear stress-strain law for reinforcing bars the effect of the physical nonlinearity on the bifurcation results was studied. In the opening lecture by Prof. Zerna the results of all these investigations carried out under axisymmetric load are already summarized. The objective of the present presentation is the nonlinear behavior under actual wind loading. In addition to wind tunnel tests on aeroelastic cooling tower models numerical investigations are carried out taking reinforced concrete properties, such as cracking and plastification of concrete, into account. The initial geometry of the unloaded shell is supposed to be perfect, however, with increasing wind load factor the geometric nonlinear effects are introduced into computations.

Buckling Tests in Wind Tunnel

As the first step of the buckling tests in wind tunnel only non-turbulent wind action was considered. In addition, the wind pressure created was constant along the height of the cooling tower models. The distribution of the internal and external wind pressures were first measured on rigid models. However, the stresses and strains induced through wind loading were measured on aeroelastic models by means of electrical strain gages. In Fig. 1 the position of the strain gages and the internal forces of a shell element are shown. The model material has a relatively great ultimate strain and behaves linearly elastic during the buckling process. In this way it is possible to use the same model in different tests. The models are manifactured casting the liquid Polyurethane between two molds. The internal mold consists of two detachable parts and the external mold of three segments.

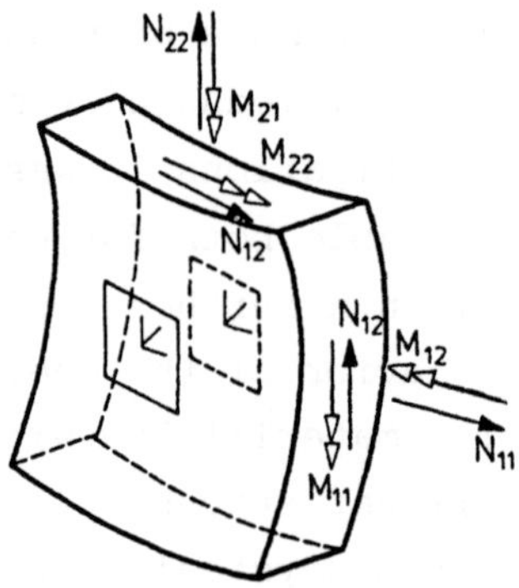

Fig. 1 Shell Element

After hardening of the model first the external mold is removed. The nearly 5 mm thick rough model is machined from the external surface to the desired thickness by means of a special lathe. In Fig. 2 the geometry of the cooling tower models tested is shown. This shape corresponds to the trend observed in recent large cooling tower shells having a large basis diameter nearly equal to the height of the cooling tower shell. The necking, as the ratio of the radii at the throat and at the base, is about 2/3. The shape of the meridian is a hypberbola with the parameters given in Fig. 2. The wall thickness is varied from 0,8 mm to 1,4 mm and is taken in some models constant along the height, in others, however, is changed, as in the case of real cooling towers. In this way it is possible to study the effect of the shell thickness on the buckling stresses and buckling behavior under wind loading. Again in some models the meridian is a circular arch giving a toroidal shape. The deviation between the toroid and hyperboloid is shown in the same figure. For both geometries the radii at the base and at the throat as well as the height are the same.

The other parameters investigated are the stiffening by rings, including the effect of the upper edge beam, and the effect of the supporting columns. The preliminary tests are carried out on cylindrical models. The coefficient for the maximum wind pressure at the windward meridian was max c_p = 1,93 and for the suction at the flanks min c_p = 0,85, the constant internal suction being c_{pi} = 0,95 is included. In Fig. 3 the buckling configurations of cylindrical and Cooling Tower shells are shown. The models at the left are without any stiffening means at the top, whereas those at the right have an upper stiffening ring.

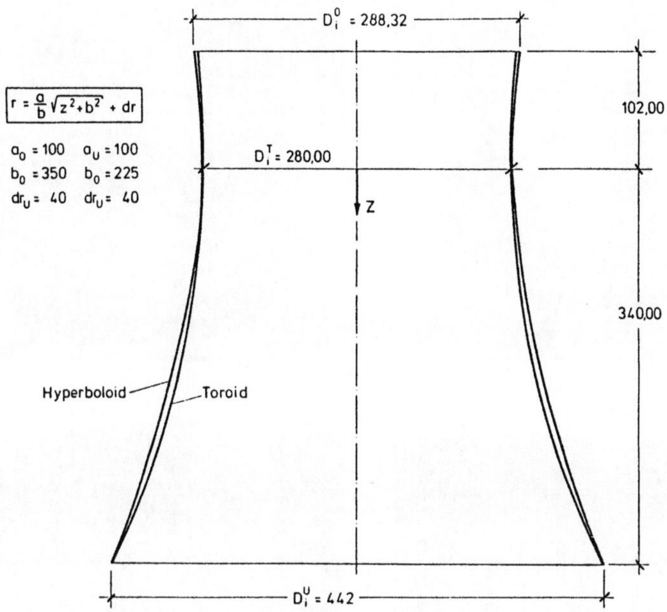

Fig. 2: Geometry of Cooling Tower Models

The effect of the depth, i.e. the stiffness, of a ring on the buckling wind pressure as observed in the tests is depicted in Fig. 4. q_o is the buckling pressure of the unstiffened model, t_M and b_R are the wall thickness of the shell and the depth of the stiffening ring, respectively. At the time of preparation of this paper the experimental work was not completed. The analysis of the test results including the strain gage measurements will yield the buckling stresses under wind loading. These stresses will be compared with the buckling stresses obtained under axisymmetric loads. In this way it will be possible to check the BSS-Approach, as presented in Ref. 1, for wind loading and to make modifications, if necessary.

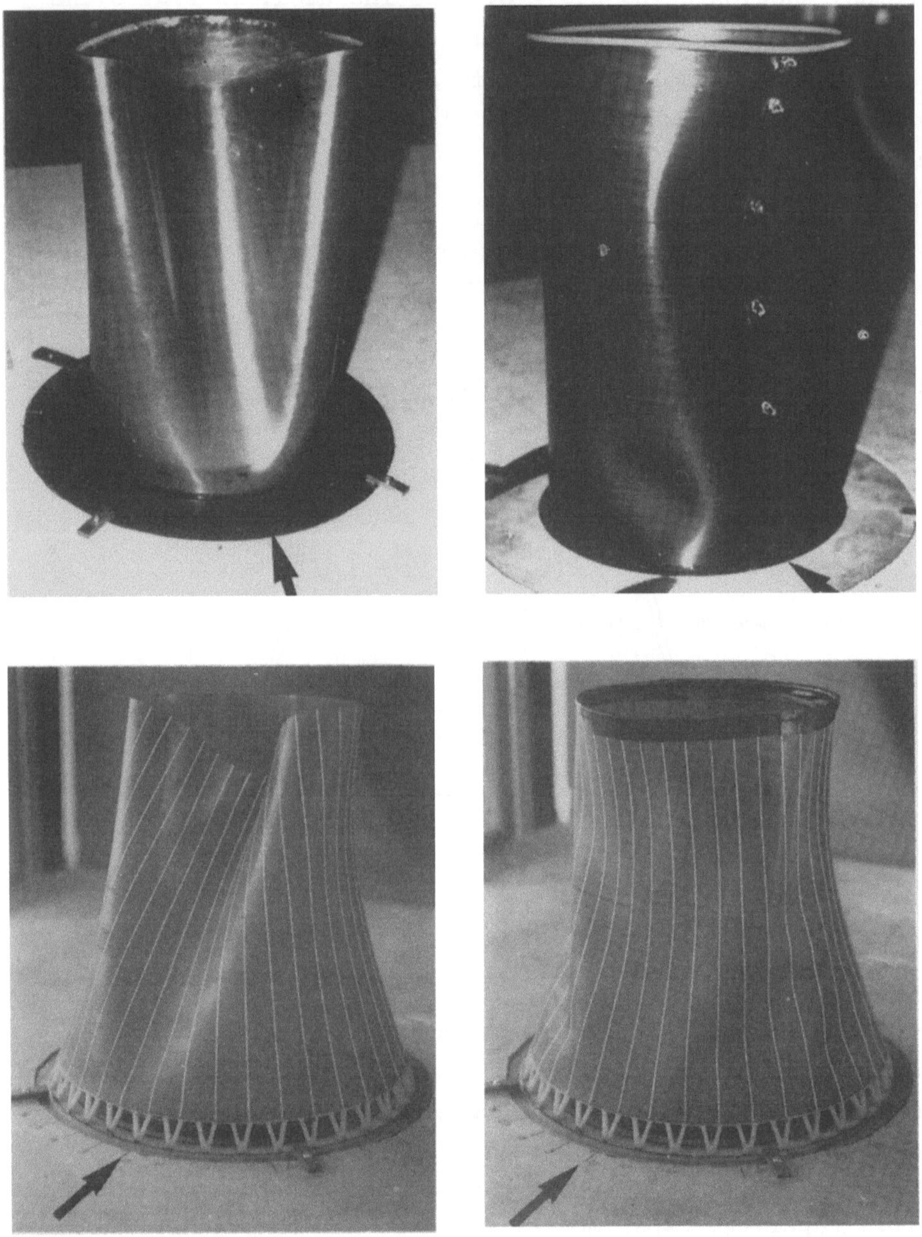

a) b)

Fig. 3: Wind buckling of Cylindrical and Cooling Tower Models
a) Without Upper Stiffening
b) With a stiff Upper Edge Beam

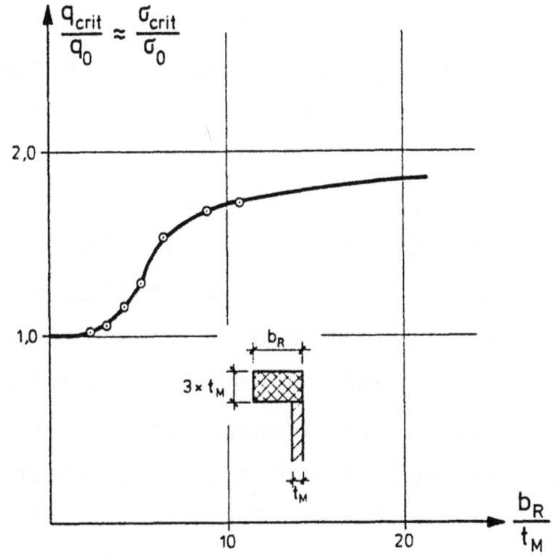

Fig. 4: Effect of the depth of a ring on the buckling wind pressure - test results on cylindrical model

Effects of RC-Nonlinearities

Wind tunnel tests provide information on the buckling behavior of shells made of linear elastic material. The question arising now is either these results can be extended to reinforced concrete shells or not, and if yes, how? Because of the very large sizes required, it is impossible to carry out buckling tests on reinforced concrete models in a wind tunnel. For this reason the only way to answer the question stated above is to investigate the problem numerically. In the case of reinforced concrete cooling tower shells failure can take place either due to instability caused through compressive stresses or due to tensile forces causing first cracking of concrete and afterwards the yielding of the reinforcing bars. These two aspects are coupled with each other and should be considered simultaneously. The recent design practise in Germany is to choose a relatively high safety factor, $\gamma_B = 5{,}0$, against buckling in order to find the shell thickness in different elevations. The amount of reinforcement, however, is obtained from the condition that the yield point of the reinforcement will be reached under dead load together with the wind load increased by a factor of $\varepsilon = 1{,}75$.

The objective of the present numerical analysis is first to find out, whether the buckling safety factor can be reduced. A reduction of γ_B results in reduction of the wall thickness and dead load. In addition, the deformations caused through wind increase, due to reduced rigidity of the shell. In the analysis an isoparametric SEMI-LOOF-Element of double curvature having 32 degrees of freedom is used. In Fig. 5 this element is shown. It is an efficient element due to its properties as explained in Refs. 2 and 3, and is developed further in order to match the requirements of the present investigation. Varying thickness along the element boundaries, different amounts of reinforcement in meridional and circumferential directions and a variable load intensity within an element are the improvements introduced. In addition in the kinematic equations the quadratic terms according to Eq. 1 are included.

$$\gamma_{\alpha\beta} = \alpha^L_{\alpha\beta} + \alpha^N_{\alpha\beta} + \Theta^3 \omega_{\alpha\beta} (\alpha, \beta = 1 \text{ or } 2) \qquad (1)$$

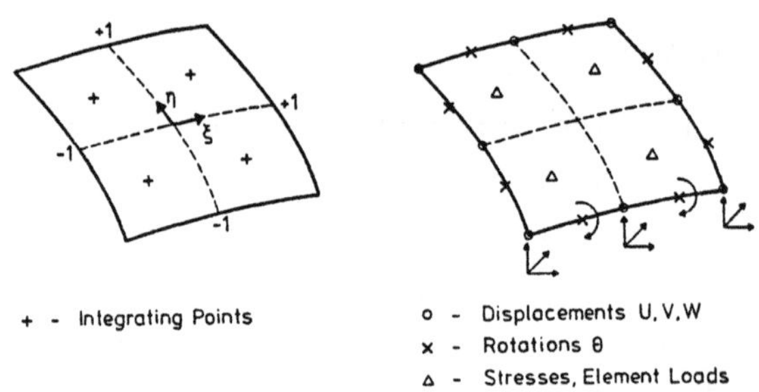

Fig. 5: "SEMI-LOOF" Shell Element with its 32 Degrees of Freedom

For concrete the biaxial stress-strain diagram is considered in the calculations [4] . Dividing each shell element in 11 layers, as shown in Fig. 7, it is possible to consider different material behaviors depending on the stress state of each layer. In computations the wind load is increased incrementally whereas the stiffness matrix of the foregoing load step is put in the equilibrium equation. The out-of-balance or residual forces resulting from the nonlinearity are added to the external nodal loads of the following step. The iterations are repeated so long

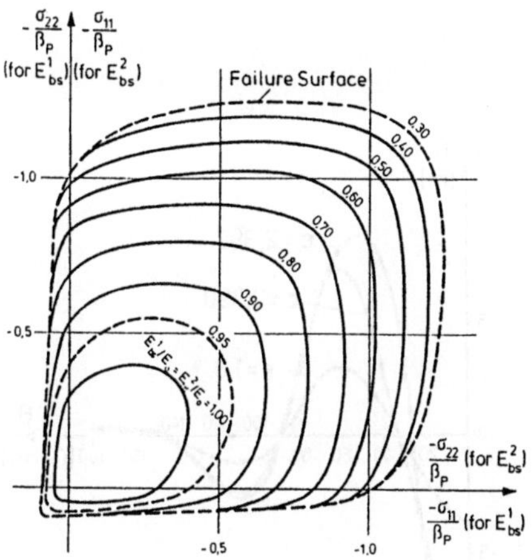

Fig. 6: Biaxial Material Law for Concrete

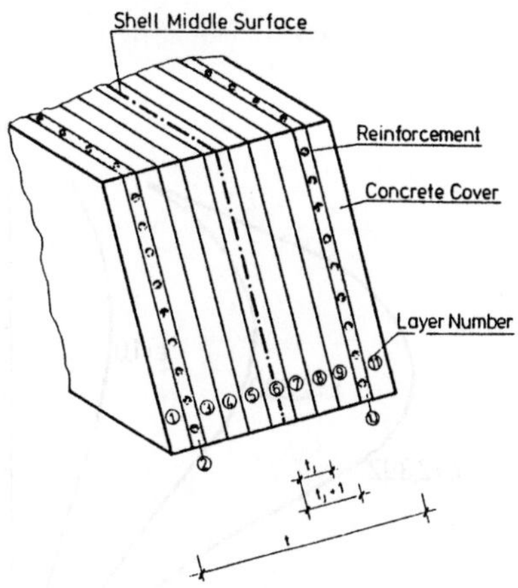

Fig. 7: Layered RC Shell Element

until the equilibrium condition is fulfiled within a certain accuracy limit. The cooling Tower shell investigated is shown in Fig. 8. The wall thickness is obtained taking the buckling safety factor once $\gamma_B = 5{,}00$ and afterwards $\gamma_B = 2{,}50$. Amount of reinforcement

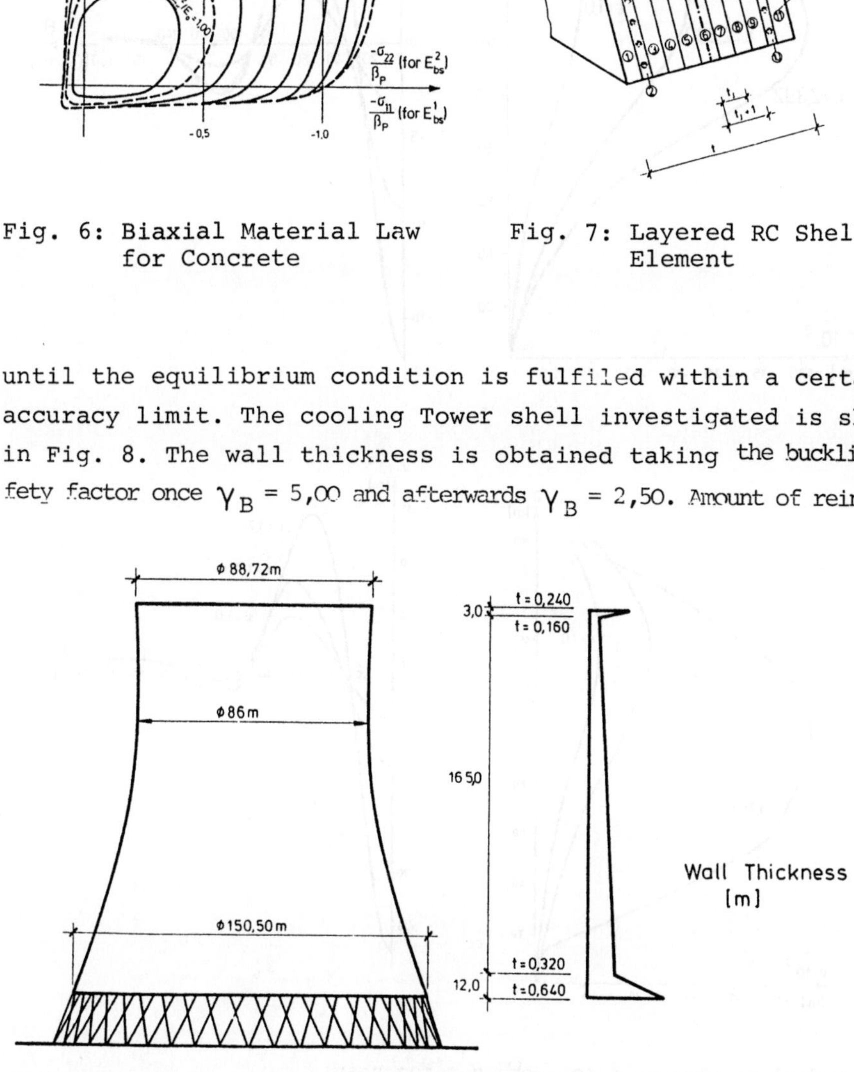

Fig. 8: Cooling Tower Shell Investigated

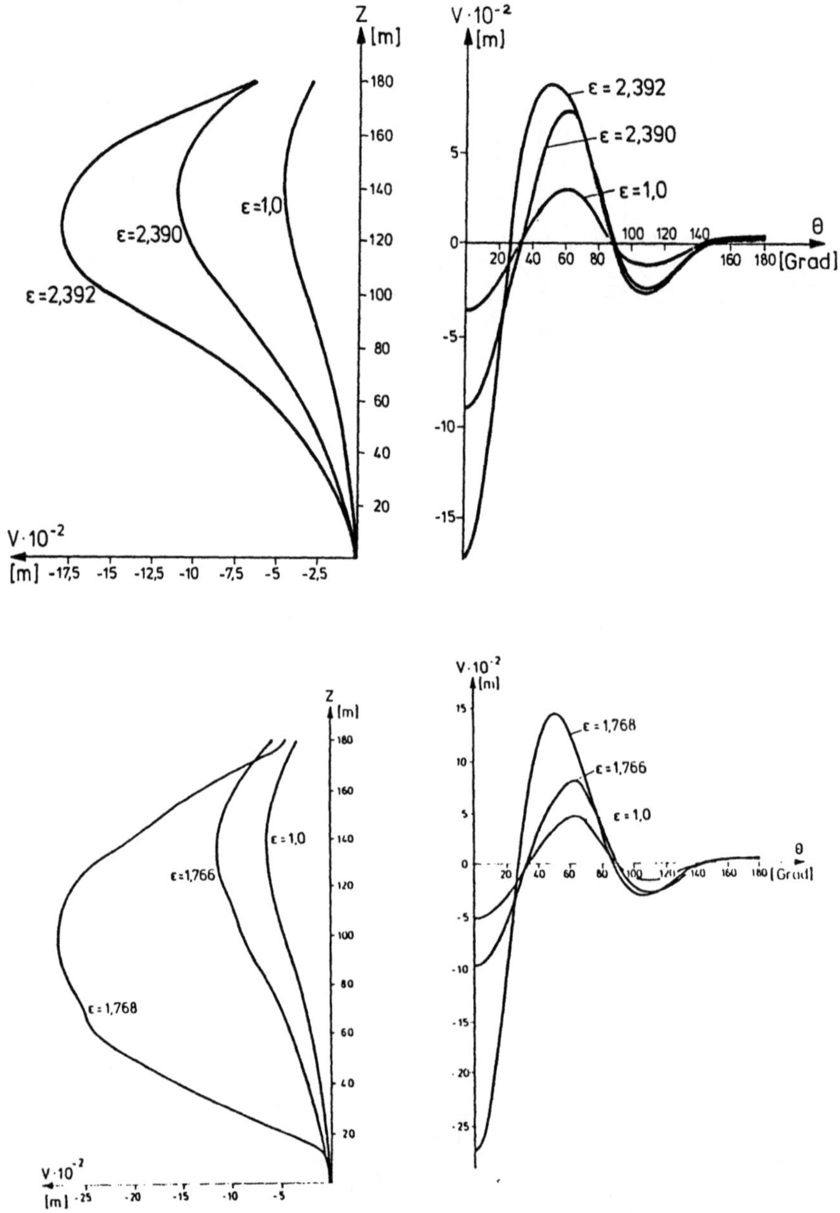

Fig. 9: Radial Displacement ($\Theta = 0^{O}$, Z = 127,5 m)

is taken as 0,3 %. As Fig. 9 demonstrates, the ultimate load of the cooling tower shell designed taking γ_B = 5,0 is reached at the wind load factor ε = 2,392, i.e. under g+2,392 W. For the cooling tower designed taking γ_B = 2,5 this limit drops to ε = 1,768, as shown in the lower part of the same figure. Finally at the left of Fig. 10 the variation of the meridional force at Element 22, the most critical element along the wind-ward meridian, is depicted with increasing wind load factors for three design buckling safety factors. In all cases failure starts as the whole concrete section is cracked at any location of the cooling tower shell.

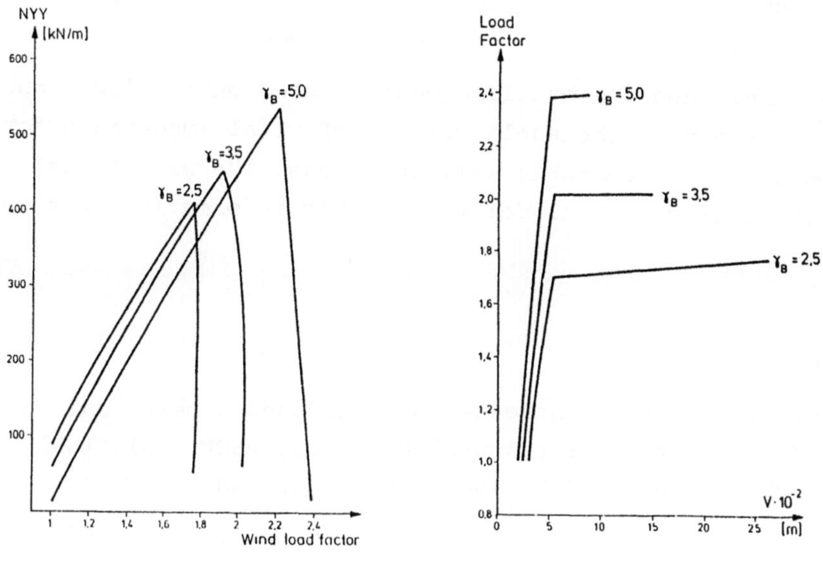

Fig. 10: Stress and Displacement of the most Critical Shell Element due to Increasing Wind Load Factor

Concluding Remarks

Following conclusions can be drawn from the parametric investigations carried out until now:

- There is a strong correlation between the buckling safety factor γ_B chosen and the wind load factor causing tensile failure. Increasing the buckling safety factor the wall thickness required and the prestressing through dead load of the cooling tower shell increase, respectively. Consequently cracking of concrete occurs at higher wind load levels.

- Complete cracking of any section is symptomatic for the failure of the whole cooling tower shell. As this stage is reached the wind load can no more be increased whereas the membrane force acting in the cracked section drops abruptly.
- For an optimal design the amount of reinforcement to be placed in the shell wall should be related to the tensile strength of the concrete section. The assumption of a certain concrete tensile strength in computations raises the wind load factor causing the failure of the shell. Up to this range the behavior of the cooling tower shell is nearly linear, as depicted in Fig. 10. For this reason it seems to be adequate to carry out an analysis according to the linear elaatic shell theory, as this is already done in the cooling tower practice.
- On the other hand the buckling design based on the local buckling resistance of the shell wall is of vital importance for the choice of a reasonable wall thickness. The wind tunnel tests in processing will deliver further information on this aspect.

References

1. Mungan, I., "Buckling Stresses of Stiffened Hyperboidal Shells," Journal of the Structural Division, ASCE, Vol. 105 No. ST 8, Proc. Paper 14775, Aug., 1979, pp. 1589-1604

2. Zienkiewicz, O.C. "The Finite Element Method," Third Edition, McGraw-Hill, London 1977.

3. Irons, B.M., "The semi Loof Shell Element", printed in Finite Elements for Thin Shells and Curved Members, John Wiley & Sons, London 1976.

4. Zerna, W., Mungan, I. and Steffen, W., "Wind-Buckling Approach for RC Cooling Towers, " ASCE Journal of Engineering Mechanics, Vol. 109, No. 3, June 1983, Proc. Paper 18032, pp. 836-848.

Comparison of Complete and Simplified Elastic Buckling Loads for Cooling Tower Shells

J. F. ABEL*, S.-C. CHANG**, AND S. L. HANNA***

Department of Structural Engineering and
Program of Computer Graphics
Cornell University
Ithaca, New York 14853
U.S.A.

Summary

By direct comparison of numerical results for a number of cooling tower geometries, various analytical design tools are evaluated for stability checks of cooling tower shells subjected to quasistatic wind loadings. The most complete and expensive method is full geometric nonlinear analysis, while the simplest and cheapest method is a bifurcation analysis for an equivalent axisymmetric loading. For the geometries considered, the equivalent axisymmetric loading approach provides a conservative estimate of the elastic stability limit.

Introduction

In normal American cooling-tower design practice, it is intended that the strength rather than the stiffness controls the proportioning and reinforcement of the shell. In other words, as loadings are increased to their factored design level, the shell is designed to fail by cracking and crushing of the concrete and yielding of the steel instead of by buckling or instability.

Because various design factors have different effects on the stiffness than they do on the strength of the shell, it is possible that a shell with adequate strength to sustain a factored design load may be subject to buckling at too low a load level. For example, the degree of upper-edge stiffening in

* Professor of Structural Engineering.
** Presently Associate Professor of Civil Engineering, National Taiwan University, Republic of China. Formerly Graduate Research Assistant, Cornell University.
*** Presently CAD Systems Development Engineer, 3D/EYE, Inc., Ithaca, NY 14850, USA. Formerly Graduate Research Assistant, Cornell University.

the form of either gradual thickening or a cornice ring may have little effect on the overall strength of the shell but will dramatically affect the buckling resistance at the rim. Therefore, an important part of design is a stability check to ensure that the elastic buckling load is significantly higher than the strength limit state.

One particularly important factor which affects stability while having little influence on strength is the vertical (meridional) microcracking that can be caused by thermal gradients through the thickness of the shell during normal operation [1]. This degradation of circumferential stiffness can sharply reduce the buckling loads [2]. In addition, there are other uncertainties or deviations from the ideal, such as geometric, load, and fabrication imperfections, which tend to have disproportionate effects on stability. Therefore, American practice calls for a minimum load factor of 1.3 for strength design of reinforced concrete cooling tower shells while the factor of safety for buckling, based on an analysis which accounts for meridional cracking, is at least 2.0 [3].

One task facing designers is a continued re-evaluation of the analysis tools used in design. As engineering analysis advances through research, new methods become available either for direct adoption in design or for re-evaluation of older techniques. In many cases, the new sophisticated methods are too complex or expensive to be used throughout the design process but nevertheless yield significant insights useful to the designer. For example, with respect to the strength design of cooling-tower shells, Mang et al. [4] have performed ultimate load predictions for a single typical cooling tower. Although their computationally expensive procedure would be prohibitive for day-to-day design, they reach an important conclusion that linear elastic nonaxisymmetric bending analysis, such as is typically used in modern design, is sufficient to estimate a lower bound to the ultimate strength.

The principal objective of this paper is to compare various buckling analysis procedures for wind loadings to ascertain which simplified methods are acceptable for design practice in terms of both economy and reliability. Although the geometrically nonlinear analysis of shells is not new and has been applied to cooling towers by several earlier investigators, e.g., Refs. [4], [5], [6], [7], and [8], the writers have developed and applied an integrated finite element geometrically nonlinear shell analysis system which employs interactive computer graphics [9]. This approach has significantly alleviated the labor necessary to prepare an analysis and has therefore permitted several analyses to be performed for each of eight different published cooling tower configurations. Full geometrically nonlinear analysis to determine limit points is used as the basis of comparison, and the results of various more simplified analyses are contrasted to these limit points.

Geometrically Nonlinear Analysis

The incremental matrix equations for the elastic, geometrically nonlinear analysis of a structure, discretized by an approach such as the finite element method, are [9] [4]:

$$[K_0 + K_\sigma(u) + K_{u1}(u) + K_{u2}(u^2) + K_p(u)]\{\Delta u\} = \{\Delta P\} \quad (1)$$

in which $\{\Delta u\}$ and $\{\Delta P\}$ are the incremental nodal displacements and loads; $\{u\}$ are the total nodal displacements; $[K_0]$ is the linear stiffness matrix; $[K_\sigma]$ is the initial-stress stiffness matrix, a linear function of $\{u\}$; $[K_{u1}]$ and $[K_{u2}]$ are the portions of the initial-displacement stiffness matrix which are linear and quadratic functions of $\{u\}$, respectively; and $[K_p]$ is the symmetrized pressure-rotation stiffness matrix, a linear function of $\{u\}$. To solve for the instability load of a structure, a designer usually employs a computer program which incorporates either Eq. (1) or a simplification of it.

The most complete and accurate, and also the most computationally expensive, method of analysis is <u>full nonlinear analysis</u> to detect a limit point. One way to accomplish this is to increment the loads from their initial, zero value and to perform Newton-Raphson iteration within each increment to satisfy equilibrium. This solution will diverge at or above a limit point, and the limiting load will be bounded from below by the load level at which the last nondivergent solution is obtained. Alternatively, one can detect a limit point by seeking, during the step-by-step analysis, the vanishing of the determinant of the composite stiffness matrix in Eq. (1).

To achieve an estimate of the load causing instability more economically, the problem described by Eq. (1) is often recast as a bifurcation or buckling problem in terms of a load parameter μ. This results in an eigenvalue problem whose lowest root μ_1 is the multiplier associated with the critical loading, i.e., $\{P_{cr}\} = \mu_1\{P\}$. Usually, Eq. (1) is linearized in the formulation of the eigenvalue problem, and one assumes that the shell is in an unstressed initial state, e.g., the dead load stresses existing prior to the application of wind load are negligible. Direct linearization leads to

$$\det([K_0] + \mu\,([K_\sigma] + [K_{u1}] + [K_p])) = 0 \qquad (2)$$

which is known as <u>linearized buckling analysis</u>. Alternatively, if the prebuckling load-deflection paths are substantially linear, as has been observed in several full nonlinear analyses [4], [5], [8], [9], $[K_{u1}]$ is also neglected, and one obtains

$$\det([K_0] + \mu\,([K_\sigma] + [K_p])) = 0 \qquad (3)$$

which is known as <u>classical buckling analysis</u>.

Although the eigenproblem approaches are computationally less expensive than full nonlinear analysis, they nevertheless require more computation than may be practical in a design process wherein many configurations are evaluated. Because of the nonaxisymmetric nature of the wind loadings, at least a full 180° of the shell must be included in the numerical model, and this entails not only significant computational time but also laborious data preparation. To alleviate this effort while still obtaining an estimate of the wind buckling load which takes into account the actual geometry and boundary conditions, equivalent axisymmetric buckling methods have been employed. The most promising of these is the equivalent axisymmetric loading analysis in which a bifurcation buckling analysis is performed for an axisymmetrical pressure equal in intensity to the external pressure plus the internal suction along the windward meridian of the shell [10]. For axisymmetric analysis, especially of ideal, negatively curved shells, Eqs. (2) and (3) provide essentially the same result. The advantage of the use of the simplified load, however, is that the analytical model can be reduced by one dimension (e.g., from surface to ring elements).

One further simplification is possible: the neglect of the pressure-rotation stiffness $[K_p]$ in Eqs. (2) and (3). Indeed, this simplification is often inadvertant since many available computer programs do not even provide the opportunity to include the pressure-rotation effect. This phenomenon is a reduction in stiffness due to the pressure continuing to remain normal to the shell surface as rotations and buckling occur. The effect is well known for the classical problem of axisymmetrically compressed circular rings, for which the neglect of this influence leads to a calculated critical pressure which is 1/3 too high [11]. If the pressure-rotation stiffness is to be neglected in cooling tower buckling analysis, the quantitative effects of this simplification for typical cooling tower shell configurations must be understood.

Numerical Investigations

The various analysis approaches have been applied to eight different cooling-tower shell configurations which have been published and used in earlier studies of stability, e.g., Ref. [12]. Unless otherwise noted, the shells investigated are hyperboloids. The first four shells are configurations tested for buckling in a wind tunnel by Der & Fidler [13]. The shells are here designated DF1 through DF4 and are the four models considered in Refs. [10] and [12]. A recent restudy of the actual models used in the experiments indicated that some of the properties, boundary conditions, and loadings presented in Refs. [12] and [13] are incorrect. The corrected values given in Ref. [10] are used in the present analyses, while all other parameters are the same as in Ref [12]. The next tower is the Ferrybridge-like shell described by Albasiny and Martin [14] and designated here A&M. It has been analyzed for buckling in Ref. [8] and is considered here with the same properties and loading as used in Refs. [8] and [12]. Mang et al. [4] [5] [15] have analyzed the Port Gibson shell, designated here as GIB. This shell is an hyperboloid above the throat and an ellipsoid below the throat. The geometry, properties, boundary conditions, and reference wind loading described in detail in Ref. [4] are employed. The Trojan cooling tower has been analyzed for buckling with a variety of stiffening, support, and load conditions by Cole et al. [2]. The configuration selected for study here and designated TROJ is the Case 2 "Simplified Trojan" of Ref. [2]. The geometry reported in Ref. [2] is incomplete, and the geometry used here, taken from the actual computer runs used by Cole et al., is an hyperbola from the upper edge at (r_1, z_1) = (124.946', 98.4') to (r_2, z_2) = (175.579', -325.922'), and a cone from (r_2, z_2) to (r_3, z_3) = (181.671', -350.175'). The additional information is reported in Ref. [2]. Finally, the eighth shell considered is the first of five studied by Yeh & Shieh [7] and subsequently by Cole et al. [12]. The properties and wind loading given in Ref. [12] are employed, with an internal suction of 0.5, except that the thickness is assumed to vary linearly from 19.2 in. at the top (z = 726 in.) to its nominal 5.12 in. at z = 581 in. and that the base restraints are of the meridional and circumferential translations.

The results of the current full nonlinear analyses are summarized in Table 1 and compared with experimental results or other available full nonlinear analyses. The writers are unaware of alternative solutions for the DF shells with corrected values of properties or for the TROJ shell. For the DF shells, the analytical results are considerably higher than the experimental, especially for the copper models (DF3 and DF4). Generally, these differences arise naturally from the

Table 1. Full Nonlinear Predictions of Critical Pressures

Cooling Tower	Mesh† (nxm)	Units	Present Work	Experiments or Other Solutions
DF1	21x13	psi	0.481	0.351 exp. [14]
DF2	21x13	psi	1.352	1.090 exp. [14]
DF3	21x13	psi	0.959	0.444 exp. [14]
DF4	21x13	psi	1.449	0.747 exp. [14]
A&M	21x13	psf	252	250-275 soln. [8]
GIB	18x12	*	19.5	≐ 20 soln. [4]
TROJ	18x10	psf	538	-
YS1	18x13	psi	3.3	2.2^{+} soln. [7]

† n = no. of quads in 180° circumferentially, m = no. of quads meridionally. Each quad = 2 triangular finite elements [9].
* Results given as multiple of design or reference load [4].
\+ Coarse mesh of flat elements; no suction.

analytical idealization of the true (imperfect) geometry, loading, and properties of the models. For the A&M and GIB shells, the agreement with other solutions is excellent, while the YS1 solution from Ref. [7] appears to be too low [12].

In Table 2, the various simplified buckling analyses results are compared to the full nonlinear buckling predictions in terms of ratios to the full nonlinear limit pressure. The linearized buckling results are all significantly lower (15 to 76%) than the full nonlinear, while the classical buckling results range from 4% below to 22% above. The equivalent axisymmetric load analysis also yields predictions on the conservative side, ranging from 21 to 47% below the full nonlinear. In Ref. [10], it is shown that the equivalent axisymmetric load method (with suction) gives good agreement with the DF experiments, and it is argued that the method is a useful compromise between accuracy and economy as it appears to give a lower-bound estimate of limiting wind pressures of cooling tower shells. Mang et al. [4] point out that the equivalent

Table 2. Wind Buckling Pressures as Multiples of the Full Nonlinear Predictions of Table 1

Cooling Tower	Linearized w/[K_p]	Classical		Equiv. Axisym. Load	
		w/[K_p]	w/o [K_p]	w/[K_p]	w/o [K_p]
DF1	0.35	1.09	1.10	0.75	0.77
DF2	0.28	1.14	1.15	0.70	0.73
DF3	0.60	1.02	1.02	0.58	0.58
DF4	0.57	0.98	0.98	0.58	0.58
A&M	0.41	1.02	1.03	0.70	0.71
GIB	0.25	0.96	-	0.53	-
TROJ	0.24	1.22	1.24	0.75	0.77
YS1	0.85	1.17	1.19	0.79	0.81

axisymmetric loading analysis is most appropriate when the circumferential compressions are most critical than the wind-induced meridional compressions which are not accounted for in the axisymmetric simplification. However, in attempting to demonstrate that the axisymmetric buckling pressure is not always a lower bound to the critical wind pressure, Mang et al. make comparisons with linearized predictions rather than full nonlinear or classical analyses and they also employ configurations (GIB with all r's proportionally scaled by 1.5 or 2.5) which are not typical of actual cooling tower shells.

Conclusions

A number of conclusions can be drawn on the basis of Table 2:

1. The relatively close agreement of the classical buckling and full nonlinear analyses indicates that the prebucking behavior of the shells considered is essentially linear.

2. The linearized buckling predictions are in every case lower than both the full nonlinear and classical buckling. This indicates that the linear and quadratic terms of the initial-displacement stiffness matrix tend to offset each other for the configurations considered.

3. The pressure-rotation stiffness effect is not signi-

ficant in the buckling of wind-loaded cooling tower shells. The differences due to this effect observed in this study range from 0 to only 3% of the full nonlinear limiting load.

4. Although both the equivalent axisymmetric pressure method (with suction) and the linearized buckling method both give wind-load buckling predictions on the safe side, the former, as a much less expensive and more widely available capability, is the more effective tool for cooling-tower designers to ensure that the elastic buckling load is significantly higher than the strength limit state. Moreover, for the DF1 and DF2 shells, the equivalent axisymmetric loading method gives closer agreement with the wind-tunnel experiments [2].

Finally, it should be restated that none of the analyses given here include the important effect of thermally induced vertical (meridional) microcracking [1]. However, as shown in Ref. [2], this effect may be simulated by any of the analytical procedures mentioned herein with an appropriately reduced circumferential stiffness provided the capability to model orthotropic properties is included within the implementation of the procedure. The use of a reduced stiffness to simulate cracking has been examined experimentally in Ref. [16].

Acknowledgements

The research reported has been partially supported by a grant to Cornell University by Research-Cottrell, Inc., of Somerville, New Jersey, and Hamon-Sobelco, S.A., of Brussels. Portions of the related work were sponsored by the U.S. National Science Foundation Grants ENG-7517400, ENG-7722568, and CME-7916818. The opinions expressed are those of the writers and not of any sponsoring organization.

References

1. Larabee, R. D.; Billington, D. P.; Abel, J. F.: Thermal loading of thin-shell concrete cooling towers. J. Struct. Div., ASCE 100 (1974) 2367-2383.

2. Cole, P. P.; Abel, J. F.; Billington, D. P.: Buckling of cooling-tower shells: Bifurcation results. J. Struct.

Div., ASCE 101 (1975) 1205-1222.

3. American Concrete Institute: Reinforced Concrete Cooling Tower Shells -- Practice and Commentary, Report ACI 334.2R, latest revision to appear 1984.

4. Mang, H. A.; Floegl, H.; Trappel, F.; Walter, H.: Wind-loaded reinforced-concrete cooling towers: Buckling or ultimate load? Engrg. Struct. 5 (1983) 163-180.

5. Mang, H.; Gallagher, R. H.; Cedolin, L; Schwinden, W. D.: Finite element instability analysis of hyberbolic cooling towers. Advances in Civil Engineering Through Engineering Mechanics, New York: ASCE 1977, 246-249.

6. Maury, J. F.: Stabilite elastique des refrigerants. Proceeding IASS Symposium on Very Tall Reinforced Concrete Cooling Towers, Paris: IASS, 1978.

7. Yeh, C. H.; Shieh, W. J.: Stability and dynamic analysis of cooling towers. J. Power Div., ASCE 99 (1973) 339-347.

8. Chan, A. S. L.; Wolf, J. P.: Cooling tower supporting columns and reinforcing rings in small and large displacement analysis. Comp. Meth. Appl. Mechs. Engrg. 13 (1978) 1-26.

9. Chang, S. C.: An integrated finite element nonlinear shell analysis system with interative computer graphics, Ph.D. Thesis, Cornell University, 1981.

10. Abel, J. F.; Billington, D. P.; Nagy, D. A.; Wiita-Dworkin, C.: Buckling of cooling towers. J. Struct. Div., ASCE 108 (1982) 2162-2174.

11. Bodner, S. R.: On the conservativeness of various distributed force systems. J. Aero. Sci. 25 (1958) 132-133.

12. Cole, P. P.; Abel, J. F.; Billington, D. P.: Buckling of cooling-tower shells: State-of-the-art. J. Struct. Div., ASCE 101 (1975) 1185-1203.

13. Der, T. J.; Fidler, R.: A model study of the buckling behavior of hyperbolic shells. Proc. Instn. of Civ. Engrs. 41 (1968) 105-118.

14. Albasiny, E. L.; Martin, D. W.: Bending and membrane equilibrium in cooling towers. J. Engrg. Mechs. Div., ASCE 95 (1967) 1-17.

15. Mang, H.; Gallagher, R. H.; Cedolin, L.; Torzicky, P.: Deformation und Stabilität windbeanspruchter Kühlturmschalen. Ing.-Archiv. 47 (1978) 391-410.

16. Hayman, B.: Flexural properties of thin reinforced concrete members. Proc. Instn. Civ. Engrs. 65 (1978) 253-269.

The Effect of Cracking on the Ultimate Load of Reinforced Concrete Cooling Towers

R V Milford (i) and W C Schnobrich (ii)

i) National Building Research Institute, Pretoria, South Africa, formerly Research Assistant, University of Illinois, USA.
ii) Professor of Civil Engineering, University of Illinois, USA.

SUMMARY.

The effect of structural integrity on the nonlinear behavior of reinforced concrete cooling towers is presented. The structure is discretized using the degenerated Lagrangian shell element, and cracked concrete is modeled using the rotating crack algorithm. It is shown that the post-cracking behavior of cooling towers is very dependent on the assumptions made which influence the load carried by the concrete in tension. This study concludes that the cracking strength of cooling towers must be investigated during the design stage, possibly to satisfy a lower factor of safety than that required at ultimate.

1) INTRODUCTION.

The safety of cooling towers under wind loading has been the subject of extensive research over the past twenty years or so, the majority of which has centered around the elastic buckling of cooling tower shells. This has resulted in several wind tunnel investigations, additional experimental investigations and numerous analytical investigations (See Refs. 1 and 2).

Recently, Mang, et al. (3) investigated the failure of reinforced concrete cooling towers, taking into account material nonlinearities, cracking of the concrete and yielding of reinforcement while including geometric nonlinear effects. This reference concluded that reinforced concrete cooling towers are not likely to buckle in the classical sense, but rather fail by cracking of the concrete and finally by yielding of the reinforcement.

The present investigation examines further the failure of reinforced concrete cooling towers, and investigates the importance of the structural integrity of the tower and the important role played by the concrete in tension. The problems in assessing the failure, or ultimate load, of the tower are discussed.

2) FINITE ELEMENT DISCRETIZATION.

The finite element used in this investigation is the 9-node degenerated Lagrangian shell element. This element is formulated as a thin shell element using explicit integration through the shell thickness. Using this approach it is possible to use a stress-resultant constitutive matrix which results in significant savings in computational effort. Material nonlinearities are accounted for using a layered material model and smeared reinforcement layers. Cracking is modeled in an average sense using the smeared crack approach.

3) MATERIAL MODEL.

The material model used is based on the biaxial orthotropic material model proposed by Liu, et al. (4). Cracking of concrete in tension is given by a maximum principal stress criterion.

Cracked concrete is modeled using the 'rotating crack model' in which the direction of the crack is not fixed after formation of the crack, but rather is assumed to be normal to the direction of the maximum principal strain. The cracks defined by this model are not cracks in the strict sense, but rather notational cracks defining the average crack direction. This model has been shown to yield significantly better results than the conventional fixed crack models, in particular for anisotropically reinforced sections. Further details can be found in Ref. 5.

It will be demonstrated in the numerical investigation to follow that tension stiffening has a significant affect on the predicted response of cooling towers. Tension stiffening has been modeled in this investigation as a gradual unloading of the concrete stress strain curve in tension, as shown in Fig. 1. In this figure β is a measure of the length of the unloading part of the stress strain curve and ϵ_t is the crack strain of the concrete.

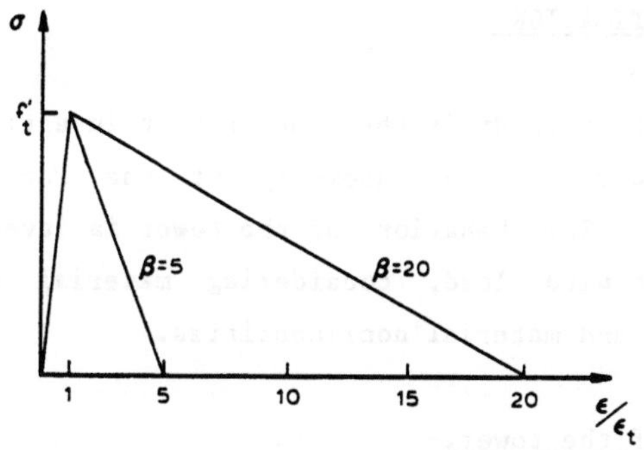

Fig. 1. Tension Stiffening Model.

There is little experimental data available on tension stiffening, in particular for lightly reinforced sections subject to inplane loading. Some measurements have been made by Vecchio and Collins (6), and these indicated significant tension stiffening effects. These results are shown in Fig. 2 together with the present model for $\beta\epsilon_t = 2.0\,\mu\epsilon$ ($\beta \simeq 20$ for $f'_t = 3.0$ MPa). These results indicate that for $\beta\epsilon_t \simeq 2.0\mu\epsilon$, the present model represents a lower bound to the experimental results but is not unduely conservative. It should be noted however that the experimental results shown in Fig. 2 were obtained on relatively heavily reinforced specimens in at least one direction (1.875%). Measurements of tension stiffening effects on specimens subject to uniaxial tension parallel to the direction of the principal reinforcement are not available.

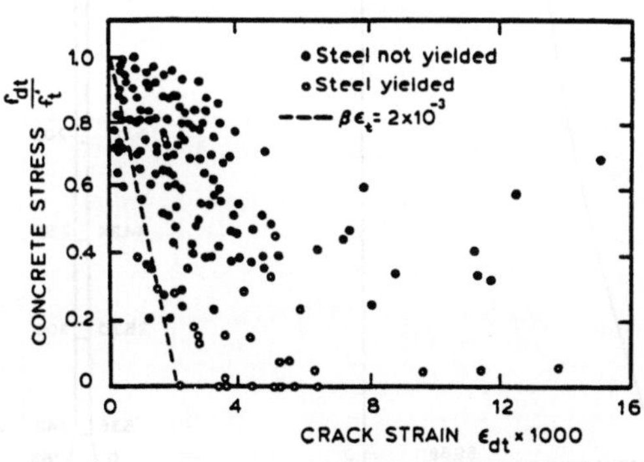

Fig. 2. Experimental Tension Stiffening Results (6).

4) NUMERICAL INVESTIGATION.

The tower used in this study is the same as that investigated in Refs. 3, 7 and 8. This tower has the geometry of the Port Gibson Tower in Mississippi, USA. The behavior of the tower is investigated under dead load and increasing wind load, considering material nonlinearities and combined geometric and material nonlinearities.

4.1) Description of the tower.

The geometry of the shell of the tower and the wall profile is given in Fig. 3. The tower stands 150.5 m (494 ft) tall and has a mininum wall thickness of 203 mm (8 in). The radius of the tower at the throat is 36.3 m (119 ft) and at the base 59.7 m (196 ft). The shell is thickened locally at the top and base to form ringbeams. Further details of the geometry of the shell can be found in Refs. 3 and 8.

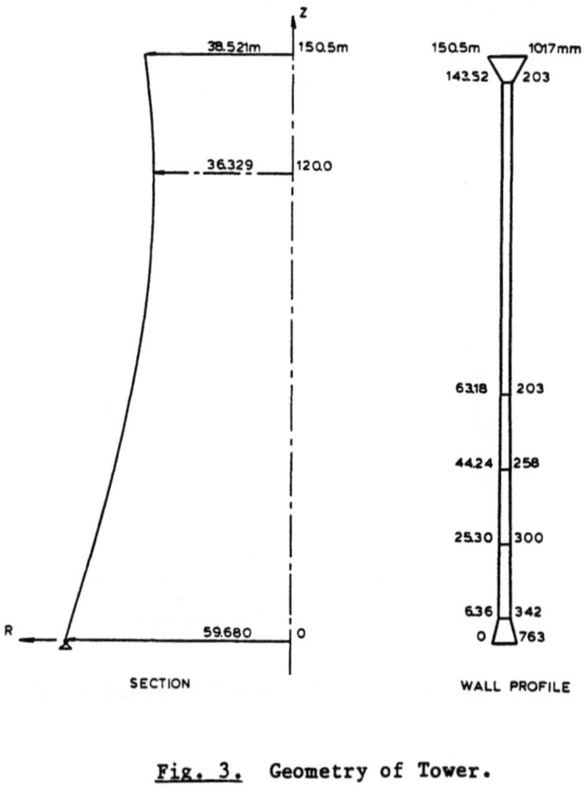

Fig. 3. Geometry of Tower.

The material properties of the concrete and reinforcement are similar to those used in Ref. 3, except for a slightly reduced tensile strength of the concrete and the reinforcement quantities. The reinforcement quantities have been taken from Ref. 7 in which the reinforcement has been designed in accordance with the ACI-ASCE recommendations (9) together with the procedure given in Ref. 10. In the upper one third of the tower the meridional reinforcement is controlled by the 0.35% minimum requirement, while the circumferential reinforcement is controlled throughout by the 0.35% minimum. The reinforcement used in this investigation (and Ref. 7) is slightly less than that used in Ref. 3.

Half of the cooling tower shell has been idealized using 90 quadratic Lagrangian shell elements- 10 elements along the circumference and 9 in the meridional direction. The simplifying assumption of a hinged base and a rigid foundation has been made.

The tower has been analyzed for combined dead and wind loading. The wind loading used is a codified psuedo-static approach in which it is assumed that the tower behaves quasistatically to a particular (design) gust load. The vertical and circumferential wind load distribution used is taken from Ref. 3, which is similar to that used in the current ACI-ASCE recommendations for exposure C open terrain.

$$p(z,\theta) = q_o H(z)\ G(\theta) \tag{1}$$

,where H(z) is the vertical distribution of the design wind pressure profile

$$H(z) = (z/10)^{2/7} \tag{2}$$

and $G(\theta)$ is the circumferential pressure distribution, including internal suction

$$G(\theta) = \sum A_n \mathrm{Cos}(n\theta) \tag{3}$$

The Fourier coefficients, A_n, are given in Refs. 3 and 8.

q_o is the design dynamic head at the reference 10m elevation. A reference windspeed of 40.2 m/s (90 mph) which is typical of a 100 year return period has been used. (Note that in Ref. 3 a corresponding reference wind speed of 42.7 m/s (95 mph) has been used.)

The meridional forces due to dead, reference wind load and combined dead plus wind load acting along the windward meridion are shown in Fig. 4. Also included in Fig. 4 is the yield strength of the meridional reinforcement, $A_s f_s$, and the cracking strength of the tower in the meridional direction, $N_{tc}+N_s$.

All the results presented in this investigation are expressed in terms of a normalized load factor, λ, obtained from the reference wind load distribution as

$$\lambda = \frac{p\ (v)}{p\ (40.2m/s)} \tag{4}$$

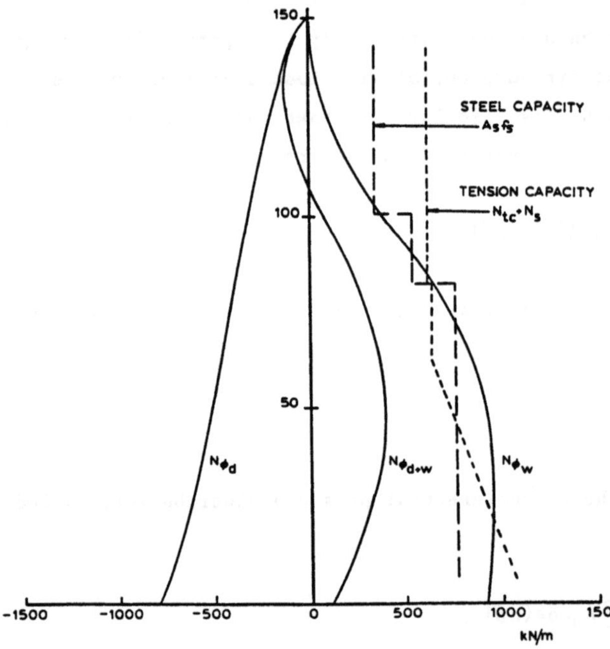

Fig. 4. **Meridional Forces and Design Strengths.**

4.2) Numerical results.

A parameter study consisting of four investigations has been carried out, namely

1. **TOWER I: Material and geometric nonlinearities included, tension strength of concrete (f'_t) equal to 3.0 MPa and the tension stiffening factor β equal to 20.**

2. **TOWER II: Material nonlinearities included but geometrically linear, f'_t equal to 3.0 MPa and β set to 20.**

3. **TOWER III: Material and geometric nonlinearities included, f'_t equal to 3.0 MPa and β reduced to 5.**

4. **TOWER IV: Material and geometric nonlinearities included, but tension strength of the concrete reduced by half to 1.5 MPa, and a tension stiffening factor, β, of 10.**

Tower I has been adopted as the reference tower for this investigation.

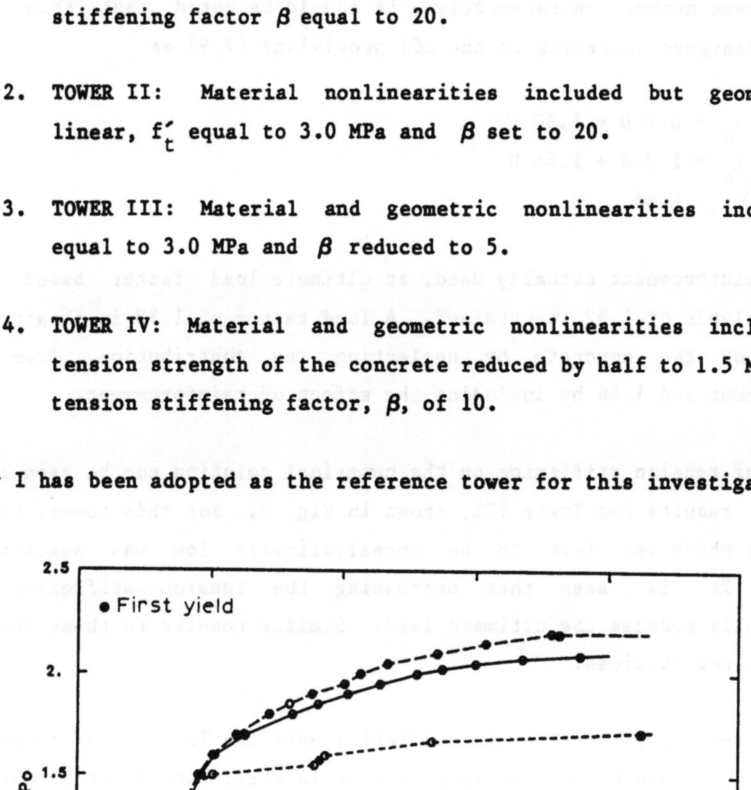

Fig. 5. Load-Displacement Curves for Windward Meridian.

The load displacement behavior for the tower, at an elevation $z = 101$ m (331 ft) on the windward meridian, is shown in Fig. 5. An ultimate load factor of approximately 2.10 is obtained for Tower I, 2.25 for Tower II, 1.725 for Tower III and 1.35 for Tower IV. The load factor at which cracking occurs is approximately 1.45 for the first three towers, and 1.0 for Tower IV.

To keep these numbers in perspective, it should be noted that this tower has been designed according to the ACI provisions (7,9) as

$$A_s f_s = 0.9\ D + 1.30\ W$$

or $$A_s f_s = 1.0\ D + 1.44\ W \qquad (4a)$$

ie $$\lambda_{u\ell t} = 1.44 \qquad (4b)$$

For the reinforcement actually used, an ultimate load factor based on a linear analysis of 1.52 is obtained. A load factor of 1.38 is obtained for cracking of the concrete by neglecting the contribution from the reinforcement and 1.46 by including the effect of reinforcement.

The role of tension stiffening on the numerical solution can be seen in the predicted results for Tower III, shown in Fig. 5. For this tower, tension stiffening which was felt to be unrealistically low was specifically chosen. It is seen that decreasing the tension stiffening used significantly reduces the ultimate load. Similar results to those found in Ref. 3 are now obtained.

As noted previously, the tension stiffening used for Tower I is relatively high, but as has been shown in Fig. 2 it is still considerably less than that observed in the Vecchio-Collins experiments.

The tension stiffening used in Ref. 3 is based on a bond slip model developed by Floegl and Mang (3,11). There is little experience by which this model can be judged, but it appears to effectively yield very low tension stiffening values.

The difference in the behavior between Tower I and Tower III can be clearly seen by examining the variation in the meridional force for integration points on the circumference, at the level at which cracking first occurs. These results are shown in Figs. 6 and 7 for Towers I and III. The

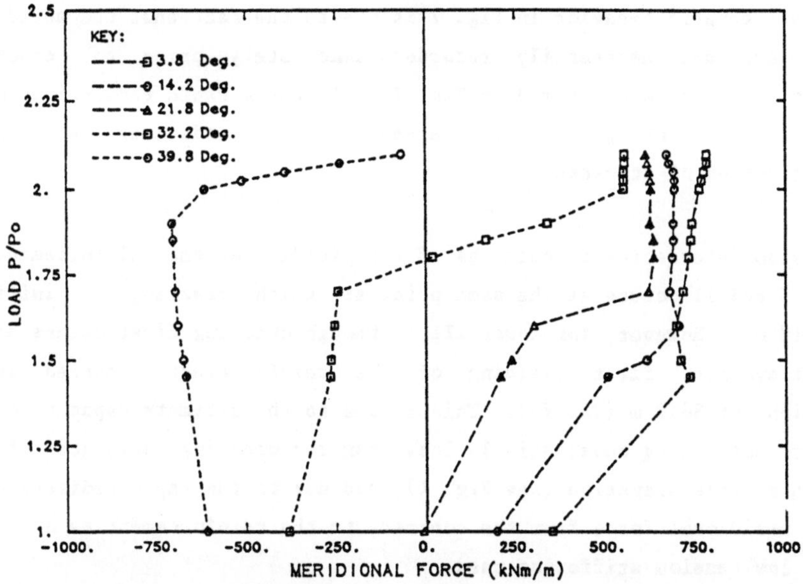

Fig. 6. Load-Meridional Force Behavior for Tower I.

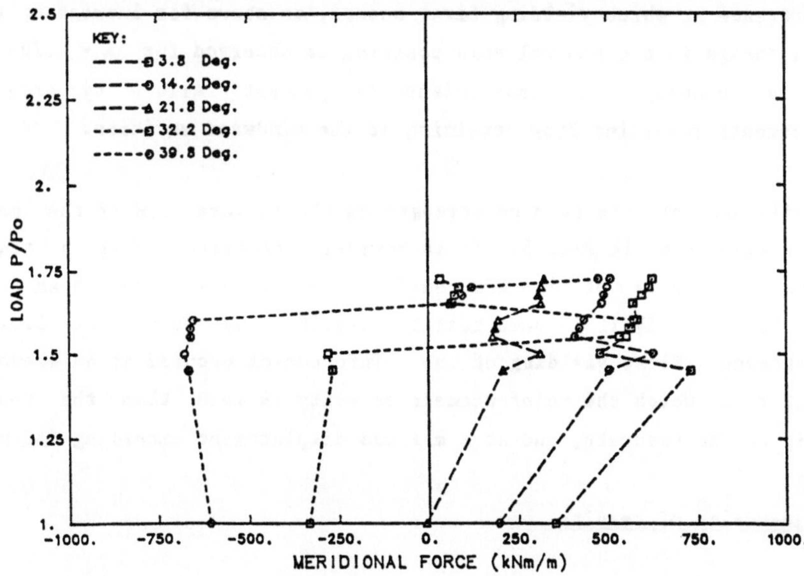

Fig. 7. Load-Meridional Force Behavior for Tower III.

somewhat erratic behavior in Fig. 7 is due to the fact that the meridional forces are not necessarily recorded immediately prior to cracking. However, the general trend in Fig. 7 indicates a rapid redistribution of stresses as cracking occurs, whereas Fig. 6 indicates a gradual redistribution of stresses.

It is also interesting to note that first yielding of the reinforcement for Towers I and II occurs at the same point at which cracking is initiated ($z=57.18$ m). However, for Tower III, although cracking first occurs at the same elevation, first yielding of the reinforcement occurred at an elevation of 86.2 m (283 ft). This is due to the ultimate capacity of the reinforcement being substantially less than the cracking strength of the section at this elevation (see Fig. 4), and due to the rapid redistribution of the meridional force from the concrete to the reinforcement as a result of the low tension stiffening used.

The inability of Tower III to compensate for the redistribution of the meridional force by a shift in the neutral axis can be seen in Fig. 8. In this figure, the distribution of the meridional force and strain along the circumference at which yielding first occurs, is shown for increasing load. A small change in the neutral axis position is observed for $\lambda = 1.725$, but this is clearly not sufficient to prevent failure by excessive displacements resulting from straining of the windward meridian.

The influence of the tension strength on the failure load of the cooling tower is also shown in Fig. 5. It is however recognised that a tension strength in excess of 1.5 MPa could reasonable be expected in an actual tower. For Tower IV it is seen that the required load factor of 1.52 is not achieved. First yielding of the reinforcement occured at an elevation of 105.2 m, at which the reinforcement capacity is less than the tensile capactiy of the concrete, and at a maximum displacement exceeding 540 mm.

4.3) <u>Discussion of results.</u>

The post cracking behavior of the cooling tower was found in this investigation to be very dependent on the assumptions made which influence the load carried by the concrete in tension. Using a relatively high tension stiffening model, the results obtained in this investigation differ

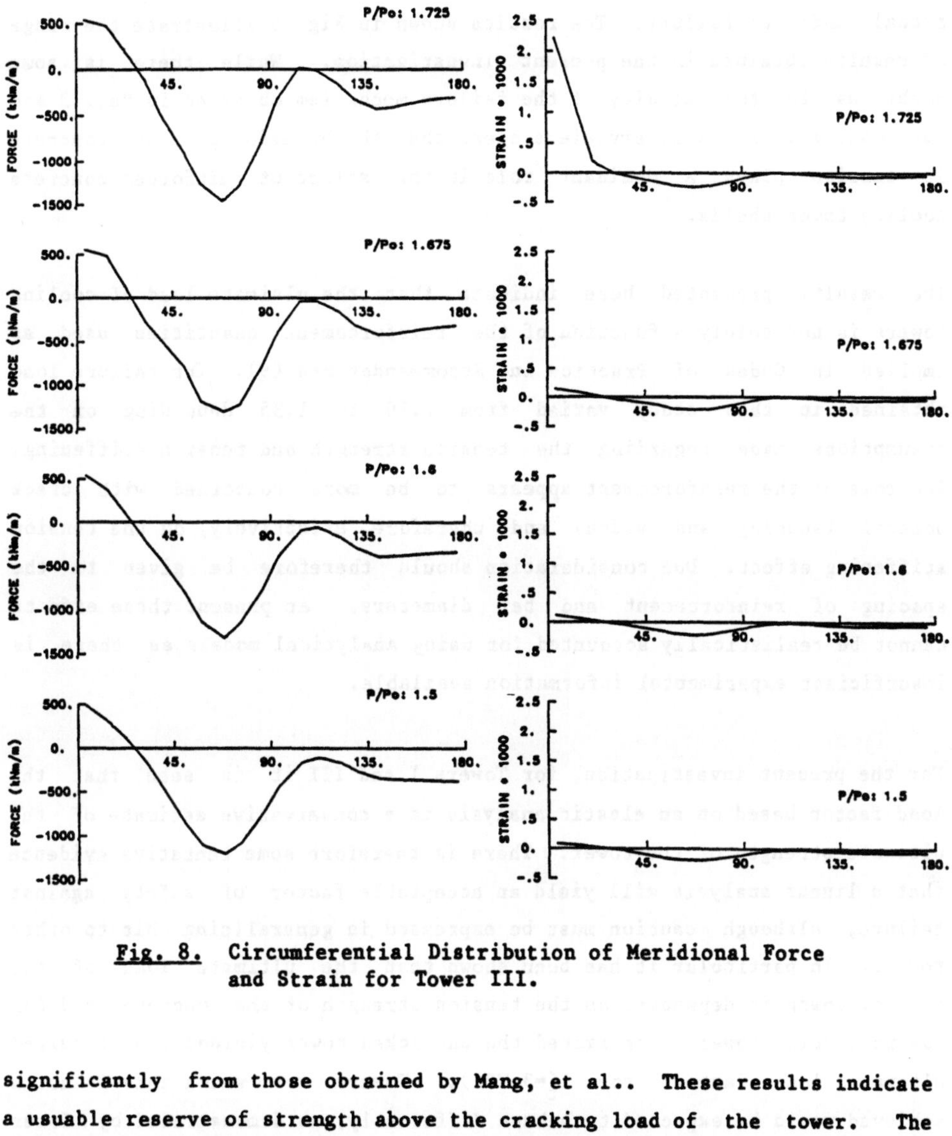

Fig. 8. Circumferential Distribution of Meridional Force and Strain for Tower III.

significantly from those obtained by Mang, et al.. These results indicate a usuable reserve of strength above the cracking load of the tower. The failure mode observed for Tower I can be regarded as a 'ductile' failure as compared to the rapid 'brittle' failure observed for Tower III. Using a low tension stiffening model, similar results to Mang, et al. have been obtained.

Reference 3 notes that at present it seems premature to make generalizations about the ultimate load based on the available numerical evidence. This is not entirely confirmed in this investigation but it has been shown that it may be premature to make generalizations about the

actual mode of failure. The results shown in Fig. 5 illustrate the range of results obtained in the present investigation. While there is some doubt as to the validity of the failure mechanism observed in Ref. 3 and for Tower III, it is nevertheless clear that the behavior of the concrete in tension plays a dominant role in the failure of reinforced concrete cooling tower shells.

The results presented here indicate that the ultimate load of cooling towers is not solely a function of the reinforcement quantities used as implied in Codes of Practice and Recommendations (9). The failure load obtained in this study varied from 2.10 to 1.35 depending on the assumptions made regarding the tension strength and tension stiffening. The role of the reinforcement appears to be more concerned with crack control (spacing and width) and therefore, effectively, on the tension stiffening effect. Due consideration should therefore be given to the spacing of reinforcement and bar diameters. At present these effects cannot be realistically accounted for using analytical models as there is insufficient experimental information available.

For the present investigation, for Towers I and III it is seen that the load factor based on an elastic analysis is a conservative estimate of the ultimate strength of the tower. There is therefore some tentative evidence that a linear analysis will yield an acceptable factor of safety against failure, although caution must be expressed in generalizing this to other towers. In particular it has been shown that the ultimate load of the cooling tower is dependent on the tension strength of the concrete, and for the particular tower investigated the uncracked tower yielded the required ultimate load factor (for f'_t=3 MPa). Towers for which this is not a achieved would be expected to behave differently, as demonstrated by Tower IV of this investigation. A linear elastic analysis will not however account for the possible large deflections above the cracking load.

Towers III and IV failed by yielding of the reinforcement in the vicinity of the throat of the tower. However, the ultimate load of Towers III and IV would have been increased, and the 'ductility' of the structure improved, if the reinforcement capacity in the vicinity of the throat of the tower had exceeded the intact strength of the concrete. In Ref. 8 it is noted that this would be good practice, and the results of this investigation support this view.

Due to the uncertainty in the post-cracking behavior of the tower, it is apparent that the cracking load must also be investigated, possibly to satisfy a lower factor of safety. A reasonable approach is to ensure that under working loads, cracking does not occur in the cooling tower shell. This requirement would be satisfied for the present tower- see Fig. 5. In this regard it is noted that requirements "to insure that neither cracking nor deflections are excessive under conditions of unfactored loading" (9) are meaningless, as the extent of cracking observed in either Tower I or Tower III cannot be predicted by a linear elastic analysis.

5) CONCLUSIONS.

The failure of an isolated cooling tower under wind loading has been examined. It has been confirmed that cracking and the subsequent redistribution of stresses plays a very important role in the post cracking behavior and failure of the tower. The results obtained in this investigation indicate that the ultimate load based on a linear elastic analysis is a conservative estimate of the actual failure load, but this conclusion is dependent on the tension strength of the concrete. Before these results can be generalized, additional investigations involving towers of differing height and wall thickness are required.

At present, as there is uncertainty in the actual post cracking behavior of the tower, is seems necessary that the cracking strength of the tower must be investigated in the design process.

6) ACKNOWLEDGEMENTS.

This research investigation was supported by the Office of Naval Research Grant N00014-83-K-0274; Nr 064-698. The first writer also wishes to acknowledge the National Building Research Institute whose support enabled him to pursue this study.

7) REFERENCES.

1. Abel, J. F., Billington, D. P., Nagy, D. A. and Witta-Dworkin, C., "Buckling of Cooling Towers", J. Struct. Div., ASCE, Vol. 108, No. ST10, Oct. 1982, pp. 2162-2174.

2. Cole, P. P., Abel, J. F., and Billington, D. P., "Buckling of Cooling- Tower Shells: State of the Art", J. Struct. Div., ASCE, Vol. 101, No. ST6, June 1975, pp. 1185-1203.

3. Mang, A. M., Floegl, H., Trappel, F. and Walter, H., "Wind-Loaded Reinforced-Concrete Cooling Towers: Buckling or Ultimate Load", Eng. Struct., Vol. 5, July 1983, pp. 163-180.

4. Liu, T. C. Y., Nilson, A. H. and Slate, F. O., "Biaxial Stress-Strain Relations for Concrete", J. Struct. Div., ASCE, Vol. 98, No. ST5, 1972, pp. 1025-1034.

5. Milford, R. V. and Schnobrich, W. C., "Numerical Model for Cracked Concrete", Int. Conf. on Computer-Aided Analysis and Design of Concrete Structures, Split, Yugoslavia, Sept. 1984.

6. Vecchio, F. and Collins, M. P., "The Response of Reinforced Concrete to Inplane Shear and Normal Stresses", Publication No. 82-03, Univ. of Toronto, March 1982.

7. Hayashi, K. and Gould, P. L., "Some Aspects of the Response of Wind-Loaded Reinforced Concrete Cooling Towers", Struct. Eng. Research Report No. 63, Dept. of Civil Eng., Washington Univ., 1982.

8. Hayashi, K. and Gould, P. L., "Cracking Load for a Wind-Loaded Reinforced Concrete Cooling Tower", J. Amer. Conc. Inst., Vol. 79, July 1983, pp. 318-325.

9. ACI-ASCE Committee 334, "Reinforced Concrete Cooling Tower Shells-Practice and Commentary", J. Amer. Conc. Inst., Vol. 74, Jan. 1977, pp. 21-31.3

10. Gupta, A. K., "Membrane Reinforcement in Shells", J. Struct. Div., ASCE, Vol. 107, No. ST1, Jan. 1981, pp. 41-56.

11. Mang, H. A. and Floegl, H., "Tension Stiffening Concept for Reinforced Concrete Surface Structures", IABSE Colloquium on Advanced Mechanics of Reinforced Concrete, Delft, Final Report, June 1981, pp. 351-369.

Nonlinear Elastic Stability of Cooling Tower Shells

U. Eckstein*, R. Harte*, W.B. Krätzig* and U. Wittek**

* Institut für Konstruktiven Ingenieurbau - Statik und Dynamik
Ruhr-University Bochum, Germany

** Lehrstuhl für Baustatik
University of Kaiserslautern, Germany

Summary

The present paper compares various design procedures to make cooling tower shells resistant against nonlinear buckling phenomena:

- a complete nonlinear computation,
- a linear eigenvalue problem including the design wind load,
- a linear eigenvalue problem for a wind load, approximated as a rotationally symmetric pressure,
- the concept of reduced-membrane-buckling and
- an experimentally founded algorithm.

Recommendations are given and verified by an example.

1. Introduction

Natural draught cooling tower shells in general are subjected to dead load g, wind load w, internal suction s and various temperature effects. Dead load as well as internal suction causes a complete state of compression in a well designed shell. Upon this state the wind load superposes fields of meridional tension stresses and hoop compression near the windward face; it further adds meridional compression and minor circumferential tension stresses in the tower flanks. Under extreme storm conditions thus the total stresses at the luff side of the shell may be changed into strong states of meridional tension, and the hoop compression will be intensified. In the tower flanks the original meridional compression will increase significantly up to 3 times the original values. All stress fields are fluctuating in time and over the shell surface due to the instationarity of intensity and direction of the storm.

All civil engineering design concepts seperate *characteristic response phenomena* from the totality of the appearing structural effects. On their basis the strength of any member is determined in order to insure its sufficient safety and serviceability. In wind-stressed cooling tower shells three major response phenomena can be isolated:

- the *tensile failure* of the meridional reinforcement near the windward face;
- the circumferential *snap-through failure* of the upper third of the shell around the stagnation point [1];
- *instability modes* in the flanks due to superposed dead load and wind compression.

In the design process these three phenomena will be met with different factors of safety (generally contained in national codes), with suitable mechanical models and appropriate numerical algorithms. All these elements of a safety concept have to be considered as a unity; they contribute equally to the structural safety of the tower. It should be emphasized that engineering design processes in general do not aim to simulate the collapse of a structure. They rather shall limit possible modes of failure, often with widely deviating probabilities of appearance, by adequate factors of safety, which make the remaining risk acceptable to the public.

The present paper mainly is concerned about the third mentioned phenomenon: *instability-modes* in the flanks of the tower due to compression. In order to exclude possible other failure-modes and because of the compression, the tower shell can be assumed *elastic* with *unlimited* tension and compression *strength*. As is well known from theoretical and experimental investigations 6 to 8 buckling-waves would appear along the circumference of the thin shell for comparable load conditions of central symmetry. One buckling wave has a relatively small extension compared with the variation of stresses. Thus the main topic of this paper is to evaluate, whether a geometrically nonlinear analysis can be substituted by time-saving classical buckling analysis, probably even for states of stress of rotational symmetry.

2. Numerical algorithms for elastic instability phenomena

Numerical algorithms for linear and nonlinear elastic instability phenomena will be derived from the *total potential energy* of the structure under consideration

$$\Pi = \Pi_i + \Pi_a = \frac{1}{2}\int_{\overset{\circ}{V}} \underline{\varepsilon}^T \underline{E}\, \underline{\varepsilon}\, d\overset{\circ}{V} - \int_{\overset{\circ}{V}} \underline{u}^T \overset{\circ}{\underline{p}}\, d\overset{\circ}{V} - \int_{\overset{\circ}{F}_t} \underline{r}^T \overset{\circ}{\underline{t}}\, d\overset{\circ}{F}, \qquad (1)$$

in which $\underline{\varepsilon}$ denotes the strains as nonlinear functions of the displacements,

- $\underline{E}$ the elasticity matrix,
- $\underline{u}$ the displacement field,
- $\underline{r}$ the displacements on the surface F of V,
- $\overset{\circ}{\underline{p}}$ the given body forces, related to elements $d\overset{\circ}{V}$ of the undeformed configuration,
- $\overset{\circ}{\underline{t}}$ the given boundary forces, related to elements $d\overset{\circ}{F}$ of the undeformed boundary surface $\overset{\circ}{F}$.

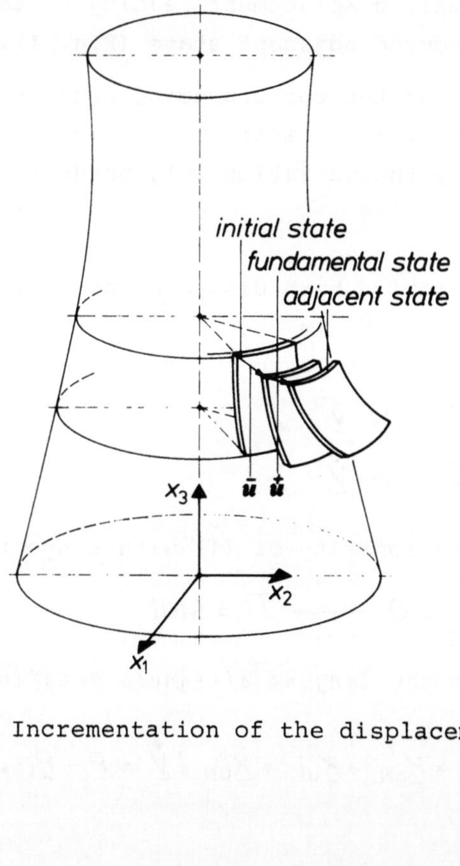

Fig. 1. Incrementation of the displacement field

The strain field $\underline{\varepsilon}$ has to accomplish the kinematic relations

$$\underline{\varepsilon} = \underline{D}_k u = (\underline{D}_{kL} + \underline{D}_{kN}(\underline{u}))\underline{u} , \qquad (2)$$

and the boundary displacements $\underline{r}$ have to fulfil the displacement boundary conditions:

$$\underline{r} = \overset{o}{\underline{r}} = \underline{R}_r \underline{u} \quad \text{on} \quad \overset{o}{F}_r . \qquad (3)$$

Herein $\underline{D}_k$ represents a nonlinear differential operator and $\underline{R}_r$ a purely algebraic one [2, 3].

In order to derive the fundamental equation of our future algorithms we *firstly* increment the displacement vector

$$\underline{u} = \bar{\underline{u}} + \overset{+}{\underline{u}} \qquad (4)$$

into $\bar{\underline{u}}$ the arbitrarily large displacement, pointing from the unloaded, stress-free and undeformed *initial state* to the *fundamental state*, beeing in equilibrium,

$\overset{+}{\underline{u}}$ the small displacement leading to the incrementally neighboured adjacent state (Fig. 1).

Substituting (4) - and the corresponding strains (2) - into the total potential energy (1) transforms Π into a functional, which is, considering the variation (6), quadratic in $\overset{+}{\underline{u}}$ and cubic in $\bar{\underline{u}}$ [9].

If we *secondly* discretize both displacement fields $\bar{\underline{u}}$, $\overset{+}{\underline{u}}$ on the element level $\bar{\underline{v}}^p$, $\overset{+}{\underline{v}}^p$ in the sense of a displacement model and assemble the element-DOFs to those of the complete structure:

$$\begin{aligned} \bar{\underline{u}}^p &= \underline{\Omega}^p \bar{\underline{v}}^p , \quad \overset{+}{\underline{u}}^p = \underline{\Omega}^p \overset{+}{\underline{v}}^p , \\ \bar{\underline{v}} &= \underline{a}\, \bar{\underline{V}} , \quad \overset{+}{\underline{v}} = \underline{a}\, \overset{+}{\underline{V}} , \end{aligned} \qquad (5)$$

the condition of stationarity of (1) with respect to $\overset{+}{\underline{V}}$

$$\delta\Pi = \delta\Pi_i + \delta\Pi_a = 0 \longrightarrow \Pi = stat \qquad (6)$$

transforms Π into the *tangent stiffness relation* of the complete structure:

$$\underline{K}_T \overset{+}{\underline{V}} = [\underline{K}_e + \underline{K}_{gL} + \underline{K}_{gN} + \underline{K}_{uL} + \underline{K}_{uN}]\overset{+}{\underline{V}} = \underline{P} - \underline{P}_i . \qquad (7)$$

In this fundamental relation the following submatrices can be distinguished:

- $\underline{K}_e$ linear elastic stiffness matrix,
- $\underline{K}_{gL}$ initial stress or geometrical stiffness matrix, linear in $\bar{\underline{V}}$ resp. $\bar{\underline{u}}$,
- $\underline{K}_{gN}$ initial stress matrix, quadratic in $\bar{\underline{V}}$ resp. $\bar{\underline{u}}$,
- $\underline{K}_{uL}$ initial displacement matrix, linear in $\bar{\underline{V}}$ resp. $\bar{\underline{u}}$,
- $\underline{K}_{uN}$ initial displacement matrix, quadratic in $\bar{\underline{V}}$ resp. $\bar{\underline{u}}$,
- $\underline{P}$ vector of applied loads,
- $\underline{P}_i$ internal nodal force vector.

Algorithms for tracing the *complete nonlinear responses* of discretized structures are usually based on combined incremental and iterative techniques: During a stepwise load - and/or displacement - incrementation the original nonlinear relationship is linearized through (7), and the linearization-error is corrected by additional equilibrium iterations using the right-hand side of (7). In our software system FEMAS, which has been used to compute all examples in section 3, this procedure is carried out by use of a symmetrized *Riks-Wempner-algorithm* with positive (negative) load-incrementation for increasing (decreasing) load-displacement-paths. Difficulties at limit points $\dot{\underline{V}}^T \underline{K}_T \dot{\underline{V}} = 0$ can be avoided by a careful selection of the respective step length.

To follow the nonlinear response of a structure with such an algorithm is a possible, but extremely *time-consuming numerical procedure*: A cooling tower shell has to be discretized into a great number of quadrilateral or triangular shell-elements and many increments are necessary to reach the collapse load [4, 5].

Especially for cooling tower shells with their long-lasting linear prebuckling state a computation due to a *classical bifurcation analysis* seems to be a promising, time-saving way. To derive the corresponding algorithm we start again with the total potential energy (1), in which the displacement field has been decomposed and discretized by

$$\underline{u} = \lambda \bar{\underline{u}} + \overset{+}{\underline{u}} \quad \text{resp.} \quad \underline{V} = \lambda \bar{\underline{V}} + \overset{+}{\underline{V}} \quad . \tag{8}$$

If the variation of Π is accomplished under the additional restriction, that no load-increase is required for a transition from the fundamental to the adjacent state

$$\delta \Pi = \delta \overset{++}{\Pi} = 0 \ , \tag{9}$$

this energetic criterion for neutral equilibrium ends up with a *quadratic eigenvalue problem*

$$[\underline{K}_e + \lambda(\underline{K}_{gL} + \underline{K}_{uL}) + \lambda^2(\underline{K}_{gN} + \underline{K}_{uN})]\overset{+}{\underline{V}}_c = \underline{0} \longrightarrow \lambda_c, \overset{+}{\underline{V}}_c, \tag{10}$$

in which the same submatrices appear as in (7). The square of λ results from the quadratic dependence of $\underline{K}_{gN}$ and $\underline{K}_{uN}$ from $\bar{V}$. λ_c denotes the critical load factor of the desired state of neutral equilibrium, and $\overset{+}{\underline{V}}_c$ the corresponding buckling mode.

In accordance with (8), assuming the prebuckling path to be linear, the quadratic terms $\underline{K}_{gN}$ and $\underline{K}_{uN}$ are neglectable, which approximates (10) by a *linear eigenvalue problem:*

$$[\underline{K}_e + \lambda(\underline{K}_{gL} + \underline{K}_{uL})]\overset{+}{\underline{V}}_c = \underline{0} \longrightarrow \lambda_c, \overset{+}{\underline{V}}_c \ . \tag{11}$$

The most common approximation of (10) in structural stability analysis is the *classical eigenvalue problem*, in which also $\underline{K}_{uL}$ has been deleted:

$$[\underline{K}_e + \lambda \underline{K}_{gL}]\overset{+}{\underline{V}}_c = \underline{0} \longrightarrow \lambda_c, \overset{+}{\underline{V}}_c \quad . \tag{12}$$

Applying equation (12) to a combination of dead load, wind and internal suction yields a simple and relatively economic numerical procedure, even if the wind load - symmetric to the wind direction - requires a discretization of the cooling tower shell by quadrilateral or triangular finite elements. Nevertheless, buckling phenomena in the flanks of cooling tower shells - if stiffened by upper and lower edge members - are local effects, as has been proved by experiments [6]. Because the number of circumferential buckling waves exceeds always 6, the solution of the eigenvalue problem (12) for rotationally symmetric load conditions - the wind distribution is approximated by its maximum pressure - lets expect reasonable results too. The application of ring-elements then will lead to the lowest possible numerical effort.

In general those procedures will deliver upper bounds of the collapse load, especially if the nonlinear load-displacement-path does not increase considerably after having passed an instability-point or -area. Lower bounds of the collapse load, taking into account the imperfection sensitivity of the structure, can be evaluated by use of the concept of *reduced-membrane-buckling* [7]

$$\lambda_c(r) = \lambda_B + r\lambda_M \, , \tag{13}$$

in which the critical load factor λ_c has been split into a membrane part λ_M and a bending part λ_B. Through energetic considerations has been proved, that all membrane contributions lose their stabilizing character in the postbuckling range due to imperfections. Thus neglection of λ_M $(r \equiv 0)$ in (12) results into a lowest bound for λ_c, where only bending energies are accompanied with the buckling phenomena [5].

Many types of shells indeed buckle at considerably lower loads compared with theoretical bifurcation results for $r=1$. It could be found from theoretical studies [8], that cooling tower shells with *negative Gaussian* curvature show a certain *restricted imperfection sensitivity*. Thus the results of experiments with large plastic models of hyperbolic shells under radial and axial loads of central symmetry [6], simultaneously corrected by numerical computations, can also be used as a design tool for buckling phenomena.

The proposed *experimental-numerical stability criterion* is shown in Fig. 2 to 4. As given in Fig. 3 it postulates, that any critical combination of the effective stresses σ_{11}, σ_{22} of a certain load combination and the buckling stresses σ_{11cr}, σ_{22cr} at any point of the middle surface is responsible for the collapse of the tower. The buckling stresses are experimen-

<u>Circumferential buckling stress:</u>

$$\sigma_{11cr} = \frac{0{,}985\,E}{\sqrt[4]{(1-\nu^2)^3}} \left(\frac{h}{R_T}\right)^{4/3} k_{G11}$$

<u>Meridional buckling stress:</u>

$$\sigma_{22cr} = \frac{0{,}612\,E}{\sqrt[4]{(1-\nu^2)^3}} \left(\frac{h}{R_T}\right)^{4/3} k_{G22}$$

E Short time modulus of elasticity
ν Poisson's ratio
h Actual thickness of the shell
R_T Shell radius at the throat
k_{G11}, k_{G22} Geometrical parameters

Fig. 2. Basic formulae for buckling stresses for hyperbolic shells

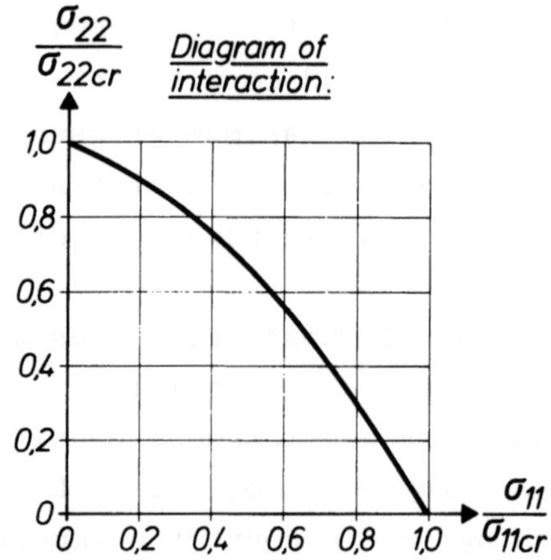

<u>Buckling safety:</u>

$$0{,}8\,\nu_B\left(\frac{\sigma_{11}}{\sigma_{11cr}}+\frac{\sigma_{22}}{\sigma_{22cr}}\right)+0{,}2\,\nu_B^2\left(\left[\frac{\sigma_{11}}{\sigma_{11cr}}\right]^2+\left[\frac{\sigma_{22}}{\sigma_{22cr}}\right]^2\right)=1$$

DIN 1045,24.3: $\nu_B \geq 5$

Fig. 3. Proposed diagram of interaction

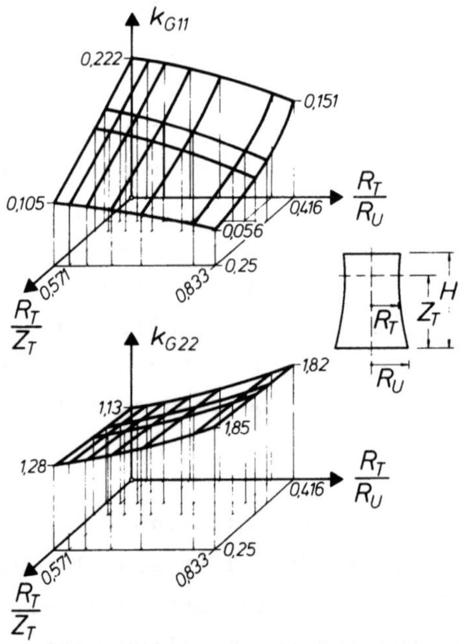

Fig. 4. Geometrical parameters k_{G11} and k_{G22}

tally derived, scaled and specified to cooling tower boundary conditions by a bifurcation theory. The parameters k_{G11}, k_{G22} in Fig. 4, which contain the influence of tower shapes other than those tested, stem also from accompanying bifurcation results.

3. Presentation of results

The various design procedures described in section 2 shall now be applied to a wide, unstiffened cooling tower of medium height. All dimensions of the tower geometry and the wall-thickness may be found in Fig. 5. Material constants, load specifications and the wind-pressure distribution are given in Fig. 6.

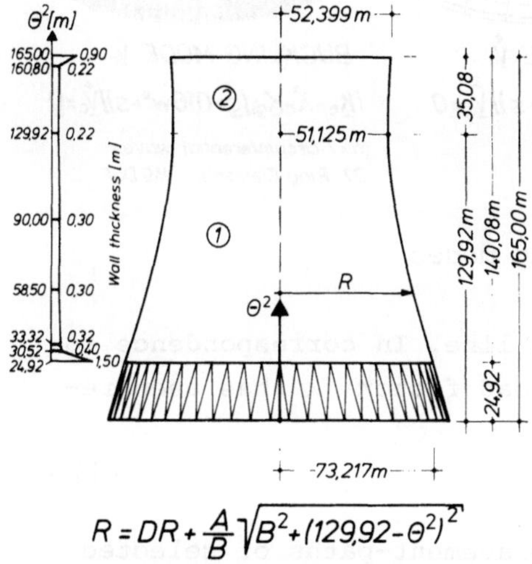

$$R = DR + \frac{A}{B}\sqrt{B^2+(129{,}92-\Theta^2)^2}$$

section	DR	A	B
①	33,077477	18,047522	52,853802
②	50,724535	0,400465	8,640975

Fig. 5. Cooling tower geometry and wall-thickness

MATERIAL: Young's Modulus: $E = 3.0 \cdot 10^7 kN/m^2$

Poisson's Ratio: $\nu = 0.2$

LOADING: Dead Load: $g(\gamma)$, $\gamma = 25\ kN/m^3$

Wind Load: $w+s$,

External Pressure: $w = 1.155\left(\frac{\Theta^2}{10}\right)^{0.22} \cdot c_p\ kN/m^2$

Internal Suction: $s = 0.525\left(\frac{165}{10}\right)^{0.22} = 0.9728\ kN/m^2$

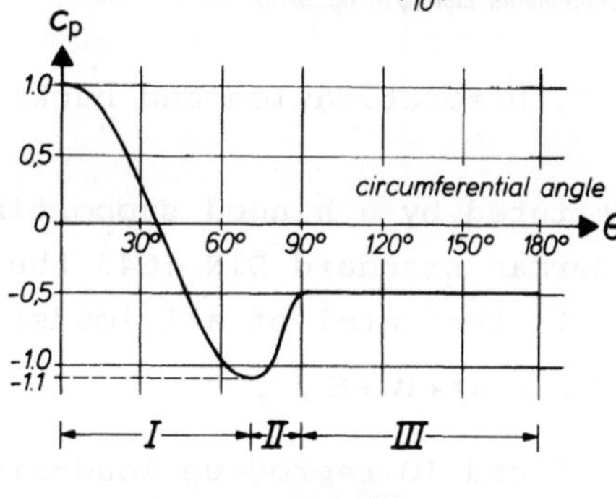

$I\quad c_p = 1.0 - 1.2\left(\sin\frac{90}{71}\Theta^1\right)^{2.239}$

$II\quad c_p = -1.1 + 0.6\left[\sin\frac{90}{22}(\Theta^1-71)\right]^{2.395}$

$III\quad c_p = -0.5$

Fig. 6. Material and load specification

We start with the evaluation of the *complete nonlinear response* of a *perfect shell* due to the algorithm following (7). Because of the symmetry of the wind loading only one half of the middle surface needs to be discretized into 120 quadrilateral shell elements NACS48 [4, 9], leading to a total of 1500 DOFs as shown in Fig. 7. The diagonal supports of the shell have been

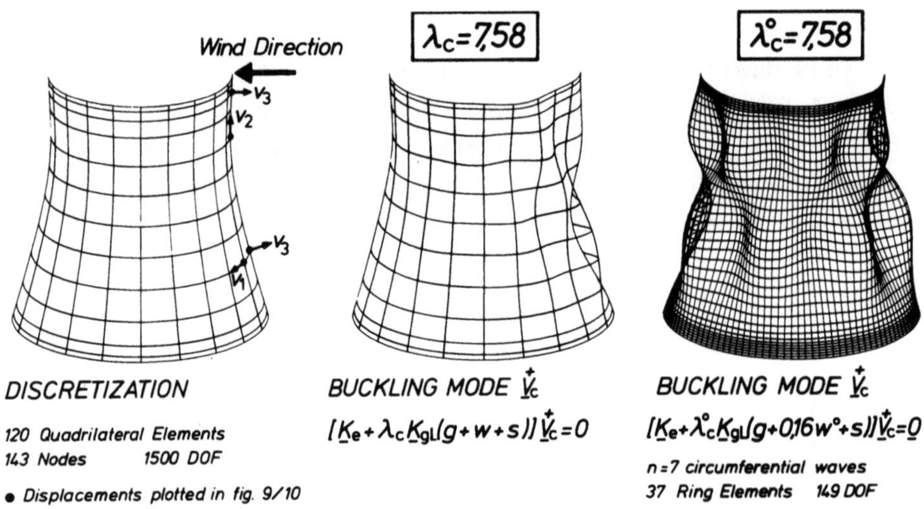

Fig. 7. Discretization and buckling modes

substituted by a hinged supporting line. In correspondence to the German standard DIN 1045 the load factor λ has been related to the total of all loads:

$$q = \lambda (g + w + s) \ . \tag{14}$$

Figs. 9 and 10 reproduce load-displacement-paths of selected DOFs: After a long linear branch a sudden weakening of the structure is followed by another branch, approximately linear and with small further increase. Fig. 8 (left and middle) shows two plots of the complete displacement field: the *load intensity factor* $\lambda = 6.0$ marks the end of the linear (prebuckling) path; and $\lambda = 7.71$ was reached immediately before interruption of the computer run.

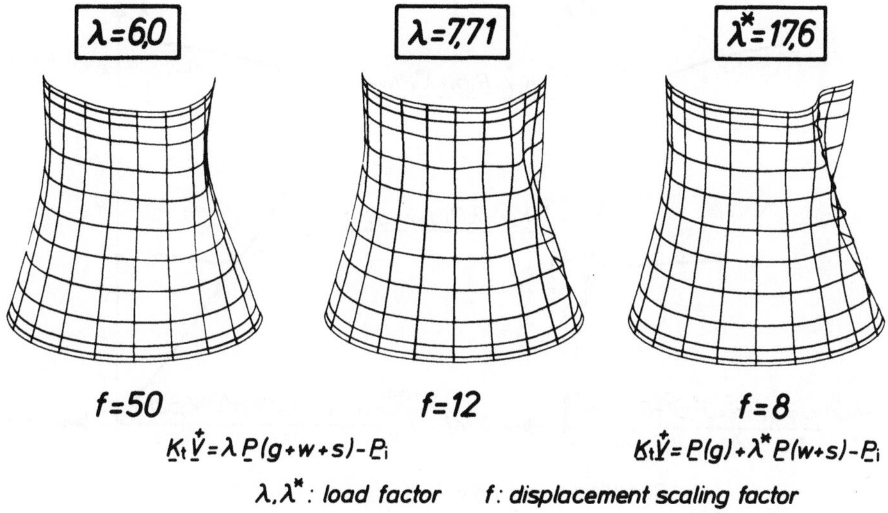

Fig. 8. Displacements of the nonlinear response

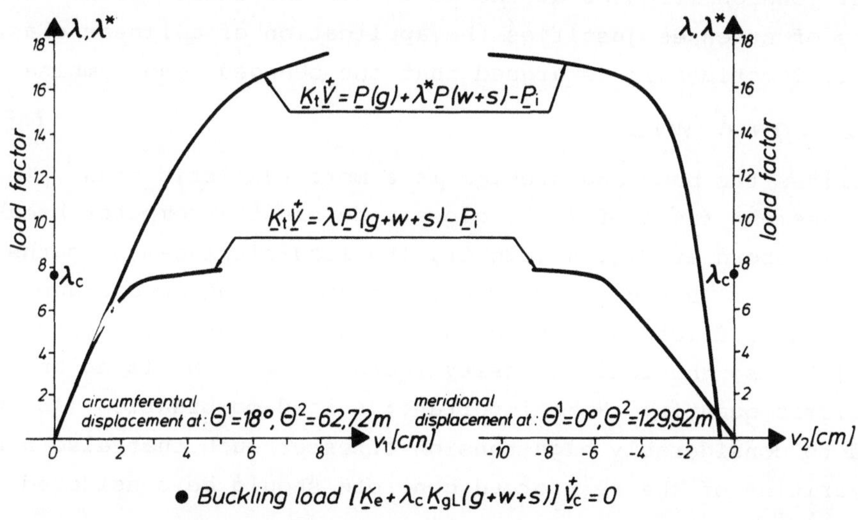

Fig. 9. Load-displacement plots

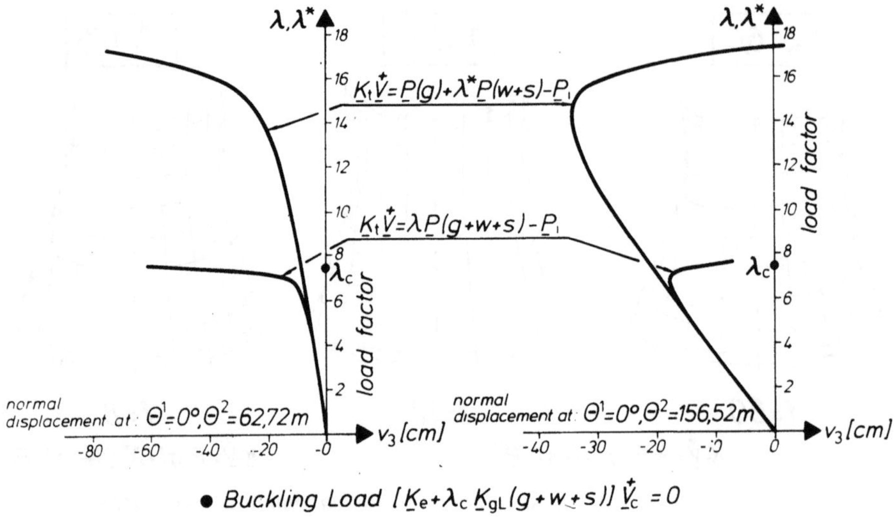

Fig. 10. Load-displacement plots

The load combination (14) ensures compression stresses - or small tension stresses at the most - in the whole shell. This state of stresses justifies the application of a linear elastic model. Sometimes it is argued that the changed load combination

$$q^* = g + \lambda^*(w+s) \qquad (15)$$

describes the buckling problem in a more realistic sense. The displacement field of (15), closely before the computer break-off, is shown at Fig. 8 (right), its load-displacement-paths again in Fig. 9 and 10. To reply the above mentioned argument: also the dead load has to be increased by a partial safety factor [10], so the load intensity factor $\lambda^* \approx 17.0$ is an unrealistic quantity. More important: a load combination (15) will lead to considerably high tension stresses such that also non-linearities of the reinforced concrete should be considered [11].

The *first simplification* of the computation procedure is the application of the *linear eigenvalue problem* (12) for the real load distribution (14). For the discretization presented in Fig. 7 a lowest *critical load factor* λ_c = 7.58 has been evaluated. The corresponding buckling mode $\overset{+}{\underline{V}}_c$, plotted on Fig. 7 (middle), is obviously nearly identical to the last displacement field of the nonlinear computer run. This explains the neighbourhood of λ_c to $\lambda = 7.71$.

A *next simplification* approximates the wind loading as rotationally symmetric. In order to find comparable results the stagnation pressure $w^o = 2.14\ kN/m^2$ has to be reduced by a reduction factor α :

$$q^o = \lambda^o (g + \alpha w^o + s) \ . \qquad (16)$$

g and s are of course rotationally symmetric. Using again the classical eigenvalue algorithm (12) the following lowest critical load factors λ_c^o could be obtained, all at $n=7$ circumferential waves:

α	0.00	0.16	0.20	0.40	0.60
λ_c^o	8.11	7.58	7.46	6.89	6.40
$\lambda_c^o(r=0)$	4.82	4.52	4.46	4.14	3.86

Obviously $\alpha = 0.16$ adjusts the critical load to the exact value of λ_c^o - for this example. Including more slender towers it can be generalized, also from earlier work, that

$$0{,}20 \leq \alpha \leq 0{,}60 \qquad (17)$$

approximates critical buckling loads for *perfect* cooling tower shells within an error-bound of ≤ 20% .

It is extremely difficult, if not impossible, to estimate buckling loads for *imperfect* shells by simple design procedures. For small imperfections up to the order of magnitude of the shell-thickness the concept of *reduced-membrane-buckling* (13) leads for $r=0$ to a lowest bound of λ_c^o as can be found in the above given table.

Finally, values for the critical load factor due to the *experimental-numerical stability criterion* of Figs. 2 to 4, which includes an unknown (small) contribution of inherent model-imperfections, range between 4.9 and 6.5 in the medium part of the shell.

4. Conclusions and Recommendations

Cooling tower shells have to resist instability phenomena as characteristic modes of failure of thin elastic structures. Necessary numerical algorithms for *perfect shells* can be based on elastic materials. The time-consuming evaluation of the complete nonlinear response can be substituted by a linear eigenvalue analysis for real wind loads, practically without any error. Errors up to 20% may appear, if the real wind load is substituted by a circumferentially constant normal pressure αw^{o} . Real, e.g. *imperfect shells* can be designed due to the reduced-membrane-buckling-concept or due to Fig. 2 to 4. Error bounds are unknown and imperfections are restricted to the order of magnitude of the shell-thickness. If imperfections grew larger, complete nonlinear computations are required; because of increasing bending effects also material nonlinearities have to be included then.

References

1. Der, T.J., Fidler, R.: A model study of the buckling behaviour of hyperbolic shells. Proc. Inst. Civ. Engrg., 41 (1968), p. 105.

2. Krätzig, W.B., Basar, Y., Wittek, U.: Nonlinear Behavior and Elastic Stability of Shells. Contribution in: Buckling of Shells, Ed.: E. Ramm. Proceedings of a State-of-the-Art Colloquium, Universität Stuttgart; Springer-Verlag, Berlin, Heidelberg, New York 1982, p. 19.

3. Krätzig, W.B., Wittek, U.: Numerical Simulation of Collapse Loads for Arbitrarily Shaped Elastic Shells. Proceedings of the 4. ASCE-EMD Specialty Conference, Purdue University, ASCE, New York 1983, p. 1087.

4. Eckstein, U.: Nichtlineare Stabilitätsberechnung elastischer Schalentragwerke. Technical Report of the Institut für Konstruktiven Ingenieurbau, Ruhr-University Bochum 1983, Nr. 83-3.

5. Krätzig, W.B., Wittek, U., Basar, Y.: Buckling of general shells - theory and numerical analysis. Contribution in: COLLAPSE - the buckling of structures in theory and practice, Ed.: J.M.T. Thompson, G.W. Hunt. Proceedings of the IUTAM-Symposium, University College London; Cambridge University Press, Cambridge 1983.

6. Mungan, J.: Buckling Stress States of Hyperboloidal Shells. Journal of the Structural Division, ASCE, Vol. 102 (1967), p. 2005.

7. Wittek, U.: Beitrag zum Verhalten der Strukturen bei endlichen Verformungen unter besonderer Beachtung des Nachbeulmechanismus dünner Flächentragwerke. Technical Report of the Institut für Konstruktiven Ingenieurbau, Ruhr-University Bochum 1980, Nr. 80-1.

8. Eckstein, U., Krätzig, W.B., Wittek, U.: Wirklichkeitsnahe Grenzbeullasten von Rotationsschalen unterschiedlicher Gauss'scher Flächenkrümmung. Technical Report of the Institut für Konstruktiven Ingenieurbau, Ruhr-University Bochum 1979, Nr. 79-5.

9. Harte, R.: Doppelt gekrümmte finite Dreieckelemente für die lineare und geometrisch nichtlineare Berechnung allgemeiner Flächentragwerke. Technical Report of the Institut für Konstruktiven Ingenieurbau, Ruhr-University Bochum 1982, Nr. 82-10.

10. Krätzig, W.B.: General Safety Concept of Reinforced Concrete Cooling Towers. Contribution in: International Symposium on very Tall Reinforced Concrete Cooling Towers, p. 59, IASS - E.D.F. - SOCOTEC, Paris 1978.

11. Mang, H., Floegl, H., Trappel, F., Walter, H.: Wind-loaded reinforced-concrete cooling towers: buckling or ultimate load? Eng. Structures 5 (1983), p. 163.

Design Specification of Tolerances for Geometric Imperfections in Concrete Cooling Towers

Charles P. Ellinas

J.P. Kenny and Partners
Burne House, London, UK

Abstract

The paper will provide a summary of the influences of both meridional and circumferenctial shape distortions on the distribution of stress in cooling tower type shells. Approximate analytical procedures will be used to evolve rational and compact procedures for setting limits on such geometric distortions. These will be shown to permit tolerances to be specified at the pre-fabrication stage.

1984
IASS-RUB

Structural Behavior and Special Problems

Two Decades of Cooling Tower Design in the United States of America

DAVID P. BILLINGTON

Department of Civil Engineering
Princeton University
Princeton, NJ 08544 USA

General Features of Built Towers

Although there are a number of natural draft cooling towers under construction in the United States in 1984, few if any are presently being designed. After two decades of intensive activity, when over 120 large-scale (over 320 ft in height) natural draft towers were put under contract, there now (1984) is a lull during which time it is useful to survey the experience derived from this period of design, construction, and research.

Of the over 120 towers put under contract, 100 had been constructed by the end of 1983 and it is some of the experience with these towers which I propose to discuss here. Of that number 56 towers are for coal-fired plants, 40 for nuclear plants, and only 4 for oil-fired ones. The distribution in heighth appears in Table 1 and the number completed per year in Table 2. If the two-decade time span is divided into three periods of roughly equal activity we can observe (see also Table 2) that during 1963-72 there were 31 towers built with the highest one 452 ft; during 1973-78, 37 towers were built with the highest one 506 ft; and during the 1979-83, 32 towers were constructed with the highest one now reaching 553 ft.

From the perspective of design, these three periods correspond very roughly to three sets of questions raised about the structural behavior of such immense structures. During the period up to 1972 most towers were designed based upon membrane theory and the failures at Ferrybridge were uppermost in the minds of designers. After 1972, finite-element methods came into more general use and this led to much concern over the validity of membrane theory as a basis for design. Moreover, between 1973 and 1978, many more complex analyses appeared especially for buckling, and dynamic behavior. New and careful wind measurements were made as well. Following 1978 the research issues had shifted towards even more complex nonlinear problems as well as imperfections and ultimate load behavior.

Lessons of Ferrybridge: 1965-1972

Prior to the failures at Ferrybridge on November 1, 1965, the design of cooling towers was based on stresses derived from the membrane theory and on working load behavior. One of the earliest technical publications in the United States gave this basis for design and in a worked example showed how little meridional reinforcement was often required because the tower dead weight gave compression stresses which nearly balanced the maximum tension calculated from wind loads. A second publication noted that "both analysis and practical experience indicate that only nominal reinforcing is needed to resist the minor stresses induced by the wind." Again the justification for such small stresses was that "in most cases, the compression forces due to weight dominate to such an extent that there is no reversal of stress. [1] Of course this use of working loads was the design basis for nearly all concrete structures in the United States prior to the 1963 ACI Building Code.

The conclusions from the 1965 Ferrybridge failures emphasized the dangers in the working load method and recommended a load factor approach to account for possible wind overloads. At the same time the Ferrybridge report ruled out buckling as a cause of failure, found the membrane theory a reasonable basis for design, and strongly urged field measurements to determine more reliable wind loadings. [2]

Therefore, during the first period and with the clear exception of the Big Sandy Tower of 1963 [3], all the towers were completed after the Ferrybridge experiences. Some had been designed before November 1965 but nearly all had been subsequently modified to some extent. One major result of Ferrybridge was the great increase in research on the behavior of cooling towers although many results did not appear before 1973. It became clear after 1965 that practices which were reasonable for small towers had been inappropriately used at Ferrybridge where the towers were the highest then built. Much of the research begun after 1965 focused on the means of predicting behavior numerically in towers higher than 500 ft and under hurricane winds or strong earthquake motions.

Design and Research: 1973-1978

In the intensive six year period of the middle 1970's the major developments consisted of the general availability of finite-element programs for cooling tower shells, new full-scale field observations, and recognition of the crucial role thermal gradients play in shell behavior.

Two similar sets of finite-element studies opened up the possibility for studying in detail the role of shell bending moments and made it possible to include more accurate representations of the geometry of the shell, ring beams, columns, and foundations. [4] Very little of this research resulted in significant changes in design. A close study, for example, of three towers (the Trojan Tower completed in 1973 and the two Martins Creek towers completed in 1974) designed before 1973, and hence without benefit of finite-element analysis, shows that the more complete analyses would have led to very little change in design. The finite-element results were, however, of great value in making it possible to answer many questions about structural behavior as towers grew in size from the 375 ft high Ferrybridge class to those of over 500 ft in height that had begun to appear regularly after 1972.

One such question was the validity of simplified analyses for earthquakes. The approach often used before 1973 consisted in simply calculating the horizontal load as a percentage of the vertical weight at each elevation. This load was then applied to the tower as a pure cantilever and the resulting membrane stresses computed by beam theory. Finite-element studies showed this method can be a reasonable basis for the shell design.[5] However, the peak response spectrum method has proved to be also relatively simple and also reasonably reliable for design.

Also of great value to design were the new field observations reported after 1972, especially those from the Ardeer failure and from the wind-load studies at Weisweiler (done before 1973), Schmeehausen, and Martins Creek. The Ardeer collapse on September 27, 1973 in a moderate windstorm (gusts up to 63 mph) raised the major issue of circumferential reinforcement and thermal cracking. Just as Ferrybridge showed how sensitive the meridional steel requirements were to slight increases in wind load, the Ardeer failure showed how sensitive hoop reinforcement was to local imperfections in the shell geometry. Professor Chilver had strongly warned the profession after Ferrybridge that circumferential reinforcement was crucial to tower safety but his warnings went unheeded in the Ardeer design.[6]

Another most important conclusion from Ardeer, and one already made by Chilver right after Ferrybridge, was the need to control vertical shell cracking arising from thermal gradients. The Ardeer report emphasized the major role of the thermal cracks in that failure and concluded that substantial circumferential reinforcement was essential. In the United States the minimum reinforcement in each direction (meridional and circumferential) was normally taken as 0.35% of the cross sectional area rather than the 0.15% used in Britain. The 1977 ACI Recommended Practice for concrete cooling towers included this minimum percentage.[7]

A related study based on field observations on one of the Martins Creek towers emphasized how vertical cracking from thermal gradients could drastically reduce circumferential shell stiffness.[8] From this conclusion it followed, as Chilver had realized much earlier, that buckling could be critical for tower safety if not enough circumferential steel were used. Chilver had even intimated that it was wrong to rule out buckling as a part of the Ferrybridge problem; and the buckling of towers was pursued during this period. However, without the benefit of a convincing nonlinear analysis, the validity of simplified methods (usually a symmetrical approximation) was debated.[9] Nevertheless the tests carried out in Great Britain immediately following Ferrybridge served as a useful guide in evaluating various results from other model tests and from numerical analyses.[10]

Finally and of probably the greatest importance were the wind measurements carried out on two separate towers in Germany at Weisweiler and Schmeehausen and in the United States on one of the Martins Creek Towers.[11] These studies included both static and dynamic wind behavior and gave rise to a number of important conclusions. First, for static wind loadings, the results from separate towers for circumferential distributions agreed remarkably well and in particular showed that as the wind velocity increased the maximum suction decreased. Therefore, for design wind velocities these results confirmed the validity of using a maximum suction coefficient of no greater than -1.0 for the types of slightly roughened towers observed.[12]

A second major conclusion was that for the dynamic loading observed there appeared to be no resonant affect that would influence design. In short, the observations and their use in time-history analyses confirmed the presently accepted quasi-static method of analyzing towers under wind loads.[13] The question of surface roughness also was discussed vigorously during this period and numerous model tests were compared to the full-scale results. At a 1977 international session on cooling tower wind loads in San Francisco, results presented from research done both in Germany, Britain and the United States, clarified the manner in which surface roughness influenced pressure distribution.[14] In particular the earlier tower geometry parameter of k/d (projecting rib height/tower diameter) was replaced by a surface parameter k/s (projecting rib height/rib spacing); and a curve of k/s versus maximum suction established. These results again confirmed the validity of using a maximum suction of -1.0 on most of the ribbed towers built in the United States.

The Safety of Towers: 1979-1983

Two issues of primary importance emerged during this period and are still under study. First came the old question raised after Ferrybridge about the validity of existing design practice as larger and larger towers are designed. The second issue, a relatively new one, related to the safety of towers already in service, especially those built during the earliest period.

Buckling capacity is a good illustration of the design-practice issue. Until very recently there had been little reliable published results on geometrically nonlinear analyses as a means for estimating the buckling capacity of new designs. During the past five years, however, such results have begun to appear, and one example will illustrate recent progress in the United States.

Carried out at Cornell University this research establishes a reliable method for nonlinear analysis, which is simple enough to be used without excessive cost or risk of error; and its primary focus is upon the establishment of other more simplified methods appropriate for design. A summary of this work appears in this Symposium Volume. [15]

In its Table 1 that paper (Ref. 15) compares its new results to those from eight earlier cooling tower examples to show that the new program is reliable. These comparisons with major earlier results are used to evaluate the new program. Although this approach seems obvious, it is often not followed in the publications intent only on presenting new methods. The high expense and complexity of computer work is part of the reason why comparisons with earlier studies were often not fully made. The new Cornell program makes use of interactive computer graphics and thus significantly reduces the time requires to prepare an analysis. It is just this concern on the visual and simplifying potentials in computer graphics that has allowed so many full geometrically nonlinear analyses to be carried out.

In turn the large number of such analyses has permitted a number of comparisons (Ref. 15, Table 2) on the basis of which the equivalent axsymmetrical load method is justified as a reasonable basis for design. However, it is also essential to include the influence of vertical cracking in such buckling estimates.

The second issue to arise recently in the United States has been the present condition of towers in service for some years. While no full survey has as yet been made, I shall report here on recent preliminary inspections of four towers having unusual significance in the United States.

The first three of these towers were completed in 1968 and 1969 at the Paradise Steam Power Plant for the Tennessee Valley Authority at Drakesboro, Kentucky. They were the only towers to exceed 400 ft in the 1960's and were the tallest (437 ft high) in the United States by 67 ft when built. Thus these three towers represented the first major height increase in the

United States above the Ferrybridge type. These towers had been in service for about 15 years, when an inspection was made in April 1984. The thin shells appeared to be in good condition both inside and out. No vertical cracking was visible externally from the ground nor internally when viewed from the fill top. Such cracking had been observed on many towers built in Great Britain up to the late 1960's.

Small vertical hairline cracks could be seen upon very close inspection in the lintel beam. These were roughly evenly spaced at about four feet and appeared to go from the free edge upward about four feet as well. They appeared to be similar to the hairline cracks normally found in reinforced concrete framed structures. There was no sign of any deterioration connected with such cracks.[16]

The fourth tower inspected was at the B. F. England Oil-fired Power Plant near Atlantic City, New Jersey. While this tower is only 208 ft high, it was the only salt water tower in the United States when completed in 1974. When inspected in January 1983, it had been in service about seven years and its structural condition in the hostile salt water environment was still good. Its design had foreseen a service life of more than 40 years in the shore area. Therefore, the designers had used a dense concrete made with type V cement, galvanized steel bars, and an epoxy coating along the horizontal construction joints. No special coatings had been applied to the exposed concrete surfaces however. [17]

As in the Paradise towers, vertical hairline cracks could be seen in the lintels. In addition there was one vertical crack visible externally at lift 3 (about 15 feet above the air inlet opening) in between each exterior vertical rib. These cracks are probably related to the thermal gradients at the level of the fill. The horizontal reinforcement is a minimum of 0.35% at that level, and thus can control the cracks there. There did not appear to be any significant structural deterioration of the tower.

Conclusions

From this brief review of cooling tower designs in the United States, we can conclude that

1. continual reference to full-scale experiences is essential to any improvements in design practice,
2. there is an encouraging trend toward research studies which establish confidence in simplified methods of analysis as a basis for design, and
3. there is a need for the continual observation of the performance of towers in service for many years.

Acknowledgments

This review is based in part upon a survey of United States cooling towers made by Research-Cottrell to whom I am indebted for its use. Also I benefited from the advice and help of E. Killam, D.K. Mahajan, and Tom Mineo of Custodis-Hamon and from the continual advice of J.F. Abel.

References

1. R. F. Rish and T. F. Steel, "Design and Selection of Hyperbolic Cooling Towers," Journal of the Power Division, ASCE, Proceedings Vol. 85, No. PO5, October 1959, pp. 89-117; and Paul Rogers, "Hyperbolic Reinforced Concrete Cooling Towers," Journal of the American Concrete Institute, Proceedings Vol. 58, No. 4, October 1961, pp. 395-405.

2. "Natural Draught Cooling Towers - Ferrybridge and After," Proceedings, Conf. held at Inst. Civ. Eng., London, June 12, 1967.

3. "Power Plant Tries a New Tower," Engineering News-Record, October 4, 1962, p. 36.

4. One complete set of finite-element programs was designed by Philip Gould at Washington University beginning in the late 1960's; see for example, P. K. Basu and P. L. Gould, "SHORE-III-Shell of Revolution Finite-Element Analysis - Program Users' Manual," Research Report No. 49, Structural Division, Washington University, St. Louis, 1977. Another set was developed by John F. Abel now at Cornell University, see for example, P. P. Cole, J. F. Abel and D. P. Billington, "Maximum Seismic Response of Cooling Towers," Report 73-SM-1, Princeton University, Princeton, NJ 08544, 1973.

5. D. P. Billington, Thin Shell Concrete Structures, McGraw Hill Book Co., Inc., New York, 2nd Edition, 1982, p. 69. The study referred to here was for the Trojan Tower.

6. Report of the Committee of Inquiry into the Collapse of the Cooling Tower at Ardeer Nylon Works, Ayrshire, on Thursday, 27 September 1973, Imperial Chemical Industries Ltd., London. See also the discussion by Chilver on p.13 of ref. 2.

7. "Reinforced Concrete Cooling Tower Shells - Practice and Commentary," Journal of the American Concrete Institute, January 1977, pp. 22-31.

8. Larrabee, R.D., D.P. Billington and J.F. Abel, "Thermal Loading of Thin-Shell Concrete Cooling Towers," Journal of the Structural Division, ASCE, Vol. 10, No. ST12, December 1974, pp. 2367-2383.

9. Cole, P.P., J.F. Abel and D.P. Billington, "Buckling of Cooling Tower Shells: Bifurcation Results," Journal of the Structural Division, ASCE, Vol. 101, No. ST6, June 1975, pp. 2305-1222, and Mungan, I., "Buckling Stress States of Hyperboloidal Shells," Journal of the Structural Division, ASCE, Vol. 102, No. ST10, October 1976, pp. 2005-2020.

10. Der, T.J., and R. Fidler, "A Model Study of the Buckling Behavior of Hyperbolic Shells," Proceedings, Inst. Civ. Eng., Vol. 41, London, September 1968, pp. 105-118 with a restudy of their results as published in J.F. Abel, D.P. Billington, D.A. Nagy and C. Wuta-Dworkin, "Buckling of Cooling Towers," Journal of the Structural Division, ASCE, Vol. 108, No. ST10, October 1982, pp. 2170-2171.

11. Niemann, H.J., Peters, H.L., and Zerna, W., "Naturzugkühltürme in Wind," Beton and Stahlbetonbau, Vol. 67, No. 6, 1972, and Niemann, H.J., "Sicherheit und Aerodynamisch Belastung von Schalenkühltürmen aus Stahlbeton," Ruhr-Universitat Bochum, 1975, and Sollenberger, N.J. and Scanlan, R.H., "Pressure Differences Across the Shell of a Hyperbolic Natural Draft Cooling Tower," Proceedings of the International Conference on Full Scale Testing of Wind Effects, London, Ontario, Canada, June, 1974.

12. Sollenberger, N.J., R.H. Scanlan and D.P. Billington, "Wind Loading and Response of Cooling Towers," Journal of the Structural Division, ASCE, Vol. 106, No. ST3, March 1980, pp. 601-621.

13. Steinmetz, R.L., D.P. Billington, and J.F. Abel, "Hyperbolic Cooling Tower Dynamic Response to Wind," Journal of the Structural Division, ASCE, Vol. 104, No. ST1, January 1978, pp. 35-53.

14. Basu, P.K. and P.L. Gould, "Cooling Towers Using Measured Wind Data"; J. Armitt, "Wind Loading in Cooling Towers"; H.-J. Niemann, "Wind Effects on Cooling-Tower Shells," all in the Journal of the Structural Division, ASCE, Vol. 106, No. ST3, March 1980, pp. 579-600, 623-641, and 643-611, respectively.

15. Abel, J.F., S.-C. Chang and S.L. Hanna, "Comparisons of Complete and Simplified Elastic Buckling Loads for Cooling Tower Shells," Proceedings of the Second International Symposium on Natural-Draft Cooling Towers, IASS, Ruhr-Universität Bochum, September 5-7, 1984.

16. This inspection consisted of walking around the base of the towers at ground level, taking the external stairs to the level at the top of the fill and closely observing the lower parts of the shell from on either side of those stairs, and finally going around the inside of the tower at the top of fill level. All inspections were with the naked eye.

17. M. E. Warner and M. R. Lefevre, "Salt Water Natural Draft Cooling Tower Design Considerations," Paper presented at the American Power Conference, Chicago, IL, April 1974, 17 pages.

Table 1

Height Range in ft.	Number of Towers Constructed
Below 300	1*
300-349	5
350-399	30
400-449	22
450-499	16
500-553	26

*This tower is only 208 feet high but it is unique in being a salt water tower.

Table 2

YEAR COMPLETED	NUMBER OF TOWERS	HIGHEST IN YEAR in ft
1963	1	320
1967	4	323
1968	7	437
1969	2	437
1970	4	370
1971	8	452
1972	5	424
1963-72	31	(452)
1973	10	500
1974	11	492
1975	3	423
1976	5	506
1977	3	499
1978	5	500
1973-78	37	(506)
1979	5	540
1980	8	514
1981	6	553
1982	6	541
1983	7	541
1979-83	32	(553)
1963-83	100	(553)

Extreme Loading Effects on Cooling Towers

H.-D. KÖPPER

Zerna, Schultz und Partner
Ingenieurgesellschaft für Bautechnik mbH
Industriestraße 27

D-4630 Bochum 7

1. Introduction

The design of reinforced concrete buildings has to ensure the reliability and durability of the structures on the one hand and has to keep a defined safety zone with regard to failures on the other. To reach this, besides the analysis of the structure that is subjected to design loads, a set of additional requirements and recommendations must be observed.

Character	Design steps and Criteria	Examples
dependent on load	Analysis of stress and strain Buckling Analysis	Dead load, live load, design wind load, thermal loads, settlement, earthquake
independent of load	General design rules Requirements Recommendations	Reinforcement - placement, minimum amount of each layer - minimum wall thickness, concrete cover

Fig.1: Design Concept

With regard to the design of cooling towers, these two aspects are shown in Fig.1 according to the German VGB-guidelines. [1] Analysis methods that are used for the design of a hyperboloidal cooling tower are usually based upon the bending theory for thin elastic shells. Nevertheless, the stress state of the shell is governed by membrane stresses, due to the above mentioned main design loads, with exception of special loadings, e.g. temperature differentials.

The additional requirements and recommendations mainly referring to the reinforcement represent an essential part of the design concept. These design rules have to cover certain approximations between the loading conditions and the analysis and shall increase the quality of the tower construction.

But there is still a further important task of these additional requirements: In cases of unexpected extreme loading, the basic load-independent design leads to the decision whether a structure resists the acting forces or has to be strengthened. For this reason the basic structural design can be designated as a surplus of load capacity for loadings that have not been quantified in the planning phase.

Experiences and the fact that spectacular damages did not occur up to now prove that the practiced design method has been sufficient in most cases. Nevertheless, it is absolutely necessary to adapt design and planning steadily to new research results and it must also steadily be checked whether the load capacity determined by basic design still meets the requirements.

In order to give an impression of character and quantity of unexpected additional extreme loads, three cases will be described in the following.

2. Differential Settlements

The intention of the planning engineer is to design the building with its structural elements that way that settlement differentials will not occur or be absolutely nigligible. Under the assumption of a complete investigation of the soil, this can be reached, as a rule, by choosing a suitable foundation. Only when subsidences because of underground mining can influence the structure, it is unavoidable to take into consideration that there will be special movements of the foundation soil. Special constructional methods and mechanical as well as structural equipments have been developed for this case, in order to limit the influences of settlements on the shell. [2]

Nevertheless, it cannot be excluded, even after intensive investigations of the soil that differential settlements can occur. As they mostly result from unrecognized irregularities or special deformation behaviour of the foundation soil, they very often are realized during construction time already, or immediately after the works are finished. So they can be registrated by normally executed surveying.

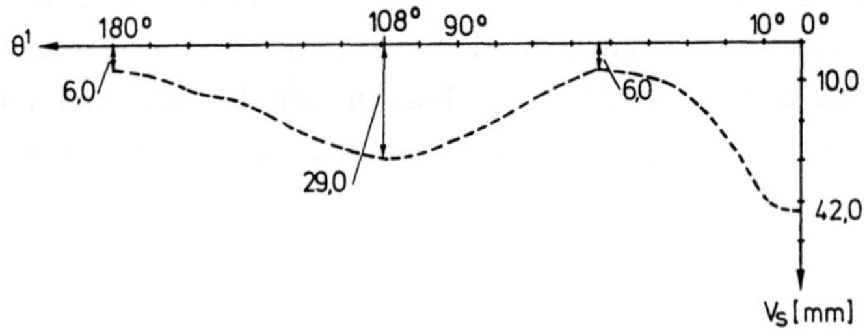

Fig. 2: Settlement Distributions

Figure 2 shows the measured settlement of a 142 m high natural draught wet cooling tower for a 750 MW power plant, and Figure 3 shows the main dimensions of the cooling tower.

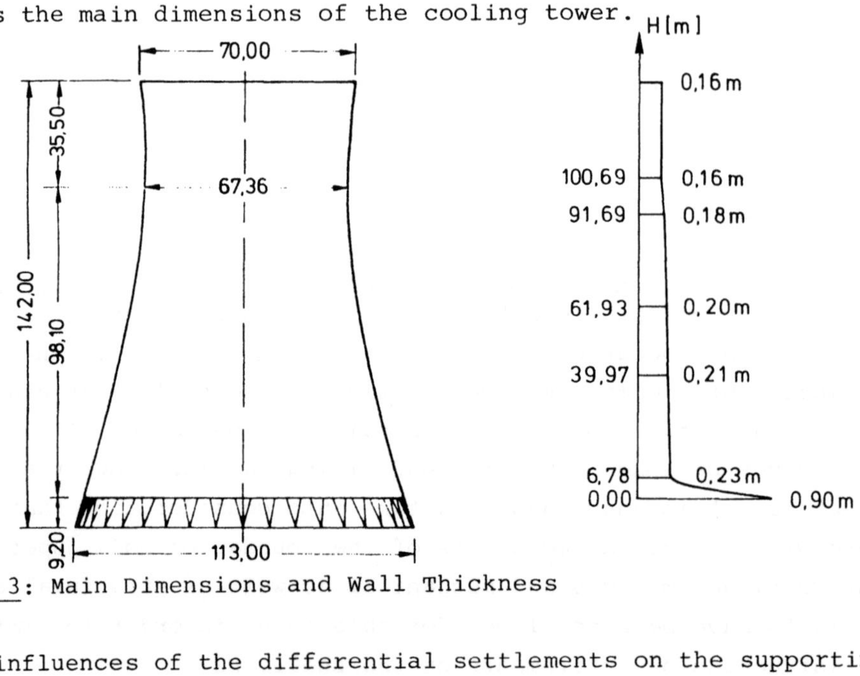

Fig. 3: Main Dimensions and Wall Thickness

The influences of the differential settlements on the supporting framework and the shell were analysed using a computer program that integrates the system of differential equations of the shell bending theory with the help of the multiplace method.[3] The distribution of measured settlement values was transformed into a Fourier-row and introduced into this investigation as a special boundary condition.[4] Some essential results are shown in Fig.4.

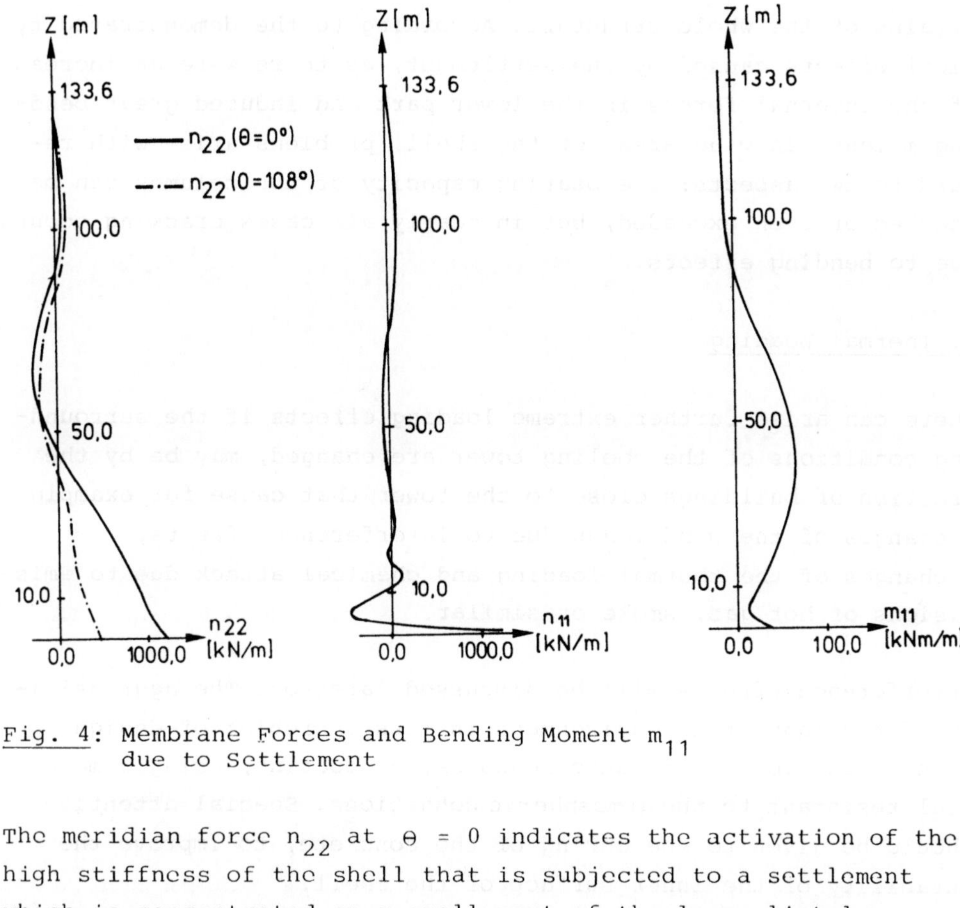

Fig. 4: Membrane Forces and Bending Moment m_{11} due to Settlement

The meridian force n_{22} at $\Theta = 0$ indicates the activation of the high stiffness of the shell that is subjected to a settlement which is concentrated on a small part of the lower lintel (deep beam effect). Within the region where long-wave settlements are measured, n_{22} is decreasing more than the ratio of extreme settlement values.

Special attention should be given to the quantities of the ring bending moments m_{11} and their distribution over a great part of the shell. In connection with the unimportant amount of the ring force n_{11} in this area these results are characteristic for inextensional bending effects.

As follows from the above mentioned results, a settlement of the here described distribution influences the state of stresses and strains of the whole structure. According to the demonstrated typical effects caused by the settlement, as there were an increase of the internal forces in the lower part and induced great bending moments in wide areas of the shell, problems arise with regard to two aspects: The bearing capacity of the columns can be reached or even exceeded, but in nearly all cases cracking occurs due to bending effects.

3. Thermal Loading

There can arise further extreme loading effects if the surrounding conditions of the cooling tower are changed, may be by the erection of buildings close to the tower that cause for example

- changes of the wind loads due to interference effects,
- changes of the thermal loading and chemical attack due to emissions of hot gaz, smoke or similar.

Interference effects will be discussed later-on. The aggressivity of the industrial environment is less a problem of design than a question of adequate measures, as coating, using a material resistant to the atmospheric conditions. Special attention should be given to the curing of the concrete, to improve the durability of the inner surface of the shell.

In the example considered now, a new heat source should be established very close to a cooling tower causing an unequally distributed thermal load on a certain part of the tower shell.

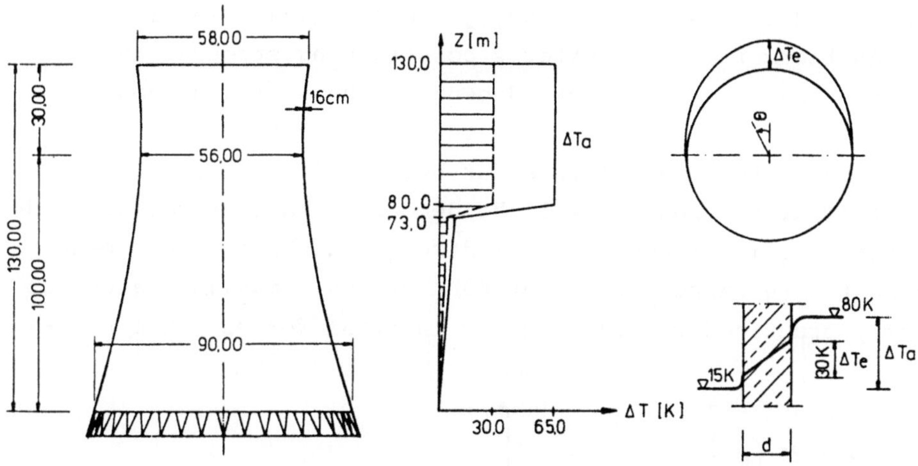

Fig.5: Main Dimensions and Thermal Loading

Figure 5 shows the main dimensions of the investigated tower and the part of the shell that was effected by this higher thermal loading. Calculations of the spreading and distribution of the hot gazes, depending on the wind velocity, show temperatures up to 80° C in the upper area of the cooling tower. Assuming a temperature of 15° C inside the tower, this leads to an absolute temperature differential of ΔT_a = 65 K and under consideration of the heat transfer to an effective temperature gradient of ΔT_e = 30 K.

Distribution in circumferential direction was assumed as cos-shaped, similar to one-sided solar radiation. For this thermal loading the internal forces and deformations were determined. The membrane force n_{11} has its maximum values at $\Theta = 0$ within that part of the shell where the loads are changing abruptly.

Bending moments in circumferential (m_{11}) and meridional direction (m_{22}) are nearly constant within the part where the thermal load is acting and they cause nearly equal strength in both directions.

All these indications concern homogeneous, linear-elastic materials. As thermal loading effects constraining stresses in the structure, the internal forces depend on the stiffness conditions. Their values decrease when the cross-section passes into the cracked status. Therefore a further non-linear calculation was performed; the results will be demonstrated considering bending moment m_{11} in circumferential direction. The greatest bending moments are reduced to about 65 % of their maximum value, as similarly practiced for the further internal forces. (See Fig.6)

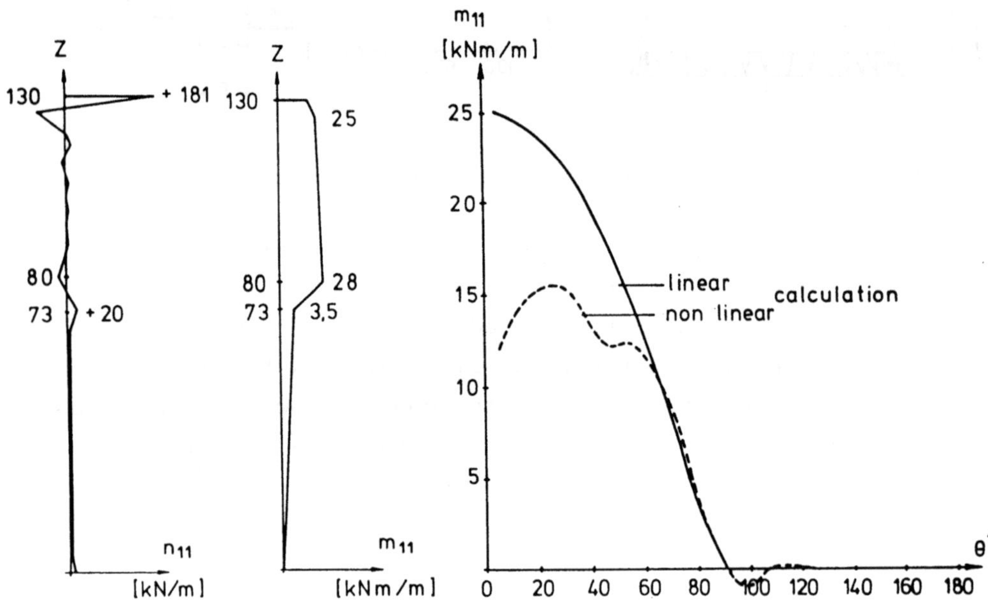

Fig. 6: Internal Forces due to Thermal Load

For the cooling tower designed according to the actual requirements and recommendations the membrane forces resulting from the investigated thermal loading do not lead to a considerable exceed of the admissible stresses. In contrast to that, the additional bending moments require an enlarged reinforcement in comparison with the basic design.

4. Interference Effects

Modern power plants using fossile fuels are composed of units with a capacity up to 750 MW. It is characteristic for the units of these power plants that they have narrow-shaped high boiler-houses with a total height up to 130 m, which is comparable with that of cooling towers.

The rising requirements of environmental protection make installations for purification (f.i. desulfurization) unavoidable. Very often the height of the buildings for these purification installations is up to 80 m, comparable to the elevation of the cooling tower throat. Following the demands of industrial processing engineering, these buildings very often must be located in the immediate neighbourhood of the cooling towers.

Furthermore, enlargements of already existing power plants by new units under the restriction that the available building ground is limited also leads to the necessity to arrange the smallest possible distances from each other.

Codes and VGB-recommendations assume wind loads that are valid for free-standing single cooling towers.
For cooling towers that are arranged in groups, these recommendations indicate empirically determined minimum distances between neighbouring cooling towers. Following these redommendations, the load assumption of a single tower may be used.

Because of the aforementioned reasons, these minimum distances cannot be kept in some cases or are later-on diminished. Resulting interference effects lead to the question, whether the decisive wind loads deviate from the wind loading according to standards.

The wind conditions (directions, velocities a.s.o.) within a densely built-up power plant area cannot be determined by analytical methods.

Realistic values of the relevant wind datas only can be verified carrying-out model tests in a boundary layer wind tunnel. It is of great importance for such model tests, which kind of datas are measured.

The most advantageous method is the measurement of strains using aeroelastic models. In this way the static as well as the dynamic structural response and the maximum load of the shell are directly obtained. In the last years there have been performed such tests for several power plants in the wind tunnel of the Ruhr-University Bochum, FRG. Some of the results shall be described in the following.

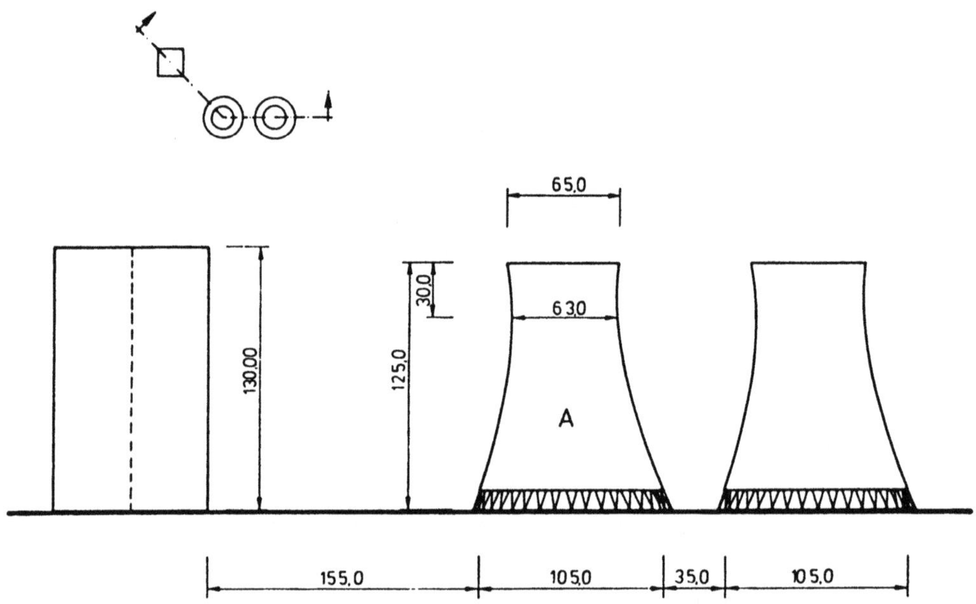

Fig. 7: Interference Effects
Investigated Building Arrangements

Figure 7 shows the arrangement and distances of two cooling towers and the boilerhouse of a power plant. The interference effects on tower A have been investigated.

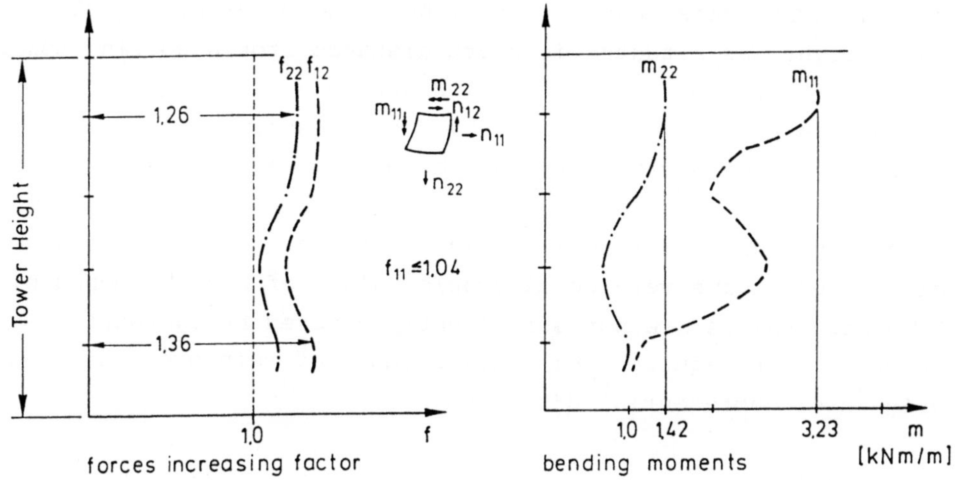

Fig. 8: Results of Interference Investigations

The values featured in Figure 8 have been measured at four different levels of this tower. They are maximum values relating to different wind directions.

The interference factors $f_{\alpha\beta}$ indicate, how far the peak membrane forces that are valid for the single tower are increased by the interference effects. Values and distribution of the additional stresses depend on the shape of the building, situated in front of the tower. At the edges of such a forestanding building a wind stream separation occurs, which causes turbulence and increased wind velocities.

The result of these effects are additional stresses in the shell. Concerning the demonstrated situation, the buildings mainly have vertical edges causing wind stream separation and are of the same height as the investigated cooling tower. Therefore, besides the shear forces n_{12}, the meridional force n_{22} increases in a higher degree then the circumferential forces n_{11} ($f_{22} > f_{11}$).

Another investigated case, where a building of great breadth and small height was situated in short distance from a cooling tower, was significated by horizontal wind stream separation at the upper edge of the building. This is, why the stresses in circumferential direction increase stronger than the meridional stresses.

Performing model tests to study interference effects there have been stated always very considerable values of bending moments. These high values are not sufficiently covered by the basic design and very often require additional amounts of reinforcement, especially in circumferential direction.

5. Conclusions

The above described cooling towers have been designed according to the actual codes and recommendations. The resulting bearing capacity is sufficient to avoid damages of the structure due to the acting of unexpected extreme loads.

On the other hand, however, the occurence of these loads leads to an exceeding of the yield strength of the reinforcement in highly stressed regions of the shell. Generally the following statements can be pointed out:

The structure resists the membrane forces, increased due to extreme loading. This results from the fact that the safety precautions of the design concept are comprising mainly the structural behaviour governed by the membrane stress state. In all considered cases the applied design concept guarantees a sufficient buckling safety [5] taking into account the increased membrane forces.

Unexpected additional loads are nearly always connected with bending moments. In many cases the bending stresses exceed the allowable values. This concerns especially loads effecting stresses in circumferential direction.

Generally valid and completing recommendations for the basic design requirements cannot be derived from few special investigations. These can only be obtained by the performance of systematic researches. Nevertheless, an experienced design engineer will always provide sufficient minimum reinforcement in circumferential direction of the cooling tower, to ensure an adequate bending stiffness of the shell, whose wall thickness is determined by membrane stresses.

References:

1. Bautechnik bei Kühltürmen, Teil 2: Bautechnik Richtlinien (BTR) VGB Kraftwerkstechnik GmbH, Essen 1980

2. Krätzig, W. B.: Tragverhalten von Naturzugkühltürmen unter Berücksichtigung bergbaulicher Einwirkungen, VGB Kraftwerkstechnik Heft 4, 1977.

3. Basar, Y.: Zur Berechnung von Rotationsschalen mit Mehrstellenverfahren. Der Bauingenieur 50 (1975), S. 41-48

4. Köpper, H.-D.: Einfluß der Randversteifung auf das Stabilitätsverhalten von Kühlturmschalen. Konstruktiver Ingenieurbau/Berichte, H. 38/39, 1981

5. Zerna, W., Mungan, I., Steffen, W.: Wind-Buckling Approach for RC Cooling Towers, J. Eng. Mech. Div. ASCE, pp. 836 ff, 1983

Repair of Cooling Towers

R.J. KRAPFENBAUER
Professor in Ordinary, Hochschule für Angewandte Kunst, Wien/Österreich
(Director of Department for Structural Science).
A-1184 Vienna, Pötzleinsdorfer Straße 94

Summary
The most important rules for the construction of cooling towers are: the concrete supports must be made of high class concrete, with dense surfaces and sufficient reinforcement. The statical analysis must be very carefully done. Also the cross section of the stanchions is essential. Failures in cooling towers usually occur either from lack of quality in the concrete or from insufficient statical analysis. The noxious effect of the water must not be underrated.

Development of Cooling Tower Construction

Looking back in the history of cooling towers, we find that the earlier species were of a cylindrical shape. In 1910 the engineer Van Itersen started to investigate shells of reinforced concrete for cooling towers that had the shape of a rotation hyperboloid. Also cooling towers in the shape of a truncated cone were examined, as it was the intention to reduce the height of the structures for reason of landscape protection.

As most of the recently built cooling towers are of the evaporating type, 70 % of the heat reduction of the water in cooling towers takes place by evaporation and so a large surface is necessary.

Most of natural draught cooling towers are therfore of the shape of a rotation hyperboloid shell resting on concrete supports. the air enters through the openings between these supports. Maximal sizes for the natural draught cooling towers are now about 130 m height and 100 m basical diameter.

The planning parameters for cooling towers are the cooling range and the occuring temperatures. Very important are also the yearly weather cycles. The prime factor that influences the performance of the power station is the temperature of cooled water leaving the tower. This affects the steam vacuum conditions in the condenser and, consequently, the heat rate and the energy output of the whole arrangement.

Construction and Safety of Cooling Towers

The particulars for the construction of Cooling Towers are of a technological, statical and structural nature.

There are three principal parts of the structure:

1. the "chimney" with the shell supports and the foundation,
2. the cooling basin with sole and walls, and
3. the "packing", a kind of grid, with the openings for the incoming and outgoing water.

The rotation shell with a negatively curved meridian curve and thin walls is most sensitive against variable settings and displacements. The supports of the shell must have a foundation that guarantees steady setting.

If the subsoil conditions are not homogeneous, a deep-reaching foundation is reasonable, possibly as far down as the horizon of the subsoil. Another solution would be an annulus-shaped foundation of reinforced concrete.

The supports for the shell must be firmly connected with the foundation, as they have to direct high tensile stresses.

Another problem is presented by the constant effect of atmospheric particles and air to which the concrete is exposed; it is likewise exposed to the temperature of the incoming air and that of the cooling water running down the inner surface of the cooling tower.

The first rule for the construction of cooling towers must therefore be: supports that are statically well analyzed, high quality concrete, and a dense concrete surface. The supports must have sufficient reinforcement. Often they are manufactured as prefabricated elements.

As the construction of cooling towers is usually built with climbing formwork, and the first ring carries the formwork for the next stage, it is very important that the concrete of this first ring has a three days' strength of 150 kiloponds/m^2 in order to carry the scaffolding jacks. The connection of the concrete to the already concreted shell asks for a great deal of plasticity of the concrete.

Possible Failures in Concrete Cooling Towers

Rather often we find that damages of cracks in the shells of supports of cooling towers come from insufficient analyses as well as from lacks of quality in the concrete and are often also a consequence of the noxious effect of water.

Cooling Towers of Austrian Mineral Oil Administration near Vienna/Austria

Two cooling towers in the gas central station of the Austrian Mineral Oil Administration, of hyperboloid shape had to undergo repairs.

Fig. 1: Cooling Towers of Gas Station of Austrian Mineral Oil Administration

They are of a diameter of 19,30 m at the bottom and of 1,10 m at the top. Their height amounts to 24 m. The thickness of the shell walls is 0,26 m at the bottom and 0,08 m at the top of the shell.

The shell rests of inclined supports that, on their part, rest on a ring-shaped foundation wall that, in its turn, rests on the soil with a broad footing. The foundation wall includes the basin, whose bottom is made of concrete. The structural firm said in its load analysis that the load from the weight proper of the shell is axially symmetrical; the wind load, it said, consists of pressures vertical to the surface of the shell; it is antimetric as to the center plane, coinciding with the wind direction, and may be approximate to every grade of accuracy, expressed by the assumption:

$$Z = \Sigma z_n \cos_n \delta$$

wherein:

Z = normal pressure per surface unit

$\cos_n$ = the angle between the considered center planes coinciding with the wind direction

n = a constant value for a certain horizontal section of the body.

As by the above assumption, a loading case for each n is fixed that may be completely taken up by the membrane forces, the total wind load may likewise be taken up by the membrane forces alone.

The analyses drew the consequence that all the loadings of the tower shell might be completely taken up by membrane forces; thus the wall of the tower would remain totally free of bending tensions.

Calculations resulted in stipulating a dynamic pressure of 0,15 t/m² with a high grade of security.

As a consequence of temperature variations, further tensions may occur in the wall of the shell. First the average wall temperature may change; this results in elongation or shortening of the height measurements. They may take place without disturbance, also elongation of shortening in the horizontal plane may take place unhindered, excepting the fixations of the shell's lower edges to the supports. By the very small number of supports and their small dimensions in the horizontal direction, their resistance against elongation or shortening of diameter was negligible, said the analysis.

Secondly, there are temperature variations in the wall, caused by the different temperatures of inner and outer air; these differences are largest in winter when they amount as far as to 45 °C (f.i. the temperature within the tower +20 °C, exterior temperature -25 °C).

As transition values of heat were assumed:

from inner air to wall	10 kg cal/m².h.°C
within the concrete wall	1,6 kg cal/m².h.°C

from wall to outer air 20 kg cal/m².h.°C

for a wall thickness D (in meters) we have thus:

temperature difference of wall = D/1.6

temperature difference of interior and exterior (1/10).(D/1.6).(1/20)

Thus, for the maximum difference of 45 °C, we have, assuming 8 cm wall thickness, a difference in the wall temperature of 12.50 °C a.s.on.

Fig. 2 Cooling Towers unter construction

The maximum pressure on the soil amounts to 0,85 kg/cm². All loadings on the foundation slab are equally distributed, with the exception of the loading on the supports. The maximum loads on the supports are 25,00 tons on the lee side; the moment above the point of support is 9,8 t. The maximum horizontal force is 10,33 t, the maximum shear tension is 5,8 kg/cm.

In the course of time, the two cooling towers showed signs of damage by wear and tear in some of the 12 supports of a diameter of 24/42 m. There was graver damage of the concrete, especially inside of some supports. Besides, cracks were found in the shells; in these cracks, there were sediments of sintered limestone. The supports of the cooling towers were under the constant influence of splashing water that alters its direction according to the wind; along the edges of these supports, further efflorescences appeared.

The damaged supports were repaired in the following way:
two new supports were manufactured and concreted to the damaged supports. The split-off concrete was removed, new reinforcement was brought in and concreted.

By an expertise of the Technical Research and Experimental Laboratory of the Technical University of Vienna, the quality of the damaged concrete, the chemical qualities of the water and the efflorescences were examined. The water had been taken from the river March and was softened according to the lime-milk process. It showed 130 mg/l of SO_3, therefore it is agressive towards concrete. This aggressivity is increased by the water temperature and the change between moistening and drying of the wall. It was recommended to reinforce the protective paint and to inspect it regularly.

In my expertise I expressed the opinion that not only the quality of the concrete and the noxious influence of the cooling water must be considered.

The edge of the cooling tower's shell is not supported along its entire length; it is supported only pointwise, i.e. on the single supports. These edge conditions seem to have been only insufficiently considered, or at least the influence of the membrane state in the area of the lower edge of the shell has not been sufficiently regarded.

Contrary to the statements of the commissioned engineering office, the transfer of loads to single supports can not longer be expressed by the membrane state alone. Edge shearing strain will remain, and this will induce edge disturbances -- e.g. bending effects - indeed they bring additional loads to the edge reinforcement.

Fig. 3 Cooling Towers in the plant

We have seen how important it is to untertake investigations on the horizontal bending moments in the shell that occur in consequence of the actual distribution of wind pressure and the uneven rise of temperature as a consequence of one-sided insolation.

For these local effects, it is desirable to take as the basis of the calculation of the wind pressure at least a three-membered Fourier series. For very high cooling towers, the wind tunnel test should be well recommended; measurements from wind tunnel tests should be taken up into the building standards. The most difficult problem is of course the erection of groups of cooling towers because of the behaviour of the wind flow between them. If one tower comes into the suction flow of another tower the situation may become critical. The cutting reactions in the shell must be carefully considered. On the whole it can be said that small failures can often occur, but measures of reconstruction can save a lot when untertaken in time.

The Effect of Unequal Settlement of Foundation on the Stress Resultants of Hyperbolic Cooling Towers and the Unequal Settlement Tolerance Limit

Lu Wenda[1], Lin Baoqing[2] and Lu Renyi[3]

1. Institute of Applied Mathematics and Mechanics, Shanghai University of Technology, Shanghai, China
2. Xian Institute of Thermal Engineering, Xian, China
3. Foundational Engineering Research Institute of Chinese Academy of Building Research, Beijing, China

Abstract

In the present paper, by using the finite element solution of hyperbolic cooling towers supported by a column system based on the curved axisymmetric finite element, the effects of unequal settlement of foundation on the stress resultants of tower shell are investigated. This partial edge effect will be spreaded from the bottom of the shell up to (1/10-1/5) height of the tower. Through the illustrative numerical example of a cooling tower being constructed, the tolerance limit of unequal settlement is given. It proved that the unequal settlement of this cooling tower from full-scale measurements satisfies that limit.

1. Introduction

In recent years, the planning and construction of hyperbolic cooling towers are approaching 150 m height and 118 m in diameter of foundation in China. The tower is a rotational thin shell supported by V-typical or X-typical column system. In order to transport the loads acting on the shell to the foundation finely, it is necessarily that vertical settlements at the bottom of shell along the circumferential direction should be equable in the main, and the in-plane displacements of base frame of column system should be small as far as possible Whereas when the power station is constructed on a flexible base, the effect of unequal settlement of foundation on the stress resultants of tower shell must be noted. In reference [2], by using the geometric approximation of meridian the analytical solutions of this problem has obtained. In the present paper, using the finite element method [3], the effects of unequal settlement of foundation on the stress resultants of tower shell are investigated. The allowable values of settlement according to the limit state design procedure are given. The numerical solutions of unequal settlement of a typical constructed cooling tower are also given. If the counteraction of dead loading is taken into consideration, the maximum relative allowable value of settlement is obtained as $\Delta U_{max} = 2.61$ mm , on the contrary, $\Delta U_{max} = 1.36$ mm . The influence of unequal settlement will be spreaded from the bottom of the shell up to (1/10-1/5) height of tower. Finally, the relative settlements of foundation measured on full-scale tower are given. According to the allowable values of relative settlement suggested by present paper, it is discovered that the vertical settlement of the 15th measured point is greater than 2.7%.

2. Finite Element Solution of Hyperbolic Cooling Tower Supported by Column System

In the present paper, the curved axisymmetric finite shell elements are used [3]. Let the shell be divided by nodal surfaces into a series of curved axisymmetric rotational shell elements. The displacement vector $\underset{\sim}{V}$ of shell's middle surface at any point in the datum coordinate system (z, θ, r) is

$$\underset{\sim}{V} = \lfloor u, v, w \rfloor^T \tag{1}$$

The transformation relationship between the displacement vector $\underset{\sim}{V}^*$ in the local coordinate system and the displacement vector $\underset{\sim}{V}$ is formed as:

$$\underset{\sim}{V}^* = \begin{bmatrix} -\sin\varphi & 0 & \cos\varphi \\ 0 & 1 & 0 \\ \cos\varphi & 0 & \sin\varphi \end{bmatrix} \underset{\sim}{V} = \underset{\sim}{T}\,\underset{\sim}{V} \tag{2}$$

The displacement vector is expanded into Fourier series in the circumferential direction. The higher order Hermite's interpolation functions are used in the meridional direction. It is defined that the Fourier series coefficients of displacement vector $\underset{\sim}{V}$, their derivatives of first order with respect to arc length s $(\frac{\partial u}{\partial s}, \frac{\partial v}{\partial s}, \frac{\partial w}{\partial s})$, and their derivatives of second order $(\frac{\partial^2 u}{\partial s^2}, \frac{\partial^2 w}{\partial s^2})$ are the generalized nodal circular displacements. The displacement functions of reference [3] are adopted. For the ith element of shell we have:

$$\underset{\sim}{V}^* = \underset{\sim}{T} \sum_{k=0}^{M} \underset{\sim}{X}_{ck} \underset{\sim}{A} \underset{\sim}{\alpha}_k \tag{3}$$

in which $\underset{\sim}{X}_{ck}$ $\underset{\sim}{A}$ = meridional, circumferential interpolation function matrix. Applying the principle of stationary total potential, the equilibrium equation of shell element of the kth harmonic term can be obtained as follows:

$$\underset{\sim}{K}_k^e \underset{\sim}{\alpha}_k = \underset{\sim}{F}_k \tag{4}$$

All the equilibrium equations are assembled, the equilibrium equation of tower shell is given as:

$$\underset{\sim}{K} \underset{\sim}{\rho} = \underset{\sim}{P} \tag{5}$$

in which the lower indice k are omitted. If the shell is divided into N nodal circles, then the order of total stiffness matrix $\underset{\sim}{K}$ of tower shell is $(8N \cdot 8N)$.

Stiffness Matrix of V-typical Column System

It is assumed that the column system is consisted of c pairs of V-typical column, and each pair of column is made up of two straight bar of uniform cross section. The lower ends may be refered to as a fixed edge. As for the location of the top ends which are connected with the bottom of shell, besides it is assumed that the bending rigidity in its own planes is infinity, the deformations of any direction have not any constraint. In order to ensure the compatibility of deformation at the conjunction between the bottom lintel and the column, we must have

$$u^* = \bar{u}^*, \quad v^* = \bar{v}^*, \quad w^* = \bar{w}^*, \quad \beta_\varphi^* = \bar{\beta}_\varphi^*, \quad \beta_\theta^* = \bar{\beta}_\theta^* \tag{6}$$

in which $\bar{u}^*, \bar{v}^*, ---, \bar{\beta}_\theta^*$ are the displacements of column at the top end which positive sense is shown in Fig. 2.. When the lower ends i, j of column are fixed, we obtain the relationship between the generalized forces and the generalized displacements of each pair of column as follows:

$$\underset{\sim}{\bar{N}} = \underset{\sim}{\bar{K}}\,\underset{\sim}{\bar{u}}^* \tag{7}$$

in which

$$\underset{\sim}{\bar{N}} = \llcorner \bar{N}_\varphi,\ \bar{N}_{\theta\varphi},\ \bar{Q}_\varphi,\ \bar{M}_\varphi,\ \bar{M}_{\theta\varphi} \lrcorner^T$$

$$\underset{\sim}{\bar{u}}^* = \llcorner \bar{u}^*,\ \bar{v}^*,\ \bar{w}^*,\ \bar{\beta}^*_\varphi,\ \bar{\beta}^*_\theta \lrcorner^T$$

Let the interval (O, 2π) be divided into C equal portions, and each pair of column be set on the left end of each portion, then the circumferential angle of the λth pair of column is as:

$$\theta_\lambda = 2\pi\,(\lambda-1)/C \qquad (= 1,\ 2,\ ---,\ C/2+1) \tag{8}$$

in which the symmetry with respect to O-π axis is taken into consideration. The principle of virtual work leads to the stiffness matrix $\underset{\sim}{K}$ of column system with respect to the kth harmonic term in the datum coordinate system as follows:

$$\underset{\sim}{K} = \frac{C}{2}\begin{bmatrix} k_{11} & & & \text{symm.} \\ k_{21} & k_{22} & & \\ k_{31} & k_{32} & k_{33} & \\ & & & \\ k_{81} & k_{82} & & k_{88} \end{bmatrix} \tag{9}$$

in which

$$k_{11} = 2EA\sin^2\alpha\sin^2\varphi/l_o + 24EI\cos^2\varphi/l_o^3\ ;$$
$$k_{22} = 2EA\cos^2\alpha/l_o\ ;$$
$$k_{33} = 2EA\sin^2\alpha\cos^2\varphi/l_o + 24EI\sin^2\varphi/l_o^3\ ;$$
$$k_{44} = 2(GJ\cos^2\alpha + 4EI\sin^2\alpha)\cos^2\varphi/l_o\ ;$$
$$k_{66} = 2(GJ\cos^2\alpha + 4EI\sin^2\alpha)\sin^2\varphi/l_o\ ;$$
$$k_{55} = k_{77} = k_{88} = 0\ ;$$
$$k_{31} = -EA\sin^2\alpha\sin2\varphi/l_o + 12EI\sin2\varphi/l_o^3\ ;$$

$$k_{41} = 12EI\sin\alpha\cos^2\varphi/l_o^2 \;;\quad k_{43} = 6EI\sin\alpha\sin2\varphi/l_o^2 \;;$$
$$k_{61} = 6EI\sin\alpha\sin2\varphi/l_o^2 \;;\quad k_{63} = 12EI\sin\alpha\sin^2\varphi/l_o^2 \;;$$
$$k_{64} = (GI\cos^2\alpha + 4EI\sin^2\alpha)\,\sin2\varphi/l_o \;;$$
$$k_{21} = k_{32} = k_{42} = k_{51} = k_{52} = k_{53} = k_{54} = 0 \;;$$
$$k_{71} = k_{72} = k_{73} = k_{74} = k_{75} = k_{76} = 0 \;;$$
$$k_{81} = k_{82} = k_{83} = k_{84} = k_{85} = k_{86} = k_{87} = 0 \;.$$

A = cross section of beam element; J = polar moment of the cross section; I = moment of inertia of the cross section; E, G = Young's modulus and shear modulus of material, representatively; l = beam length. The terms of the order of h/r^2, h/rr_θ, h/r_θ^2 compared to unity are neglected, which express the effect of the rotation of the cross section of beam element on the membrane rigidity of column.

Boundary Loading of Self-Equilibrating

Let the stiffness matrix of column system $\hat{\underset{\sim}{K}}$ to be assemble on the corresponding location of the last shell element. To solve the assembled equation of equilibration we obtain the finite element solution, in which the effects of column system are taken into consideration. Call it a particular solution. So that the stress resultants with continuous distribution on the base of the bottom lintel are given as follows:

$$\bar{\underset{\sim}{N}} = \sum_{k=0}^{M} \underset{\sim}{S}_k\, \bar{\underset{\sim}{N}}_k \tag{10}$$

in which

$$\bar{\underset{\sim}{N}} = \llcorner \bar{N}_\varphi,\ \bar{N}_{\theta\varphi},\ \bar{q}_\varphi,\ \bar{M}_\varphi \lrcorner^T \;;\quad \bar{\underset{\sim}{N}}_k = \llcorner \bar{N}_{\varphi k},\ \bar{N}_{\theta\varphi k},\ \bar{q}_{\varphi k},\ \bar{M}_{\varphi k} \lrcorner^T$$

$$\underset{\sim}{S}_k = \ulcorner \cos k\theta,\ \sin k\theta,\ \cos k\theta,\ \cos k\theta \lrcorner \quad \bar{q}_\varphi = \bar{Q}_\varphi + \frac{1}{r}\frac{\partial \bar{M}_{\theta\varphi}}{\partial\theta}$$

The mean stress resultants transfered from bottom lintel are obtained as:

$$\bar{\underset{\sim}{N}}_\lambda = \frac{1}{2\Omega}\int_{\theta_\lambda-\pi/c}^{\theta_\lambda+\pi/c} \sum_{k=0}^{M} \underset{\sim}{S}_k\, \bar{\underset{\sim}{N}}_k\, d\theta \tag{11}$$

in which the Ω is defined as one-half of the central angle corresponding to the width of column's top. According to the formula of (11), we obtain four groups of rectangular wave with different amplitude corresponding to the $\bar{\underset{\sim}{N}}$. Each group has four waves. If the symmetry with respect to O-π axis is taken into account, then only (C/2 + 1) waves need to expand into Fourier series. So that we have

$$\underset{\sim}{N}_\lambda = - \begin{bmatrix} \bar{N}_{\varphi 1} + \bar{N}_{\varphi \frac{c}{2}+1} \\ 0 \\ \bar{q}_{\varphi 1} + \bar{q}_{\varphi \frac{c}{2}+1} \\ \bar{M}_{\varphi 1} + \bar{M}_{\varphi \frac{c}{2}+1} \end{bmatrix} + \frac{2\Omega}{\pi} \sum_{\lambda=2}^{c/2} \begin{bmatrix} \bar{N}_{\varphi\lambda} \\ 0 \\ \bar{q}_{\varphi\lambda} \\ \bar{M}_{\varphi\lambda} \end{bmatrix}$$

$$+ \sum_{k=1}^{M} \left[\frac{2\sigma_k}{\pi k} \underset{\sim}{S}_k \begin{bmatrix} \bar{N}_{\varphi 1} \sin k\Omega + \bar{N}_{\varphi \frac{c}{2}+1} (\sin k\pi - \sin k(\pi-\Omega)) \\ \bar{N}_{\theta\varphi 1} (1-\cos k\Omega) + \bar{N}_{\theta\varphi \frac{c}{2}+1} (\cos k(\pi-\Omega) - \cos k\pi) \\ \bar{q}_{\varphi 1} \sin k\Omega + \bar{q}_{\varphi \frac{c}{2}+1} (\sin k\pi - \sin k(\pi-\Omega)) \\ \bar{M}_{\varphi 1} \sin k\Omega + \bar{M}_{\varphi \frac{c}{2}+1} (\sin k\pi - \sin k(\pi-\Omega)) \end{bmatrix} \right.$$

$$\left. + \frac{2\sigma_k}{\pi k} \underset{\sim}{S}_k \sum_{\lambda=2}^{c/2} \begin{bmatrix} \bar{N}_{\varphi\lambda} (\sin k (\frac{2\pi}{c}(\lambda-1)+\Omega) - \sin k (\frac{2\pi}{c}(\lambda-1)-\Omega)) \\ \bar{N}_{\theta\varphi\lambda} (\cos k (\frac{2\pi}{c}(\lambda-1)+\Omega) - \cos k (\frac{2\pi}{c}(\lambda-1)-\Omega)) \\ \bar{q}_{\varphi\lambda} (\sin k (\frac{2\pi}{c}(\lambda-1)+\Omega) - \sin k (\frac{2\pi}{c}(\lambda-1)-\Omega)) \\ \bar{M}_{\varphi\lambda} (\sin k (\frac{2\pi}{c}(\lambda-1)+\Omega) - \sin k (\frac{2\pi}{c}(\lambda-1)-\Omega)) \end{bmatrix} \right]$$

In order to accelerate the convergence of Fourier series, we introduce a Sigma factor:

$$\sigma_k = \sin (k\pi/M) / (k\pi/M) \tag{13}$$

So that, the boundary loadings of self-equilibrating can be obtained as:

$$\hat{\underset{\sim}{N}} = \bar{\underset{\sim}{N}}_\lambda - \bar{\underset{\sim}{N}} \tag{14}$$

Let this boundary loadings act on the base of bottom lintel of tower shell. Meanwhile, the upper and lower boundary conditions are considered as free edges. The solution is called "complementary function". Let the complementary function superpose on the particular solution, then we obtain the general solution of hyperbolic cooling tower supported by column system. This solution should ensure that the base of bottom lintel between two pairs of V-typical column appear as "forces-free state".

3. Model of Unequal Settlement of Foundation

Assuming that a pair of V-typical column appears settlement, which be denoted as the first of column and the origin of coordinates be set on it, i.e. $\theta_1 = 0^{\circ}$; then the coordinates of near-column are given as follows:

$$\theta_{\lambda} = 2\pi (\lambda-1)/C \qquad (\lambda = 1, 2, ----, c)$$

The column will be subject to a tensile force P_c as a result of its settlement. Assuming that tensile force is uniform distribution along top surface of column, then its meridional pressure can be given as:

$$p = P_c / L / H \tag{15}$$

in which $L = 2R_s \Omega$ denoted the width of column; H = thickness of bottom lintel; R_s = radius of shell base. Assuming that the relative displacement induced by bottom lintel is along the meridian, and the total force of column is subjected on the bottom lintel in tangential direction. In order to establish the model of unequal settlement of foundation, the shell base is simplified as a semi-infinite plate. In reference [2] the model of unequal settlement of five columns has been investigated. When the first column is subjected tensile force, then the second and Cth column will be subjected compressive forces $\gamma_1 p$ as well as third and (C-1)th column will be subjected tensile forces $\gamma_2 p$. These proportional factors, γ_1 and γ_2, can be obtained from equilibrium equations as follows:

$$\gamma_1 = 2 (1-\cos (4\pi/C))/(\cos (2\pi/C) - \cos (4\pi/C)) \tag{16}$$

$$\gamma_2 = \gamma_1 - (1/2) \tag{17}$$

It must be noted that the shell base will be ensured to be subjected a "force-free state", if γ_1 and γ_2 satisfy the Equ. (16) and (17).

Assuming that the meridional relative displacement induced by the 1st and 2nd columns is

$$\Delta u = R_a \, \delta u \tag{18}$$

in which $\delta u = \delta_1 + \delta_2$; R_a = radius of the shell's throat.

Applying the Boussinesq's solution, the relationship between the force subjected on the top surface of column and the relative settlement can be obtained with no difficulty as:

$$\bar{p}_c = \frac{\pi h}{l\bar{u}(m, l)} \, \delta u \tag{19}$$

in which

$$\begin{aligned} \bar{u}(m,l) = & -2(1+\gamma_1)\ln(1/2)+(1+2\gamma_1+\gamma_2)((2m/l+1)\ln(m+1/2) \\ & -(2m/l-1)\ln(m-l/2))-(\gamma_1+\gamma_2)(4m/l+1)\ln(2m+1/2) \\ & -(4m/l-1)\ln(2m-l/2))+\gamma_2((6m/L+1)\ln(3m+l/2) \\ & -(6m/l-1)\ln(3m-l/2)) \end{aligned} \tag{20}$$

$m=\frac{2\pi R_s}{R_a C}$ = dimensionless distance between the centre of top surface of column and the centre of top surface of near column along the circumferential arc;

$l = L/R_a = 2R_s\Omega/R_a$ = dimensionless width of the top surface of column;

R_s = radius of the shell's base;

$p_c=\frac{P_c}{R_a l E}$= dimensionless force in a unit of length.

4. Column's Forces Expansion in Fourier Series

The meridional membrane forces acting on the bottom lintel of the shell's base which are induced by unequal settlement of foundation are given as follows:

$$\bar{N}_{\varphi 1} = ER_a\bar{p}_c \ ; \quad \bar{N}_{\varphi 2} = -\gamma_1 ER_a\bar{p}_c \ ; \quad \bar{N}_{\varphi 3} = \gamma_2 ER_a\bar{p}_c \ ;$$

$$\bar{N}_{\varphi C-1} = \gamma_2 ER_a\bar{p}_c \ ; \quad \bar{N}_{\varphi C} = -\gamma_1 ER_a\bar{p}_c \ ;$$

$$\bar{N}_{\varphi\lambda} = 0 \ , \quad \lambda = 4, 5, ---, C-2. \tag{21}$$

We expand this group of rectangular wave into Fourier series as:

$$\begin{aligned} \bar{N}_{\varphi}(\theta) = {} & \frac{\Omega}{\pi}\bar{N}_{\varphi 1} + \frac{2\Omega}{\pi}\sum_{\lambda=2}^{3}\bar{N}_{\varphi\lambda} \\ & + \sum_{k=1}^{M}\frac{2\sigma_k}{\pi k}(\bar{N}_{\varphi 1}\sin k\Omega + \sum_{\lambda=2}^{3}\bar{N}_{\varphi\lambda}(\sin k(2\pi(\lambda-1)/C+\Omega) \\ & - \sin k(2\pi(\lambda-1)/C-\Omega))\cos k\theta \end{aligned} \tag{22}$$

in which $\sigma_k = \sin(k\pi/M) / (k\pi/M)$.

Substituting the formula of (21) into the formula of (22) we obtain:

$$\bar{N}_{\varphi\theta} = E\ R_a\ (1-2\gamma_1+2\gamma_2)\ \frac{\Omega h}{l\bar{u}(m,l)}\ \delta u \equiv 0$$

$$\bar{N}_{\varphi\theta} = E\ R_a\ (1-2\gamma_1\cos(2k\pi/C) + 2\gamma_2\cos(4k\pi/M))\ \frac{2\sigma_k h\sin k\Omega}{kl\bar{u}(m,l)}\ \delta u \tag{23}$$

5. Boundary Conditions

In the base of shell, the meridional membrane force is equal to the force $N_{\varphi}(\theta)$ transported from column. So that the boundary condition can be given as:

$$N_{\varphi k} = \bar{N}_{\varphi k} \tag{24}$$

The horizontal resultant of five column forces acting on the base of shell is obtained as follows:

$$\Delta H_Y = L\cos\Phi_s(1-2\gamma_1\cos(2\pi/C)+2\gamma_2\cos(4\pi/C))\bar{N}_{\varphi 1} \equiv 0 \quad (25)$$

There, the membrane shear force $\bar{N}_{\theta\varphi}$ and the transversal generalized shear force $\bar{q}_\varphi$ must satisfy the following equilibrium equations:

$$\sin\Phi_s(1+2\cos(2\pi/C)+2\cos(4\pi/C))\bar{q}_\varphi+2(\sin(2\pi/C) \quad (26)$$

$$+\sin(4\pi/C))\bar{N}_{\theta\varphi} = 0$$

$$\sin\Phi_s(\sin(2\pi/C)+\sin(4\pi/C))\bar{q}_\varphi-(\cos(2\pi/C)+\cos(4\pi/C))\bar{N}_{\theta\varphi} = 0$$

Because the determinant of coefficient does not equal to zero, so that we have

$$\bar{N}_{\theta\varphi} = 0\,, \quad \bar{q}_\varphi = 0 \quad (27)$$

Then the boundary conditions are given as:

$$N_{\theta\varphi} = 0 \quad (28)$$

$$q_\varphi = 0 \quad (29)$$

Because the algebraic sum of moments induced by the column's forces on the base of shell are equal to zero, then we may assume that

$$M_\varphi = 0 \quad (30)$$

in the base of bottom lintel. Eqs. (24), (28), (29) and (30) are the total boundary conditions on the base of tower shell. By using the principle of virtual work:

$$\int_0^{2\pi} \delta\underset{\sim}{\alpha}^T \underset{\sim}{F}\, d\theta = \int_0^{2\pi} \delta\underset{\sim}{\bar{u}}^T \underset{\sim}{\bar{N}}\, r\, d\theta$$

then we obtain the boundary generalized forces of nodal circle as:

$$\underset{\sim}{F}_k = r\llcorner \bar{N}_{\varphi k},\ \bar{N}_{\theta\varphi k},\ \bar{q}_{\varphi k},\ -\bar{M}_{\varphi k}\cos\varphi,\ 0,\ -\bar{M}_{\varphi k}\sin\varphi,\ 0,\ 0\lrcorner^T \quad (31)$$

Let these generalized forces act to the base of bottom lintel, and the upper edge as well as lower edge are assumed to be free boundary conditions.

6. Numerical Solutions of Computation

Suppose there is a reinforced concrete hyperbolic cooling tower with a height of 105 m. The radius of throat is R_o = 21.9 m; radius of shell's base is R_s = 38.74 m; maximum thickness of shell is h_{max} = 0.6 m; minimum thickness h_{min} = 0.16 m; the height of column is h = 7.8 m; length of column is l_o = 9.95 m; area of column section is $b \cdot b = (0.55\ m)^2$; and the inclination of column is $\alpha = 75.9^o$. The equation of meridian of middle surface of tower shell is expressed as $r^2 = b^2 + az^2$, in which a = 0.176 m and b = 21.9 m. The thickness varies with power law which exponent η = 3.0. Young's modulus of concrete is $E = 2.7 \cdot 10\ t/m^2$, and Poisson's ratio is μ = 0.167. The design strength of steel is $R_g = 3.4\ t/cm^2$. Density of reinforced concrete is $\gamma = 2.4\ t/m^3$.

Fig. 4.-7. show the variations of membrane forces with respect to the tower height corresponding to relative settlement Δu = 2.19 mm, in which $\theta = 0^o, 8.18^o, 16.36^o$ represent the first, second and third columns, respectively; $\theta = 4.09^o$ and 12.27^o are denoted centre between two near-columns. It will be seen that the partial edge effect induced unequal settlement of foundation will be spreaded up to (1/14-1/6) heigth of tower. When the relative settlement Δu = 1.26 , the membrane compressive force induced by dead loading can be cancelled out. According to the limit state design procedure, meanwhile this cancel is taken into consideration, the maximum relative settlement is obtained as 2.16 mm. If the cancel of dead loading is not taken into consideration, then Δu_{max} = 1.36 mm. In table 1. the allowable values of relative settlement controlled by circumferential membrane force and obtained by using working stress design procedure [1] are also given. It will be seen that the allowable values of unequal settlement of foundation are controlled by the meridional membrane force N_φ .

Type \ Controlled factors \ Maximum allowable values		ΔU_{max} (mm) Limite state design procedure	ΔU_{max} [1] (mm) Working stress design procedure
Cancel out by dead loading	N_φ	2.61 (2.48)*	2.23
	N_θ	5.91 (5.58)*	----
No cancel out by dead loading	N_φ	1.36 (1.29)*	1.20
	N_θ	4.37 (4.13)*	----

*are the maximum allowable values of relative settlement in the vertical direction

Table 1. Maximum Allowable Values of the Relative Settlement

7. Settlement of Foundation Measured on Full-Scale Tower

In 1981-1982, a full-scale measurement of foundation settlement of a hyperbolic cooling tower was done at Matou Power Station. Fourty-four measured points were arranged with equal interval along the circumferential direction. The mean value of settlement is equal to 5.3 mm. The final values of settlement of total measured points are shown in Fig.8.. The maximum relative settlement between two near measured points is 2.0 mm which appears at the 15th and 16th measured points. If the effect of the settlement of the 17th point on the 15th point is taken into account, then the settlement of the 15th measured point exceeds about 2.7%. Hence the computing method and allowable values of unequal settlement of foundation suggested by present paper are of avail.

Conclusion

1. The partial edge effects induced by unequal settlement of foundation will be spreaded up to (1/14-1/6) height of tower, but the edge effects of shell supported by column

system induced by dead loading will be only spreaded up to 1/20 height of tower.

2. The allowable values of unequal settlement of foundation are controlled by the meridional membrane force N_φ in main. If the effect of dead loading on the settlement is not taken into account, the allowable value will be decreased by 50%. Such case is in danger for the cooling tower design practice against wind loading, in particular, while the relative settlement appears near to the stagnation point on oncoming flow.

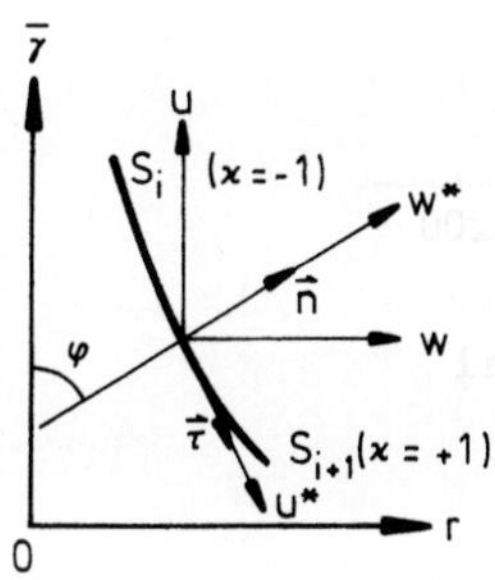

Fig. 1. Transformation Relationship of Coordinate System

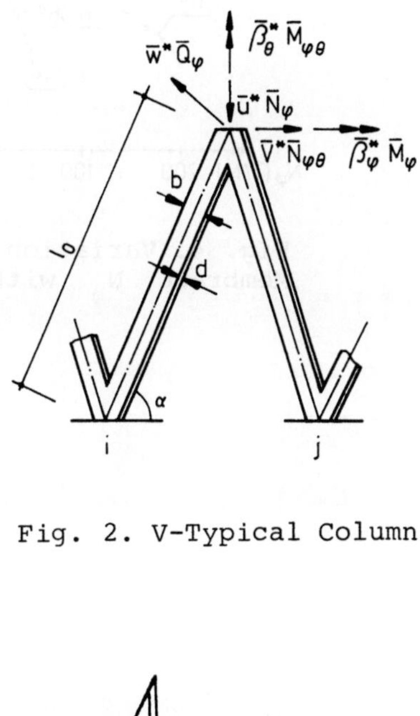

Fig. 2. V-Typical Column

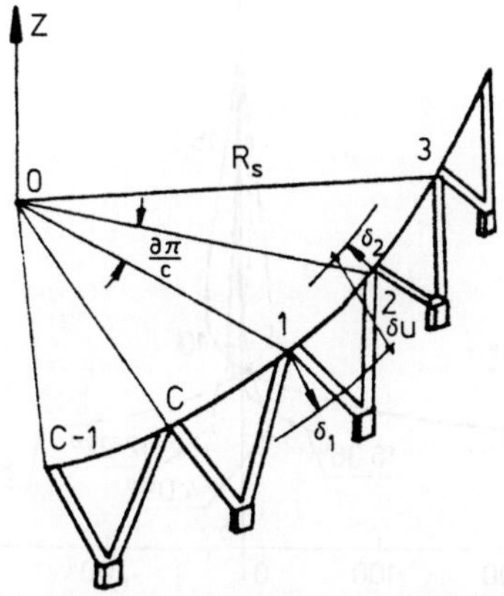

Fig. 3. Model of Unequal Settlement of Foundation

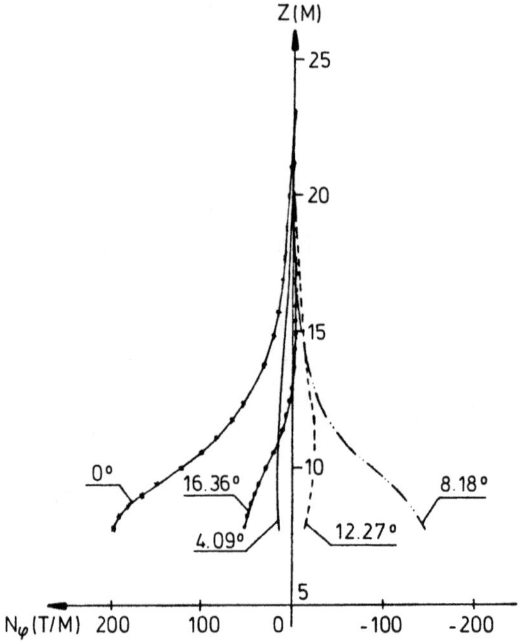

Fig. 4. Variation of Meridional Membrane N_φ with Height

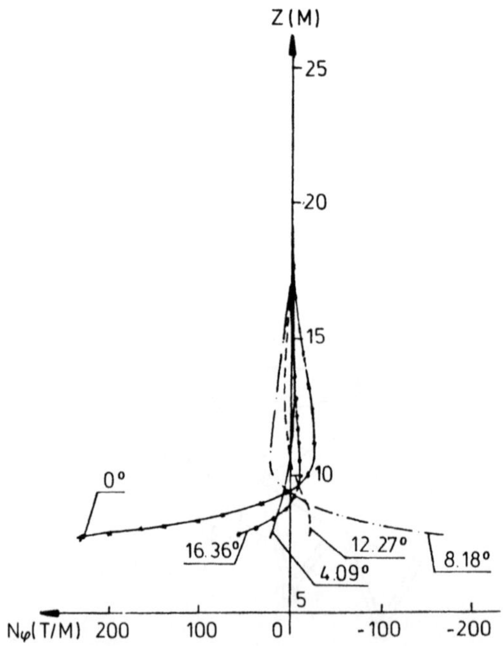

Fig. 5. Variation of Circumferential Membrane N_φ with Height

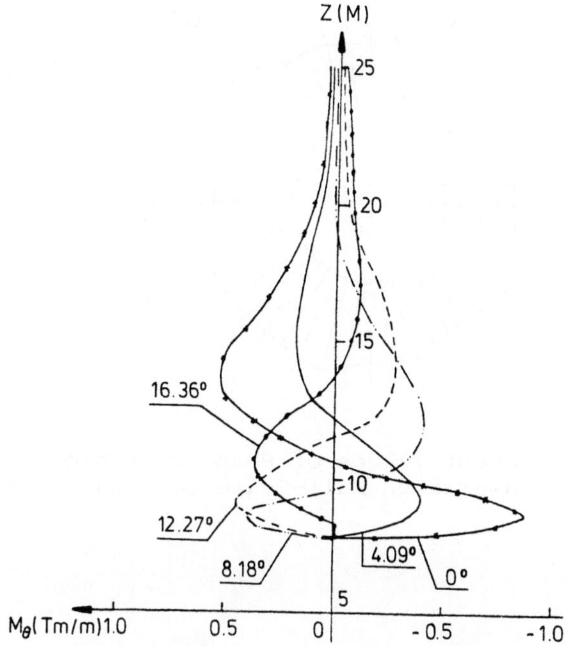

Fig. 6. Variation of Meridional Bending Moment M_φ with Height

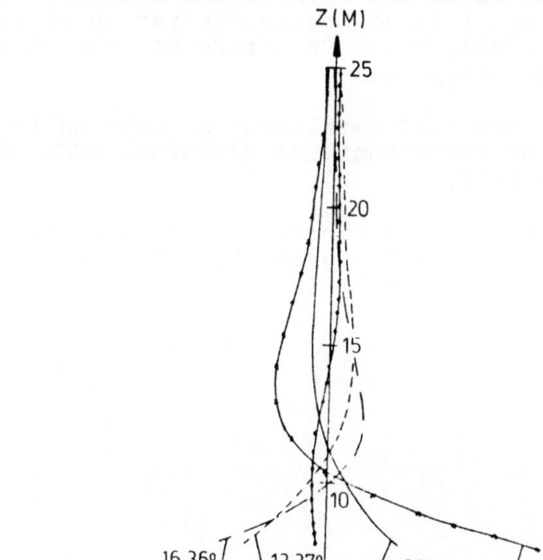

Fig. 7. Variation of Circumferential Bending Moment M_θ with Height

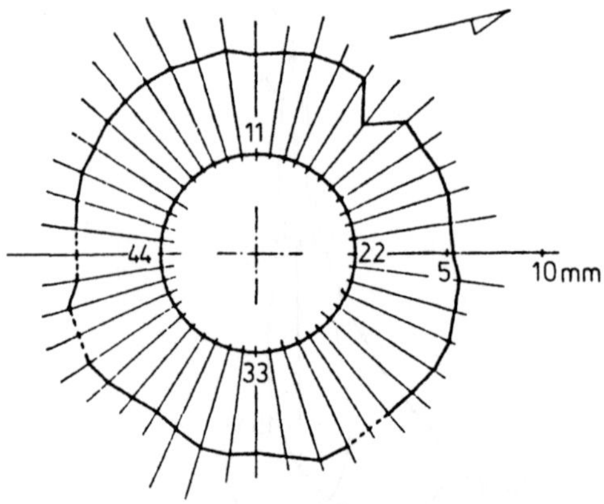

Fig. 8. Settlement Values of Foundation along Circumference Measured on Full-Scale Cooling Tower

<u>Reference</u>

1. The Effect of Unequal Settlement of Foundation of Natural Draft Cooling Towers on the Deformation and Stress. Power Station Technology of Steam-Engine Generation of Large Capacity, Title No. 354, Water Structure No. 036, Translation from Japanese into Chinese.

2. Gould, P.L.: Differential Settlement of Hyperbolic Cooling Towers. Journal of the Structural Division, ASCE, St. 10, P. 2207, October 1972.

3. Wenda, L. and Jingxin, P.: Finite Element Solution of the Hyperbolic Cooling Tower Supported by Column System. Research Report of Xian Institute of Thermal Engineering, June 1976, China.

Analysis of Hyperbolic Cooling Towers with Local Imperfections

P.L. GOULD,* K.J. HAN,** and G.S. TONG**

*Department of Civil Engineering
Washington University
St. Louis, Missouri 63130
U.S.A.

**Department of Civil Engineering
University of Houston
Houston, Texas 77004
U.S.A.

Summary

Two methods to analyze hyperbolic cooling towers with local imperfections are presented. One method is based on the finite element method. For this purpose, a specialized finite element program, which can model any arbitrary imperfections while retaining the advantage offered by the basically axisymmetric nature of the shell, was developed. The other method is an approximate procedure which may be implemented with a purely axisymmetric analysis capability.

Through numerical studies, the two methods are compared. A cooling tower shell with a bulge-type imperfection is examined under dead load and wind load conditions. It is concluded that the finite element model presented is effective for the analysis of such shells, while the equivalent load method may be adequate for some cases. Also, it is shown that both meridional and circumferential stress resultants may be radically influenced by a small bulge imperfection.

Introduction

The presence of geometric imperfections disturbs the pattern of stresses in hyperbolic cooling towers and results in higher than usual compressive as well as tensile stresses. As such, many investigators have studied the effects of imperfections in these shells over the last two decades [1-9]. Due to the complexity of the analysis involved, however, these studies were often limited to two idealized imperfection geometries: meridional and circumferential imperfections. The former refers to the geometry where the shell is assumed to remain axisymmetric and the imperfection is mainly due to deviations in the meridional shape, whereas the latter assumes that the changes of curvature are appreciable in the circumferential direction but negligible in the meridional direction. Thus, each case treats curvature errors in only one of the two principal directions of the surface. While these idealizations facilitate the analysis and enable the examination of the effects of imperfections, an actual imperfection would seldom be of either form. A realistic estimate of the effects of imperfections can only be obtained when the curvature errors in both directions are simultaneously included in the analysis.

In this paper, two methods to analyze hyperbolic cooling towers which have bulge-type imperfections with curvature errors in both principal directions are presented. One method uses a finite element model which is developed specifically for the analysis of shells of revolution with arbitrary local deviations. The other is the equivalent load method where the effects of imperfections are simulated by the application of ficticious normal loads on the perfect shell. The theoretical backgrounds of the two methods will be briefly discussed in the following sections, and the two methods will be compared through numerical examples. The advantages and the disadvantages of each method will be pointed out. Also, some characteristics of bulge imperfections which are not evident in the case of single directional imperfections will be examined.

Finite Element Model

Hyperbolic cooling towers may be analyzed by the finite element method using rotational shell elements. Once a bulge imperfection is created, however, the structure is no longer a perfect shell of revolution, and can not be represented by rotational elements alone. One alternative is to model the structure by general shell elements of triangular or quadrilateral shape. If the entire shell is modeled by such elements, however, the analysis may be very costly. Moreover, this type of analysis does not take advantage of the fact that a large part of the shell remains axisymmetric even though the bulge has locally annulled the axisymmetric nature of the shell.

The finite element model presented in this paper combines general shell elements and rotational shell elements in a single analysis (Figure 1). The general shell elements are specified over a limited region around the imperfection, whereas the rotational shell elements are utilized in the remaining part where the shell is axisymmetric. Since the rotational element is computationally more efficient, this model takes the advantage of the prevailing axisymmetry. At the same time, the general element provides the required generality in modeling the nonaxisymmetric portion. While appealing in concept, two complications arise in the implementation of the model. One is in the incompatibility between the two different types of elements, and the other is in the coupling of Fourier harmonics. The sources of and the remedies for these problems will be discussed in the following paragraphs.

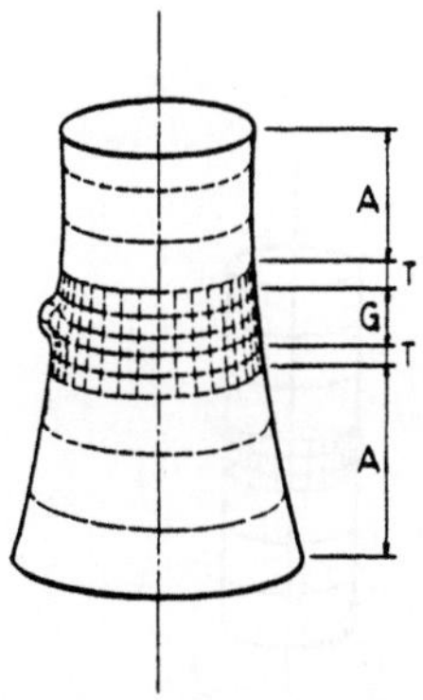

Figure 1. Combined F.E. mesh

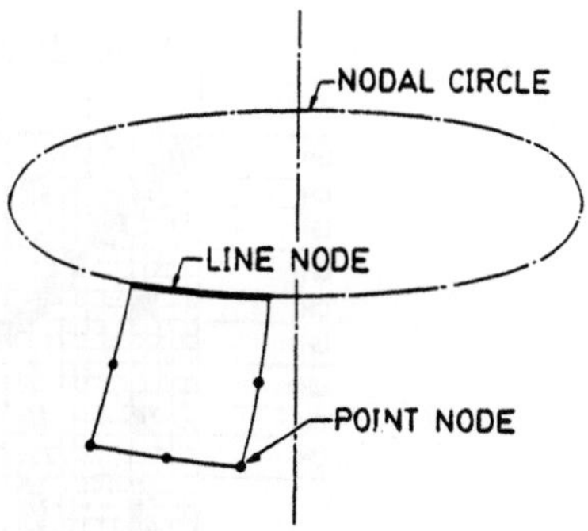

Figure 2. Transitional shell element

The incompatibility is due to the fact that the two types of element use different functions to express displacement and stress fields. The general element employs polynomial shape functions, while the rotational element uses trigonometric functions along the circumferential direction. Thus, if the two elements are juxtaposed, they will not produce identical displacement fields along the common boundary. To overcome this deficiency, a new type of element, a transitional shell element, is introduced [10]. This element has a line node on one side of the boundary and point nodes on the remaining boundaries (Figure 2). Trigonometric functions are used to express displacements along the line node, while the point nodes have polynomial shape functions which are identical to those of the general shell element. By specifying a layer of transitional shell elements between the rotational and the general shell elements (Figure 1), compatibility is achieved.

The coupling of Fourier harmonics is due to two facts [11]: (1) the shell is locally nonaxisymmetric; and (2) the general shell element has point nodes which do not rely on Fourier series to express circumferential variations. Figure (3) shows a typical form of the global stiffness matrix obtained in this finite element model. The shaded area has nonzero terms, while the remaining terms are zeros. Item (1), mentioned above, is responsible for the non-zero off-diagonal terms, marked "CH" in the figure, and item (2) produces the terms marked "P."

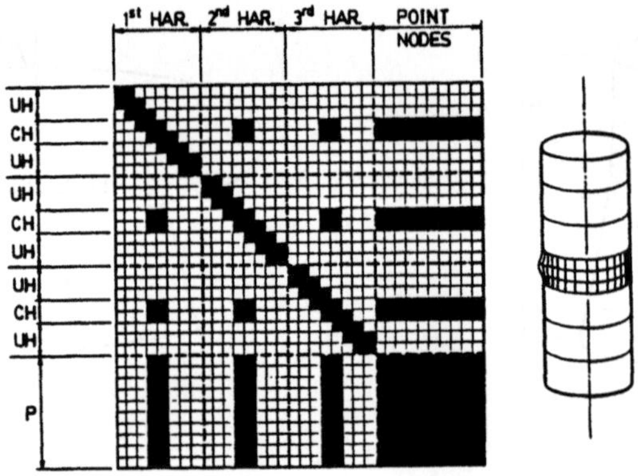

Figure 3. Stiffness matrix for a shell of revolution with a bulge imperfection

The presence of non-zero off-diagonal terms which couple harmonics makes it impossible to use the superposition of individual harmonic analyses technique which is commonly used for shells of revolution. However, a close examination reveals that the coupling is effected through only the two nodal circles which bound the region of the imperfection. As far as the remaining nodal circles are concerned, each harmonic component is completely decoupled from the other harmonics. Thus, separate harmonic analyses are possible for these nodes. Taking this advantage, the solution is carried out in three steps [11]. First, the stiffness matrix is assembled for only general and transitional shell elements. At this stage all the terms related to point nodes, marked "P" in the figure, are eliminated by the static condensation. This leaves only the terms which are marked "CH" in the figure. Second, the stiffness matrix for the rotational elements is assembled, followed by the condensation of the uncoupled harmonic terms, marked "UH" in the figure. This step is carried out for each harmonic separately as only uncoupled terms are eliminated. Again, this leaves only the terms marked "CH." Finally, the remaining terms of the first and the second steps are combined and simultaneously solved. Thus, the entire global matrix is never assembled. Instead, only a small portion of the matrix highly populated with non-zero terms is involved at one time. As a result, unnecessary manipulation of zero terms is avoided in large part.

Equivalent Load Method

This is an approximate method which was first suggested by Munro et al [12]. Instead of directly analyzing a shell with geometric imperfections, this method uses the perfect shell geometry but with additional loads which produce a similar stress distribution as that of the imperfect shell. The accuracy of the method was established for the case of axisymmetric imperfections by comparisons of numerical results with a more rigorous analysis. Later, this procedure was applied for circumferential imperfections by Ellinas et al [8]. However, the validity of the result for circumferential imperfections was not closely examined. One of the advantages of this method is that it requires only the capability to analyze shells of revolution even if the actual shell is nonaxisymmetric. In this paper, the application of this procedure for bulge-type imperfections is studied. The accuracy and the limits of the method are examined by numerical results.

The intensity of the equivalent load [12] is obtained by

$$P(s, \theta) = X_s N_s + 2X_{s\theta} N_{s\theta} + X_\theta N_\theta \qquad (1)$$

where P is the normal pressure representing the effects of imperfections; N_s, $N_{s\theta}$, and N_θ are meridional, in-plane shearing, and circumferential stress resultants, respectively; and X_s, $X_{s\theta}$, and X_θ are the corresponding curvature errors. As the stress resultants in equation (1) are unknown functions at the outset, the analysis must be carried out iteratively. In the imperfect region, the shell is first considered to be subjected to the same in-plane stress resultants as present in the perfect shell under the design load. These stresses give the first order equivalent load as

$$\Delta^1 P = X_s N_s^P + 2X_{s\theta} N_{s\theta}^P + X_\theta N_\theta^P \qquad (2)$$

When the first order equivalent load is applied to the perfect shell, the first corrections to the stress resultants and couples are obtained. These corrections, in turn, generate the second order normal pressure which may be expressed as

$$\Delta^2 P = Y_s N_s^P + 2\, Y_{s\theta} N_{s\theta}^P + Y_\theta\, N_\theta^P \qquad (3)$$

The second order equivalent load, when applied to the perfect shell, produces the second corrections to the stresses. This iterative process can be continued for as many cycles as is necessary. Finally, the stress resultants and couples in the imperfect shell are obtained by adding all of the corrections of stresses to the stresses in the perfect shell.

In the application of equation (2) for a bulge imperfection, all six quantities appearing on the right hand side are variable functions of position over the region of the imperfection. Of these, X_s, $X_{s\theta}$, and X_θ are related to the bulge geometry. The remaining three terms, N_s, $N_{s\theta}$, and N_θ, are functions of the loading on the perfect shell, and must be changed when different loads are applied. Since there may be several loading cases in the complete analysis of a shell, the exact representation of these stresses for each loading may be too demanding in terms of computational effort. If the extent of the imperfection zone is small in comparison to the overall dimensions of the shell, however, the following simplifications may be made: the stresses acting in the perfect shell do not vary significantly within the zone and may be considered to be uniform. This assumption considerably simplifies the whole analysis if it is recognized that equation (2) may be degenerated into three separate loads, $X_s N_s^P$, $2X_{s\theta} N_{s\theta}^P$, and $X_\theta N_\theta^P$. Since the stress resultants in these expressions are now constants, they may be considered as being equal to unity when the loads are applied. Then, actual magnitudes can be multiplied to the stresses thus obtained. A similar argument may be made with regard to the second order equivalent load, equation (3), and so on. Thus, only three separate analyses are required in each cycle regardless of the number of loading cases involved in the entire analysis. It should be noted, however, that the above-mentioned assumption, while simplifying the whole analysis, also limits the range of the applicability of the method to only localized imperfections.

Each of the three loads in equation (2), $X_s N_s^P$, $2X_{s\theta} N_{s\theta}^P$, and $X_\theta N_\theta^P$, represents the effects of meridional, twisting, and circumferential imperfections, respectively. When the imperfection is uni-directional, e.g., a meridional or circumferential imperfection, the first order equivalent load is given by one of the three terms, and the second order load is negligible. This is because there is no interaction between different curvature errors. When a bidirectional bulge imperfection is considered, however, there are interactions which can not be represented by the first order load. These interactions are simulated only by the application of the second and higher order loads. Thus, the first order equivalent load alone is not sufficient for a bulge imperfection. As the iterations progress, however, the magnitudes of the equivalent loads tend to diminish, and their variations within the imperfection zone become local in nature. Thus, based on numerical studies, iterations up to the second cycle seem to be sufficient for a typical cooling tower geometry.

In the computation of the second order equivalent loads, $Y_{s\theta}$ and Y_θ are found to be much smaller in magnitude than Y_s. Furthermore, shearing and circumferential stress resultants are generally smaller in magnitude than meridional forces for a typical cooling tower under the usual design loads. The result is that small quantities are multiplied by other small numbers, while large quantities are multiplied by large numbers. Thus, the multiples of small numbers become negligible, and only the load given by $Y_s N_s^P$ is found to be meaningful for the second order equivalent load. In summary, only four separate analyses, three for the first order load and one for the second order load, are necessary for a bulge imperfection in a typical cooling tower shell regardless of the number of loadings involved.

Numerical Examples

The cooling tower shell shown in Figure (4) with the bulge imperfection shown in Figure (5) is analyzed by the two methods discussed in this paper. The center of the bulge is located at the elevation z = 92.2 feet, and the vertical height and the horizontal "width" are 41.6 feet and 30°, respectively. Except for the results shown in Figure (8), the maximum amplitude of the bulge is 1 foot, which is twice the thickness of the shell in the area. The loadings considered are the selfweight load and a wind load. The circumferential distribution of the wind pressure is based on reference [13], with a wind speed of 90 mph at 30 feet above the ground. For simplicity, the wind pressure is assumed to be constant over the height of the shell. In the equivalent load method, only four separate loading terms, three terms in the first order equivalent load and one term in the second as discussed previously, are included in the analysis. The direct analysis of the imperfect shell by the finite element method is carried out using the RONALD finite element program [14] which is based on the formulation presented in this paper. The analysis of the axisymmetric shell for the equivalent load method is carried out using the SHORE-III finite element program [15].

In Figure (6), meridional and circumferential stress resultants due to selfweight loading are plotted along the meridian which passes through the center of the bulge. As the effect of the imperfection diminishes away from the bulge, results are shown only for one and one-half times the height of the bulge above and below the center of the bulge. The agreement between the two methods is reasonably good, though the results obtained by the equivalent load method are somewhat conservative. Similar agreements have been found for the other stresses as well as along the other meridians. Also shown in the figure are the results obtained when only the first order equivalent load is included in the analysis. This clearly shows that the first order equivalent load alone is not sufficient and that

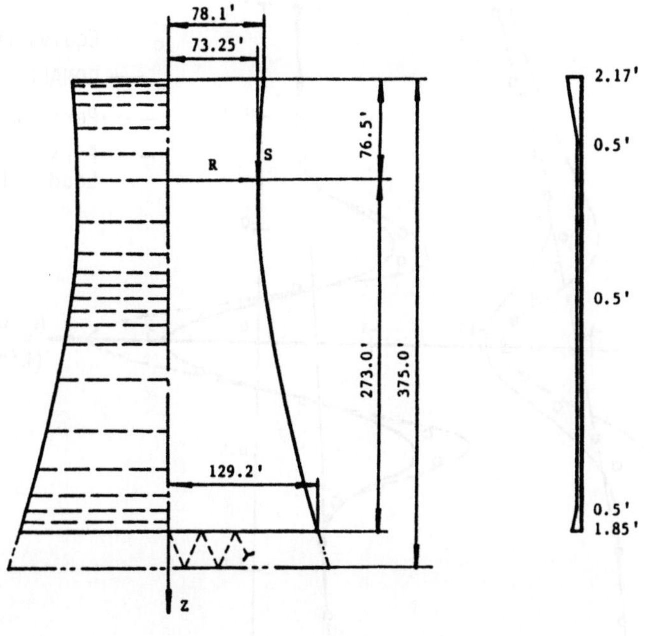

Figure 4. Geometry of the cooling tower analyzed

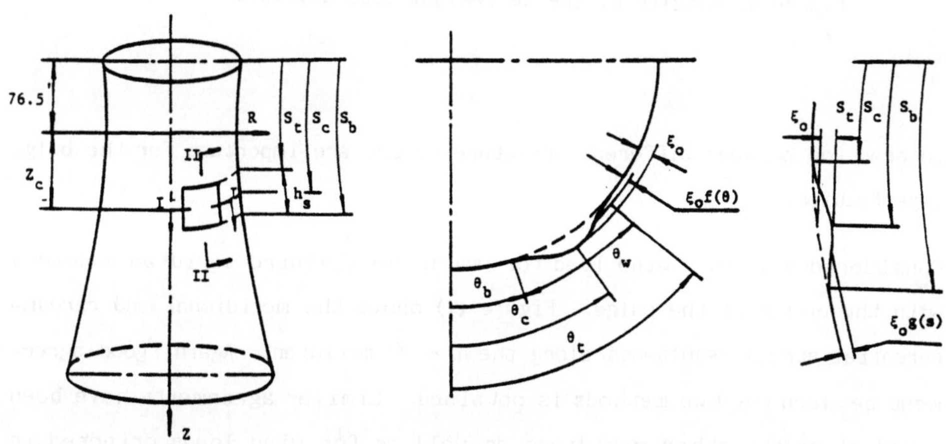

Figure 5. Geometry of the bulge imperfection; (a) Overall description; (b) Section I-I; (c) Section II-II

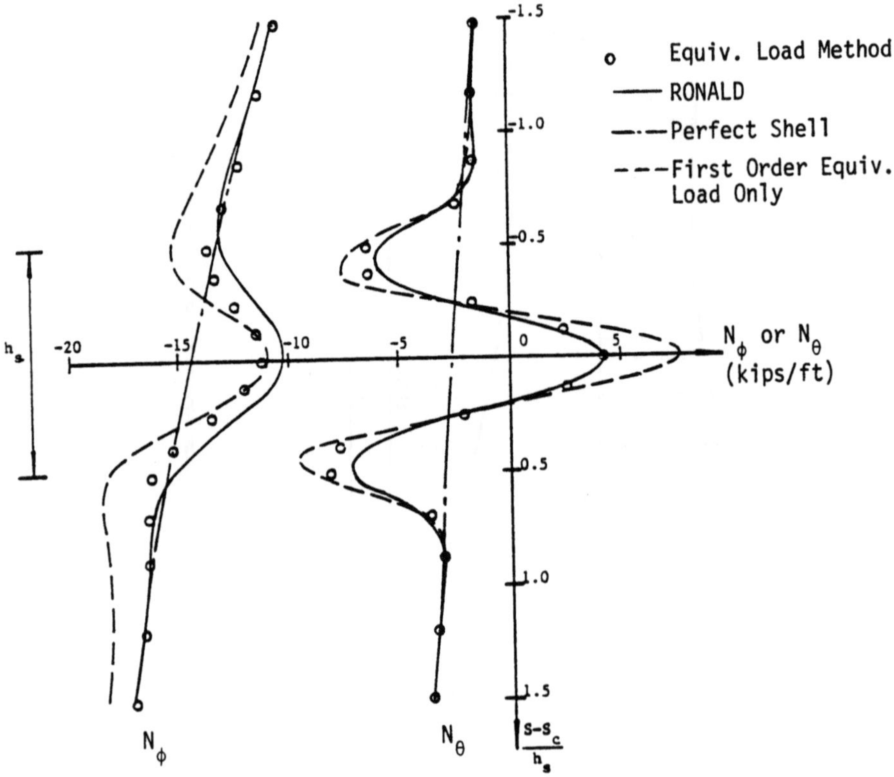

Figure 6. Results of the selfweight load analysis

interaction between different curvature errors are important for the bulge imperfection.

Considered next is a wind load for which the windward meridian concides with the center of the bulge. Figure (7) shows the meridional and circumferential stress resultants along the $\theta = 2^0$ meridian. Again, good agreement between the two methods is obtained. Similar agreements have been found along the other meridians as well as for wind loads oriented in different directions with respect to the location of the bulge. Shown in Figure (8) are the results when the maximum amplitude of the bulge is changed to 0.5 feet and then to 2.0 feet. As can be seen, the agreement is

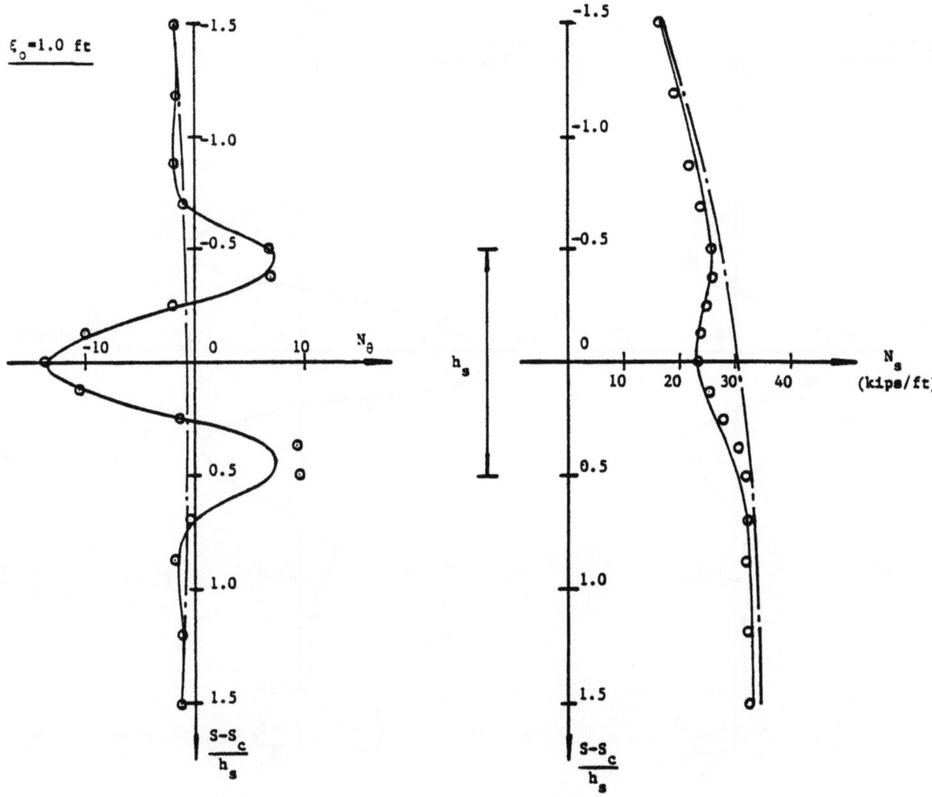

Figure 7. Results of the wind load analysis (Bulge at θ = 0°)

even better when the amplitude is reduced to the thickness of the shell, while there is an appreciable disagreement when the amplitude is increased to four times the thickness of the shell. This shows that the equivalent load method may be applied for deviations of small magnitude, but is not suitable for large deviations form the perfect geometry. While not being elaborated in this paper, this is due to the fact that the equivalent load method uses the perfect shell geometry for the analysis [16].

Meridional and circumferential moments due to the wind load are plotted in Figure (9). The agreement is not as good as in the case of in-plane

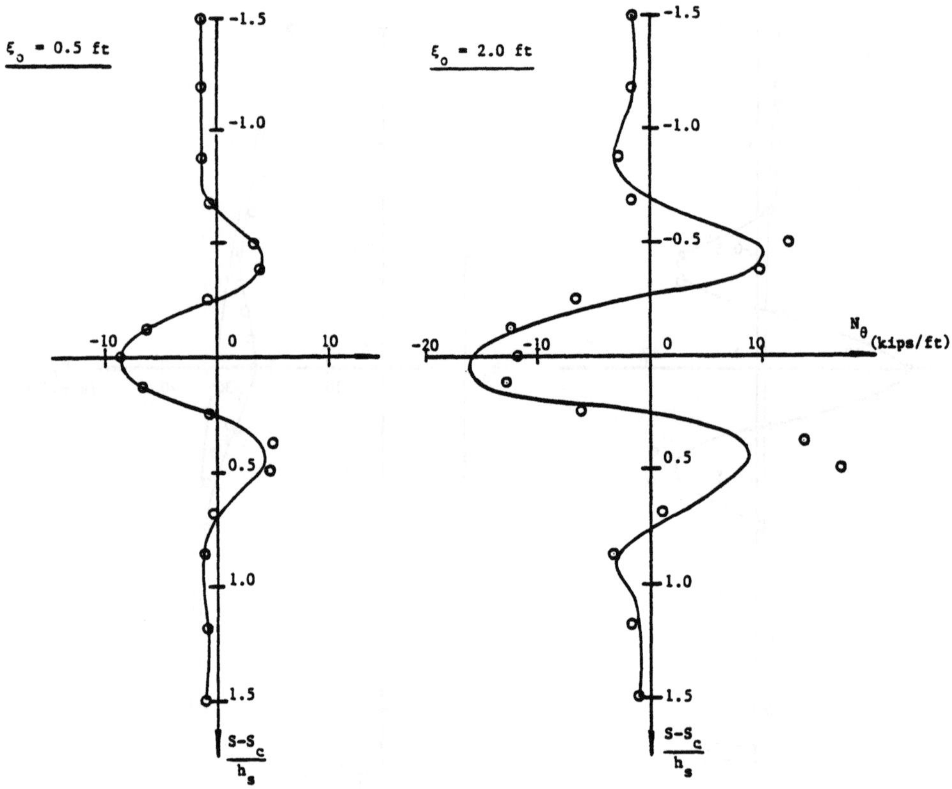

Figure 8. Comparisons of N_θ for different amplitudes of the bulge

forces. However, the shape of the variations are very similar between the two analyses. Considering that moments are smaller than in-plane forces by an order of magnitude, this discrepancy is probably due to precision errors in the numerical computation.

Conclusion

From the preceeding discussion, the following conclusions may be made:

(1) Both the finite element model and the equivalent load method provide satisfactory results for small dimple-shaped imperfections.

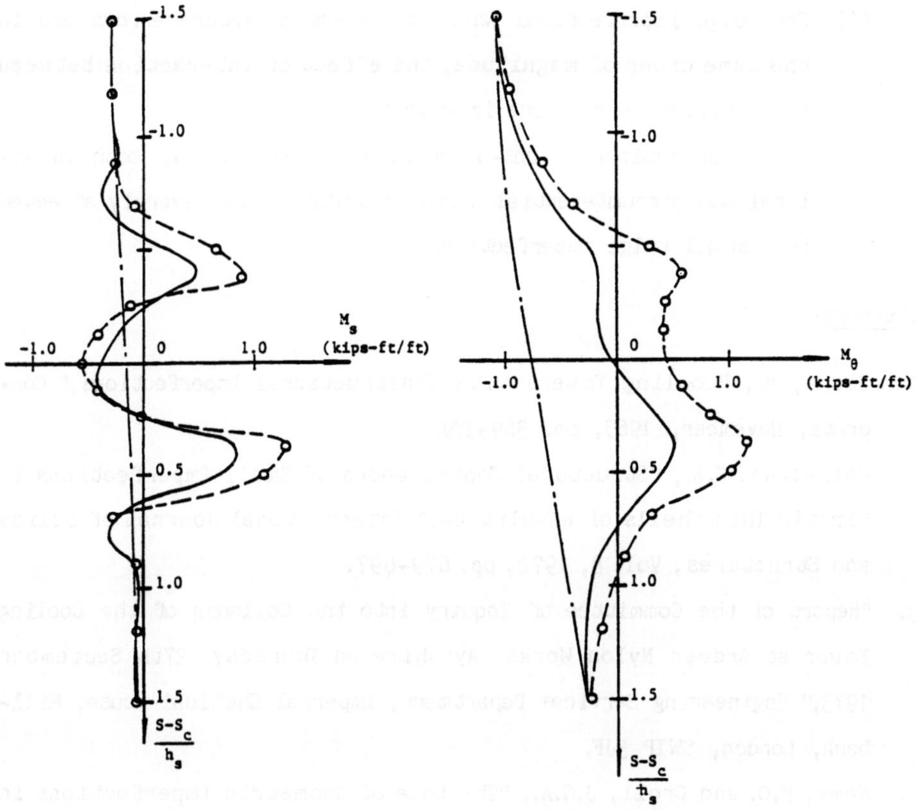

Figure 9. Stress couples due to the wind load (Bulge at $\theta = 0^o$)

(2) The equivalent load method uses axisymmetric geometry for the analysis. Thus, many existing programs can be used to analyze shells with nonaxisymmetric geometry.

(3) While the finite element method may be used for any arbitrary imperfection geometry, the equivalent load method is limited to imperfections with small amplitudes.

(4) The extent of the imperfection may not be too large when the equivalent load method is used. In such cases, the computational effort required is too great for the method to be a practical tool.

(5) While the finite element method requires one analysis for each loading, the equivalent load method requires four analyses regardless of the number of loadings involved.

(6) For bulge imperfections where different curvature errors are in the same order of magnitude, the effect of interaction between the curvature errors are important.

(7) Unlike predominantly uni-directional imperfections, both meridional and circumferential stress resultants are severely affected by a small bulge imperfection.

References

1. Soare, M., "Cooling Towers with Constructional Imperfections," Concrete, November, 1965, pp. 369-379.
2. Calladine, C.R., "Structural Consequences of Small Imperfections in Elastic Thin Shells of Revolution," International Journal of Solids and Structures, Vol. 8, 1972, pp. 679-697.
3. "Report of the Committee of Inquiry into the Collapse of the Cooling Tower at Ardeer Nylon Works, Ayrshire on Thursday, 27th September 1973," Engineering Services Department, Imperial Chemical House, Millbank, London, SWIP 3JF.
4. Kemp, K.O. and Croll, J.G.A., "The Role of Geometric Imperfections in the Collapse of a Cooling Tower," The Structural Engineer, January 1976, No. 1, Vol. 54, pp. 33-37.
5. Al-Dabbagh A. and Gupta, A.K. "Meridional Imperfection in Cooling Tower Design," Journal of the Structural Division, ASCE, June 1979, pp. 1089-1102.
6. Han, K.J. and Gould, P.L., "The Effect of Geometric Imperfection on Cooling Towers," bulletin of the International Association for Shell and Spatial Structures, Vol. XX-2, August 1979, pp. 63-68.
7. Croll, J.G.A., Kaleli, F., and Kemp, K.O., "Meridionally Imperfect Cooling Towers," Journal of the Engineering Mechanics Division, ASCE, Vol. 105, No. EM5, Oct. 1979, pp. 761-777.
8. Ellinas, C.P., Croll, J.G.A., and Kemp, K.O., "Cooling Tower with Circumferential Imperfections," Journal of the Structural Division, ASCE, ST. 12, Dec. 1980, pp. 2405-2433.

9. Gupta, A.K. and Al-Dabbagh, "Meridional Imperfection in Cooling Tower Design: Update," Journal of the Structural Division, ASCE, August 1982, pp. 1697-1708.

10. Han, K.J. and Gould, P.L., "Line Node and Transitional Shell Element for Rotational Shells," International Journal for Numerical Method in Engineering, Vol. 18, pp. 879-895, June 1982.

11. Han, K.J. and Gould, P.L., "Shells of Revolution with Local Deviations," International Journal for Numerical Methods in Engineering, Vol. 20, pp. 305-313, February 1984.

12. Croll, J.G.A., Kaleli, F., Kemp, K.O., and Munro, J., "A Simplified Approach to the Analysis of Geometrically Imperfect Cooling Tower Shells," Engineering Structures, Vol. 1, No. 2, January 1979, pp. 92-98.

13. "Reinforced Concrete Cooling Tower Shells--Practice and Commentary," ACI-ASCE Committee on Concrete Shell Design and Construction, ACI-ASCE Committee 334, ACI Journal, January 1977, pp. 22-31.

14. Han, K.J. and Gould, P.L, "User's Manual of RONALD Program," Structural Engineering Research Report No. 64, Structural Division, Washington University, St. Louis, MO, September, 1982.

15. Basu, P.K. and Gould, P.L., "User's Manual of SHORE-III," Research Report No. 49, Structural Division, Washington University, St. Louis, MO, Sept. 1977, Revised 1981.

16. Han, K.J. and Tong, G.S., "Simplified Analysis of Hyperbolic Cooling Towers with Localized Bulge Imperfections," in preparation.

A Study of the Effect of Geometric Imperfection on the Hyperbolic Cooling Tower Shells

P. Srinivasa Rao[*], P.K. Aravindan[*], Manohar A. Mata[**], B.S. Srinivas[*]

[*] Structural Engg. Laboratory,
Indian Institute of Technology,
Madras, India

[**] M/s Engineers India Ltd,
New Delhi, India

Abstract

This paper gives the results of investigations carried out on the effect of geometric imperfections on the behaviour of 114.4 m high hyperbolic tower built in India in the year 1978.

After the critical review of the existing methods of analysis, the paper deals with the analysis of the tower. The thickness of the shell is 150 mm in most part of the shell. The shell is reinforced with 0.776% of steel in the meridional direction and 0.527% of steel in the circumferential direction in a single layer. It has geometric imperfections which are mostly inward with a maximum amplitude of 325 mm. The zone of imperfection is between 91.64 m and 101.96 m from the base.

The tower is analysed by Finite Element Method taking into account the circumferential as well as meridional variation of the imperfection, for both dead and wind loads. The wind load assumed is as per I.S.Code.

From the analysis it is found that the most affected stress resultants are the hoop stress and the meridional moment. In fact the hoop stress in the imperfect shell is almost 5 time as much as that of the perfect shell. Also the meridional moment is 3 times that in the perfect shell. Variation of imperfection in the circumferential direction is not found to have a pronounced effect on circumferential and meridional forces but influences the corresponding moments.

As a part of the study the results of all the existing methods of analysis are compared with the results of the present method of investigation. It is found that Soare's and Banarjee's methods give much higher values of hoop stress. But Kemp and Croll's Dabbagh and Gupta's methods gave results slightly higher than those given by Finite Element Method.

Optimization of Large Cooling Towers

K. T. ALOWOOJA and P. L. GOULD

Department of Civil Engineering
Washington University
St. Louis, Missouri 63130 USA

I. FURZER

Department of Chemical Engineering
University of Sydney
Australia

Abstract

The optimal shapes and heights of hyperboloidal cooling towers and the magnitude of possible savings achievable through optimization are considered.

The study is based on an advanced prediction of power rejection as a function of height, diameter and depth of packing. The entire practical range of shell geometries is examined and both self-dead load and wind loading are included. The structural analysis was carried out, to a large extent, using a microcomputer-based membrane theory formulation and was supplemented by an advanced finite element analysis for the effects of wind.

INTRODUCTION

The thermal design of hyperbolic cooling towers permits combinations of tower diameter and height which provide a constant value of the recooled water temperature. The most suitable economic design requires an examination of both the thermal and structural design problems to find the optimal values of the tower diameter and height. The problem can be presented as an optimization problem using a suitable cost objective function. Optimization based on primarily structural consideration were presented by Gould [1] and by Reinschmidt and Narayanan [2]. Furzer [3] has approached the problem from the thermal aspect and has presented a simple cost function taking into account the separate costs of packing, shell and basin. A more detailed cost function including the basin wall and the ring beam is proposed in this paper.

To limit the dimensions of the optimization problem, it is necessary to include a number of constraints on the thermal variables. The most significant constraint is the constant heat rejection of the tower of 1500 MW under constant inlet and outlet water temperatures with constant wet and dry bulb air temperatures. This would be equivalent to an electrical power output of approximately 1000 MW. The constant temperatures are T(IW) = 40°C, T(OW) = 27°C, T(WB) = 15°C and T(DB) = 18°C. It has also been necessary to keep constant the type of packing, the depth of counterflow packing and the height of the air inlet. The packing selected consists of corrugated asbestos - cement sheets spaced on 43 mm centers with the heat transfer and air flow resistance parameters given by Lowe and Christie [4]. This packing is positioned above the air inlet in the counterflow position with an open crossflow region underneath.

From the structural standpoint, all significant components are included, except the supporting columns at the base. Instead, the shell wall is assumed to rest directly on the ring beam. In addition, certain dimensions are selected as typical values to reduce the number of variables.

THEORY

Thermodynamic Model

An improved thermodynamic model of a hyperbolic cooling tower has been given by Furzer [5]. The model considers the horizontal and vertical components of the air velocity in the spray region below the counterflow packing and in the counterflow packing. The conservation equations of mass and momentum provide dimensionless differential equations which can be solved for the appropriate boundary conditions. A finite element model is made of the heat transfer processes in both the spray and packing regions with the given air velocities. A final overall iteration is necessary to close all balances. A computer package IMPACT accepts as data the geometrical parameters of the tower and packing and provides as output the recooled water temperature. The output is flexible and provides the vertical air velocity leaving the packing as shown in Figure 1 and the radial variations in water temperature, Figure 2, as the water leaves the packing and also as it enters the pond. The large radial temperatures should be noted, with a central zone providing only

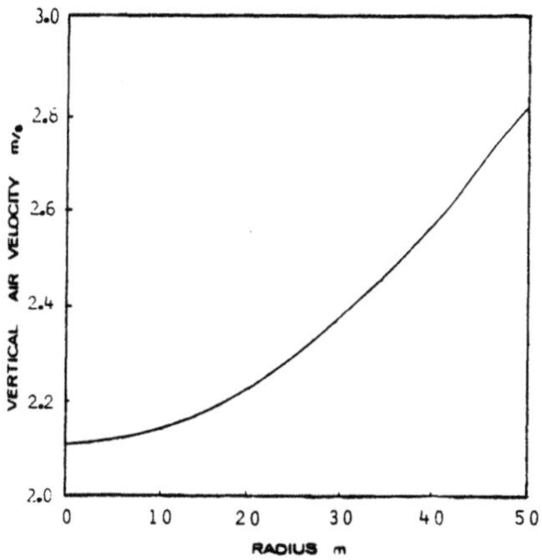

Fig. 1

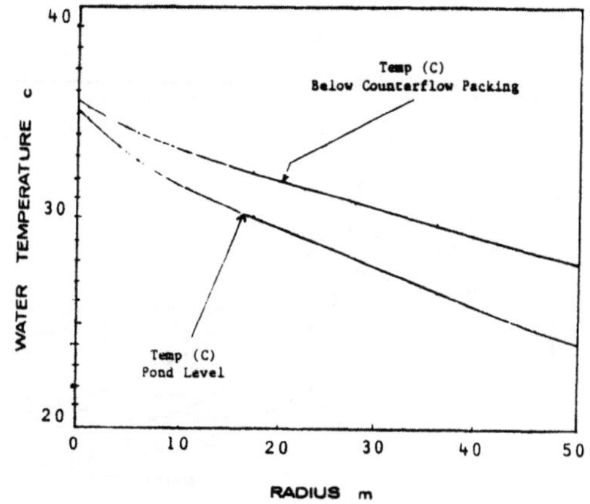

Fig. 2

marginal cooling. Figures 1 and 2 have been obtained for a 100 m diameter tower, 300 m high and containing 2 m of packing.

Structural Model

Candidate geometries for optimization were analyzed using the membrane theory of thin shells, considering dead load only. A segmental model [6] was used to account for the variation in shell thickness with height, and was executed quickly and economically with an APPLE-II PLUS microcomputer. From the meridional stress resultant, the shell weight is easily computed. In addition, representative candidate cases for optimum design were analyzed for dead and wind loading by the finite element method, using the SHORE-III computer program [7,8]. This allowed the shell reinforcement to be properly proportioned and introduced into the cost function.

GENERAL APPROACH

Cost Contours

Shown in Figure 3 are thermal design curves [9] for hyperbolic cooling towers that, for a specified cooling duty, give different sets of D, base diameter, H, height, and Z_P, depth of packing, for a constant height of air opening, H_A = 10 m. A given cooling duty is defined as rejecting power with fixed inlet and outlet water temperatures [9]. Any point on the curves labeled Z_P = 3 m, 2 m, 1.5 m and 1 m on Figure 3 represents a design

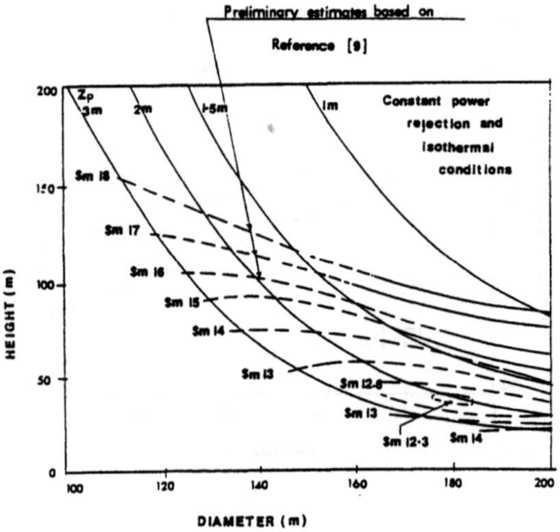

Fig. 3. Optimization study cost contours

(thermal) point with inlet water temperature, T(IW) = 40°C, outlet water temperature, T(OW) = 27°C and a power rejection of 1500 MW associated with an electrical output of 1000 MW for an overall 40% station efficiency. In this work, the thermal curve for the depth of packing Z_P = 2 m in Figure 2 was chosen for all investigations.

The dashed lines in the figure refer to preliminary cost values [9] which do not pertain directly to this study.

Procedure

The objective of this study is to investigate the different tower geometries on Figure 3 in order to provide improved calculations of the shell thickness and the quantities of concrete and reinforcing bars required, leading to more accurate cost estimates.

The study was executed as follows:

1. Determination of shell geometric parameters.
2. Establishment of shell wall thickness profile.
3. Investigation of least weight of each group of shells starting from a cylindrical case and proceeding to the maximum possible angle of inclination using the simple membrane theory and considering only dead load stresses.
4. Cost comparison of shells analyzed by membrane theory.

5. More refined analysis including wind load using the finite element method. On this basis, the cost comparison in 4 is adjusted to account for any additional reinforcement required.

OPTIMIZATION

Geometry

The surface geometry of a hyperboloid of revolution (Figure 4) is defined by

$$\frac{R^2}{a^2} - \frac{Z^2}{b^2} = 1 \tag{1}$$

in which R_o = the horizontal radius; Z = the vertical coordinate; a = the throat radius, and

$$b = \frac{aT}{\sqrt{t^2-a^2}} = \frac{aS}{\sqrt{s^2-a^2}} \tag{2}$$

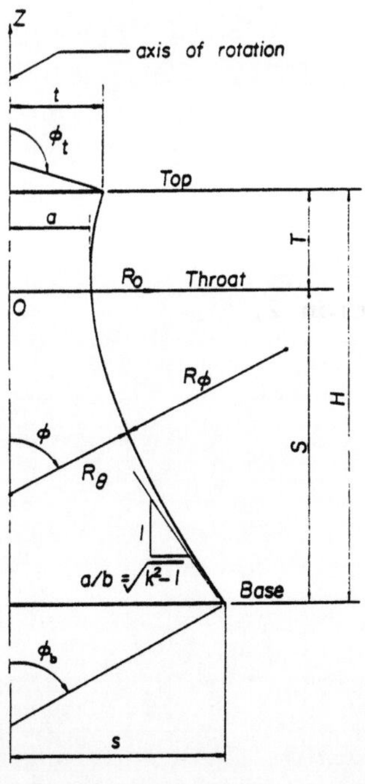

Fig. 4. Hyperboloidal shell geometry

in which s = the base radius; t = the top radius; and S and T = the vertical distances from the throat to the base and top of the shell, respectively. Another useful parameter is

$$k = \sqrt{1 + \frac{a^2}{b^2}} \tag{3}$$

which indicates the deviation of the shell profile from the degenerate case of a cylinder, k = 1, with increasing k corresponding to a more pronounced curvature of the meridian. The ratio $\frac{a}{b}$ in equation 3 is the slope of the asymptote of the generating hyperbola. The principal radii of curvature, R_ϕ and R_θ, and other relevant parameters are shown on Figure 4. The base radius is evaluated as

$$s = R_o(\phi_b) = R_\theta \sin\phi_b \tag{4}$$

Therefore,

$$s = \frac{a\sqrt{k^2-1}}{\sqrt{k^2 \sin^2\phi_b - 1}} \sin\phi_b \tag{5}$$

Also, from equation 3,

$$b = \frac{a}{\sqrt{k^2-1}} \tag{6}$$

Equating equation 6 to the rhs of equation 2,

$$\frac{a}{\sqrt{k^2-1}} = \frac{aS}{\sqrt{s^2-a^2}} \tag{7}$$

from which

$$a^2 = s^2 - S^2(k^2-1) \tag{8}$$

With S, s, and ϕ_b known and solving equations 5 and 8 simultaneously,

$$a = \sqrt{s^2 - Ss(\arctan\phi_b)} \tag{9}$$

$$k = \sqrt{\frac{s}{S}\arctan\phi_b + 1} \tag{10}$$

and, from equation 2,

$$t = \frac{1}{S}\sqrt{T^2(s^2-a^2) + a^2S^2} \tag{11}$$

For any given shell height, H, and base diameter, D, the corresponding throat radius, a, the top radius, t, and parameter k may be obtained using equations 9-11. Here, the vertical distance from the throat to the base, S, and to the top, T, are taken as 0.90 H and 0.10 H, respectively, which are in the practical range.

Selection of Base Angle

As shown on Figure 5, ϕ_b is the angle that a line drawn normal to the shell middle surface at the base makes with the Z axis. Also shown is the angle of inclination of the formwork

$$\Phi = \frac{\pi}{2} - \phi_b \tag{12}$$

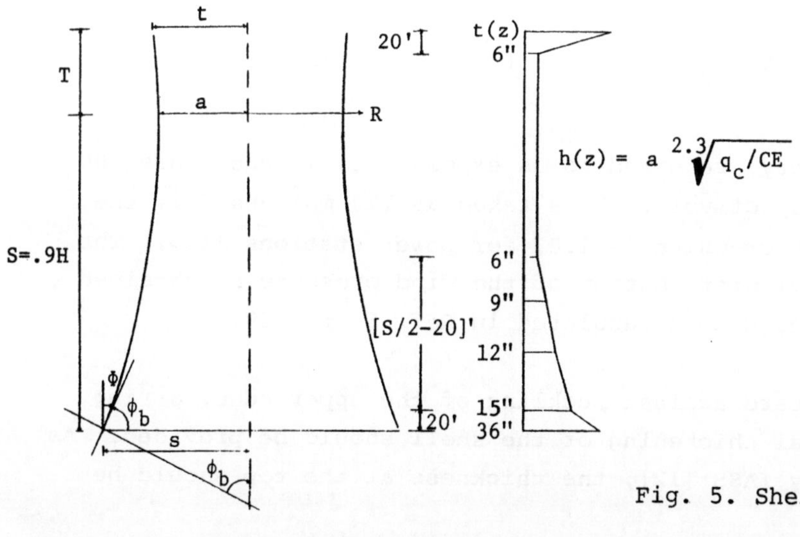

Fig. 5. Shell profiles

From equations 4 and 5,

$$\phi_b = \sin^{-1} \frac{s}{[a^2 + k^2(s^2-a^2)]^{1/2}} \tag{13}$$

The base angles ϕ_b used for each set of D and H range from 80° to 70° with increments of 0.5°. The limit $\phi_b = 70°$ is based on the maximum inclination possible with commercial formwork systems, $\phi \simeq 20°$. With ϕ_b set, shell parameters a, k, and t, as defined by equations 9-11, may be calculated to produce, for each set of D and H, twenty-one different hyperboloid of revolution surface geometries [10].

Determination of Shell Wall Thickness Profile

Generally, wind velocity causes external pressure on structures, which can be expressed at any elevation z by

$$q_z = 0.00256\ K_z(IV)^2 \qquad \text{(psf)} \tag{14}$$

The velocity exposure coefficient, K_z, can be obtained as follows:

$$K_z = \begin{cases} 2.58\left(\dfrac{z}{z_g}\right)^{2/\alpha} & \text{for } z \geq 15 \text{ feet} \\[2ex] 2.58\left(\dfrac{15}{z_g}\right)^{2/\alpha} & \text{for } z \leq 15 \text{ feet} \end{cases} \tag{15}$$

For open country, referred to as exposure C, z_g and α are 900 feet and 7 respectively. V is taken as 110 mph and I is the importance factor which is 1.07 for power stations [11]. The circumferential distribution of the wind pressure is obtained using harmonics 0-5 as tabulated in Ref. 6, p. 139.

In order to guard against buckling of the upper edge, a ring beam or gradual thickening of the shell should be provided. As recommended by IASS [12], the thickness at the top should be

$$h(\phi_t) = a\left(\frac{q_c}{CE_c}\right)^{1/2.3} = 12a\left(\frac{50.90401K_z}{E_c}\right)^{1/2.3} \quad (16)$$

in which

$$q_c \geq \gamma_b q_z = 5q_z \times \frac{1}{144} \text{ (psi)} \quad (17)$$

where $\gamma_b \geq 5$, is the required buckling factor; the important coefficient C is assumed to be 0.052; and E_c is Young's modulus for concrete.

The thickness at the base $h(\phi_b)$ is assumed to be 36 inches for every shell, while the minimum thickness of a hyperboloidal shell wall should not be less than [12]

$$h_{min} = \sqrt{\frac{R_o^2 H}{400}} \quad (18)$$

Since the minimum wall thickness for cooling tower shells with two layers of reinforcement may not be less than about 5.5 inches [12], the minimum wall thickness for all shells investigated is taken as 6 inches (considering the shortest shell of throat radius, a = 321.51 feet and height, H = 82.02 feet). On Figure 5, the general thickness profile of the shell wall is shown.

Membrane Theory Results

As stated previously, the variation of ϕ_b from 80° to 70° provided a subset of different surface geometries for each set of "D's" and "H's". Using fourteen segments, the dead load meridional stress resultant for each shell, $N_\phi(\phi)$, was evaluated taking the unit weight of concrete to be 150 pcf. Then the weight of the shell was computed as

$$W_S = 2\pi s N_\phi(\phi_b)\sin\phi_b \quad (19)$$

It was found that shorter hyperboloidal shells generally weighed much less than comparable taller ones, even with consideration of the thickness profile of Figure 5 and minimum vertical and circumferential shell wall reinforcement, A_S = .35% × gross cross-sectional area of the wall. In each group, the maximum ϕ_b corresponded to the lightest tower. However, additional components enter into the weight determination and cost comparisons, as discussed in the following sections.

Additional Components

In addition to the shell wall, a ring beam, basin wall and ground slab, as shown on Figure 6, are considered with minimum reinforcement. As mentioned earlier, the columns, shown in Figure 6, were not explicitly included.

A most important component in the cost comparison is the packing, which is assumed to cover the entire base area of the shell. The specific characteristics of the packing were given previously.

Capital Cost Function

The capital cost function given by Furzer [9] is modified as follows:

$$C_T = C_P \frac{\pi}{4} D^2 Z_P + [C_C + UC_R][V_B + V_{RB}] + C_C \frac{W_S}{\gamma_C} + C_{RS} \tag{19}$$

where

V_{RB} = volume of ring beam concrete

$= \pi(r_B^2 - r_A^2) d_B$

V_B = volume of water basin concrete

= volume of concrete ground slab and basin wall

$= \pi r_A^2 h_s + \pi(r_d^2 - r_c^2) h_w$

$\pi(r_B^2 - r_A^2)$ = surface area of ring beam (see Figure 6)

$\pi(r_d^2 - r_c^2)$ = surface area of basin wall (see Figure 6)

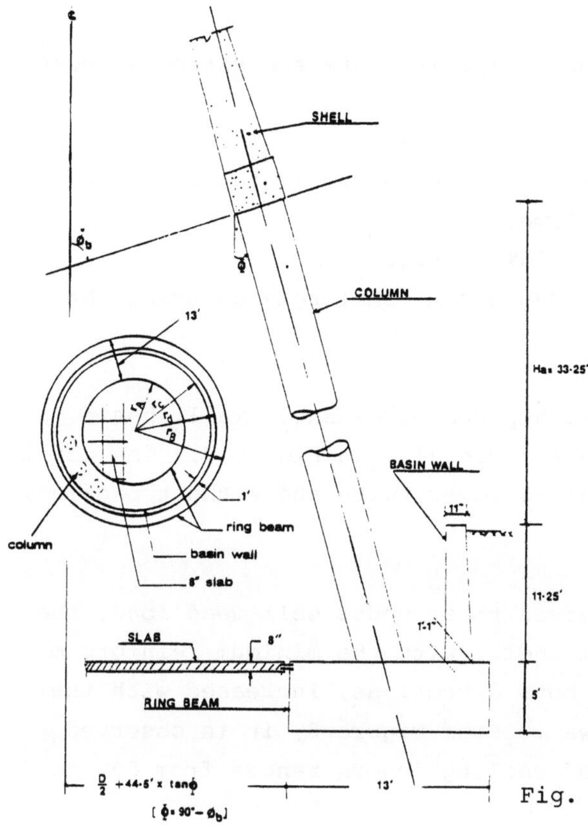

Fig. 6. Cooling tower basin

C_{RS} = cost of reinforcing bars for shell wall

C_R = cost of reinforcement per lb; taken as 45¢

C_C = cost of concrete/cf; taken as \$11.11

C_P = cost of packing/cf; taken as \$2.8317

γ_C = unit weight of concrete = 150 pcf

γ_S = unit weight of steel = 494 pcf

h_s = thickness of floor slab = 8 in.

d_B = depth of ring beam = 5 ft.

h_w = height of basin wall = 11.25 ft.

$U = .0035\gamma_S$ (pcf)

D = shell base diameter

Z_P = depth of asbestos packing = 2 m = 6.562 ft.

W_S = weight of concrete in shell wall

Cost Comparison

Items included in this cost comparison are the estimated cost of:

1. Hyperboloidal shell wall with minimum 0.35% reinforcement.
2. Cold water basin with reinforcement estimated as 0.35% of the total concrete volume.
3. Ring foundation with 0.35% reinforcement.
4. Packing with constant depth 2 m completely covering the base area.

The cost variations of the hyperboloidal shell wall, basin, and ring foundation are shown individually in Figure 7. The total cost variation of the cooling tower, with and without packing, is shown in Figure 8.

From Figure 7, it is observed that, under self dead load, the cost of the hyperboloidal shell, with the minimum reinforcement provided for the wall in both directions, increased with the height of the cooling tower. From Figure 8, it is observed that the optimum height of cooling towers ranges from 500 to

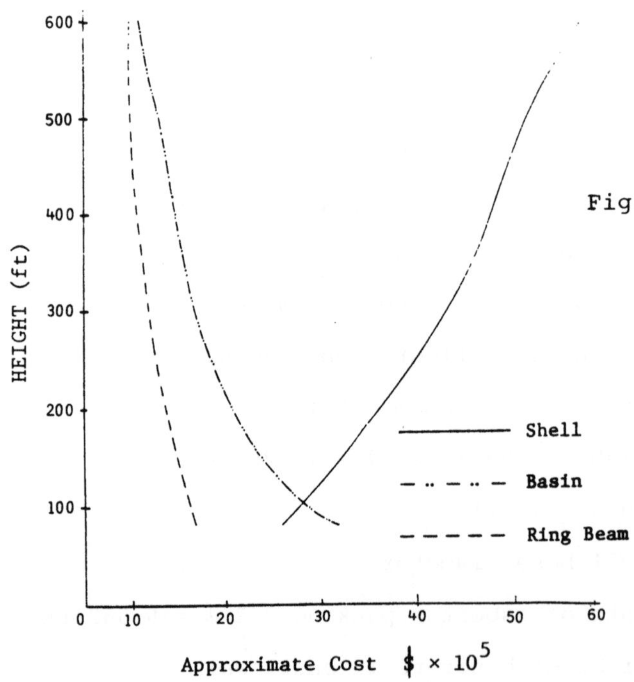

Fig. 7. Cost comparison of ring beams, basins and shells

600 feet with the cost of the asbestos packing included, and from 160 to 200 feet if the cost of the packing is neglected.

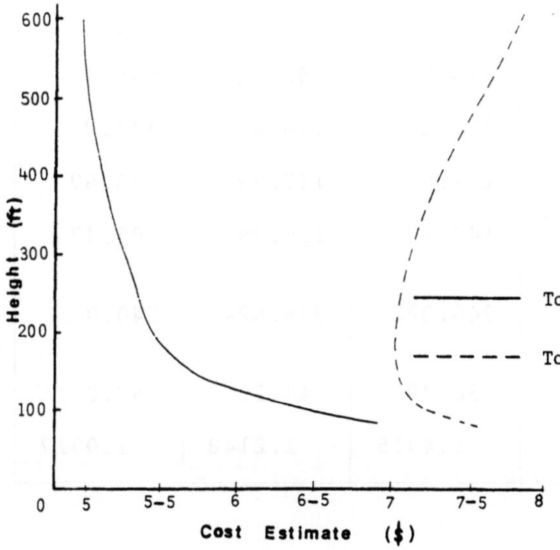

Fig. 8. Cost comparison of cooling towers

IMPROVED COST COMPARISON

Basis

The previous section dealt with hyperboloidal shell cost comparisons with reinforcement determined by the shell thickness profile. In this section, the cost of three hyperboloidal shells analyzed by the finite element method for factored dead load plus wind load corresponding to V = 100 mph are compared. The shell dimensions are given in Table 1 and the detailed design calculations, carried out in accordance with ACI-ASCE 334 [13], are given in Ref. 10.

The calculations reveal that the top half of each hyperboloidal shell requires only minimum reinforcement in both directions. In the lower half of the shell, the vertical reinforcement is controlled by the stress resultants while the circumferential reinforcement remains at the minimum required by the wall thickness. Vertical reinforcement for intermediate shell geometries was computed by interpolation.

Table 1. Description of Analyzed Shells

Description (units in ft)	Tower I	Tower II	Tower III
Height of shell	384.88	442.91	600.
Base radius, s	218.34	210.8	190.7
Throat radius, a	141.93	117.69	93.67
Top radius, t	143.12	119.28	95.47
Vertical distance from throat to base of shell, S	346.392	398.624	540.0
Vertical distance from throat to top of shell, T	38.49	44.29	60.0
Top thickness, $h(\phi_t)$	1.4415	1.2148	1.0017

Results

The items included in this revised cost comparison are the same as those previously considered, with the only major difference being the revised cost of reinforcement for the hyperboloidal shells. This is principally the vertical reinforcement in the lower half of the shell.

Figure 9 shows the cost of adequately reinforced shells and the cost of cooling towers, with and without packing, plotted for various cooling tower heights. From this figure, it may be observed that the curve that compares the revised total cost of the cooling tower with the cost of the packing included follows a somewhat different course, and particularly shows that the optimum heights of cooling towers are lower than previously indicated in Figure 8 where only the self-weight was considered. This is, of course, due to the increased influence of wind on the taller towers.

CONCLUSIONS

1. Generally the weight of a hyperboloidal shell is very sensitive to the base angle of inclination. All thermally com-

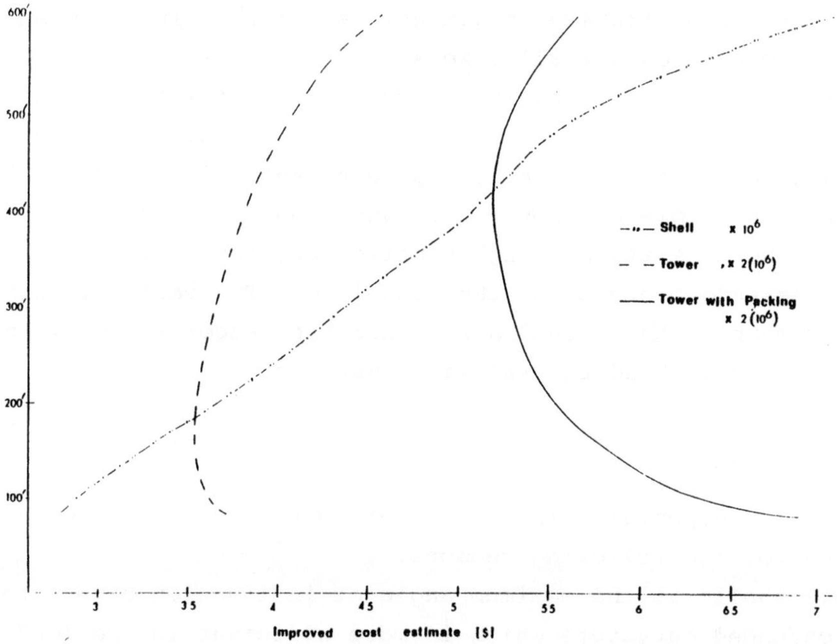

Fig. 9. Improved cost comparison of cooling towers

parable shells had the least weights for the maximum angle of inclination possible for commercial formwork.

2. The top thickness of the hyperboloidal shell provided, as recommended by IASS, decreased with increased shells height. This is because the top thickness is dependent on the throat radius of the shell, and tall shells having more pronounced curvature at $\phi_b = 70°$ have a drastically reduced throat radius, a. It would have been expected that the top thickness be increased as the shell height progressed. The IASS recommendation should be followed with caution.
3. The ratio of the height of shell, H, to the vertical distance from the base to throat of shell, S, was chosen as 0.9 because the investigations showed that lesser ratios, e.g. 0.8 or 0.75, would produce heavier weight even for the same angle of inclination.
4. The estimated cost of the hyperboloidal shell increased rapidly with height, with or without inclusion of the reinforcement, while the cost of the basin and ring beam reduced with height. However, it should be noted again here

that the same transverse dimensions for the ring beam and basin were used for all towers.

5. The total cost of a cooling tower decreased with increasing height only when the cost of packing was included. However, again the packing depth is kept constant at 2 m and is allowed to completely cover the base area of every shell.
6. Tall shells sustained high tensile stresses from the effects of pressure produced by the wind load. The vertical reinforcement of these shells was increased accordingly, which more than doubled the reinforcement cost.

RECOMMENDATIONS

Tall shells, especially from 400 feet up to 500 feet, are recommended for the following reasons:

1. Tall shells at the maximum angle of inclination have more pronounced curvature which is more efficient in the load resistance mechanism.
2. Usage of short shells is discouraged since they require enormous land space for construction.
3. The optimum tower heights shown on Figure 9 may be more economically maintained since the area covered is smaller.
4. The results could be generalized by considering various depths of packing to provide the same power rejection. Also, in regions with less severe wind conditions, the influence of the added reinforcement would not be so drastic.

References

1. Gould, P.L.: Minimum weight design of hyperbolic cooling towers. J. Struct. Div., ASCE; V. 95; No. 572 (Feb. 1969) 203-208.

2. Reinschmidt, K.F.; Narayanan, R.: The optimum shape of cooling towers. Computers and Structures; Vol. 5, (May 1975) 321-325.

3. Furzer, I.A.: Very large hyperbolic cooling towers. Mech. and Chem. Engg. Trans. I.E. Aust.; Vol. MC8; No. 2 (Nov. 1972) 122-124.

4. Lowe, H.J.; Christie, D.G.: Heat transfer and pressure drop data on cooling tower packings. International Heat Transfer Conference; Denver, CO (1962) V, 993.

5. Furzer, I.A.: An improved thermodynamic analysis of hyperbolic cooling towers. 6th International Heat Transfer Conference; Toronto, Canada (1978).

6. Gould, P.L.: Static analysis of shells. Lexington (1976).

7. Basu, P.K.; Gould, P.L.: SHORE-III shell of revolution finite element program-theoretical manual. Structural Engineering Research Report No. 48, Dept. of Civil Engineering, Washington University, St. Louis, MO (Sept. 1977).

8. Basu, P.K.; Gould, P.L.: SHORE-III shell of revolution finite element program-user's manual. Structural Engineering Research Report No. 49, Dept. of Civil Engineering, Washington University, St. Louis, MO (Sept. 1977).

9. Furzer, I.A.: Sub-optimization of a hyperbolic cooling tower for a 1000MW power station. Dept. of Chemical Engineering, University of Sydney, NSW 2006, Australia.

10. Alowooja, K.T.: Sub-optimization of a hyperbolic cooling tower for a 1000MW power station. Thesis; Dept. of Civil Engineering, Sever Institute of Technology, Washington University, St. Louis, MO (Aug. 1983).

11. Minimum design loads for buildings and other structures. American National Standards Institute, Inc., New York (March 10, 1982).

12. Recommendations for the design of hyperbolic or other similarly shaped cooling towers. International Association for Shell Structures Working Group No. 3, Brussels (1977).

13. Reinforced concrete cooling tower shells: practice and commentary. ACI-ASCE Committee 334; American Concrete Institute Journal; Vol. 74 (Jan. 1977) 22-31.

A New Look at the Ultimate Behavior of Hyperbolic Cooling Towers

A.K. Gupta

North Carolina State University,
Raleigh, North Carolina, USA

Abstract

The cracking criterion conventionally used in the nonlinear analysis of reinforced concrete shells is not consistent with the limit state used in the design of reinforcement. This limit state has been verified by recent experiments. A new algorithm has been developed which can successfully predict the initial crack direction and the limit crack direction. A finite element computer program to perform nonlinear analysis of membrane reinforced concrete shells, which incorporates the new algorithm, has been developed. The program is used to analyze a hyperbolic cooling tower. Important results and conclusion have been discussed.

1984
IASS-RUB

Stress Analysis

Stress Analysis of Cooling Towers with Ribs and Discrete Supports Using F.E.M.

DAJUN WANG

Department of Mechanics, Peking University, China

JIAN CHEN, DONGYING JU

Hebei Institute of Mechanical and Electrical Engineering, China

Summary

The stress analysis of a natural-draught cooling tower with ribs is treated by finite element method. Two cases usually encountered in engineering, viz. ribs arranged densely along meridians and parallel circles as well as arranged sparsely along parallel circles, are considered. Influence of the supports and their foundations to the stress distribution is treated. In order to show the local characteristic near the discrete supports, a second iterative computation is made.

The stress analysis for a shell of revolution with ribs and discrete supports can be used in treating an assembled type of cooling tower, as well as in selecting the model of a tower. The ribs improves the stability of the tower, it also enhances the strength of the tower locally.

The stress distribution in the tower is influenced by the structural behaviors of the supports and its foundations (including prestressed annular beam and maybe piles), particularly by the settlement of the foundations.

The discrete distribution of elastic supports of the tower greatly affects the stress in the bottom part of the tower, so it is necessary to consider the discrete distribution of the supports to get a better approximate value of the real stress in this area.

In the following the formulation and programming are given to solve these problems using the results of Ref.[1]

The stiffness matrices of ribbed shell of revolution

A shell of revolution andits coordinate systems are shown in Fig. 1. The displacements of the middle surface of the shell in the coordinates $\{\Pi\} = \{\underline{k}, \underline{e}, \underline{r}^{\circ}\}$ are denoted by $\{V\} = \{u, v, w\}$. The shell is divided into several elements by parallel circles which are also called nodal circles. For the displacement functions, trigonometric polynomial is employed in the Θ

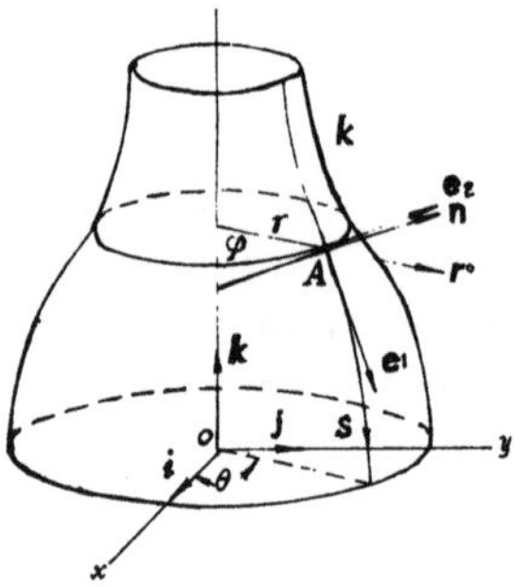

Fig. 1

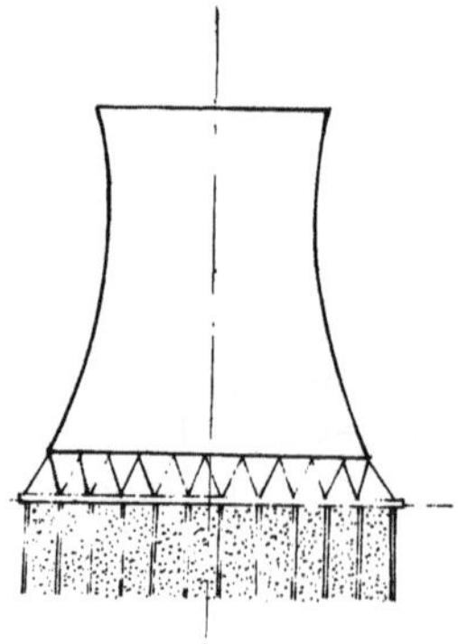

Fig. 2

direction but Hermitian polynomial in the S direction. For the majority of engineering problems the stress in a shell is continuous, hence only the order 2 Hermitian polynomial for u and w, the order 1 polynomial for v are used. Thus, the components of the generalized displacements are the values of the coefficients of the trigonometric polynomial of u, v, w, $\partial u/\partial s$, $\partial v/\partial s$, $\partial w/\partial s$, $\partial^2 u/\partial s^2$, $\partial^2 w/\partial s^2$ at the nodal circles. It is important to notice the following facts. The expressions for the displacements can be divided into two groups: the terms of symmetric deformation (in which case u and w have only terms in cos $k\theta$, while v only in sin $k\theta$) and the terms of anti-symmetric deformation. Because those two groups of terms can be solved separately, and the latter can be obtained from the former by a proper rotation in the direction θ, so one needs only to treat the former. The stiffness matrix of the element corresponding to the kth harmonic is

$$[KE_k] = [A]^T \int [B_k]^T [D] [B_k] rds [A] \quad (1)$$

where [A] is a 16 × 16 matrix concerning Hermitian interpolation, $[B_k]$ is a 6 × 16 matrix operator connecting the components of

the deformations of the shell with the displacement components.
(1) The stiffness matrix of the element of a shell with dense ribs. Let Young's modulus and Poisson's ratio of the shell be denoted by E and μ respectively. Suppose there are ribs arranged densely and uniformly along meridians and parallel circles, let E_mI_m, G_mJ_m, k_1, F_1, b_1 and η_1 denote the section flexural modulus of torsion, the shear coefficient, the section area, the distance between two ribs and the eccentric distance of a meridian rib respectively, let E_cI_c, G_cJ_c, k_2, F_2, b_2 and η_2 denote those of a parallel rib respectively. These dense ribs are treated as orthogonal anisotropic elastic body, which elastic matrix is

$$[D]=\begin{bmatrix} \frac{Eh}{1-\mu^2}+\frac{E_mF_1}{b_2} & \frac{E\mu h}{1-\mu^2} & 0 & \frac{E_mF_1\eta_1}{b_2} & 0 & 0 \\ & \frac{Eh}{1-\mu^2}+\frac{E_cF_2}{b_1} & 0 & 0 & \frac{E_cF_2\eta_2}{b_1} & 0 \\ & & Gh+\frac{1}{2}\left(\frac{F_2k_2G_c}{b_1}+\frac{F_1k_1G_m}{b_2}\right) & 0 & 0 & 0 \\ & & & \frac{Eh^3}{12(1-\mu^2)}+\frac{E_cI_c}{b_2} & 0 & 0 \\ & & & & \frac{Eh^3}{12(1-\mu^2)}+\frac{E_cI_c}{b_1} & 0 \\ & \text{Symmetric} & & & & \frac{Gh^3}{12}+\frac{1}{4}\left(\frac{G_mJ_m}{b_2}+\frac{G_cJ_c}{b_1}\right) \end{bmatrix} \quad (2)$$

Substituting expression (2) into expression (1), we get the stiffness matrix of the element of a shell with dense ribs.
(2) The stiffness matrix of a parallel rib arranged sparsely. Let E_cF_c, G_cJ_c, E_cI_s and E_cI_n denote the section extension modulus, the section modulus of torsion, the section flexural modulus about <u>e</u> and the one about <u>n</u> respectively, the stiffness matrix of a rib is

$$[K_k]_c = [B_k]_c^T\,[D]_c\,[B_k]_c \quad (3)$$

$$[D]_c = \begin{bmatrix} E_cF_2 & E_cF_2\eta_2 & 0 & 0 \\ & E_cI_s & 0 & 0 \\ & & E_cJ_n & 0 \\ \text{Symmetric} & & & G_cJ_c \end{bmatrix} \quad (4)$$

where $[B_k]_c$ is a matrix connecting the components of deformation with those of displacements of the rib. The generalized displacements are the values of the coefficients of trigonometric polynomial of u, v, w, $\partial u/\partial s$, $\partial v/\partial s$ and $\partial w/\partial s$. Expanding the stiffness matrix $[K_k]_c$ into a 8×8 matrix and adding it to the corresponding elements of the stiffness matrix of the element of the shell, we get the coupling between the rib and the shell.

Discrete supports of the shell

In general, a cooling tower is supported on many closely spaced, uniform and discrete sloping columns which are founded on a prestressed annular beam, there may be some piles located under the beam as shown in Fig. 2.

(1) Stiffness matrix of supporting columns. The stiffness matrix $[K]_b$ of each sloping column in local coordinates may be obtained if translation and rotation displacements of the top and bottom edges of the column in three directions are defined as the generalized displacements. Denote $[T_1]$, $[T_2]$ as the coordinate transformation matrices of a pair of sloping column from local coordinates of the column to the coordinates II . $[B_k]_c$ is a matrix connecting the generalized displacements of the shell with those of the sloping columns. Then the stiffness matrix of a pair of sloping column is

$$[K_k]_{cb} = [B_k]_c^T([T_1]^T[K]_b[T_1]+[T_2]^T[K]_b[T_2])[B_k]_c \quad (5)$$

Assuming that n pair of sloping column are spread into a continuous and uniform distribution, then the stiffness matrix will be

$$[K_k]_{cc} = (n/2\pi)[K_k]_{cb} \quad (6)$$

(2) Stiffness matrix of the foundation of columns. Similar to the sparse parallel rib the stiffness matrix of prestressed annular beam $[K_k]_a$ can be obtained. The piles are treated as beams of infinite length located on elastic foundations. The top of piles are coupled with the displacements of annular beam. The stiffness matrix of a pile is

$$[K]_p = \begin{bmatrix} k_e & 0 & 0 & 0 & 0 & 0 \\ & 4C^3EI_p & 0 & 0 & 0 & 2EI_pC^2 \\ & & 4C^3EI_p & 0 & -2C^2EI_p & 0 \\ & & & k_t & 0 & 0 \\ & \text{Symmetric} & & & 2CEI_p & 0 \\ & & & & & 2CEI_p \end{bmatrix} \quad (7)$$

in which EI_p is sectional flexural modulus, k_e, k_t and k_f are the spring coefficient of elastic foundation, $c=(k_f/4EI_p)^{\frac{1}{4}}$. Assuming the spring coefficients of the foundation under the annular beam in the direction $\underline{k}$, $\underline{e}$ and $\underline{r}^o$ are k_1, k_2 and k_3 respectively, then the stiffness matrix of the foundations of the columns is

$$[K_k]_f=[K_k]_a+[k]_p+\lceil k_1,k_2,k_3,\ 0,0,0 \rfloor \quad (8)$$

(3)Treatment of Discrete Allocation of Supporting Columns

In the case that the cooling tower is supported on the top of n bars of columns which are allocated along a circle at equal interval a simple method is presented for the consideration of the effect of the discrete allocation. The main of this method are: 1. First at all, the columns are treated as a continous support, and the generalized forces on the supporting boundary are calculated. 2. Then the generalized forces which acted in between the two neighbouring middle span of a column uniformly allocated to the columns in their statical equivalents. In the case of a fairly numerous columna, this re-allocation does not induce notable error. 3. The supporting boundary is considered as a free boundary, and the mechanical quantities on this boundary are four generalized forces mentioned above while on the other boundary of the shell the displacements calculated in the first step are used to carry out the second iterative computation. The results obtained under this boundary can be considered as those supported on discrete columns. Because the differences of the solution between the discrete support and the continous support are notable only in the region near the supporting boundary, it is not necessary to compute for the entire shell, only the domain in which notable error is expected needs to be computed.

Example 1. The stresses in a cooling tower under dead load and vertical uniform load of 0.8 T/m on the top of the tower. The size and material data of the tower are: $r^2=b^2+az^2$, here a=0.1833 and b=26.91; at the top and the bottom z=17.50m and -102.7m respectively; thickness h changes as exponential function, h_{min}= 0.4m, h_{max}=0.75m; Young's modulds $E=2.7x10^6 T/m^2$; Poisson's ratio =0.167; specific gravity of concrete =2.4. The bottom is supported on 50 pairs of sloping column; the length of the column is 9.66m, cross-section is a square with sides of length 0.65m. The results of computation are shown in Fig 3, 4 and 5. T_1 and T_2 in Fig 3 are the membrane forces in tower when the boundary with columns are regarded as continous supports.

Fig 4 and 5 show the distribution of T_1 and T_2, respectively, from the middle point of a column to that of a span below the level of -85.6m. It can be seen that the stresses computed for the two cases that the columns are regarded as continous supports and as discrete supports have only small differences beyond a distance of about the same length of the span between two columns from the bottom.

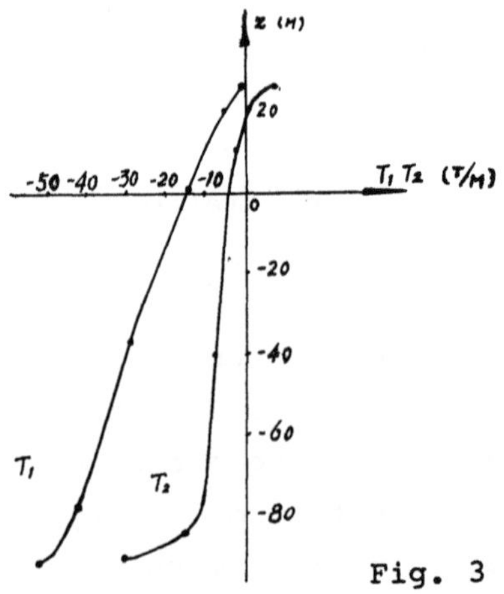

Fig. 3

Example 2. The stresses in tower caused by the settlement of the foundation of the columns. The values of displacements and stresses in tower are listed in following form for the case that the foundations of the columns have a uniform displacement w=0.1m in direction $\underline{r}°$.

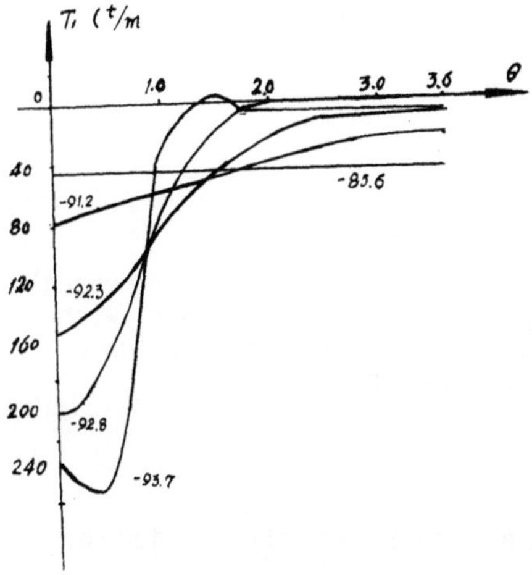

Fig. 4

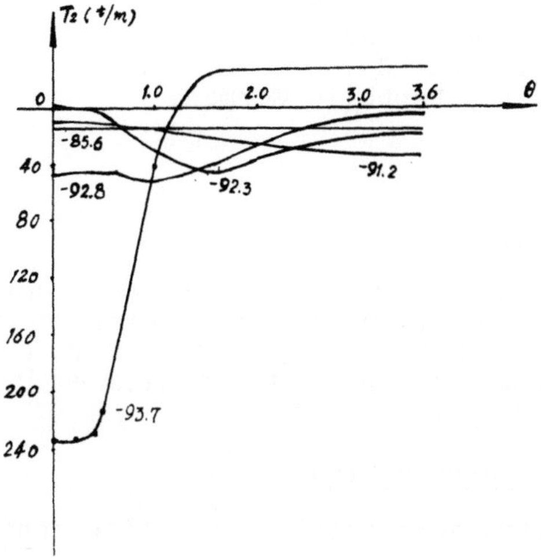

Fig. 5

Z (m)	w^* (m)	u^* (m)	T_1 (T/m)	T_2 (T/m)	M_1 (T·m/m)
129.0	0.0061	0.0363	0.0	0.0	0.0
100.0	-0.0007	0.0368	0.0	0.0	0.0
34.0	-0.00111	0.0351	0.003	-0.026	-0.001
26.0	-0.0115	0.0350	0.022	0.244	-0.009
14.0	-0.0142	0.0348	-2.199	49.761	-15.953
9.6	-0.0053	0.0347	0.745	357.293	-35.711

References

1. Jike Wu; Dajun Wang; Mingwu Yuan. Stresses Analysis of Shells of Revolution, Irrigation and Electric Power Publisher, 1979, Beijing.
2. O.M.E. Shafee; P.K.Basu; B.J. Lee & P.L. Gould. SHORE IV, Finite Element Program for Dynamic and Static Analysis of Shells of Revolution, PB.82-147919.
3. Jian Chen; Dongying Ju; Dajun Wang. Calculation of Stresses for Ribbed Shell of REvolution Made of Composite Materials, Presented at Symposium on Strength of Structure, Chinese Aerospace Society, Oct., 1983.

Economical Calculation for the Optimal Design of General Shells of Revolution for Natural-Draft Cooling Towers by Means of Membrane Theory

W. KRINGS

E. Heitkamp GmbH
Langekampstraße 36
4690 Herne 2
Federal Republic Germany

Summary

By means of a fast integration procedure the membrane equations for general shells of revolution can be solved. The application of the procedure with regard to the determination of the most economical shell shape and support framework are presented and explained in this contribution.

1. Introduction

Designing a shell in practice confronts the structural design engineer with the problem of establishing - for a few given dimensions - a precise calculation of the shell's centre surface, the wall-thickness curve and the percentage of reinforcement. In doing so, all the structural requirements as well as the specific requirements as to the available shuttering and scaffolding systems have of course to be fulfilled. All this is to be done in such a way that the resulting construction costs remain as low as possible.

This problem, together with proposals for its solution, has already been dealt with in [1] and [2]. Krätzig [1] keeps the wall-thickness curve constant and varies the shape of the shell's centre surface. Harnach [2] starts from the assumption of a constant shell centre surface and determines the required wall-thickness curve. Both make use of the characteristics procedure which is not suitable for shells with a positive Gaussian curvature.

Therefore, we will present a design procedure which is generally suitable for shells of revolution of any shape for reinforced concrete natural-draft cooling towers. This means that shells of revolution of any Gaussian curvature can be optimized with regard to the costs, including, of course, also the newly built ATS-shapes (see fig. 1) (Atmospherical, Thermal, Stabilized).

Fig.1. ATS natural-draft cooling tower

The wall-thickness curve, the shape of shell's centre surface and the percentage of reinforcement are determined in such a way as to obtain the minimum of costs for the shell and the shell supports.

The solution method used here for the required statical calculations is a numerical integration scheme which was already mentioned by Flügge [3] in 1934. This method is simpler, more universal and more advantageous than the well-known characteristics procedure in establishing the required grid. To the author's knowledge this method has never before been applied to the calculation of cooling tower shells. It has, however, been referred to in short in [5].

2. Membrane theory solution method for shells of revolution

The three conditions of equilibrium in membrane theory can be described by two partial differential equations and one algebraic equation (1).

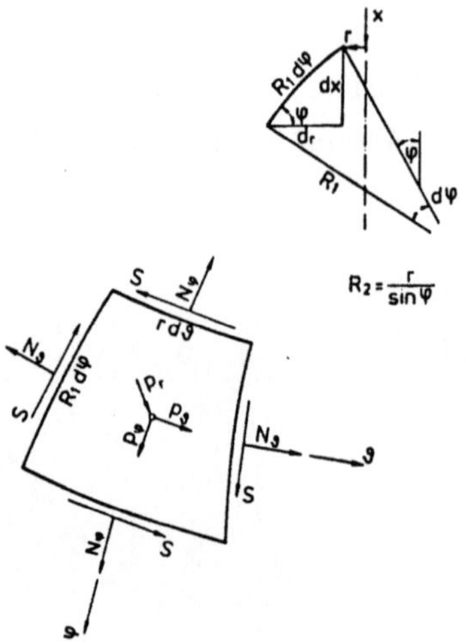

Fig.2. Shell of revolution element with membrane forces

$$\frac{\partial(N_\varphi \cdot r)}{R_1 \cdot \partial\varphi} + \frac{\partial S}{\partial\vartheta} - N_\vartheta \cdot \cos\varphi + p_\varphi \cdot r = 0$$

$$\frac{\partial(S \cdot r)}{R_1 \cdot \partial\varphi} + \frac{\partial N_\vartheta}{\partial\vartheta} + S \cdot \cos\varphi + p_\vartheta \cdot r = 0 \qquad (1)$$

$$N_\vartheta = - R_2 \cdot \left(p_r + \frac{N_\varphi}{R_1} \right)$$

The designations used for coordinates and membrane forces are represented in fig. 2.

By developing the circumferential loads in Fourier series (2) the system is decoupled.

$$N_\varphi = \sum_n N_{\varphi n} \cdot \cos(n \cdot \vartheta)$$

$$N_\vartheta = \sum_n N_{\vartheta n} \cdot \cos(n \cdot \vartheta) \qquad (2)$$

$$S = \sum_n S_n \cdot \sin(n \cdot \vartheta)$$

$$p_\varphi = \sum_n p_{\varphi n} \cdot \cos(n \cdot \vartheta)$$

$$p_\vartheta = \sum_n p_{\vartheta n} \cdot \sin(n \cdot \vartheta) \tag{2}$$

$$p_r = \sum_n p_{rn} \cdot \cos(n \cdot \vartheta)$$

$$n = 0,1,2,\ldots$$

For each Fourier element there will be two ordinary differential equations and one algebraic equation (3).

$$\frac{d(N_{\varphi n} \cdot r)}{R_1 \cdot d\varphi} + n \cdot S_n - N_{\vartheta n} \cdot \cos\varphi + p_{\varphi n} \cdot r = 0$$

$$\frac{d(S_n \cdot r)}{R_1 \cdot d\varphi} - n \cdot N_{\vartheta n} + S_n \cdot \cos\varphi + p_{\vartheta n} \cdot r = 0 \tag{3}$$

$$N_{\vartheta n} = - R_2 \left(p_{rn} + \frac{N_{\varphi n}}{R_1} \right)$$

$$n = 0,1,2,\ldots$$

With the substitutions proposed by Rabich in [4], the transition to the vertical coordinate x and the elimination of the ring force $N_{\vartheta n}$ we obtain the differential equation system for each Fourier element as follows:

$$\frac{dU_n}{dx} = - \frac{n}{r^2} \cdot V_n - A_n$$

$$\frac{dV_n}{dx} = - \frac{n \cdot r}{R_1 \cdot \sin^3\varphi} \cdot U_n - B_n \tag{4}$$

with the substitutions:

$$dx = R_1 \cdot \sin\varphi \cdot d\varphi$$

$$U_n = r \cdot \sin\varphi \cdot N_{\varphi n}$$

$$V_n = r^2 \cdot S_n \qquad (5)$$

$$A_n = \frac{r}{\sin\varphi} \cdot (p_{\varphi n} \cdot \sin\varphi + p_{rn} \cdot \cos\varphi)$$

$$B_n = \frac{r^2}{\sin\varphi} \cdot (\frac{n \cdot p_{rn}}{\sin\varphi} + p_{\vartheta n})$$

The ordinary differential equation system (4) can be numerically integrated, irrespective of the shape of the centre surface of the shell of revolution. This is a decisive difference with regard to the characteristics method!

In principle, every numerical integration scheme can be used. The scheme chosen here is the well-known Runge-Kutta procedure. The integration should reasonably proceed from top to bottom as the present initial value problem implies that the membrane force functions U_n and V_n at the upper boundary are known. If shown in the form of a matrix (4) becomes:

$$(\frac{d...}{dx} = ...')$$

$$\underline{z}_{n(x)} = \begin{bmatrix} U_n \\ V_n \end{bmatrix}_{(x)} \qquad (6)$$

$$\underline{z}'_{n(x)} = f_{(\underline{z}_n, x)} \qquad (7)$$

(4) and (7) are equivalent. For a Runge-Kutta step from x to $x+\Delta x$ we obtain:

$$\underline{k}_0 = f(\underline{z}_n, x)$$

$$\underline{k}_1 = f(\underline{z}_n + \underline{k}_0 \cdot \frac{\Delta x}{2}, \; x + \frac{\Delta x}{2})$$

$$\underline{k}_2 = f(\underline{z}_n + \underline{k}_1 \frac{\Delta x}{2}, \; x + \frac{\Delta x}{2}) \qquad (8)$$

$$\underline{k}_3 = f(\underline{z}_n + \underline{k}_2 \cdot \Delta x, \; x + \Delta x)$$

$$\underline{z}_n(x+\Delta x) = \underline{z}_n(x) + \frac{\Delta x}{6}(\underline{k}_0 + 2 \cdot \underline{k}_1 + 2 \cdot \underline{k}_2 + \underline{k}_3)$$

The definite membrane force distribution is obtained by superposing the individual series elements and their subsequent conversion to the meridian force N_φ and the shear force S. The circumferential force N_ϑ can be determined by means of the last formula of (1).

$$N_{\varphi(\vartheta)} = \frac{1}{r \cdot \sin\varphi} \cdot \sum_n U_n \cdot \cos(n \cdot \vartheta)$$

$$S_{(\vartheta)} = \frac{1}{r^2} \cdot \sum_n V_n \cdot \sin(n \cdot \vartheta) \qquad (9)$$

$$N_{\vartheta(\vartheta)} = -R_2 \cdot (p_{r(\vartheta)} + \frac{N_{\varphi(\vartheta)}}{R_1})$$

3. Suggestions for an economical programming of the calculation method presented

The exactness of the numerical integration depends on the step width Δx . The smaller the step chosen the more precise the results. For ordinary cooling towers it will, however, be sufficient to choose the step width from values between 2 and 4 m.

The differential equations to be integrated contain terms depending only on the geometry of the shell's centre surface. They need, of course, not be recalculated for every Fourier element. It is sufficient to calculate them once beforehand and then use them again for each individual integration or for each new case of loading.

After the integration has been performed for each Fourier element the solutions must be superposed for the circumference in accordance with formula (9). It is reasonable to choose an equidistant spacing for the circumference, i.e. the step width $\Delta\vartheta$. For the purpose of evaluating the above formulas the terms

$$\cos(n \cdot m \cdot \Delta\vartheta) \qquad \text{and} \qquad \sin(n \cdot m \cdot \Delta\vartheta)$$

have to be calculated. The recalculation of these terms for each case would be a waste of valuable computation time. It is therefore recommended to calcultate $\sin(\Delta\vartheta)$ and $\cos(\Delta\vartheta)$ only once and to determine all the other sine and cosine values recursively by means of the addition theorems as follows:

$$CO = \cos(\Delta\vartheta) \quad ; \quad SI = \sin(\Delta\vartheta)$$

$$\sin(i \cdot \Delta\vartheta) = \sin((i-1) \cdot \Delta\vartheta) \cdot CO + \cos((i-1) \cdot \Delta\vartheta) \cdot SI$$

and

$$\cos(i \cdot \Delta\vartheta) = \cos((i-1) \cdot \Delta\vartheta) \cdot CO - \sin((i-1) \cdot \Delta\vartheta) \cdot SI$$

4. Application of the solution method

It is well known that the above differential equation for the forces of membrane theory apply analogously also to the deformations and also to the calculation of strainless states of deformation of the shell's centre surface. This can, of course, also be done with the solution method presented here.

The shape of the shell's centre surface is described by the generatrix which may be composed, for example, of hyperbolas ellipses, polynomials, trigonometric functions, exponential functions, etc. If membrane theory is to retain its validity the generatrix must not present any jumps, breaks or jumps in the 2nd derivation. For all types of functions mentioned above test calculations have been carried out. Cooling-tower construction mainly makes use of hyperbolas and circles which are also used in the following considerations of our contribution.

5. Design

5.1 General

The dimensioning of the structural elements of a natural-draft cooling tower is based on the input values provided for by the thermal design engineers who are responsible for the design of the cooling tower's thermal features. Their calculations as well as the trickling equipment provided for are to ensure that the warm water entering the tower is cooled down to the temperature required.

The detailed input values provided for by the thermal design engineers which are to be adhered to are the following (fig. 3):

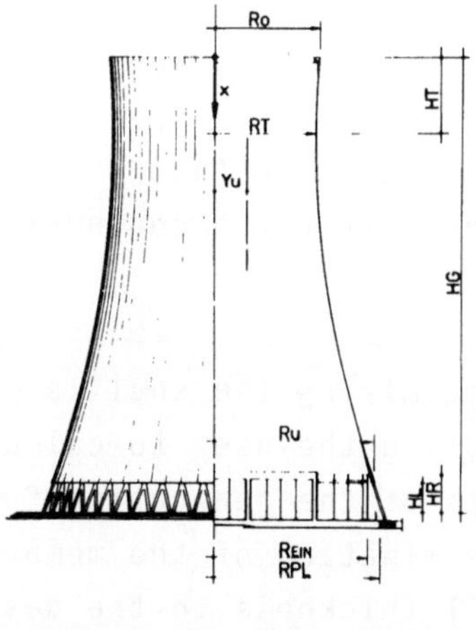

Fig.3. Input values from thermal design engineers

1. Total height: HG
2. Height of the air intake: HL
3. Radius at the waist: RT
4. Radius at upper edge of shell: RU
5. Radius at lower edge of shell: RO
6. Radius of column framework at upper edge of basin RPL
7. Radius of the trickling equipment REIN
8. Height of trickling equipment HR

The structural design engineer's task now consists in designing the cooling-tower geometry in such a way that, on the one hand, the required thermal design values are adhered to and, on the other hand, the most economical form can be chosen.

The range within which variations in the design of the optimal geometry are possible is a rather narrow one, as can be seen from the input values. It is further narrowed by the limits due to the working technique chosen and mainly the manufacturer-specific characteristics of the climbing scaffolds used for the construction of the cooling-tower shell.

As will be explained in detail later on, practically the only free variables remaining are the position of the waist H_T and the shift between the lower axis of the hyperbola and the rotational axis Y_U .

The importance of such an optimization, however, results from the fact that the state of stress of shells reacts in a very sensitive way to changes in the centre surface's curvature [1] with the corresponding consequences for the consumption of building material.

To fulfil the requirement of optimizing the shell's geometry we have developed a programme enabling the user to calculate in a very short time as many variants of the generating function as possible, ranging from the determination of the membrane forces and of the required minimum wall-thickness to the design and the evaluation of quantities and costs.

The flowchart in fig. 4 shows how this programme works.

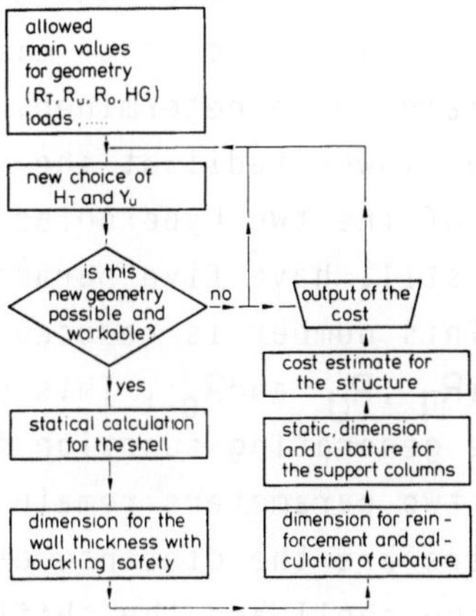

Fig.4. Optimization method

The individual steps will be explained in the following:

5.2 Cooling-tower shell

The following procedure is chosen:

Shell centre surface (generatrix):

For the area above and below the waist, the generating function of the shell's centre surface can be described by two hyperbolic functions with four parameters each:

$$r_{(x)} = Y + \frac{A}{B}\sqrt{(x-H_T)^2 + B^2} \tag{10}$$

If the given value of R_0 is smaller than R_T the upper part of the cooling tower will be described by the circular equation:

$$r_{(x)} = Y + \sqrt{-(x-H_T)^2 + B^2} \tag{11}$$

At the waist a curvature jump is then inevitable and leads to the jump in the ring force which is impossible in membrane theory so that, strictly speaking, membrane theory loses its validity in such case. Comparison calculations with bending theory have shown, however, that this case can be tolerated.

In the upper and lower equation the height of waist H_T is identical and the first derivation of $r_{(x)}$ at the waist disappears in case of the hyperbola and the circle. In the case of two hyperbolas only seven parameters and in the case of circle and hyperbola only six parameters remain to be determined. Due to the requirement that the upper and lower radii at the waist must be equal and that - in case of the two hyperbolas - the curvature radii are the same, we still have five parameters left that are to be determined. This number is reduced by three further values of the given radii R_T, R_U and R_O. This means that for a complete description of the generating function for the centre surface of the shell only two parameters remain to be chosen freely. These two parameters are the distance of the waist R_T from the upper edge of the shell and the shift value Y_U of the lower hyperbola with regard to the axis of revolution. With the two free parameters H_T and Y_U the above requirements and conditions the generating function of the shell's centre surface is fully determined.

Determination of membrane forces and wall-thickness curve

As to the relevant literature and the guidelines for the structural design of cooling towers [6], calculations in accordance with membrane theory are reasonable and precise enough for a state-of-the-art determination of membrane forces. This implies, however, that only a statically determined calculation is required where the distribution of membrane forces is dependent only on the loads and not, as in the case of statically indetermined structures, also on the stiffness. The calculation of the membrane forces, resulting in particular from wind loads, and the determination of the required wall thickness are therefore decoupled.

The wall-thickness curve with regard to the height must now be chosen in such a way that the structural dimensions do not lie below the required minimum of 16 cm and that the buckling of the shell is avoided with a safety factor of 5. The justification of a sufficient buckling safety employs the buckling criterion recommended in [6]. The buckling criterion is based on the fact that shell buckling is primarily dependent on the local state of stress. This circumstance is used here and leads to the following programme sequence:

For a chosen centre surface of the shell the determination of the membrane forces is effected from top to bottom following the grid of the numerical integration steps. Due to the statical determination and the local buckling criterion the required wall thickness for the respective height can at once be definitely calculated with the most unfavourable point on the circumference being considered as relevant. The load cases to be examined and to be superposed are the dead load, the inner suction and the wind load. The wind load along the circumference is sufficiently described by about 11 Fourier elements.

In parallel to the calculations of membrane forces and stability, the concrete quantities and the shuttering surfaces are determined.

Reinforcement

With the membrane forces for each height section the necessary bar cross-sections are calculated on the basis of the relevant dimensioning formulas and the required reinforcement is fixed.

Empirical formulas are used to calculate the supplements for overlapping, S-hooks, etc., with the result that the simultaneous determination of the steel quantities is obtained.

Choice of the optimal shell geometry

The calculation procedure described above is performed in accordance with the explanations given in chapter 2 for the different combinations of the two shell parameters H_T and Y_U as may be required by the design engineer.

The complete calculation of a shell geometry including the determination of membrane forces, the dimensioning, the computation of quantities and the multiplication of the concrete and steel quantities with the height-depending unit costs, i.e. the cost estimation procedure, is performed by a Siemens computer with a core storage time of 1 to 2 seconds depending on the cooling tower height. This is a considerably short time!

Thus, the design programme presented enables the user to completely calculate - during the design phase - as many variants as required and to select the most favourable variant from the point of view of costs in a fully automatic procedure.

5.3 Column framework

In basically the same way the column framework for each variant can be statically determined, dimensioned and evaluated with regard to quantities and costs.

The starting values used for such purpose are, on the one hand, the membrane forces of the shell already mentioned and, on the other hand, additional data concerning the number and the geometry of the columns. Here, too, the possibility to provide for variations is granted, which means that the number of columns may be varied within an assumed range and that the most economical solution for a specific purpose can be chosen.

The computing time of 1 to 2 seconds mentioned above includes, of course, the complete calculation of the column framework.

5.4 Cost estimation

In each calculated variant the concrete and steel quantities are assessed for the shell and the column framework. After multiplication of these quantities with possibly height-depending unit costs the total costs for each variant are obtained.

6. Example

As a conclusion an optimization procedure will be demonstrated for the cooling tower represented in fig. 5. All important input data provided for by the thermal design engineers as well as the loads can be found there. As already mentioned on several occasions all the usual requirements of [6] are, of course, fulfilled.

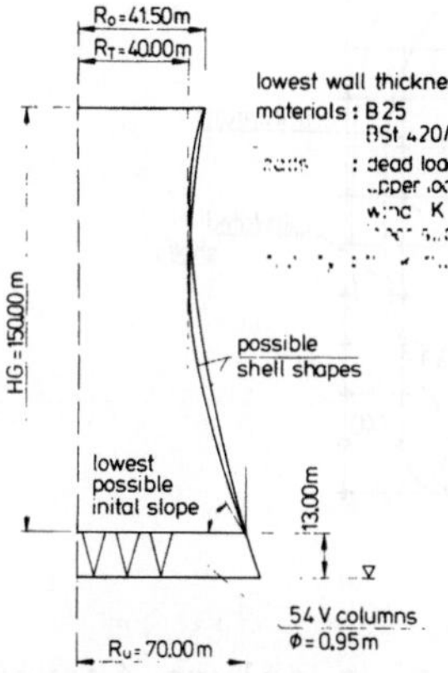

Fig.5. Preliminary values for the design of a shell (example)

The distance H_T of the waist from the upper edge of shell is varied from 20 m to 40 m in steps of 2 m and the shift value of the lower hyperbola Y_U from 15 m to 35 m in steps of 2.50 m. The estimation of the concrete costs for the shell and the columns is based on a factor of 0.002. For the steel quantities of the shell a corresponding value of 0.011 and for those of the columns a value of 0.020 has been chosen.

Impossible geometries and geometries hardly realizable with a climbing scaffold are recognized as such by the programme and rejected. Of the 11·9 = 99 combinations required due to the choice of H_T and Y_U 54 are abandoned with 45 cooling-tower shells remaining to be calculated.

The wind load has been described by 11 Fourier elements and a maximum step width of 5 m chosen for the numerical integration. Along the circumference the step width chosen corresponds to an angle of 3.33°. The evaluation stops after having reached an angle of 90° from the point of wind attack.

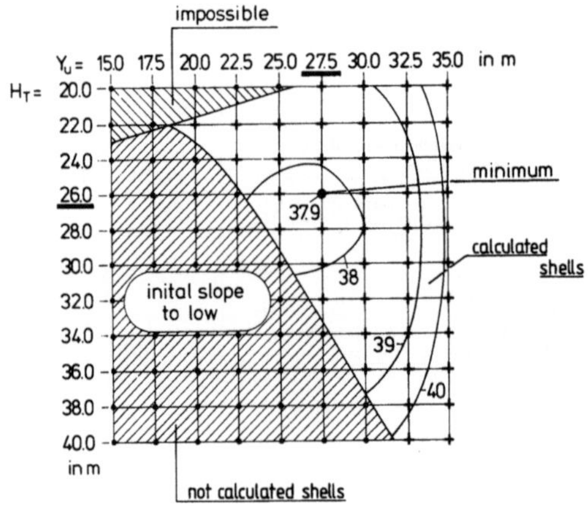

Fig.6. Cost analysis for shell plus column framework

Fig. 6 shows the results of the cost analysis. The most favourable variant is obtained for an H_T of 26.0 m and a Y_U of 27.5 m. The pertaining wall-thickness curve can be inferred from fig. 7.

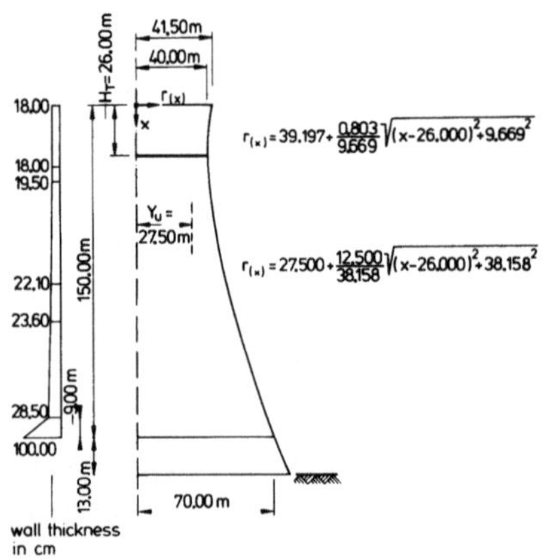

Fig.7. Optimal shell

7. Conclusions

The procedure presented in this contribution enables the user to determine in a fast and reliable manner the most economical shape and wall thickness of a natural-draft cooling tower. The calculation of any type of generating function is possible. In this contribution preference has been given to the calculation with hyperbolas and circles. Even shells having a positive Gaussian curvature can thus be calculated in accordance with membrane theory. The procedure described has been in practical use now for three years.

References

1. Krätzig W.:
Kühlturmformen besonderer Wirtschaftlichkeit in:
Konstruktiver Ingenieurbau/Berichte, Heft 1,
Vulkan-Verlag, Essen 1970

2. Harnach R.:
ANKE - Ein Programmsystem für den Automatischen Naturzugkühlturm-Entwurf. Techn. wiss. Mitteilungen des Inst. für Konstruktiven Ingenieurbau Ruhr-Universität Bochum, November 1977

3. Flügge W.:
Statik und Dynamik der Schalen.
Springer-Verlag Berlin, 1934

4. Rabich R.:
Statik der Platten, Scheiben, Schalen in:
Ingenieurtaschenbuch Bauwesen, Band I,
Teubner Verlagsgesellschaft Leipzig 1963

5. Form J. und Krings W.:
CAD-Anwendungen in einer Baufirma, insbesondere beim Bau von Kühltürmen
VDI-Berichte Nr. 492, 1983

6. Bautechnik bei Kühltürmen
VGB-Richtlinie, VGB-Kraftwerkstechnik GmbH,
Essen 1980

Current Developments in the Cooling Tower Analysis at Purdue University

T.Y. YANG
SUNIL SAIGAL
RAKESH K. KAPANIA

School of Aeronautics and Astronautics
Purdue University
West Lafayette, IN 47907 U.S.A.

Summary

This paper describes some current developments towards linear and nonlinear responses of column-supported cooling towers. Linear elastic seismic and wind response analyses have been performed. For seismic loads, deterministic methods are used and for wind loads, both deterministic and probabilistic methods are used. Probabilistic responses are obtained using both frequency- and time-domain analyses, based on quasi-steady aerodynamic theory and also ARIMA (Auto-Regressive Integrated Moving Averages) models. For a realistic wind load model, the results obtained using the frequency- and time-domain analyses are compared. The materially and geometrically nonlinear formulations and solution procedures have been developed for two types of finite elements; a 6 d.o.f. beam element and a 48 d.o.f. quadrilateral curved shell element. A series of examples are studied to evaluate the present developments. These developments can be applied to obtain nonlinear seismic responses of column-supported cooling towers. A brief description of the research being planned is also given.

Linear Elastic Analyses

After the collapse of the Ferrybridge cooling towers in England in 1965, a number of investigators have conducted studies to better understand the dynamic response behaviors of the cooling towers. Earlier studies were conducted for the cases of fixed bases [see, for example, Reference 1]. The presence of discrete column-supports, however, significantly alters the condition of the physical restraint at the base and thereby greatly influences the dynamic response. This makes an accurate modeling of the column supports of utmost importance.

A number of efforts were made to model the base columns. Gould, Sen, and Suryoutomo [2] and Basu and Gould [3] used an equivalent rotational-type shell element to model the columns. Gran and Yang [4,5] suggested an alternative approach by using quadrilateral flat plate elements to model the shell and discrete column elements to model the columns. The advantage is that the columns were modeled more accurately, but the shell is obviously modeled less efficiently as compared to that by using rotational-type shell elements [2, 3]. There appears to be a need for using more efficient shell elements by including proper curvatures. The modeling using quadrilateral shell elements and discrete column elements has the advantage and capability of obtaining detailed response results, for both linear and materially nonlinear cases, in the vulnerable region of shell-column joints.

The column element used in linear study is a sophisticated 3-D column element. The element contains 8 d.o.f.'s at each node, namely: three orthogonal displacements, u, v, and w in the η, ξ and ζ directions, respectively; two strain components $\partial u/\partial\eta$ and $\partial v/\partial\xi$; two rotations $\partial w/\partial\eta$ and $\partial w/\partial\xi$; and a rotation $\partial v/\partial\eta$ about the ζ-axis. The curvilinear coordinates η, ξ, and ζ are along the meridional, circumferential, and normal-to-surface directions, respectively, of the shell.

The response behavior of a cooling tower due to earthquake loads is, in general, predominantly membrane except in the vicinity of the base where bending behavior is also significant. This is vividly illustrated in Fig. 1 [4], which shows the variation of the meridional bending moment along the height (θ=0) for a cooling tower when subjected to the N-S component of the El-Centro earthquake. It would be advantageous to use different types of elements ranging from "general" to "membrane" to model the shell optimally.

A highly efficient high-order quadrilateral membrane element was developed by Gran and Yang [6]. The element shown in Fig. 2, is a doubly-curved quadrilateral shell of revolution element defined by lines of principal curvature. An array of 8 mapping nodes specifies a cubic variation of R and Z in the meridional direction. The element possesses 9 d.o.f.'s at each of the four vertices: u, $\partial u/\partial\xi$, $\partial u/\partial\eta$, $\partial^2 u/\partial\xi\partial\eta$; v, $\partial v/\partial\xi$, $\partial v/\partial\eta$; $\partial^2 v/\partial\xi\partial\eta$, and w. The performance of this element was evaluated by a series of example analyses [6].

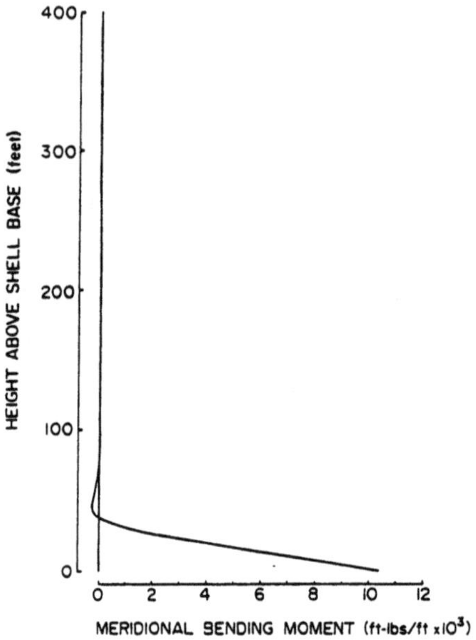

Fig. 1 Longitudinal distribution of meridional bending moment.

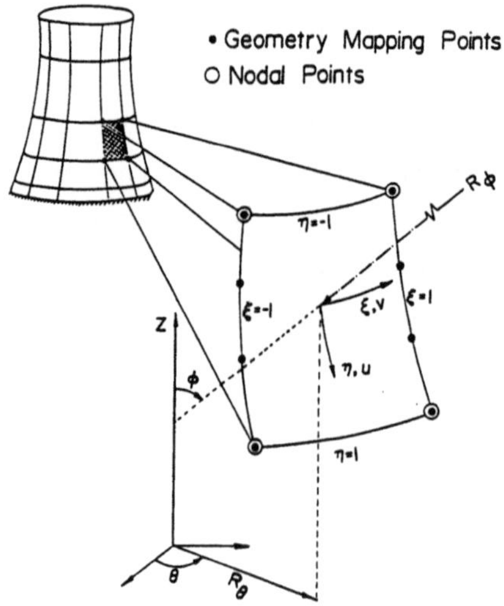

Fig. 2 Geometric representation of the "membrane" and "general" shell elements.

The membrane element was used in combination with a doubly-curved quadrilateral general shell element to model the cooling tower. The general element has the same geometry as the membrane element and possesses 12 d.o.f.'s at each of the four vertices: u, $\partial u/\partial\xi$, $\partial u/\partial\eta$, $\partial^2 u/\partial\xi\partial\eta$ and similar for v and w. The detailed formulation of the element along with the evaluative analyses is given elsewhere [7].

The modeling of the cooling towers using the above two types of elements requires the use of a transition element which is compatible with the general shell elements at the lower edge and with the pure membrane element at the upper edge [7]. This element has the same geometry as that of the membrane and the general shell elements.

To fill between a fine and a coarse layer of quadrilateral membrane elements, a triangular membrane filler element was developed [7]. The element possesses 7 d.o.f.'s at each node, namely:

w, u, $\partial u/\partial\xi$, $\partial u/\partial\eta$, v, $\partial v/\partial\xi$, and $\partial v/\partial\eta$. Two additional d.o.f.'s u and v are specified at the centroid and are partitioned out at the element level before assemblage.

Natural Frequencies and Seismic Responses

The above elements were used to predict the natural frequencies and the seismic response of an example cooling tower, previously analyzed in References 3 and 4. Both fixed and flexible cases were considered. The results for natural frequencies and maximum relative displacements were found to be in good agreement with the previously available results. Details of the finite element modeling and the various results are given elsewhere [7,8]. For the seismic response analysis, modal superposition method was used. As the eccentric modes are known to dominate the seismic response, only these modes were considered. To avoid the expensive task of computing the Duhamel's integral for time-history responses, maximum relative displacements were obtained using the response spectrum method [8].

Wind Responses

Both the static (due to mean wind velocity) and the dynamic (due to fluctuating wind) components of the wind responses were studied. The dynamic component was studied using both the deterministic and probabilistic methods. For deterministic responses, the results were obtained using both direct integration and modal-superposition methods. Dynamic and quasi-static responses of a cooling tower, alternative results for which are already available [9], were obtained. Figure 3 shows the load and also the time-history of the normal displacement at the throat at $\theta=0$. The present results are in good agreement with those given in Reference 9. It is noted that for the wind loads, higher (breathing) modes become important. For these modes, the shell no longer behaves primarily as a membrane and therefore the general shell element must be used throughout the shell height.

For the probabilistic responses, both frequency- and time-domain analyses were performed. A finite element formulation and Gaussian quadrature procedure were presented for studying the

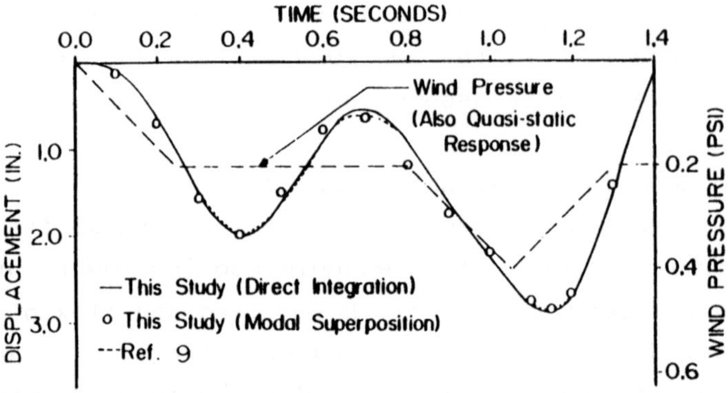

Fig. 3 Dynamic response of a cooling tower subjected to an artificial wind load.

stationary random response of shells. Details of the formulation and the evaluative examples are given elsewhere [10]. Results were also obtained using Monte-Carlo simulation techniques, based on quasi-steady aerodynamics and on recently developed ARIMA time-series models [11]. Five different wind simulation models, all with different correlations assumptions in the meridional and circumferential directions, were tested [12]. For the most realistic model, results were obtained using both frequency- and time-domain analyses. Figure 4 shows the circumferential variation of mean and RMS values of the meridional stress resultant at the base of an example cooling tower for a wind velocity of 102 fps and a pressure variance of 12.1 psf^2 at the windward meridian. The variation of the variance along the circumference was that given by Ruscheweyh [13].

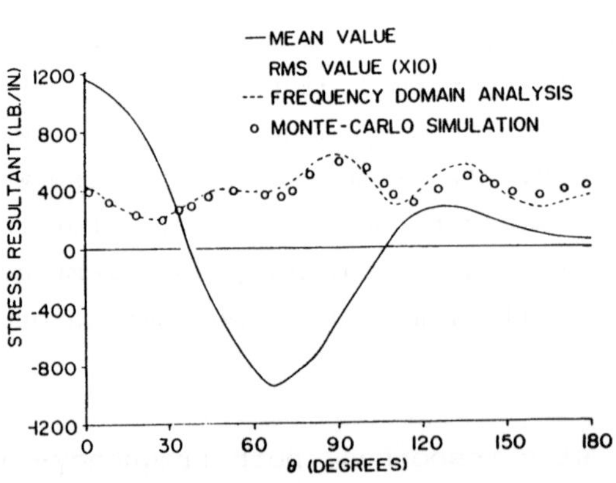

Fig. 4 Circumferential variation of mean and RMS values of meridional stress resultant at shell base.

The RMS values in Fig. 4, as obtained by the two methods, are in good agreement with each other.

Materially and Geometrically Nonlinear Analyses.

Past experience has shown that the base columns and the adjoining shell region of a column-supported cooling tower may yield under the action of severe earthquake loads [5]. In such situations, an elastic-inelastic analysis is desirable. To perform such an analysis, a 6 d.o.f. nonlinear column element and a 48 d.o.f. nonlinear doubly-curved quadrilateral shell element were developed. The details of the formulation for the two elements are given in References 14 and 15, respectively. For reinforced concrete columns, the material nonlinearity was considered by assuming the concrete as a nonlinearly elastic-strain hardening material in compression and a linearly elastic-brittle fracture material in tension. The spread of plastic zones for both the column and the shell elements were allowed using the concept of layered element model.

A step-by-step loading combined with Newton-Raphson iteration procedure is employed for tracing the nonlinear behavior. For dynamic analyses, the numerical integration schemes were employed to solve the incremental linearized equations of motion. Wilson-θ and Newmark's methods were interchangeably used. The structure was loaded in incremental steps (of load or time) and iterations were performed at every or desired step to guarantee equilibrium [16].

Evaluative Analyses

The formulation for the nonlinear beam and shell analyses were evaluated for a series of examples [14,15,16]. Two examples, pertinent to the static and dynamic responses of shells, respectively, are presented here.

(i) Cylindrical Roof Subjected to Uniform Gravity Loading

The cylindrical panel under consideration and its finite element modeling are shown in Figure 5. This problem was previously analyzed by Parisch [17]. The present results for gravitational load intensity versus the vertical deflection at midpoint A of free edge is plotted in Figure 5 along with the previous results [17]. A good correlation among the three sets of results for load deflection behavior is seen.

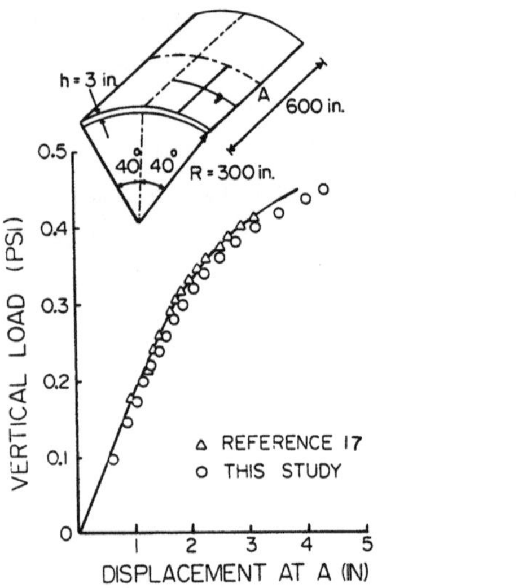

Fig. 5 Large displacement of an elastic-plastic cylindrical panel.

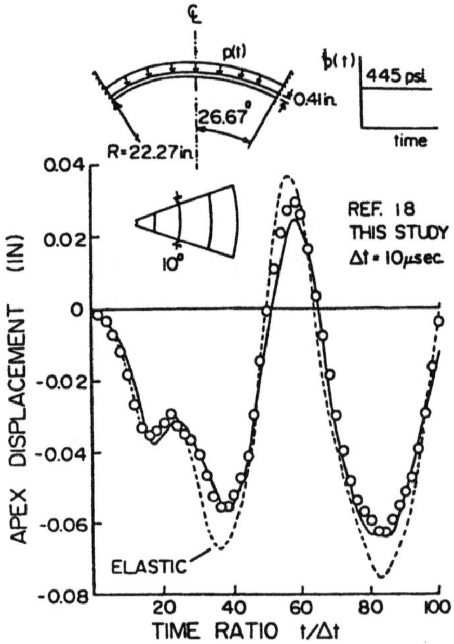

Fig. 6 Large displacement, elastic-plastic dynamic response of a spherical cap.

(ii) Clamped Spherical Cap Under Uniformly Distributed Step Load

The spherical cap under consideration and its finite element modeling are shown in Figure 6. The material was assumed to be elastic-linear strain hardening obeying Von Mises yield criterion. Due to the axisymmetric nature of the problem, a 10° segment of the cap was modeled using four elements. This problem was previously analyzed by Nagarajan and Popov [18]. The response time-history of the normal displacement at the apex of the cap, using Newmark's method with $\Delta t = 10 \mu sec.$, is plotted in Figure 6 along with the previous results [18]. A good correlation among the two sets of results is seen.

Elasto-viscoplastic Analysis

In recent years, elastic-visco-plastic formulations have been used to predict elastic-plastic behavior of structures [see, e.g.,

Reference 19]. In this quasi-static approach, stress states beyond the yield surface are allowed to develop. Then, through postulation of strain rate laws, steady states corresponding to stress states reaching the yield surface (within a tolerance) are obtained. Elasto-viscoplastic analyses were performed for a series of examples for which elastic-plastic results were previously obtained [15]. An illustrative example is given here.

Cylindrical Shell Roof Under Uniform Gravity Loading

The cylindrical panel, shown in Fig. 5, was reanalyzed using elastic-viscoplastic approach. Only material nonlinearity was considered. This problem was previously analyzed by Cormeau [19]. The present results for gravitational load intensity versus the vertical deflection at midpoint A of the free edge are plotted in Fig. 7 along with the previous results [19]. A good correlation among the three sets of results is seen. It is noted that the present results were obtained using an initial time step of 0.1 hour. The gravity load was applied in seven increments. Both instantaneous and steady-state displacements are shown in Fig. 7.

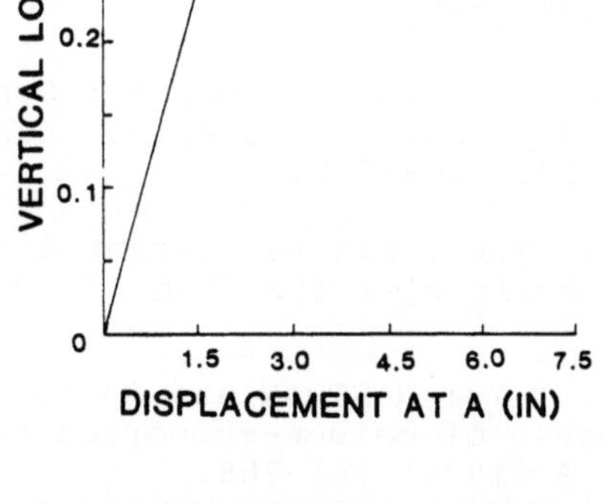

Fig. 7 Small displacement elastic-viscoplastic analysis of cylindrical shell roof.

Areas of Current Research

Two aspects being active pursued are: (i) realistic material modeling and (ii) the effects of geometric imperfections.

The material model for the concrete includes the effects of progressive cracking of concrete in tension, nonlinear response under multiaxial state of stress, and the post-fracture behavior. The concrete is considered as elastic, plastic, strain hardening, and fracture material in both compression and in tension. Stress release across a crack under a tensile type of fracture and limited plastic flow under a complete stress release after crushing under a compressive type of fracture will be implemented.

The effect of the imperfections on the seismic and wind responses are also being studied. To accurately model the imperfections, variable order polynomials and B-spline surfaces are being used. To accurately represent the rigid body modes, the 48 d.o.f. general shell element has been reformulated using the Cartesian displacement components and their derivatives as the element d.o.f.'s. To broaden the applications of this element to arbitrary shells, the strain displacement and the constitutive relations as expressed in the tensorial form have been used. To study the behavior of shells beyond critical points, modified Riks-Wempner method is being used. Efforts will also be made to study the large amplitude vibrations of cooling towers.

References

1. Carter, R.L.; Robinson, A.R.; Schnobrich, W.C.: Free vibration of hyperboloidal shells of revolution. J. Engng. Mech. Div. 95 (1969) 1033-1052.

2. Gould, P.L.; Sen, S.K.; Suryoutomo, H.: Dynamic analysis of column-supported hyperboloidal shells. Earth. Engng. and Struct. Dyn. 2 (1974) 269-279.

3. Basu, P.K.; Gould, P.L.: Finite element discretization of open-type axisymmetric elements. Int. J. Num. Meth. Engng. 14 (1979) 159-178.

4. Gran, C.S.; Yang, T.Y.: NASTRAN and SAPIV applications on the seismic response of column-supported cooling towers. Comp. and Strut. 8 (1978) 761-768.

5. Gran, C.S.; Yang, T.Y.: Refined analysis of seismic response of column supported cooling towers. Comp. Struc. 11 (1980) 225-231.

6. Gran, C.S.; Yang, T.Y.: Doubly curved membrane shell finite element. J. Engng. Mech. Div. 105 (1979) 567-584.

7. Yang, T.Y.; Kapania, R.K.: Shell elements for cooling tower analysis. J. Engng. Mech. 109 (1983) 1270-1289.

8. Yang, T.Y.; Kapania, R.K.; Saigal, S.: Linear and nonlinear dynamic response analysis of complex shell structures. Eighth World Conf. on Earth. Engng. San Francisco, Calif. July 21-28, 1984.

9. Suryoutomo, H.: Gould, P.L.; Basu, P.K.: Direct dynamic analysis of shells of revolution using high-precision finite elements. Comp. Strut. 7 (1977) 425-433.

10. Yang, T.Y.; Kapania, R.K.: Finite element random response analysis of cooling tower. J. Engng. Mech. 110 (1984).

11. Reed, D.A.; Scanlan, R.H.: Time series analysis of cooling tower wind loading. J. Strut. Engng. 109 (1983) 538-554.

12. Yang, T.Y.; Kapania, R.K.: Wind response of cooling tower using Monte-Carlo simulation. ASCE Specialty Conf. on Probl. Mech. and Strut. Reliability, Jan. 11-13 (1984) 240-243.

13. Ruscheweyh, H.: Wind loading on hyperbolic natural draught cooling towers. J. Wind Engng. Ind. Aero. 1 (1975/1976) 335-340.

14. Yang, T.Y.; Saigal, S.: A simple element for static and dynamic response of beams with material and geometric nonlinearities. Int. J. Num. Meth. in Engng. (to appear).

15. Yang, T.Y.; Saigal, S.: A curved quadrilateral element for static analysis of shells with geometric and material nonlinearities. Int. J. Num. Meth. in Engng. (to appear).

16. Yang, T.Y.; Saigal, S.: Materially nonlinear, static, and dynamic response of shell structures of revolution using quadrilateral elements. Second Int. Conf. on Num. Meth. for Nonl. Probs., Barcelona, Spain, April 9-13 (1984).

17. Parisch, M.: Large displacements of shells including material nonlinearities. Comput. Meth. Appl. Mech. Engng., 27, (1981) 181-214.

18. Nagarajan, S.; Popov, E.P.: Elastic-plastic dynamic analysis of axisymmetric solids. Comp. and Strut. 4 (1974) 1117-1134.

19. Cormeau, I.: Elastoplastic thick shell analysis by viscoplastic solid finite elements. Int. J. Num. Meth. Engng. 12 (1978) 203-227.

An Evaluation Method of Load Combination of Windstorm and Low Atmospheric Temperature on Cooling Towers

HUANG WEIMOU

Department of Science and Technology of Ministry of Water Resources and Electric Power, P.R.C.

Summary

The maximum windstorm will never occur with the lowest atmospheric temperature at the same time. Assumming that both the windstorm and the low atmospheric temperature which occur simultaneously can be treated as unrelated normal distributed random variables, a combined probability of both the windstorm and low atmospheric temperature has been obtained. Through the comparison of Beijing in China, it was shown that the present method may be used to describe the pattern of combined appearance of windstorm and low atmospheric temperature with a good coincidence with the historical records. It can be used after the tower's service life and the damage ratio are given.

Introduction

For many years researches have been done in China to develop better analytical methods for cooling tower applications, an evaluation method of load combination of windstorm and low atmospheric temperature has also been studied.

Both the windstorm and low atmospheric temperature are phenomena of nature, they occur at random in time, space and intensity but rules can be found after a certain period. The occurrence of these two phenomena is closely relative to many factors such as the radiation of the sun, the earth rotation and revolution and the different conditions of seas and continents etc. If all those factors unchange every year, then the occurrence of windstorm and low atmospheric temperature may be considered as repeated exams. It is possible to find out their eigen values and probability of occurrence by applied probability theory.

Sample Selections

The windstorm occurs in a certain district may be defined as cold current, typhoon, thunderstorm, hurricane, hailstorm etc, and all of them are relative to the behaviors of different air masses. There is a transition

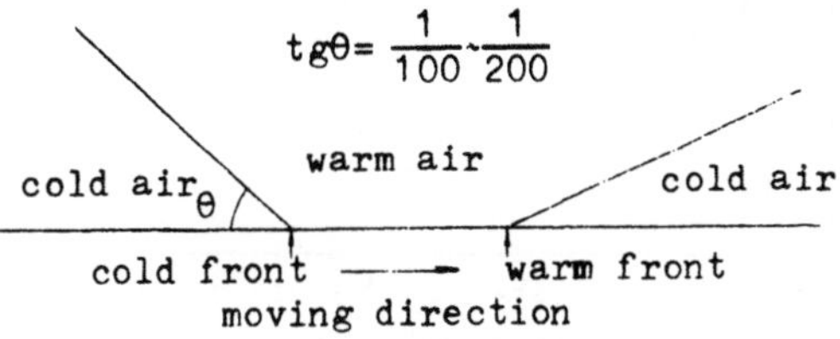

Fig.1. Sketch of warm and cold front (vertical profile)

band called the front surface existed between two air masses. A strong wind happens while the front moves across the district. In that case the warm and cold air of two different air masses mix seriously not only in horizontal but also in vertical directions, the air temperature of that district is higher than that of the cold air mass.

In the cold winter, when there is no cloud or only a few clouds which can prevent the earth from dispersing the heat to the sky, low atmospheric temperature will occur, but the lowest will only occur at no wind weather while the warm current at high level couldn't come down to mix with the cold air near the ground. Therefore, a certain mistake will surely occur if we use the maximum windstorm and the lowest air temperature together in a calculation. The only right way is to use the values recorded at the same time. Hence three situations should be considered individually as following:

1. the annual maximum windstorm and its simultaneous air temperature,
2. The annual maximum windstorm in winter (Dec. Jan. Feb. three months) and its simultaneous air temperature,
3. The annual lowest air temperature and its simultaneous wind velocity.

Calculation Method

Consider the wind velocity and its simultaneous air temperature as two random variables V and T, and make a plane right angle coordinate with t and v as the vertical and horizontal axeses (Fig.2). The probability of the random event (V,T) which will fall into the region R will be as following:

$$P(V,T\subset R)=\iint_R f(v,t)dvdt \tag{1}$$

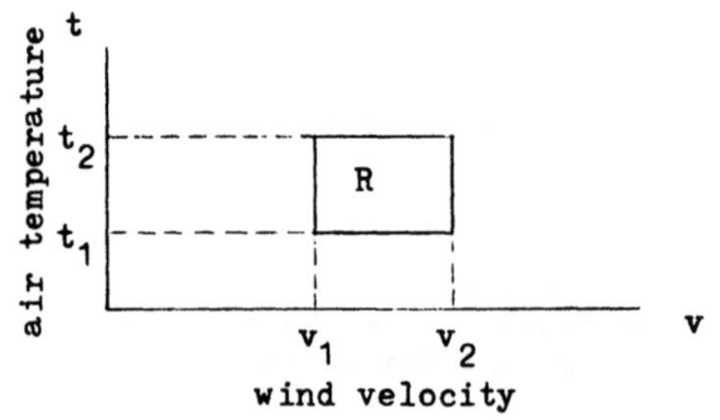

Fig.2. Random event (V,T) falls into the region R

Where f(v,t) is the density of probability. While V and T are unrelative with each other, we can get:

$$P(V,T \subset R)=\int_{v_1}^{v_2} f(v)dv \cdot \int_{t_1}^{t_2} f(t)dt \tag{2}$$

The relation between V and T is still undertermined or unknown nowadays, but after we made studies of some districts in China, we are sure that the relative coefficient of these two random varibles may be treated as zero and then the formula (2) may be used in practice. Since we are interested only in the event $(V>v,\ T<t)$, we get:

$$P(V>v,\ T<t)= P(V>v)\cdot P(T<t) \tag{3}$$

Where $P(V>v,\ T<t)$, $P(V>v)$, $P(T<t)$ are being called the functions of distribution.

Function of Wind's Distribution

As the condition will only have a little change every year, a normal distribution for wind should be acceptable in practice and because of windstorm would never be of minus, some distributions have been recommended. The most useful one is the critical distribution. In this case, the function of wind's distribution is:

$$\left.\begin{aligned} &P(V\leqslant v)= \exp(-\exp(-y)) \\ &y= a(v-u) \end{aligned}\right\} \tag{4}$$

Where a is the disperse factor of maximum values, u is the mode of the function of probability density. Both parameters can be determined by least

-square method [1], Gumbel gave:

$$\left.\begin{aligned} a &= \frac{\pi}{\sqrt{6}}\cdot\frac{1}{\sigma_v} \\ u &= \bar{v} - 0{,}45005\,\sigma_v \end{aligned}\right\} \qquad (5)$$

Substitute this values into (4), we get the value v of $P(V>v)$ as:

$$v = \bar{v}(F\cdot C_v + 1) \qquad (6)$$

Where

$$F = -\frac{\sqrt{6}}{\pi}(0{,}57722 + \ln(-\ln(P(V>v)))) \qquad (7)$$

$$C_v = \sqrt{\frac{\sum_{i=1}^{n}(v_i - \bar{v})^2}{n-1}}\cdot\frac{1}{\bar{v}} = \frac{\sigma_v}{\bar{v}} \qquad (8)$$

Where $\bar{v}$ is the mean, σ_v is the r.m.s. Use $v=\bar{v} + n\sigma_v$, and $\sigma_v = C_v\cdot\bar{v}$, we can find out n=F, and get the probability expressed in terms of $\bar{v} + n\sigma_v$. The relation between the normal and the critical is shown in Fig.3.

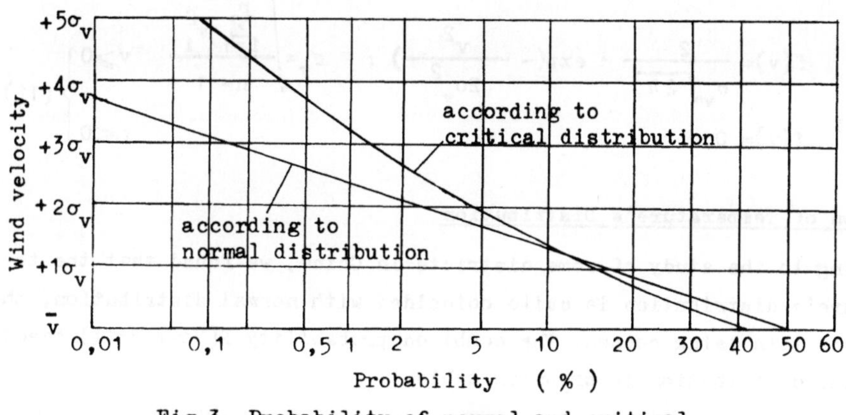

Fig.3. Probability of normal and critical

The difference of these two probabilities is considerable. When $P(V>v)=$ 0,02, the normal is $v= \bar{v}+2{,}05\sigma_v$, the critical is $v= \bar{v}+2{,}58\sigma_v$, when $P(V>v)= 0{,}001$, the normal one is $\bar{v}+3{,}12\sigma_v$, the critical one is $\bar{v}+4{,}94\sigma_v$, the differences are $0{,}53\sigma_v$ and $1{,}82\sigma_v$.

Since we have never got a windstorm which could have a greatest ten minute average wind velocity more than $\bar{v}+3\sigma_v$, for example, in the north-eastern part of China, so called Chifeng in Liaoning province, the maximum ten minute average within 1951 to 1981 is 33,3m/s, the mean is 18,5m/s, σ_v=5,66m/s. If we had to consider a wind velocity of $\bar{v}+4{,}94\sigma_v$=46,46m/s, we would have to require a two times kinetic energy than that of the 33,3 m/s to produce such a windstorm. Therefore, we should have to require a one half shorter distance of the original air masses or a more than two times thermal energy received from the sun. Since both these two cases look so impossible to achieve and a 0,001 probability is not a too harsh requirement for engineering, the normal distribution will be a suitable selection. (For another time intervals, the problem will remain the same.) A function of density is then being considered:

$$\left.\begin{aligned} f(v) &= \frac{1}{\int_0^{\infty} \exp\left(-\frac{(v-\bar{v})^2}{2\sigma_v^2}\right) dv} \cdot \exp\left(-\frac{(v-\bar{v})^2}{2\sigma_v^2}\right) && v \geqslant 0 \\ f(v) &= 0 && v < 0 \end{aligned}\right\} \quad (9)$$

And $$P(V>v) = \int_v^{\infty} f(v)\,dv \quad (10)$$

When $\bar{v} > (3 \sim 4)\sigma_v$, the coefficient of (9) may remain as $\frac{1}{\sigma_v \sqrt{2\pi}}$. As for the lowest atmospheric temperature condition, we use:

$$\left.\begin{aligned} f(v) &= \frac{2}{\sigma_v \sqrt{2\pi}} \cdot \exp\left(-\frac{v^2}{2\sigma_v^2}\right) ; \quad \sigma_v = \sqrt{\frac{\sum_{i=1}^{n} v_i^2}{n-1}} && v \geqslant 0 \\ f(v) &= 0 && v < 0 \end{aligned}\right\} (11)$$

<u>Function of Temperature's Distribution</u>

According to the study of some districts in China, we found that the temperature's distribution is quite coincided with normal distribution, the normal is then being chosen. The combined probability $P(V>v, T<t)$ based on normal distribution is shown in table 1.

<u>Damage Ratio</u>

Defined that both v and t are the critical load acted on cooling towers, the annual probability is $p = P(V>v, T<t)$, the service life of tower is N years, we can have a guarantee ratio W as:

$$W = (1-p)^N \quad (12)$$

And $$P = 1 - W = 1 - (1-p)^N \quad (13)$$

May be called the damage ratio. And p can be calculated as:

Table 1 Combined Probability $P(V>v, T<t)$

t \ v	$\bar{v}-3\sigma_v$	$\bar{v}-2\sigma_v$	$\bar{v}-\sigma_v$	$\bar{v}$	$\bar{v}+\sigma_v$	$\bar{v}+2\sigma_v$	$\bar{v}+3\sigma_v$
$\bar{t}+3\sigma_t$	0,99730	0,97593	0,84021	0,49933	0,15844	0,02272	0,00135
$\bar{t}+2\sigma_t$	0,97593	0,95502	0,82220	0,48863	0,15505	0,02223	0,00132
$\bar{t}+\sigma_t$	0,84021	0,82220	0,70786	0,42067	0,13348	0,01914	0,00114
$\bar{t}$	0,49933	0,48863	0,42067	0,25000	0,07933	0,01138	0,00068
$\bar{t}-\sigma_t$	0,15844	0,15505	0,13348	0,07933	0,02517	0,00361	0,00021
$\bar{t}-2\sigma_t$	0,02272	0,02223	0,01914	0,01138	0,00361	0,00052	0,00003
$\bar{t}-3\sigma_t$	0,00135	0,00132	0,00114	0,00068	0,00021	0,00003	0,00000

Table 2 The comparison of Beijing in winter max. wind condition

t (°C) \ v (m/s)		$\bar{v}-3\sigma_v$	$\bar{v}-2\sigma_v$	$\bar{v}-\sigma_v$	$\bar{v}$	$\bar{v}+\sigma_v$	$\bar{v}+2\sigma_v$	$\bar{v}+3\sigma_v$
		7,57	10,52	13,48	6,44	19,40	22,36	25,31
$\bar{t}+3\sigma_t$	5,06	* 21,94 22	21,47 22	18,48 19	10,99 11	3,49 4	0,50 1	0,03 0
$\bar{t}+2\sigma_t$	1,33	21,47 21	21,01 21	18,09 19	10,74 11	3,41 4	0,49 1	0,03 0
$\bar{t}+\sigma_t$	-2,41	18,48 19	18,09 18	15,57 16	9,25 9	2,94 3	0,42 1	0,03 0
$\bar{t}$	-6,14	10,99 11	10,74 11	9,25 9	5,50 5	1,75 1	0,25 1	0,02 0
$\bar{t}-\sigma_t$	-9,88	3,49 4	3,41 4	2,94 4	1,75 2	0,55 0	0,08 0	0,01 0
$\bar{t}-2\sigma_t$	-13,61	0,50 0	0,49 0	0,42 0	0,25 0	0,08 0	0,01 0	0 0
$\bar{t}-3\sigma_t$	-17,35	0,03 0	0,03 0	0,03 0	0,02 0	0,01 0	0 0	0 0

Note.*The values shown above are the frequency of occurrence calculated =probability × number of recorded years. The values shown beneath are the actual number of times of recorded years.

**The number of recorded years used in calculation is 22 years, from 1954 to 1975. Plotted in Fig.4. σ_v=2,96m/s, $\bar{v}$=16,44m/s, σ_t=3,74°c, $\bar{t}$= -6,14°c.

***There are also the winter lowest temperature condition plotted in Fig.4. From 1954 to 1972. σ_v=1,15m/s, $\bar{v}$=0m/s, σ_t=1,47°c, $\bar{t}$=-15,16°c.

$$p = 1 - W^{\frac{1}{N}} \tag{14}$$

Example

An example of Beijing made for winter situations is shown in table 2 and Fig.4. A good coincidence between the calculated frequency of occurrence and the actual number of times has been found. When the velocity of windstorm is of a higher value, the probability decreases rapidly while the velocity increases slowly, and so is the lower air temperature. This characteristic shows a very high wind velocity or a very low air temperature would hardly occur. An isogram of probability is given in Fig. 4.

Conclusions and Prospects

The probability of load combination of windstorm and low atmospheric temperature can be obtained through two unrelated normal distributed probabilities, which are of windstorm and low atmospheric temperature individually, times together, but the samples of these two factors should be simultaneous. After the service life and the allowable damage ratio are given, the required annual probability can be obtained, the isogram would be of help to find out the controlled combination in design. A further research to simplify the load combination method is beneficial.

In this paper, we have used the normal distribution to approach the wind's distribution, but this approach will only be suitable for samples in which the number of negative maxima could have a value of one half or at least 0,04 of the number of the whole maxima in a zero-mean stationary Gaussian process. [1]. If there are only positive maxima then we should choose the Rayleigh distribution as the most suitable selection in practice [2].

References

1. Clough, R.W.; Penzien, J.: Dynamics of structures. 1975.
2. Li Guohao: Dynamics of engineering structures in earthquake. Shanghai Science Press 1980.
3. Chen Kai; Wang Yuxiao: The applied theory of probability and statistics. 1982.
4. П.Н.ТВЕРСКОЙ et al.: Meteorology. 1951.

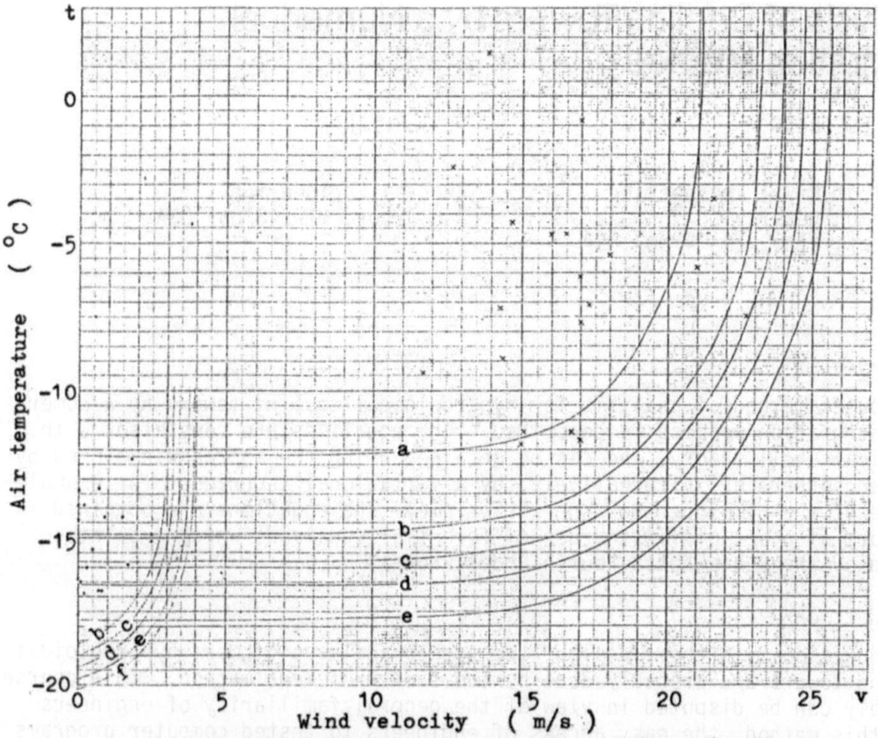

Fig.4. An example of Beijing made for winter situations
× Annual maximum wind velocity condition
. Annual lowest air temperature condition
a-p=0,05, b-p=0,01, c-p=0,005, d-p=0,0025, e-p=0,001
f-p=0,0005

Analysis of Cooling Towers by a Numerical Integration Method

J. E. GOLDBERG

School of Civil Engineering
Purdue University
West Lafayette, IN 47907 USA

D. V. PATHAK

Bechtel Engineering
Ann Arbor, MI 48106 USA

Summary

A computer method of analysis for hyperboloidal cooling towers by a numerical integration process is described. The equations are consistent with classical shell theory, and can be obtained using the approach employed by Love and others or, alternatively, by a variational technique, for example by use of the Hellinger-Reissner Principle. The equations are obtained in the form of a set of first-order differential equations which are integrated by the Runge-Kutta Fourth-Order Process.

Introduction

In connection with their design, the stress analyses of most hyperboloidal cooling towers are probably done by the finite element method. This course scarcely can be disputed in view of the general familiarity of engineers with this method, the easy access of engineers to tested computer programs based upon this method, and the generally good results obtained through the use of these programs with reasonable effort on the part of the engineer.

In addition to the finite element method, other methods for handling cooling tower problems as well as problems of other shell configurations exist or may be constructed in a straightforward manner. Despite the popularity and wide use of the finite element method, it may in fact be advisable to give consideration to the use of these other methods in connection with the analysis and design of hyperboloidal cooling towers and other shell-type structures. These other methods might be considered for general use as substitutes for or alternatives to the finite element method or, at least, in supportive applications. In the supportive applications, these other methods might be used either regularly to check the output of the finite element program -- and thus the condition or state of the computer code -- or occasionally for this same purpose, including situations where the structure and its response are particularly significant and important because of the cost of the structure or the consequences of poor performance. The latter reason may also justify consideration of these other methods as substitutes or alternates to the finite element methods. It is, in fact, entirely possible that a careful comparison which takes into account total cost, accuracy and reliability would provide sufficient justification for using another method as an alternative to the finite element method at least in certain special circumstances. The method which is often called the Numerical Integration Method is suggested as a viable candidate for consideration as an alternative or, at least, in a supportive role to the more commonly used methods. This method, which was devised during the early and middle fifties, has been applied with excellent results to a number of problems of shells of revolution as well as to a number of other shell and plate problems [1,-,15].

In this paper the application of the Numerical Integration Method is described. The pertinent set of governing differential equations may be developed through the application of a variational principle. Other standard approaches for obtaining the desired equations, however, are also available. The equations are obtained in a first-order form in which the dependent variables are taken to be the complete set of tractions and displacements. The equations include the effects of pre-existing stresses due to gravity loading or other axisymmetric or mildly asymmetric state. A numerical solution to this set of equations is obtained readily by means of any of several well-known numerical algorithms. A technique is described whereby the computer effort in obtaining the numerical solution to the set of equations can be reduced by almost fifty percent in comparison with the usual procedures.

The equations and numerical technique are intended to handle asymmetrical as well as axisymmetrical problems of hyperboloidal cooling towers. Among these problems are:

- Modes and frequencies of free vibration
- Response to steady and gusty winds
- Response to blast
- Response to earthquakes
- Stress and deformation due to axisymmetric and asymmetric temperature fields and shrinkage
- Moderate nonlinearities and imperfections
- Circumferentially stiffened shells

The method has proven to be reliable and accurate. It has proven also to be efficient and economical, not only in terms of computer time but also in terms of the engineer's time and effort. When using this method, the procedure takes on an almost automatic character. In view of the fact that it avoids the task of establishing an element mesh as required by the finite element method, this method not only saves much of the engineer's time but also avoids the compromises and sometimes arbitrary judgments involved in establishing the element mesh with an acceptable combination of element size and accuracy.

Theory and Derivation of Equations

The desired equations can be obtained in the conventional way, that is by formulating the equilibrium equations and the stress-displacement relations in the appropriate coordinate system such as done by Love [16]. Alternatively, the equations can be obtained through the use of a variational principle, for example the Hellinger-Reissner theorem [16,17,18].

If the conventional approach is used, equilibrium of a differential element is considered, with the forces and couples acting at the middle surface being taken in the usual way. Lines α and β are two orthogonal curvilinear coordinate lines on the middle surface, α being along a meridian, and x and y are the two tangents to these coordinate lines. A third direction z is chosen such that x, y and z form a right-hand triad. The in-plane forces are denoted by N_1^* , N_{12}^* , N_2^* and N_{21}^* where N_{12}^* and N_{21}^* are membrane shears. The transverse shears are denoted by Q_1^* and Q_2^*. Similarly moments and twisting couples are denoted by M_1^* , M_2^* and M_{12}^* and M_{21}^*. The three components of the applied forces in the local coordinate directions are denoted by X, Y and Z and the components of the applied couples are m_1 and m_2 . The equations of equilibrium for a general shell of revolution are given by Love [16].

The hyperbolic cooling tower is essentially a shell of revolution. The meridian and the parallel circle are chosen as two coordinates which are denoted by ϕ and θ, respectively. The two principal radii of curvature are denoted by R_1 and R_2 . The radius of the parallel circle is r . Thus, for the shell of revolution

$$\alpha = \phi \ ; \quad \beta = \theta \ ; \quad A = R_1 \text{ and } \quad B = r \tag{1}$$

where A and B are Lamé parameters; and from the Gauss-Codazzi relations

$$\frac{\partial A}{\partial \beta} = 0 \quad ; \qquad \frac{\partial B}{\partial \alpha} = R_1 \cos \phi \tag{2}$$

The final state of stress in the cooling tower is assumed to be sum of a symmetric state of stress (e.g., due to the dead load) and the general state of stress (e.g., superimposed stresses due to the wind). Thus the forces and moments can be written as

$$N_1^* = \bar{N}_1 + N_1 \tag{3}$$

with similar expressions for the remaining stress resultants and stress-couples where, $\bar{N}_1$, etc. represent the pre-stress stage and N_1 , etc. represent the additional internal tractions set up during the additional deformations.

Substitution of relations (1),(2) and (3) into the equilibrium equations of Love yields a set of nonlinear equations of equilibrium. These nonlinear equations are linearized by the following assumptions:

a. The nonlinear terms which are products of superimposed tractions such as N_1 and the deformations are small and can be neglected.
b. The pre-stress state of stress is axisymmetric, i.e.,

$$\bar{N}_{12} = \bar{M}_{12} = \bar{Q}_2 = \bar{M}_{21} = \bar{M}_{21} = \bar{N}_{21} \equiv 0 \tag{4}$$

c. The pre-stress deformations are small and do not alter the geometry of the shell significantly. Hence the original configuration of the shell is used to calculate geometric properties of the middle surface.

These linearizing moves are made here partly for simplifying the description of the method, and it may be pointed out that the linearizing simplifications need not be as drastic as indicated above. The membrane shear N_{12} and transverse shear Q_1 in the equations are replaced by the corresponding boundary tractions as follows:

$$N_{12}^{O} = N_{12} + \frac{M_{21}}{R_2} \quad ; \quad N_{21}^{O} = N_{21} + \frac{M_{12}}{R_1} \tag{5}$$

$$V_1 = Q_1 - \frac{\partial M_{21}}{r\partial\theta} \quad ; \quad V_2 = Q_2 + \frac{\partial M_{12}}{R_1\partial\phi}$$

The strain-displacement relations are obtained on the basis of small deformation theory and the stress-strain relations are those developed by Novozhilov. Proceeding, first-order partial differential and supporting equations which follow are obtained.

$$s = R_1 d\phi$$

$$r = R_2 \cos\phi$$

$$\varepsilon_2 = \frac{\partial v}{r\partial\theta} + \frac{u\cot\phi}{R_2} - \frac{w}{R_2}$$

$$\frac{\partial u}{\partial s} = \frac{w}{R_1} + \frac{N_1}{B} - \nu\varepsilon_2$$

$$N_2 = \nu N_1 + Eh\varepsilon_2$$

$$\frac{\partial w}{\partial s} = \beta_1 - \frac{u}{R_1}$$

$$K_2 = \frac{\partial^2 w}{r^2\partial\theta^2} + \frac{\partial v}{R_2 r\partial\theta} + \frac{\cot\phi}{R_2}\beta_1 \tag{6a}$$

$$M_2 = \nu M_1 - \frac{Eh^3}{12} K_2$$

$$\frac{\partial\beta_1}{\partial s} = -\frac{M}{D} - \nu K_2$$

$$\frac{\partial v}{\partial s} = \frac{N_{12}^{O}}{C^*} - \frac{G}{C^*}\left[\frac{\partial u}{r\partial\theta} - \frac{v\cot\phi}{R_2}\right]$$

$$- \frac{2F}{C^* R_2}\left[\frac{\partial}{r\partial\theta}\left(\beta_1 - \frac{u}{R_1}\right) + \frac{\partial}{\partial s}\left(\frac{1}{r}\right)\frac{\partial w}{\partial\theta} + v\frac{\partial}{\partial s}\left(\frac{1}{R_2}\right)\right]$$

$$H = F\left[\frac{\partial}{r\partial\theta}(\beta_1 - \frac{u}{R_1}) + \frac{\partial}{\partial s}(\frac{1}{r})\frac{\partial w}{\partial\theta} + \frac{1}{R_2}\frac{\partial v}{\partial s} + v\frac{\partial}{\partial s}(\frac{1}{R_2}) - \frac{1}{R_1}\frac{\partial v}{\partial s}\right]$$

$$\frac{\partial(N_1 r)}{\partial s} = -\frac{\partial(N^o_{12} r)}{r\partial\theta} + \frac{2}{R_2}\frac{\partial H}{\partial\theta} + \frac{V_1 r}{R_1} + N_2\cos\phi$$

$$+\bar{N}_2\left[\frac{\partial}{\partial\theta}(\frac{\partial v}{\partial s}) - \sin\phi\,\beta_1\right]$$

$$+\bar{Q}_1 r\frac{\partial\beta_1}{\partial s} - Xr \qquad (6b)$$

$$V_2 = -\frac{2H\cos\phi}{r} + \frac{\partial M_2}{r\partial\theta} + \bar{M}_1\left[\frac{\partial}{\partial s}(\frac{\partial v}{\partial s}) + \frac{1}{R_1}(\frac{\partial w}{r\partial\theta} + \frac{v}{R_2})\right] - Zr$$

$$\frac{\partial(N^o_{12} r)}{\partial s} = V_2\sin\phi - N^o_{12}\cos\phi + \frac{2H\cos\phi}{R_2} - \frac{\partial(N_2 r)}{r\partial\theta}$$

$$+\bar{Q}_1 r\left[\frac{\partial}{r\partial\theta}(\beta_1 - \frac{u}{R_1}) + \frac{\partial w}{\partial\theta}\frac{\partial}{\partial s}(\frac{1}{r}) + \frac{1}{R_2}\frac{\partial v}{\partial s} + v\frac{\partial}{\partial s}(\frac{1}{R_2}) - \frac{1}{R_1}\frac{\partial v}{\partial s}\right]$$

$$-\bar{N} r\left[\frac{\partial}{\partial s}(\frac{\partial v}{\partial s}) + \frac{1}{R_1}(\frac{\partial w}{r\partial\theta} + \frac{v}{R_2})\right] - Yr$$

$$\frac{\partial(V_1 r)}{R_1\partial\phi} = -\frac{\partial V_2}{\partial\theta} - \frac{N_1 r}{R_1} - \bar{N}_1 r\frac{\partial\beta_1}{\partial s} - N_2\sin\phi - \bar{N}_2\left[\frac{\partial}{\partial\theta}(\frac{\partial w}{r\partial\theta} + \frac{v}{R_2}) + \cos\phi\,\beta_1\right]$$

$$\frac{\partial(M_1 r)}{\partial s} = 2\frac{\partial H}{\partial\theta} + M_2\cos\phi + V_1 r + \bar{M}_2\left[\frac{\partial}{\partial\theta}(\frac{\partial v}{\partial s}) - \sin\phi\,\beta_1\right]$$

where,

$$B = \frac{Eh}{1-\nu^2} \quad ; \quad G = \frac{Eh}{2(1+\nu)}$$

$$F = \frac{Eh^3}{12(1+\nu)} \quad ; \quad D = \frac{Eh^3}{12(1-\nu^2)} \qquad (6c)$$

$$C^* = G + \frac{2F}{R_2^2} - \frac{2F}{R_1 R_2}$$

To reduce the system of partial differential equations to a system of ordinary differential equations corresponding to the n-th terms of a Fourier series, the following are introduced:

$$\begin{Bmatrix} w \\ u \\ N_1 \\ M_1 \\ V_1 \\ N^O_{12} \\ \varepsilon_2 \\ \beta_1 \\ N_2 \\ M_2 \\ K_2 \end{Bmatrix} = \Sigma_n \begin{Bmatrix} w_n \\ u_n \\ N_{1n} \\ M_{1n} \\ V_{1n} \\ N^O_{12n} \\ \varepsilon_{2n} \\ \beta_{1n} \\ N_{2n} \\ M_{2n} \\ K_{2n} \end{Bmatrix} \cos n\theta \qquad \begin{matrix} v \\ V_2 \\ H \end{matrix} = \Sigma_n \begin{Bmatrix} v_n \\ V_{2n} \\ H_n \end{Bmatrix} \sin n\theta \tag{7}$$

This transforms the partial differential set into the following general n-th ordinary differential set. A similar set is obtained when cos nθ and sin nθ are interchanged.

$$\varepsilon_{2n} = n\frac{v_n}{r} + \frac{u_n \cos\phi}{r} - \frac{w_n}{R_1}$$

$$\frac{du_n}{ds} = \frac{w_n}{R_1} + \frac{n_{1n}}{B} - \nu\,\varepsilon_{2n}$$

$$N_{2n} = \nu N_{1n} + Eh\,\varepsilon_{2n}$$

$$\frac{dw_n}{ds} = \beta_{1n} - \frac{u_n}{R_1}$$

$$K_{2n} = \frac{-n^2 W_n}{R_1} + \frac{n\,v_n}{R_2\,r} + \frac{\cos\phi\,\beta_{1n}}{r} \tag{8a}$$

$$M_{2n} = \nu\,M_{1n} - \frac{Eh^3}{12}\,K_{2n}$$

$$\frac{d\beta_{1n}}{d\phi} = -\frac{M_{1n}}{D} - \nu\,K_{2n}$$

$$\frac{dV_n}{ds} = \frac{N^O_{12}}{C^*} - \frac{G}{C^*}\left[-n\frac{u_n}{r} - \frac{v_n\cos\phi}{r}\right]$$

$$-\frac{2F}{C^*R_2}\left[-\frac{n}{r}\frac{dW_n}{ds} - nW_n\frac{d}{ds}\left(\frac{1}{r}\right) + V_n\frac{d}{ds}\left(\frac{1}{R_2}\right)\right]$$

$$H_n = F\,[-\frac{n}{r}\frac{dw_n}{ds} - nW_n\frac{d}{ds}(\frac{1}{r}) + \frac{1}{R_2}\frac{dv_n}{ds} + V_n\frac{d}{ds}(\frac{1}{R_2}) - \frac{1}{R_1}\frac{dv_n}{ds}]$$

$$\frac{d\,(N_{1n}r)}{ds} = -n\,\frac{(N^o_{12n}r)}{r} + 2n\,\frac{H_n}{R_2} + \frac{(V_{1n}r)}{R_1} + N_{2n}\cos\phi$$

$$+ \bar{N}_2[\,n\frac{dVn}{ds} - \sin\phi\;\beta_{1n}\,] + \bar{Q}_1 r\,\frac{d\beta_{1n}}{ds}$$

$$V_{2n} = -\frac{2\cos\phi}{r}M_n - n\frac{M_{2n}}{r} + \bar{M}_1\;\frac{d^2Vn}{ds^2} + \frac{1}{R_1}[-\frac{nWn}{r} + \frac{Vn}{R_2}]$$

$$\frac{d(N^o_{12n}r)}{ds} = V_{2n}\sin\phi - N^o_{12n}\cos\phi + \frac{2H_n\cos\phi}{R_2} + n\,N_{2n}$$

$$+ \bar{Q}_1 r\,[-\frac{n\beta_{1n}}{r} + \frac{nu_n}{r\,R_1} - n\,W_n\frac{d}{ds}(\frac{1}{r}) + \frac{1}{R_2}\frac{dv_n}{ds}$$

$$+ v_n\frac{d}{ds}(\frac{1}{R_2}) - \frac{1}{R_1}\frac{dv_n}{ds}\,] \tag{8b}$$

$$- \bar{N}_1 r\,[\,\frac{d^2v_n}{ds^2} + \frac{1}{R_1}[-\frac{nw_n}{r} + \frac{Vn}{R_2}]\,]$$

$$\frac{d(V_{1n}r)}{ds} = -nV_{2n} - \frac{N_{1n}r}{R_1} - \bar{N}_1 r\,\frac{d\beta_{1n}}{ds} - N_{2n}\sin\phi$$

$$- \bar{N}_2\;[-\frac{n^2w_n}{r} + n\frac{v_n}{R_2} + \cos\phi\;\beta_{1n}\,]$$

$$\frac{d(m_{1n}r)}{ds} = +\,2nH_n + M_{2n}\cos\phi + V_{1n}r$$

$$+ \bar{M}_2 \quad + n\,\frac{dVn}{ds} \quad - \sin\phi\;\beta_1 n$$

For the symmetric case, n = 0 and the set reduces to the set of six first-order ordinary differential together with three simplifying definitions:

$$\bar{\varepsilon}_2 = \frac{\bar{u}\cos\phi}{r} - \frac{\bar{w}}{R_2}$$

$$\frac{d\bar{u}}{ds} = \frac{\bar{w}}{R_1} + \frac{\bar{n}_1}{B} - \nu\;\bar{\varepsilon}_2$$

$$\bar{N}_2 = \nu\bar{N}_1 + Eh\,\bar{\varepsilon}_2$$

$$\frac{d\bar{w}}{ds} = \bar{\beta}_1 - \frac{\bar{u}}{R_1} \tag{9a}$$

$$\bar{K}_2 = \bar{\beta}_1\,\frac{\cos\phi}{r}$$

$$\bar{M}_2 = \nu\bar{M}_1 - \frac{Eh^3}{12}\bar{K}_2$$

$$\frac{d\bar{\beta}_1}{ds} = -\frac{\bar{M}_1}{D} - \nu\,\bar{K}_2$$

$$\frac{d(\bar{N}_1 r)}{ds} = \frac{\bar{V}_1 r}{R_1} + \bar{N}_2 \cos\phi - Xr$$

$$\frac{d(\bar{V}_1 r)}{ds} = \frac{-\bar{N}_1 r}{R_1} - \bar{N}_2 \sin\phi - Zr \tag{9b}$$

$$\frac{d(\bar{M}_1 r)}{ds} = \bar{M}_2 \cos\phi + \bar{V}_1 r$$

Numerical Procedure

The differential equations have been put in the first-order form, having in mind the problem of nonuniform shells, to use one of the several available and dependable quadrature processes. Excellent results have been obtained with the Runge-Kutta fourth-order process, and this process is recommended as being efficient and accurate, but other good processes are also available. The integrations are to be carried out for each circumferential harmonic which contributes significantly to the result, keeping in mind the required or desired accuracy. As indicated by the first-order form of the derived ordinary differential equations, the path of integration will be along an antinodal meridian.

In the general problem of thin shells of revolution, four conditions must be satisfied at the initial boundary and four more remain to be satisfied at the terminal boundary. For a meaningful solution, the effect or influence of the initial conditions must be retained, at changing -- and generally decaying magnitudes -- along the path of integration, until the terminal boundary is reached. A simplistic technique for accomplishing this is to divide the total path into segments of such length that the effect of the conditions at the beginning of the segment is retained, at changing magnitude, until the far end of the segment is reached by integration and the free parameters introduced at the beginning of the segment can be evaluated. A step size which provides the desired accuracy must be used within the segment. Nine integrations over the segment are necessary, eight of these with the homogeneous form of the equations and one with the nonhomogeneous form which includes the loading terms. The results for that segment can be put into the form of an 8 x 8 stiffness matrix plus a set of fixed-edge tractions due to the nonhomogeneous terms. This is to be done for each segment. A tridiagonal matrix can then be set up for the entire shell system, each element of this system being the 8 x 8 stiffness matrix of the associated segment. This tridiagonal system matrix premultiplies a column vector of the displacements at the ends of the system and at the junctions of the segments, and the right-hand side of this system matrix equation is essentially a vector made up of the fixed-edge tractions of the several segments. The solution provides the displacements at all segment boundaries. Substitution into the segment stiffness matrices leads to the set of tractions at each segment boundaries. It should be noted that the set of differential equations reduces from eight to six in the axisymmetric case and for the zeroth-harmonic of the general loading asymmetric case.

A considerably more efficient method was developed by the senior author in the early sixties and published several years later [12]. In this method, the number of integrations of the homogeneous equations is reduced from eight to four in the asymmetric case and from six to three in the axisymmetric case. In each case, a nonhomogeneous set must also be integrated.

In this improved method, four integrations of the homogeneous equation are made starting at an end of the shell, using four linearly independent vectors as starting values of the unknown variables. These vectors may be, but need not be, made up of zeroes and ones. A fifth integration, of the nonhomogeneous equations, is also made. At the far boundary of this segment, four linearly independent vectors of displacements, including the rotation of the edge in the meridional plane, these vectors being made up of zeroes and ones. It should be noted that all eight of the differential equations are integrated simultaneously. The tractions at this far edge are evaluated from the results of the four integrations which have been made. In this way, a 4 x 4 stiffness matrix is established at this edge. This stiffness matrix, post-multiplied by a 4 x 4 matrix of linearly independent displacements, again made up of zeroes and ones, provides a complete set of initial values at the start of the second segment, and four integrations are done on the eight differential equations. At the far end of the second segment, and a stiffness matrix is established at this edge in the same way as was done at the interior edge of the first segment. It may be noted that this matrix is the stiffness matrix for the entire portion of the shell from the beginning of the shell to the second internal boundary. Continuing in the same way sequentially over the other segments, the final boundary -- which is the far boundary of the last segment -- is reached. At that location, the prescribed boundary conditions are imposed on the solution. Now, the values of the complete set of dependent variables at the terminal boundary are known. Having these values and using the matrix equation which coupled these terminal values to the values at the initial boundary of the last segment, the values at the initial boundary can be determined. Proceeding in the same way, the initial values can be determined for the preceding segment and so on for every segment back to the beginning of the shell. Having the complete set of initial values, values at any desired points within the segments can be determined by integration. Alternatively, if the coordinate system were to be reversed, a single backward integration, segment by segment, would provide information on displacements and tractions wherever desired.

Commentary and Conclusion

It is perhaps unlikely that the deformations due to axisymmetric loadings such as dead load will modify the geometry tc a significant degree so that the effect of the superimposed loads will be changed significantly. This question, however, can be faced head on simply by modifying the geometry by the amounts of the deformations due to the initial axisymmetric loading due to gravity and possibly to construction. In the same vein, if a superimposed asymmetric loading has an appreciable Fourier component corresponding to n = 0, it might be advisable to modify the geometry additionally to include the changes due to that harmonic. In some cases including the case of mildly nonlinear problems, an incremental procedure in which the products of tractions and associated displacements obtained in the preceding increment are treated as nonhomogeneous terms, i.e., as loads, may prove to be satisfactory for engineering purposes. Other techniques of this type may be used.

The question of how to treat the lattice supporting structure at the base of the shell must be considered. There are at least two obvious techniques. In both of these, the elasticity of the soil and the foundation ring should be taken into account in both techniques. One technique is to replace the set of lattice struts by an equivalent shell, cylindrical but preferably conical. In the other technique, the struts are treated as elastic members which impose discrete loads on the base of the shell.

If the cross section of the ring at the top of the tower has compact proportions, it is likely that it can be handled in a simple way. The 4 x 4 stiffness matrix of the ring with respect to the attachment to the ring can be readily calculated by what may be called ring theory . It may, in fact, be advisable to take the attachment as the starting point for the integrations. With the attachment to the supporting lattice structure as the terminal boundary, alternate models of the supporting structure and soil may be used and compared without re-doing the forward integration from the top boundary down for each model.

When the tower is provided with a canopy extending out from and attached to the lower edge of the shell, the shell equations presented here may be used and integrated from the outer free edge to the attachment to the shell, and its stiffness matrix at that circle added either to the stiffness matrix of the lattice structure or of the shell.

References

1. Goldberg, J.E., and Bogdanoff, J.L., "Analysis of Coned Discs Under Symmetrical Conditions," Report 57-7, 1957 Midwest Applied Science Corp., West Lafayette, IN.
2. Goldberg, J.E., Bogdanoff, J.L., and Marcus, L., "Analysis of Conical Shells by Electronic Computer," Presented at the National Meeting of ASCE, Cleveland, OH, May 1959.
3. Goldberg, J.E., Bogdanoff, J.L., Oshira, H.O., and Alspaugh, D.W., "Stress and Deformation in Symmetrically Loaded Shells of Revolution," Midwest Applied Science Corp., Report 61-1, April 1961.
4. Goldbert, J.E., "Computer Analysis of Shells," Proceedings --Symposium on the Theory of Shells to Honor L. H. Donnell, University of Houston Press, 1967, pp. 5-22.
5. Goldberg, J.E., "Analysis of Moderately Thick Shells of Revolution," Proceedings, Third Canadian Congress of Applied Mechanics, ed. P. G. Glockner, Calgary, May, 1971, pp. 243-244.
6. Goldberg, J.E., and Alspaugh, D.W., "Stress and Displacement in Pressurized Cylindrical Shells Subjected to Localized Loads," Midwest Applied Science Corp., Report 64-3, March 1964.
7. Goldberg, J.E., Baloch, M.H., and Tang, Y.K., "Analysis of Moderately Thick Shells of Revolution," First International Conference on Structural Mechanics in Reactor Technology, Vol.3, Part G, Sept.1971,pap.G1/3

8. Goldberg, J.E., and Bogdanoff, J.L., "Static and Dynamic Analysis of Conical Shells under Symmetrical and Unsymmetrical Conditions," Proceedings of the Sixth Symposium on Ballistic Missile and Aerospace Technology, Los Angeles, 1961, Academic Press, New York, p. 219.
9. Goldberg, J.E., Bogdanoff, J.L., and Alspaugh, D.W., "On the Calculation of the Modes and Frequencies of Pressurized Conical Shells," Journal of Aircraft, AIAA, Vol. 1, No. 6, 1964, pp. 372-374.
10. Goldberg, J.E., Bogdanoff, J.L., and Alspaugh, D.W., "Stresses in Spherical Domes under Arbitrary Loading Including Thermal and Shrinkage Effects," Proceedings of the IASS Symposium, Warsaw, Sept., 1963.
11. Goldberg, J.E., Bogdanoff, J.L., and Marcus, L., "On the Calculation of the Axisymmetric Modes and Frequencies of Conical Shells," J. Acoust. Soc., Vol. 32, 1960.
12. Goldberg, J.E., Setlur, A.V., and Alspaugh, D.W., "Computer Analysis of Non-Circular Cylindrical Shells," Proceedings of the International Symposium on Shell Structures in Engineering Practice, IASS, Budapest, 1965.
13. Goldberg, J.E., Setlur, A.V., and Pathak, D.V., "Stability of Ring-Stiffened Cylindrical Shells with Closed Ends under External Pressure," IASS Pacific Symposium on Hydromechanically Loaded Shells, Part I, Honolulu, Hawaii, USA, October 1971.
14. Goldberg, J.E., Tang, Y.K., and Pathak, D.V., "Analysis of Moderately Thick Flanged Reactor Pressure Vessels," 2nd International Conference on Structural Mechanics in Reactor Technology, Berlin, Germany, September 1973, paper G 1/3.
15. Goldberg, J.E., Pathak, D.V., and Roschke, P.N., "Thermal Stress Analysis of Curved Box Girder Bridges," IASS Symposium on Shell Structures and Climatic Influences, Calgary, Canada, eds. P.G. Glockner and Ghalia., July 1972.
16. Love, A.E.H., A Treatise on the Mathematical Theory of Elasticy, 4th Ed. Dover Publication, 1944.
17. Hellinger, E., "Der Allgemeine Ansatz der Mechanik der Kontinua," Encylopädie der Mathematichen Wissenschaften, Vol. 4, Part 4, (1914) pp. 602-94.
18. Reissner, E., "On a Variational Theorem in Elasticity," Journal of Mathematics and Physics, Vol. 29, No. 2, July 1950, pp. 90-95.
19. Naghdi, P., "On the Theory of Thin Elastic Shells," Quarterly Applied Mathematics, Vol. 14, No. 4, 1957, pp. 364-380.

1984
IASS-RUB

Durability

Long Time Measurements of the Effects of Water and Temperature on Cooling Towers

Dr.-Ing. E.h. U. Stoffregen

HOCHTIEF Aktiengesellschaft vorm. Gebr. Helfmann
P.O. Box 10 17 62, 4300 Essen 1

1. Purpose of the measurement campaign

I do not think it too bold to say that such factors as load application, design and other conceptual criteria for reinforced concrete structures of very high natural-draught cooling towers have by now been investigated to a satisfactory degree and that they have stood the test in respect of the stability of these structures.

As to the questions concerning the durability of our structures, I am afraid, they haven't yet been solved quite satisfactorily. No doubt, anybody involved in construction, is aware of the same and in quite a number of cases the consequences of experience gathered have been taken into account in our technical rules and regulations. Yet, again and again, we encounter a phenomenon whose cause we can detect only partially or not yet at all. Naturally, it then becomes difficult, to find the right measure for design and execution.

You will know that in the Federal Republic the examination of inner shell surfaces of natural-draught cooling towers that have been in operation for many years, has revealed erosion of varying degrees on the concrete although, provenly, owing to a very carefully designed concrete mix and to proper workmanship, an excellent concrete quality had been achieved.

Very soon, it was generally admitted that it was insufficient to investigate the possible causes of such erosion by way of laboratory tests alone. It became necessary to determine on the spot, the effects of temperature, water, vapour and condensates on the respective concrete surface, taking due account of the prevailing operational and environmental conditions.

Complementary to measurement results obtained earlier elsewhere, HOCHTIEF decided to carry out a measurement campaign on a natural-draught cooling tower in the Saarland which we had recently completed. The programme proposed by HOCHTIEF and realized with kind permission and expert support by the Owner and Operator the Saarbergwerke AG, Saarbrücken, is to be the subject of this paper.

2. Description of the test programme

The test programme conducted on the natural-draught cooling tower of the new-built coal-fired Bexbach power station started in August 1983 and still continues. The power station was completed in spring of 1983 and commissioned in the autumn of the same year.

Figure 1 on the right shows the dimensions of the cooling tower. Wall thicknesses range between 90 and 16 cm in the upper shell portion.

The Concrete mix design of the shell is as follows:

370 kg per cubic metre of early strength, highly sulphate resistant Portland Cement PZ 45 FHS
Aggregate: Rhine gravel
Water/cement ratio: 0.5

The lower portion of the inner shell surface was provided with a protective coating consisting of a prime coat and a tack coat of a two-component epoxy resin product.

The cooling system used is by Hamon-Sobelco. The capacity of the associate power station block is about 750 MW and the water flow is about 14.000 kg per second or about 50.000 cubic metre per hour. The cooling range is about 12 Kelvin.

Measurements are being taken above the accessible drift eliminator level at plus 15 m, and at the level of the inner circumferential platform for the aviation warning light system at plus 96 m. In the measurement control stand at plus 15 m the results of all measurements taken are being registered and stored.

The following equipment is used:

- Central Computer PDP 11/23 of Digital Equipment Company
- Measurement and control system (Multiplexer) μMac-4000 of Analog Devices Company.

2.1 Examination of shell condensates

For the purpose of assessing the stresses on the inside shell surfaces from the cooling process effects, for the shell section above the drift eliminator, only the composition of the condensate is determining. Therefore, condensate samples have regularly been taken and their chemical composition been examined in our laboratory.

In addition, condensates from the shell wall and of the vapors have been taken continuously in order to determine their pH value. These measurements are carried out by means of an electronic pH-value sensor which is permanently installed at both measuring levels (Figure 2).

The water collected in the catch basins is pumped computer-controlled towards the sensors for determination of the pH value. The water running down the shell wall was automatically measured at hourly intervals.

The deposition on the condensors placed within the cooling air flow, despite freshwater cooling, is so small, that it took a relatively long time to obtain sufficient condensate quantities. We discovered that algae had polluted the conduits and basins when sufficient quantities were finally obtained. Thus, it was only possible to make measurements at irregular intervals and after prior cleaning.

Apart from the chemical composition of the condensate which may vary according to local circumstances,also the quantity of deposition on the inner shell surface is of interest for our tests. In order to obtain this information, we have measured the condensate quantity at some of the measuring points, on the catch basin mentioned, and in special test areas which had been sealed against water penetration from above and from the sides.

These measurements, are certainly not automatically transferable to the overall shell surface, but allow to obtain an idea on the magnitude of the amount of condensate involved.

2.2 Temperature measurements

For measuring the temperature stresses within the shell, four temperature measuring points have been installed on each measuring level. Figure 2 shows the position of the measuring points within the structure. On both levels, there are two measuring points on the south and two on the north side of the shell. Five resistance thermometers on each point served to measure the air temperatures inside and outside the shell, i.e. at a distance of approximately 5 cm from the wall as well as at the wall centre and 2 cm behind the wall surface.

The measured values inside the wall allow to compute the surface temperatures and heat flow under stationary conditions. For determining the warm and cold water temperatures more measuring points are provided within the riser and the cold water outlet structure.

At the start of the measurement campaign, these temperatures were taken manually at intervals of approximately two weeks. Once the automatic control of the system was installed and functioning, a central computer called in the measured values continuously every ten minutes, storing them on a magnetic disk until later evaluation.

2.3 Other measurements

Complementary to the tests so far mentioned and which are directly relative to the cooling tower shell, the following additional measuring points were provided:

- hygroscopic moisture contents in the vapour and the outside air
- temperature of the vapour
- velocity within the cooling air flow.

Weather data required for evaluating the measured values have been obtained from nearby measuring stations of the German Meteorological Office.

3. Results of Condensate examinations

Continuous control of the pH values of the water film running down the inner shell surface yielded results which almost invariably lay between pH = 7.5 and 8.5. As expected, the lower values were registered on the measuring level at + 15 m. Since the shell is coated here, the composition of the water collected corresponds more closely to the pure condensate.

The range of fluctuations observed was extremely small, in very rare cases extreme values of pH = 6.5 resp. 9.0 were measured. A dependance on the seasons or the daytime was not found to exist, the temperature dependance of the pH value being automatically compensated by the sensor.

The pH values of the condensate from vapour measured at irregular intervals did not show any other results either, with values varying between pH = 6.0 and 7.5, as one will expect it from condensates.

As far as the pH value is concerned, these results do not constitute any particular stress for the concrete, even if maintained over a prolonged period of time.

In order to further evaluate the condensate, water samples were taken and analysed. Three types of samples were involved:

A) Condensate from the cooling air flow
These samples consist of recondensed vapour and deposition of water carried by the vapour. Thus, their composition largely corresponds to that of the water depositing on the concrete surface.

B) Samples from the condensate after watering the wall
For this purpose, approximately 3 m^2 large surfaces were sealed off against water penetration from above and from the sides. These samples yielded information on the wall watering processes.

C) Samples from the water film running down the wall.

Table 1 contains the average values of these chemical analyses for the three types of samples:
Samples A indicate a very soft water with low conductivity.

The contents of aggressive carbonic acid is clearly below the limit values of the German Standard Specification DIN 4030. Sulphates are practically non-existant and sulphites were not detected at all.

Samples B show that water of such softness dissolves components from the concrete. The contents of calcium and sulphate is clearly higher than in the A-samples and the conductivity increases fourfold. The higher $KMnO_4$ consumption is an indicator of organic substances, presumably algae. The overall hardness is deduced from the increased contents of calcium and magnesium.

The lower values of the C-samples are an indicator of the dissolving process being smaller on average in their case. Consequently, no enrichment takes place on the flow path along the shell surface.

Shortly said, in our case, the values measured are extremely low. Even over prolonged periods of time, they can have no aggressive effect on the concrete.

4. Temperature measurements results

Results of measurements taken during the period between October 1983 (after commissioning of the cooling tower) and April 1984 are being presented here. During the time of year concerned, the shell is subjected to heavy temperature stresses, which are of considerable importance for the design.

It was one essential aim of the measurement campaign to determine in how far the regulations in respect of temperature design laid down by the VGB (TECHNISCHE VEREINIGUNG DER GROSSKRAFT-WERKSBETREIBER e.V.) (Technical Association of Large Power Plant Operators) in BAUTECHNIK BEI KÜHLTÜRMEN (Cooling Tower Construction Code) indicate accurate values. According to these regulations the shell design must be based on steady-state conditions with inside and outside temperatures of + 30° C resp. - 15° C. Heat transfer by convection may be taken into account with 20 $W/m^2/K$, the same values applying to the inside and outside. The temperature difference within the wall then only depends upon the wall thickness.

Figure 3 shows a typical daily variation line of the five thermocouples of one measuring point in Winter 1983/84. One clearly recognizes short-term fluctuations of the vapour temperature during sectional switch-off of filling surfaces.

Within the shell, fluctuations decrease from the inside towards the outside and have ceased almost entirely in the wall centre. Temperatures at the outer rim mainly follow the daily variation line of air temperature.

On Figure 4 and 5 , the maximum temperature differences between vapour and outside air Δ T, as well as within the wall Δ t, taken on both measuring levels, have been traced, i.e. in each instance the pairs of values max Δ T with the appurtenant Δ t and vice versa. The values indicated are daily peak values observed between December 1983 and February 1984, measures being taken every ten minutes. Also included in the diagrams are the values resulting from the VGB regulations, being inserted as straight lines. From the diagrams, two things are clearly perceptible:

a) Generally, maximum air and wall temperature differences do n o t occur simultaneously.

b) The maximum temperature differences measured within the wall clearly agree with the VGB computing code. It is important to note that the indicated values strictly represent daily peaks. The maximum hourly averages are definitely of lower values at this stage.

Consequently, in the case considered, the VGB regulations provide accurate values. On the other hand, it becomes obvious from the example of the daily variation line in Figure 3 that the ratio between temperature jumps inside and outside, and the temperature differences within the wall deviate from these calculative assumptions. By no means is it the same inside and outside of the tower. Among others, this ratio is dependent upon the thermal conductivity, the thickness and surface roughness of the wall, as well as upon the flow conditions and the properties of the medium responsable for the convective heat transmission, which deviate from each other inside and outside the tower.

In Figure 6 , the measured ratios between heat transfer and heat conduction are evaluated statistically. Since the thermal conductivity is not precisely known, the non-dimensional Biot number was chosen to serve as observation variable. The hourly average values of 10-minute measurements over the indicated observation period were evaluated, without taking into account values measured under solar radiation.

The following calculative assumptions were applied:

$$q_L = \frac{\lambda}{D}\frac{d\vartheta}{d\xi} \; ; \; \xi = \frac{x}{D}$$

$$q_K = \alpha_a \cdot \Delta\vartheta_a$$

with $q_K = q_L$ $(\xi = 0)$ it follows

$$Bi^a = \frac{\alpha a\ D}{\lambda} = \frac{1}{\Delta\vartheta_a}\left[\frac{d\vartheta}{d\xi}\right]_{\xi=o}$$

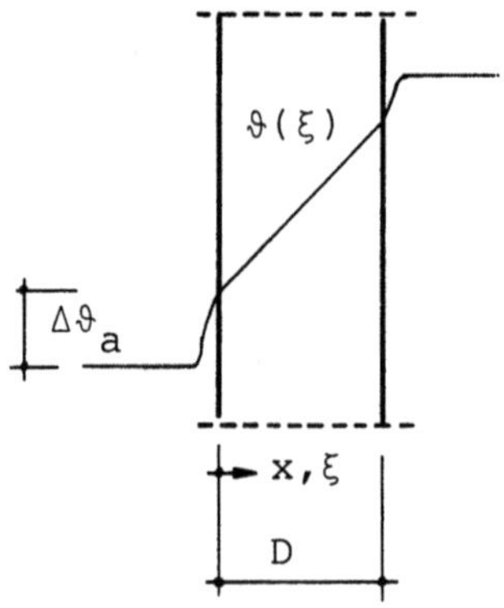

where:

q_L = heat flow through heat conduction in the wall

q_K = heat flow through convection

$\Delta\vartheta_a$ = jump of temperature within the boundary layer

$Bi^{a,i}$ = Biot number (outside resp. inside)

α_a = heat transfer coefficient outside

The temperature gradient on the surface is replaced by the difference quotient of measured values in the wall centre and at the outer rim, which are hardly influenced by the short-term fluctuations of the inside air.

The result is a skew distribution function (lower limitation through natural convection) with the average value of $Bi_m^a = 1.00$ and a 95 % fraction of $Bi_{95}^a = 1.65$.

A similar view cannot be adopted for the inside of the tower, since the heat transferred to the wall is transmitted not only through convection, but major portions come from the evaporisation heat which is freed during condensation of the vapour on the inner surface. The exact condensate quantity is unknown, though, as is the exact temperature gradient near the surface which is required to quantify the heat flow. This circumstance is expressed by a temperature jump on the inner surface which is very much smaller than that on the outer surface. This can easily be recognized on Figure 3 . If, for the purpose of comparison, one still proceeds to determine an "apparent" inner heat transfer coefficient, the resulting Biot numbers fluctuate between 1.0 and 3.0.

For steady-state conditions, Figure 7 allows to compare the temperature differences in the wall with the VGB regulations in dependance of the heat transfer coefficients measured by us. It is revealed that for shell areas with wall thickness between 16 and 20 cm an underdesign of up to maximum 10 % might be possible. But since one has to base on the assumption that maximum heat transfers do not occur simultaneously with maximum air temperature differences, we are of the opinion that there is no cause for proposing a modification in the VGB regulations.

Since the wind velocities have a determining influence on the heat transfer, we have used the average hourly wind velocities recorded by the weather station of Saarbrücken-Ensheim during the observation period and have plotted them on Figure 8 . It is revealed that an adequately large wind velocity range has been covered by our measurements.

Switch-off procedures

Figure 9 illustrates graphically the wall and air temperatures taken at measuring level + 15 m resp. + 96 m during a scheduled block switch-off in March 1984. The relevant wall thicknesses of the shell are d = 21 resp. 27 cm. They clearly reveal the differences in the cooling behaviour with a centre-point temperature difference of maximum 4.5 Kelvin. And yet, this switch-off was a scheduled, "gentle" procedure, as one can see from the decrease in the inside air temperature.

Conditions deteriorate with an abrupt block switch-off. Figure 10 shows such an event measured mid-February 1984.
Within a few hours only, it causes temperature changes of more than 5 K in the shell centre.

5. Conclusion

Evaluation of the results of the measurement campaign which has been in progress for more than 9 months now, investigating the effects of temperature changes during varying operational conditions, as well as the quality of the condensate and of the cooling water, gave no indication of any particular danger to the reinforced concrete shell.

We found the design specifications as laid down in the VGB regulations confirmed in respect of the temperature stresses to which the shell is subjected. Obviously, there is no need to aggravate the existing calculative assumptions for these structures in our climatic zones.

The permanent watering of the inner wall surfaces of the shell by the condensate measured at this location will most probably, not result in any deterioration of the concrete surface quality for a long time to come.

Obviously, this statement cannot serve as an example for other cooling towers and other locations. In our case, the chemical analysis of the condensate yielded such low concrete-aggressive substances, that the optimistic prognosis seems justified for this specific cooling tower, a prerequisite being, though, that conditions under which the measured results were obtained do not change.

As to this, an authoritative forecast cannot be made.
In the introduction to my paper, I had touched on the question of the durability of the cooling towers built by us. It requires that the quality of the structure cannot follow alone the rules of stability. In respect to the profitability and in the interest of lower constructional maintenance cost later on, it seems quite sensible to allow higher initial capital investment for less sensitive structures.

Therefore, I can only recommend, that prior to any decision taken for the construction of a cooling tower, the operational conditions to be expected in the long run, are meticulously defined. Only this allows to make the right decision with respect to the material to be used, so that the desirable durability is ensured.

Figures

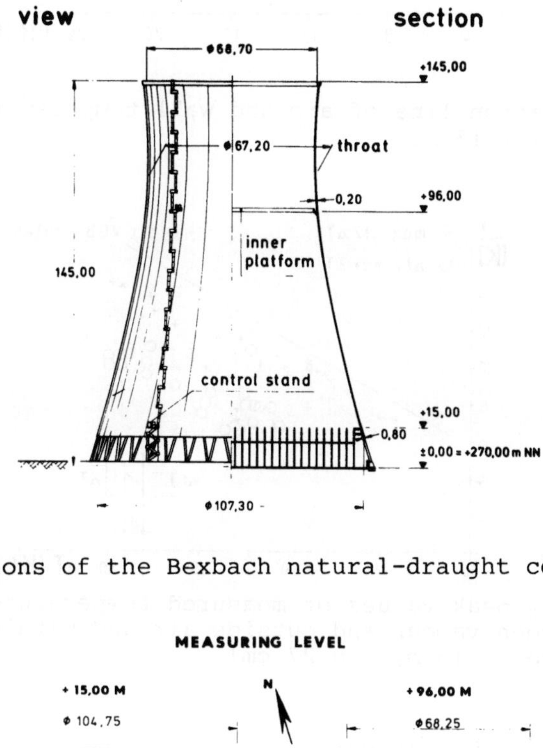

Fig. 1. Dimensions of the Bexbach natural-draught cooling tower.

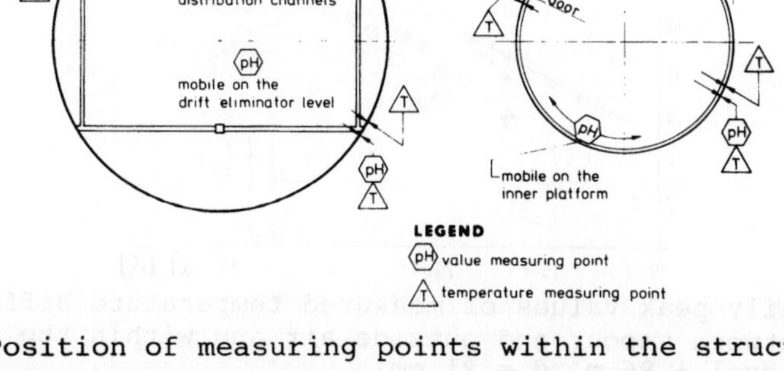

Fig. 2. Position of measuring points within the structure.

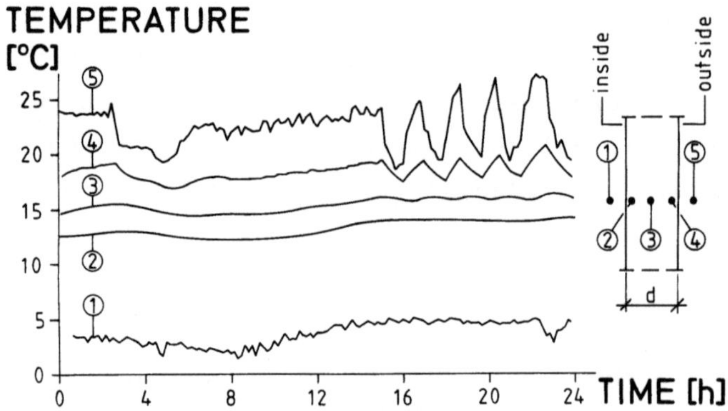

Fig. 3. Variation line of air and wall temperatures taken at level + 15 m.

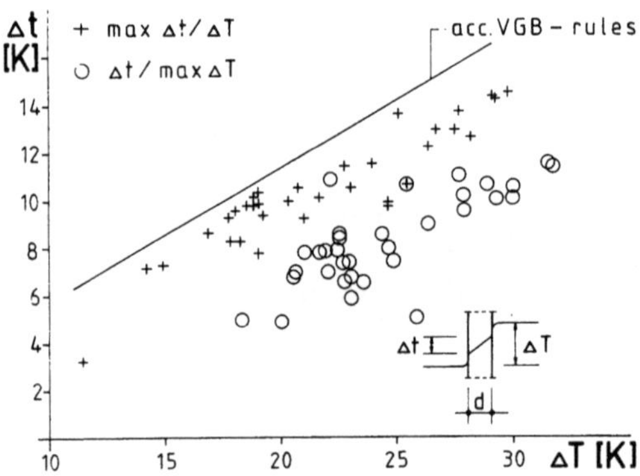

Fig. 4. Daily peak values of measured temperature differences between vapour and outside air and within the wall (level + 15 m, d = 27 cm)

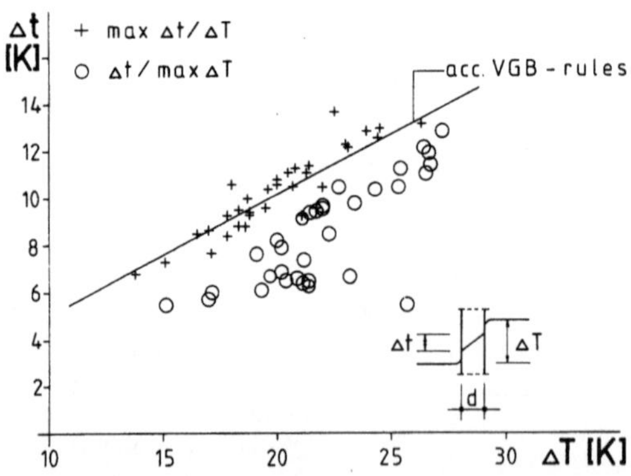

Fig. 5. Daily peak values of measured temperature differences between vapour and outside air and within the wall (level + 96 m, d = 21 cm)

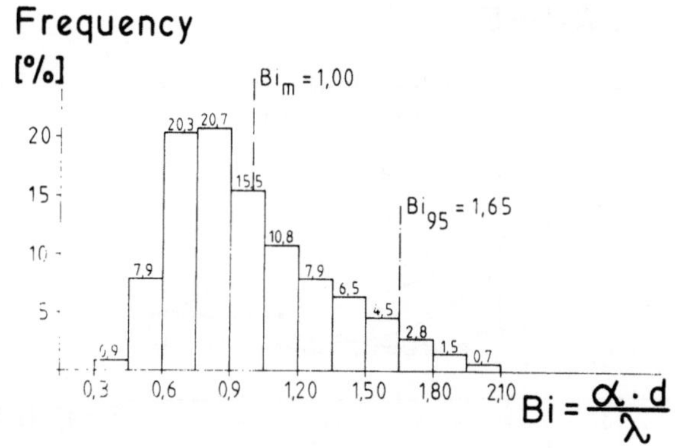

Fig. 6. Statistical distribution of outside heat transfer.

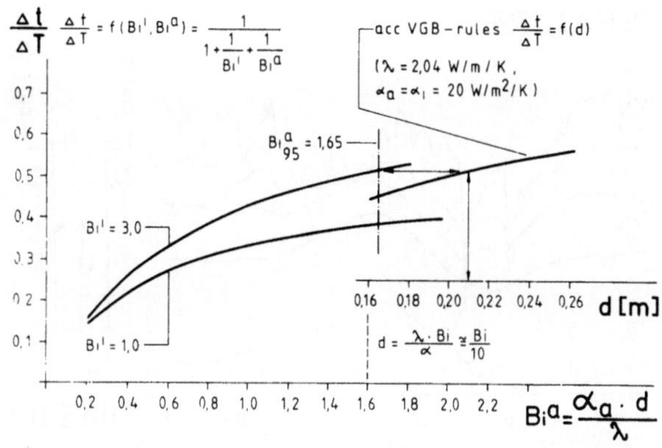

Fig. 7. Comparison of wall temperature differences in relation to the inner and outer heat transfer coefficient with the VGB regulations.

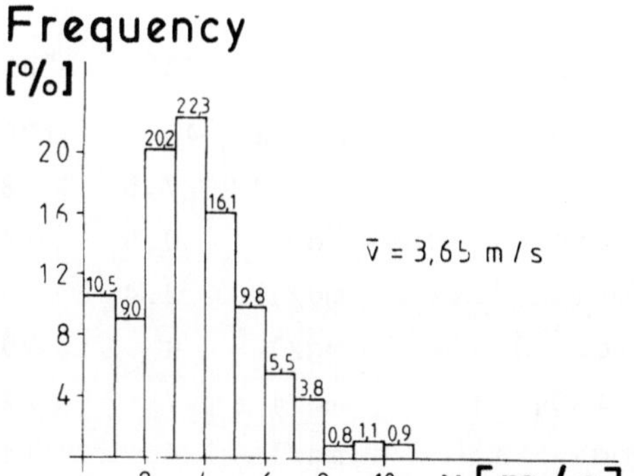

Fig. 8. Statistics of average hourly wind velocities recorded at the weather station of Saarbrücken-Ensheim.

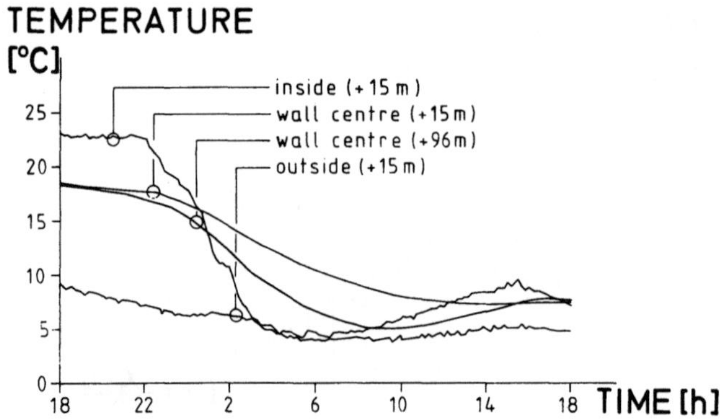

Fig. 9. Air and wall centre temperatures measured during a scheduled block switch-off.

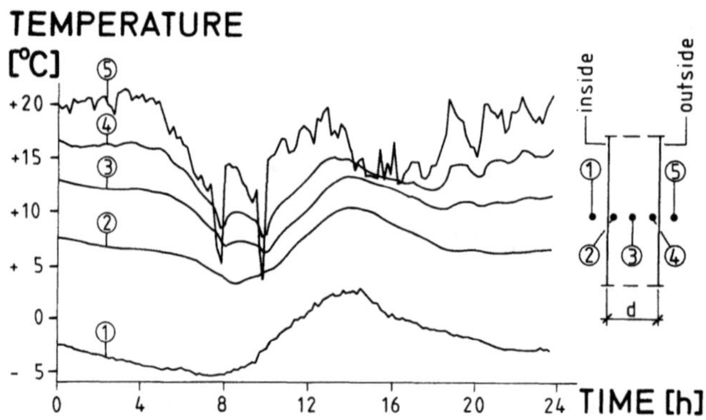

Fig. 10. Air and wall temperatures measured during an abrupt block switch-off.

Table 1: Chemical analyses of condensates

Type of sample		"A"	"B"	"C"
Conductivity	μS/cm	45	170	135
pH value		7.5	8.1	8.2
Hardness, total (DIN)	°dH	0.5	4	2.5
CO_2-aggresive	mg/l	2.4	-	-
Calcium (Ca^{++})	mg/l	4	29	18
Sulphate (SO_4^{--})	mg/l	7	22	8
$KMnO_4$-used	mg/l	4	14	5

Durability of the Inner Surface of Cooling Tower Shells

O. Lehmkämper

Dr.-Ing., Rheinisch-Westfälisches Elektrizitätswerk AG,
Essen, West Germany

Summary

In the last few years there has been increasing damage at the inside walls of concrete cooling tower shells. Low surface strengths, extensive sand formation, spallings of concrete and corrosion of the embedded reinforcement characterize the situation. In several areas high carbonation depth endangers the corrosion protection. It is assumed that damage is essentially attributable to the effect of acid vapour condensate and frost. The required restoration work is the subject of this paper. Quality assurance and additional surface protection systems are supposed to improve the durability of new cooling towers.

Introduction

In recent years defects and damage at the inside walls of the reinforced-concrete shells of wet natural-draught cooling towers have been observed more frequently. The characteristics and the extent of the damage vary in the individual cases.

This paper deals with cooling towers that were put into service between 1968 and 1975 (Fig.1.). These are counterflow cooling towers with a height of 90 to 129 m and base diameters between 68 and 100 m. Their minimum wall thickness amounts to 14 cm. Their inner and outer nets of reinforcement run in an orthogonal fashion in a ring and meridional direction. Carry-over eliminators of very high efficiency have been installed above the film-type packs and the hot-water distribution pipes.

Manufacturing of the shells

The shells have been concreted in rings by means of climbing formwork systems in 1.20 m high sections. The formwork consisted of about 60 cm high formwork panels which were connected in a polygon-like fashion with metal plates at the formwork jacks. Depending on the weather conditions the concreting progress amounted to a maximum of 1.20 m per day.

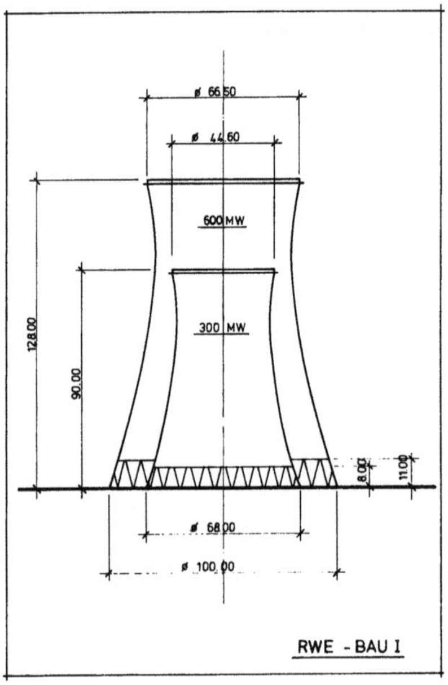

Fig.1. Dimensions of the defective cooling towers

The concrete remained in the formwork for at least 36 hours. Following the stripping the concrete was sprayed with a curing agent.

Following is a list of the most important characteristics of the concrete formula:

Concrete strength class	:	B 25
Type of cement	:	HOZ 35 L (partly NW-HS)
Cement content	:	340 - 360 kg/m³
Water-cement ratio	:	$<$ 0,5
Aggregate	:	Rhine gravel of medium grading curve.

Reinforcing bar steel was used. The prescibed minimum concrete cover was 2 cm.

Identified damage and defects

In all the affected cooling towers sand deposits can be observed at the surface which is touched by vapour (Fig.2.).

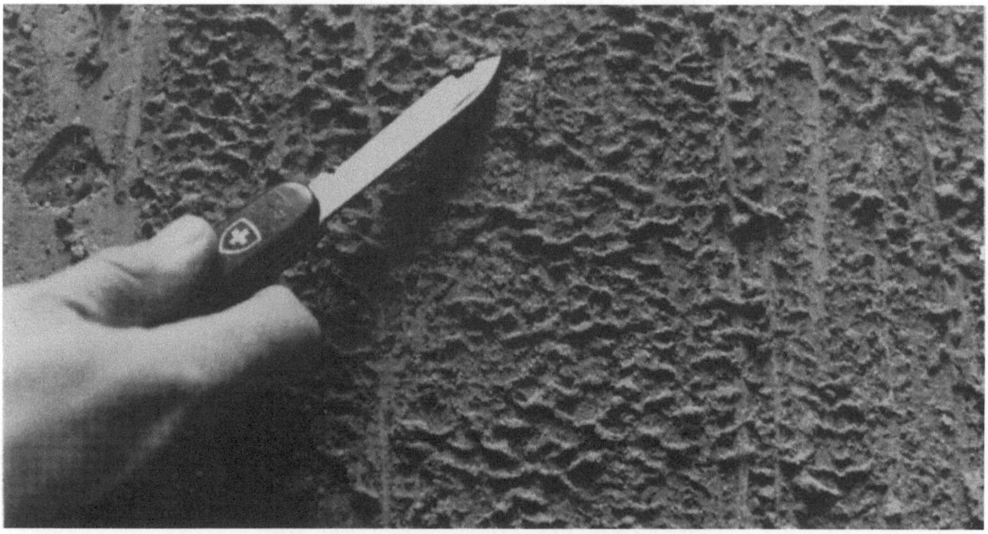

Fig.2. Sand formation on concrete surfaces

This sand is carried from higher altitudes by means of condensate running down the shell and is deposited on the shell. The release of the cementinous material results in the concrete surface loosing its strength. The fine grain components can be rubbed of manually resulting in a coarse-grained concrete surface.

The coarse aggregate particles have been laid bare in many places (Fig.3.). At the edges of these zones the concrete has come off the underlying layer in scales of up to 10 mm in thickness. These scales can be removed mechanically without problems. Areas that do not show any defects have also been damaged in some places already. This can be detected by moving a hammer along the surface. Another sign of damage is corrosion of the reinforcement (Fig.4.) which lies bare or already begins scaling off its concrete cover. These damaged places are distributed at random in the cooling tower.

In some cooling towers the carbonation depth has been measured in ring and meridional direction (Fig.5.). It is striking that the values attain their maximum directly above the carryover eliminators and become clearly smaller towards the upper edge. The distribution over the circumference does not show any regularities.

Fig.3. Scaling concrete layers

Fig.4. Corrosion of the reinforcement

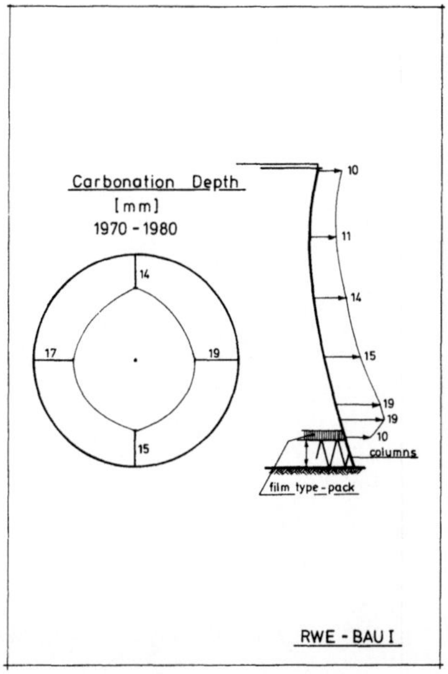

Fig.5. Carbonation depths

Cause of the damage

In 1979 and 1980 extensive investigations were carried out to find out the cause of the damage. Measurements of the compressive strength, capillary porosity and water penetration depth at drill cores from the concrete of the shells yields more favourable values than required. The measurement of the freezing resistance at drill cores, however, points to a correlation with the degree of damage in the cooling towers. As the frost resistance declines the degree of damage increases.

The composition of the condensed vapour, which precipitates in large quantities on the inner cooling tower surface and drains off, is essential for the chemical action upon the cooling tower shell. Fig.6 shows the values of chemical analyses of the vapour condensate carried out in three cooling towers. The demineralized vapour has very low degrees of hardness, shows pH-values in the acidic range and contains aggressive carbonic acid. The condensate can thus be classified as highly aggressive according to DIN 4030. This constantly renewed, flowing attack on the concrete results in the separation of lime and silicic acid. This explains the high rates of sand formation as well as the low strength of the concrete layers that are close to the surface.

Chemical Analysis of the Vapour	
pH	4.5 - 5.6
Carb. Hardness °d	0 - 0.3
Hardness °d	0.9 - 1.4
CO_2 aggr. mg/l	10 - 20

RWE - BAU I

Fig.6. Chemical vapour analysis

The biochemical attack by algae and other micro-organisms is still largely open to investigation. Measurements in cooling towers are being made in order to obtain quantitative information on the entire effects of physical, chemical and biological action. In the context of the series of publications, this subject is reported on by Stiller and Stoffregen.

The mode of operation and the cooling system of the cooling towers have proved to influence the degree of scaling decisively. Starting and shut-down as well as low-load operation favour the formation of ice at the inside walls of the cooling towers in winter. This results in increased sand formation as well as spallings of the already damaged, water-saturated concrete surfaces.

Corrosion of the reinforcing bars can be attributed to the fact that either the prescribed minimum concrete cover has not been respected or that the carbonation depth has already reached the reinforcement bars.

Today it is impossible to determine in retrospect what effect the kind of follow-up remedial treatment of the formwork concrete has on the damage.

It is undisputed, however, that the strength and the resistance of the concrete surface largely depend on the quality of the follow-up treatment.

Cooling tower restoration

Cooling tower restoration is aimed at preventing a damage and carbonation from deteriorating. The required work is carried out during the inspection periods of approx. 2 months.

Prior to every cooling tower restoration a concrete diagnosis is carried out to determine the degree of damage. Thus the visual degree of damage is described and documented by photographs. The depth position of the reinforcement, the carbonation depth and the surface strength are ascertained and evaluated statistically at 4 to 5 measuring points above the carry-over eliminators. The decision for restoration depends on the diagnosis.

A working platform with an elevator serves to transport all the required material through the cooling tower gate (Fig.7.).

Fig.7. Construction site with working platform

In order to protect the sprinkling grates and to ensure safe access and inspection, planks on an elastic support are placed on the drip-water receivers (Fig.8.).

Fig.8. Planks placed on carry-over eliminator level

Fairlead mobile scaffolding equipment is used to carry out the repairs on the cooling tower surfaces (Fig.9.). Horizontal access is ensured by an upper and lower elevator cage. The upper elevator cage runs on top of the cooling tower, the lower one on a guide rail fastened with dowels. The working platforms are up to 8 m long.

A high-pressure water-jet procedure is applied to remove the scaled and brittle concrete layers (Fig.10.). Meanwhile, rotating water-jets have come to be used working with a pressure of up to 300 bar. The jet plates are fastened to an unmanned platform controlled and operated from the centre of the cooling tower (see Fig. 8.). By varying the water pressure, the travel speed and the number and kind of jets fitted, the desired degree of cleaning can be adjusted.

The exposed reinforcement bars are sand-blasted and given a double protective coating. Afterwards the cooling tower surface is coated with epoxy resin

Fig.9. Fairlead working platform

Fig.10. Water-jet controlled from the centre

systems (Fig.11.). These coatings have proved suitable as protective system for several years.

Fig.11. Coating from the working platform

Precautions

According to what is known so far the following quality characteristics are relevant for the durability of new cooling tower shells to be built:

- A smooth concrete surface without shrink holes;
- A type of concrete of high frost resistance, as impermeable to water as possible;
- Curing of the concrete surfaces to optimize strength and imperviousness;
- Adherence to the prescribed construction tolerances, in particular with respect to the minimum concrete cover.

These exacting requirements can only be met by a quality assurance programme. For this purpose, the required steps must already be coordinated with the manufacturer at the preliminary planning stage. Some planning objectives have been attained during the construction of the cooling tower for the

Ibbenbüren power station, unit B (Fig.12.).

Fig.12. Inner surface with pigmented coating

The climbing formwork system worked with a very high degree of precision. The formwork panels were connected both with the concrete and among themselves so that rock pockets and faults between the concreting sections were largely avoided. A concrete laboratory at the construction site continuously verified the required values of the concrete formula. Following the formwork removal the concrete surfaces were immediately sprayed with a curing agent giving additional protection. A second pigmented coating was rolled on at the inside to improve the diffusion resistance against CO_2 and the UV resistance. The nominal dimensions as well as the concrete cover were constantly checked.

This list contains just a few important points of the quality assurance programme. Owing to the cooperation between the company in charge of the construction, the test engineer and the builder, the endeavours resulted in a construction of very high quality (Fig.13.). It is to be hoped that no damage will occur during its useful life.

Fig.13. Cooling tower after completion

Long Term Deterioration of Concrete Cooling Towers

J.G.A. Croll

University College London,
Department of Civil and Municipal Engineering,
London, U.K.

Abstract

Over their operational lives, a number of concrete cooling towers in the U.K. have developed serious levels of cracking and geometric distortion. After many years of operation, this damage has at times been found to be severe that very costly remedial action has had to be undertaken. In extreme cases, towers have had to be replaced.

This paper will describe the forms that this progressive growth of damage can take and will put forward possible explanations for its occurence that may have important consequences for current cooling tower design and construction.

Special Systems for Treatment of Fresh Concrete Surfaces Combined with Long Term Protection

Dr. Fritz M. Depke

MC-BAUCHEMIE Müller GmbH + Co, Chemische Fabriken, Bottrop

Summary

The surfaces of natural draught coolers have to be both sound and well protected in order to withstand the varying types of often extreme attacks to which they are subject. The first step towards achieving this end is to ensure adequate curing. The main point outlined in this essay is the possibility of curing and at the same time protecting the concrete in one single operation actually while the building is under construction. For this purpose, two particular types of coating are available.

Introduction and Subject

The great requirement for energy makes it necessary to increase the capacity of power stations considerably. As a consequence there is a rising demand on the capacity of cooling plants. For some years natural draught coolers have been used to meet this demand. The reason is to be found in the expected greater economy and efficiency due to lower wear and shorter interruptions for maintenance work - and thus lower operating cost.

The special problems of statics, design and construction as well as foundation or aerodynamics are not dealt with in this essay. The essay is only concerned with concrete as a building material and with its resistant qualities. The natural draught - especially in consideration of the usual locations of the power stations - continuously sucks in aggressive industrial gases which are a great strain on reinforced concrete. In addition the concrete is affected by the influence of weather and temperature on its outer side and by the discharged cooling water on its inner surface.

If one takes into account that the wall thickness of the cooler shell in the upper two thirds was approx. 14 cm in most of the older coolers and is hardly more than 20 cm in the considerably larger constructions of today, and that the covering of the reinforcement is limited and hardly meets the usual requirements, it becomes evident that only a very high quality of the concrete can guarantee a long-lasting stability of the cooler jacket. The quality of the concrete always lies in the category B 300 and as a rule a waterproof concrete is required.

A material that has been successfully applied in this field is a concrete admixture which is, at the same time, a plasticizing and hydrophobic agent. This material has been used by a major construction company to build more than a dozen natural draught coolers in Germany and also in Europe with a very high quality of concrete (1).

The hydrophobic effect, in addition to the general waterproofness, lies in the fact that the material, owing to its specific composition, also repels water and moisture that are not under pressure together with any noxious matter that may act on the concrete, while the waterproofness of the concrete is, of course, examined under pressure.

Concrete Curing

The more a concrete is affected by aggressive substances and the thinner the walls, especially in proportion to the total size of the building, the more important is the condition of its surface, which is even significant if "normal concrete" is used. The condition, however, considerably depends on the curing of the concrete. The traditional method of wetting concrete constructions still used today is not only uneconomical but it is also very much limited in time for climbing formwork so that it is unsatisfactory and ineffective and should not be considered. The same is true of the application of mats and foils. This consideration leads to the preferred application of so-called chemical curing agents. Such a method has been known for years in other fields of concrete construction and has proved its efficiency. The cheaper products based on wax-type raw materials soon disintegrate due to the influence of the weather or due to mechanical abrasion and are thus unable to meet the higher requirements in the construction of coolers. They may stain the concrete or may considerably impair a later treatment of the surface, including painting work which might be desirable. Nevertheless the use of these simple curing compounds was a first step towards obtaining better surfaces in draught coolers, since there was a considerably increased resistance to aggressive environmental elements.

In order to cope with these problems, our company produced materials which were based on plastic dispersion agents and which no longer had all these disadvantages.

The first types of these products, however, had not yet such a high waterproofing coefficient as the wax-containing products, and so more research and development work was necessary before the desired effects could be achieved with specific chemical products and certain admixtures as far as possible. The first natural draught coolers which were treated in this way already showed a considerable increase in the quality of their surfaces.

Such a single treatment, which primarily aimed at sufficiently retaining the water in the fresh concrete and at a hydration on the surface which was as complete as in the concrete itself, did not produce long-term results. This is not surprising if you compare the method with a single unpigmented dispersion paint coat on a house, e. g. on plaster, and with its durability and resistance.

The increasing pollution, especially in greatly industrialized areas where the new power stations were erected caused far-sighted construction companies to consider whether a longer-lasting protection would not be more reasonable in the long run. As a result of the negative experience made with unprotected coolers it has now become standard practice to prescribe anti-corrosive coatings.

Fig. 1. Application of the second coat (pigmented) on the internal surface of the cooler during the construction phase

Owing to the size of the constructions and because they were often not easily accessible, the major question was, of course, whether such a permanent protection could not be applied while the unit was still under construction and the cooler was still comparatively accessible, although there was the maxim that concrete should be at least four to eight weeks old before it was painted because otherwise no permanent protection could be guaranteed.

Long-term protection

This consideration was taken up in our company, and on the basis of our previous experiences we came to the conclusion that, in contrast to the usual opinion, better results could be obtained, not only for an initial but also for a permanent adhesion, if specific solvents were used instead of the usual dispersion materials.

This, of course, is not possible with just any solvent or any combination of solvents, but the individual components have to meet specific requirements. This also applies to the binders used. Among the one-component plastics used, it was above all the vinyl and acrylic copolymers that proved to be most suitable.

As to the solvents we only wish to point out that they must be a mixture of such solvents as are compatible with water and such solvents as cannot be mixed with water, plus some special additives.

The first applications of this new material known by the name MC-Schutzüberzug were, however, not in the construction of coolers but of chimneys, because there was a need for an immediate external protection to be applied while the building was still under construction (2). Then the same major company having introduced the hydrophobic admixtures bravely proceeded to apply this method of long-term protection in the construction of coolers, i. e. natural draught coolers. Here again successful results were achieved.

The material was produced as a colourless substance, i. e. unpigmented, and as a pigmented substance - in several concrete

Fig.2. Natural draught cooler under construction. Internal surface: two pigmented coats Schutzüberzug 702 External surface: single non-pigmented cout as a curing compound.

colours - and it was applied, depending on the purpose and circumstances, once, twice or even three times in a single operation (however, not in such thick coats on both sides of the concrete at the same time, but then not more than twice on the inner and outer surfaces).
A test at one of the first chimneys thus protected - carried out by the Institute of Building Material Technology and Material Examination of the University of Hannover (3) - showed amongst other things a very good protection against carbonation.

This protection was sufficient for general purposes, but could not be applied where there were very high levels of corrosion or pollution. As a result we developed another protective coating on the basis of an epoxy resin. This required the development of a suitable hardener. Here, too, it was shown that the use of solvents, as described above, brought better results than solvent-free products, which can be easily produced with epoxides.

This special epoxy primer, called MC-DUR 1277 WV, which can be painted with different types of epoxy resins or, according to new findings, also with selected light-resistant polyurethanes as a finish coat, was also first applied on chimneys, which must be specially protected in their upper areas because of the strong acid attack (4). Meanwhile, however, it has also been successfully applied in the construction and operation of natural draught coolers. When, some years ago, the repair of the interior surfaces of natural draught coolers became an acute requirement, two special products mentioned above were tested by the University of Dortmund. The examination showed excellent results with regard to the penetration and permanent adhesion both on fresh and on older moist (not waterlogged) concrete and the epoxy material was successfully applied, first as a test and then with slight corrections, on the final coatings of several coolers in succession.

The present considerations about the discharge of flue gases, which are desulphurized but still contain far too much sulphur dioxide for the concrete (because it turns to sulphurous and sulphuric acid under the influence of oxygen and moisture) have made epoxy appear an interesting product also for new natural draught coolers. The cooler in the model power station at Völklingen (Saar Territory), which is fully operated in this way, was therefore immediately coated with the primer MC-DUR 1277 WV on the inside and with a two-layer coating in the one-component protective coat on the outside, while the building was still being constructed (also with climbing formwork) by the construction company Heitkamp. Two final coatings with a solvent-containing "normal", i. e. usual epoxy resin were

Fig.3. Carrying out repair work on the cooler: MC-DUR 1277 WV, unpigmented primer

subsequently applied from an elevator cage. This was done because the client and the construction company wanted to be on the safe side and make sure that, after a number of adhesion tensile tests on all areas of the cooler and in all operating phases, the adhesion both of the primer on the concrete and of the finish coat on the primer would be durable.
This expectation was fully satisfied.

Conclusion

There are therefore today a number of possibilities for a safe curing of concrete used in the construction of natural draught coolers. Above all, there is the chance of combining curing

Fig.4. First and second epoxy finishing coats applied on top of MC-DUR 1277 WV primer (Cooler of the model power station)

with long-term protection without longer interruptions. This long-term protection can be adapted to the expected level of attack so that its application results in a greater economy and efficiency. It is not a subject of this essay, but it may be interesting to know that this method is now used also in other fields, apart from slip and climbing form, especially in such fields where a sufficient protection against chemical aggression in addition to curing is required immediately after the formwork has been removed.

But also in the construction of the cooler itself, the above-mentioned measures and products are not only used for the cooling jacket but also, in an adapted way, for the supports, the water-trough and for other concrete parts and they have proved to be fully reliable.

As a summary, it may be said that a process has been developed which provides concrete with an additional stability and resistance against various types of chemical attack.

References

1. MC-Report No. 7, ed. by MC-Bauchemie, Essen/Bottrop 19

2. Korrosionsschutz im Schornsteinbau, p. 172 in the Congress Book of the 3rd Intern. Conference on Chimneys in Munich in 1978, ed. by Vulkan-Verlag Essen 1979

3. Test Report I-2/83-Scho/Hm of April 27, 1983 from the Institute for Building Material Technology and Examination of the University of Hannover

4. DIN 1056 E and Technical Leaflet "Coating of Reinforced Concrete Chimneys" of the Deutsche Beton-Verein, Wiesbaden 1983

Physical, Chemical and Microbiological Influences of the Environment on Thin Cooling Tower Shells

Dr.-Ing. W.Stiller

Philipp Holzmann Aktiengesellschaft
Hauptniederlassung Düsseldorf
Münsterstrasse 291

4000 Düsseldorf

1. Introduction

In most cases waste heat from power stations are released into the atmosphere continuously by means of cooling towers. Those giant-like buildings which decisively characterise the outside image of a power station, are extremely thin structures with reinforced concrete walls of a merely 14 - 16 cm in strength, concrete surfaces of 20,000 to 100,000 m^2 and ratios between surfaces and volume of up to 10.

The layman properly knows reinforced concrete as an endurable and reliable building material which is able to resist influences from weather for decades. The constructor, however, bears in mind the relation between stresses caused by the jet stream during the cooling process and the low thickness of the shell. Some damages exclusively on the inner surfaces of the cooling towers were discovered by accurate surveillance of these structures, carried out by the operating company. These damages were observed first after the extreme fall of temperature between Dec.30th 1978 and Jan. 1st 1979. In some parts of Northrhine Westphalia temperatures fell by 27°C to -20°C, which caused the inner surfaces partly to sand and to show shallow chippings. The stability of the structure, of cause, was not endangered. Damage analyses at the cooling towers, which were erected early- or in the middle of the past decade from slag-rich blast furnace cement, showed extraordinary depths of carbonatisation in a range of 1 - 1,5 cm with a concrete covering of the reinforcement bars of 2.5 cm. (Fig. 1)

In order to obtain data about the progress of the damages and the actual scale of stress within the material of colling tower shells, a research- and development project was carried out, kindly supported by the German Ministry for Research and Technology (BMFT). We moreover owe our gratitude to the construction department of the "RWE" for their active and kind assistance.

Since the end of 1981 the research- and development projekt has been carried out at the cooling tower of the brown coal power station at Niederaussem. The natural-draught cooling tower, analysed, works after the worldwide known, and well-tried Hamon-Sobelco system. The objectives of the research- and development project are to record continuously the scale of environmental and operational effects and to find extreme data from those readings by means of a comprehensive measuring programme, which will, by far, surmount the extent of comparable research programmes. (Fig. 2)

In the following the matter will be about realization and results, so far, of the measuring programme without digging too deep into the comprehensive fields of physics, chemistry, and micro-biology.

2. Realisation of the Measuring Campaign

2.1 Measuring of Temperature

The different thermal effects on the shell have been measured in this programme using 70 measure spots on the inner and outer surfaces of the shell. Thermocouples which allow to "spotmeasure" were used as sensors. Trouble-free installation of measuring leads on the shell's surface proved to be especially difficult.

2.2 Measuring of Humidity

To judge the actual stress on the inner surface of the cooling tower it is important to find out for how long and to what extend the surface is humid or completely wet. Thus humidity was measured continuously in the 4 main meridians, by using humidity sensors produced by Novasina.

2.3 Measuring of Strain

The dominant forces of wind cause the shell to strain. Because there is very little data in the literature as to the extent of strains, at certain spots, covered strain gauges with an enlarged measured section were applied, partly in rose-shape arrangements.

2.4 Meteorological Recordings

The co-effect of climatic elements is an important stress-factor for the shell. Thus wind direction and -force, atmospheric pressure, sunshine hours, and humidity of the environment were recorded. The actual quantity of polluted air entering the cooling tower was measured by means of wind meters which were installed on the lower brink of the tower.

2.5 Measuring of Immission Rates

When judging the effect of polluted air on materials, a relation between immission and the resulting effects has to be constructed. [I] .Therefore determination by means of the IRMA-method was used. A paper felt body which must be kept moist, is being exposed to the polluted air. The pollutants diffuse into it and their quantity can be recorded by means of a chemical analysis.

2.6 Composition of the Cooling Water

Whereas the cooling water and its components are known as to their effects on the materials in the cooling system, there is very little data about the composition of the condensed water on the cooling towers's shell [2] . Thus analyses of the circuit water were set in relation to the condensed water from special places on the shell.

The readings for the temperature gradient ΔT, occurring at a stationary heat flow, showed an annual average of 12 ^{o}C with extreme readings of about 25-30 ^{o}C in winter. This gradient is extremely dependant on the vapor temperature of the tower and tends to rise when atmospherical temperatures fall. [4] . (Fig. 3)

In operation, the change in average shell temperature due to sun rays, is a neglectable factor and can only be sensed in midsummer in our latitudes. Extreme differences in temperature between structure components of different thickness occur when the cooling tower changes its average temperature during the starting and cooling down periods.

Within the shell considerable stresses build up, which will become apparent in sufficiently reinforced sections when the power station is being put into stepwise operation.

The switch-off process is different. The cooling tower is left on its own. Those thin (14 - 16 cm), heavily soaked sections of the tower will soon release their stored heat to the surrounding air. The lower sections of the shell with a wall diameter of up to 70 centimeters maintain the operation temperature in their cores. Owing to different degrees in temperature strains, stresses will build up within the shell, which may lead to an accumulation of vertical cracks in the lower quarter of the shell. Stresses created this way will, of course, superimpose one another by wearing - being a consequence of the instationary temperature field which arises during the cooling down process. A critical case of temperature load will occur in a cooling tower, when a station in regular load service would be shut down at extreme winter conditions. Water, soaked into the concrete surface will then freeze immediately causing chipping on the surface.

2.7 Microbiology

It is already known that blue algae settle in the upper area of the inner surface of the cooling tower shell. This living coating can be removed like a hide, causing, however, fine particles of concrete to remain on the coating's backside. Besides algae, micro-organisms can be found to a considerable extent on the inner surface of the shell. For the research- and development project Professor Bock from the University of Hamburg has been working on the problem of microbiology. His institute has aquired considerable experience as to damages in the concrete of sewage canals in the city of Hamburg. Theses damages occurred partly owing to "biological attacks" [3] .

The readings measured at the cooling tower - 5,000,000 so far - have been recorded and stored in a central measure equipment. The statistical analysis and the extreme data finding have been carried out in our Duesseldorf branch.

3. Results

3.1 Measurement of temperature

The concrete shell is object to different thermal influences from operation process and environment. The temperatures measured are dependant from the time. Only extreme readings have any significance for the construction.
Finding of extreme readings is based on the following three temperature values:

. temperature gradient in the shell

. changes in average shell temperature during the sunlight hours

. differences in temperature owing to varying wallthickness.

3.2 Measurement of Humidity

A continuous measurement of humidity - a process which caused many problems - leads to the result that the inner surfaces are constantly spread by vapor with a humidity of 85 - 100 percent. So on the interface to the concrete shell humidity condenses, creating thus a thin film of water. On only a few days a year when, owing to sun rays the outer surface is warmer than the inner one the shell will dry.A visible sign being vapor formation which in those periods will be minimal if air humidity as well is low [4] .

3.3 The Influence of the composition of Cooling Water on Shell Concrete

The cooling water used in the cooling system is basic with a pH-value of about 9. Generally this value is not harmful for cemented concrete. "Acid corrosion" with the typical structure of exposed aggregates concrete cannot occur. Condensed water, however, which collects on the shell shows a chemical composition fundamentally different from that of the cooling water. Merely in the sprinkler area and up to 3 - 5 meters above the sprinkler assembly cooling water components can be found in larger quantities.
The average pH-values were 5 - 6 but also values of 4.2 were measured, which means that the water film trickling down on the inner surface of the shell proves to be acidic.
(Fig. 4 and 5)

It moreover shows that the salt content in the condensate decreases drastically. It makes the changes in chemical composition apparent reminding us that distilled water is harmful to concrete, owing to the fact that it vigorously solves the salt in the concrete [5] .
Moreover the condensate contains considerable quantities of lime-aggressive carbon acid. (Fig. 6). (The dissolving effect of it towards concrete is known [6]) . It is moreover interesting that the pH-value rises in downward direction measured on the inner surface of the towers. These changes

occur owing to microbiological influences. From these data it becomes apparent immediately that the shell concrete must be protected against slightly acidic condensate to avoid "dissolving corrosion".

3.4 Air-Chemical Research

In the analysis of immission rates the components SO_2, Cl, F, and NO_x were researched. The 1982 recordings supply characteristic high values for all winter-months and lower values for the summer-months. As to effects on materials peak values of pollutants may generally be neglected. In contrast to effects on living nature, only longrange average values of pollutants are of any interest [1] .
A review of annual average values of the Niederaussem-power-station showed definitely that according to present data, harmful effects of immissions (SO_2-compounds, in particular) can be proved. It is a known fact that, owing to higher air-stream speed, immission rates rise at higher altitudes.
Thus thorough considerations will be necessary whether the upper sections of the tower's outer surface, as well, must be protected more efficiently against pollutants from the air.

3.5 Microbiological Influences

The inner surface of a cooling tower is animated. It is inhabited by a large number of blue-algae and bacteria (Fig. 7).
According to Mr Parker it is a surface temperature of about 30 $^\circ$C and high air humidity that make out ideal living conditions in those regions. Blue-algae predominantly settle in the upper regions of the shell. They do not produce concrete-destroying acids. The researched bacteria groups are different. They produce anorganic acids in a biological process.
In this respect, nitrifying bacteria, producing nitric acid, are an important factor [6] . Acid production is a function of the ambient temperature. The optimal temperature is at about 30 $^\circ$C. At 10 $^\circ$C there exists hardly any life on the shell.

As life needs a certain time to develop, acid production mainly occurs in towers, which are operated in the regular load service. Acid quantities produced by microbiology are naturally low. Owing to the great extension of the inner surface, however, a certain quantity problem arises anyway. Those acid producing organisms, which were detected by the University of Hamburg will corrode the concrete surface in the long run. Corrosion will be caused by a continuous leach-out process in the concrete by means of a constant ion exchange. The pre-damaged surfaces will then be more exposed to the weather, especially to frost, than an undamaged surface. Regarding the formation of micro-organisms, the characteristics of subsequently sealed surfaces were of some interest. In the course of the research programme it could be observed that the influence of microbiology was diminishing considerably. After two years in operation, minor traces of blue-algae were found in the waist-section of the cooling tower, acid producers, however, were not detected [7] .

4 .Summery

Not only the results of static calculations are relevant for the construction of a concrete shell. The measuring campaign proved that, moreover so far often neglected influences from physics, chemistry, and the operational process are important factors. For a long-term protection of cooling towers in regular load service, sealing is the right way. Because the sealing is being applied immediately while errecting the tower, it will not have any negative effect on economy and elegance of the concrete structure.

References

[1] Luckat, S.: Die Einwirkungen von Luftverunreinigungen auf die Bausubstanz des Kölner Domes
Kölner Domblatt 38/39, 1979,

[2] Held, H.D. : Kühlwasser, Vulkan-Verlag, Essen (1977)

[3] Bock, E. : Bedeutung der Mikroorganismen bei der Korrosion von Abwasserkanälen
Tiefbau Sonderheft zum 4. Statusseminar "Bauforschung und Technik"; Gütersloh 1983

[4] Baer, E. : Thermodynamische Untersuchungen am Naturzugkühlturm C des Kraftwerkes Neurath
VDI Fortschrittsbericht 15,7; VDI-Verlag Düsseldorf 1979

[5] Krenkler, K. : Chemie des Bauwesens; Springer Verlag 1980

[6] Wierig, H.J. : Zum Verhalten von Beton gegenüber physikalischen, chemischen und biologischen Angriffen.
Vortrag VGB Bautagung, Bautechnik in Wärmekraftwerken 1982

[7] Bock, E. : Abschlußbericht über das Forschungsvorhaben: Korrosion an hochbelasteten Stahlbetonkonstruktionen in industrieller Atmosphäre; wird 1984 veröffentlicht

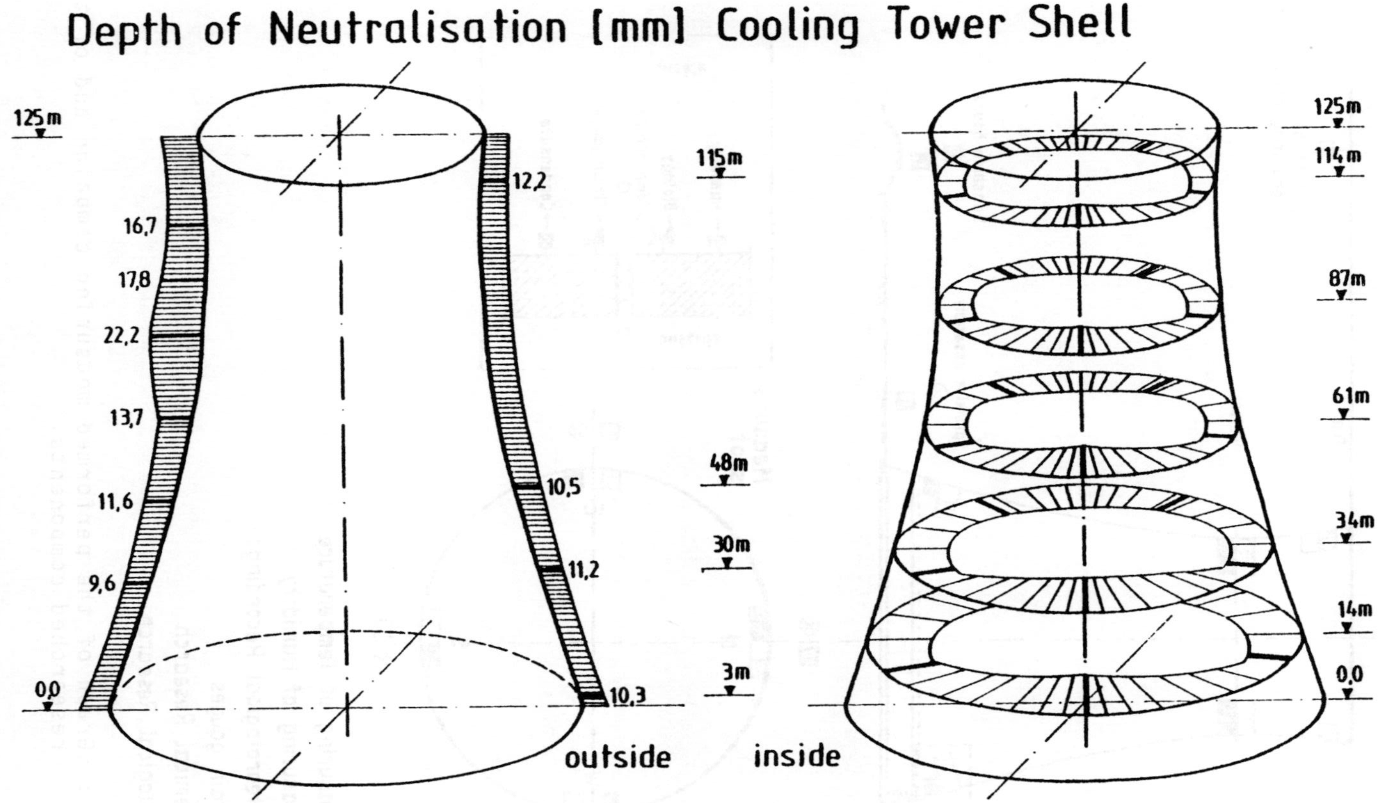

Fig.1 : The depths of neutralization as a scale for the carbonation of the concrete are in the range of 10 - 20 mm at the inner and outer surfaces of the shell.

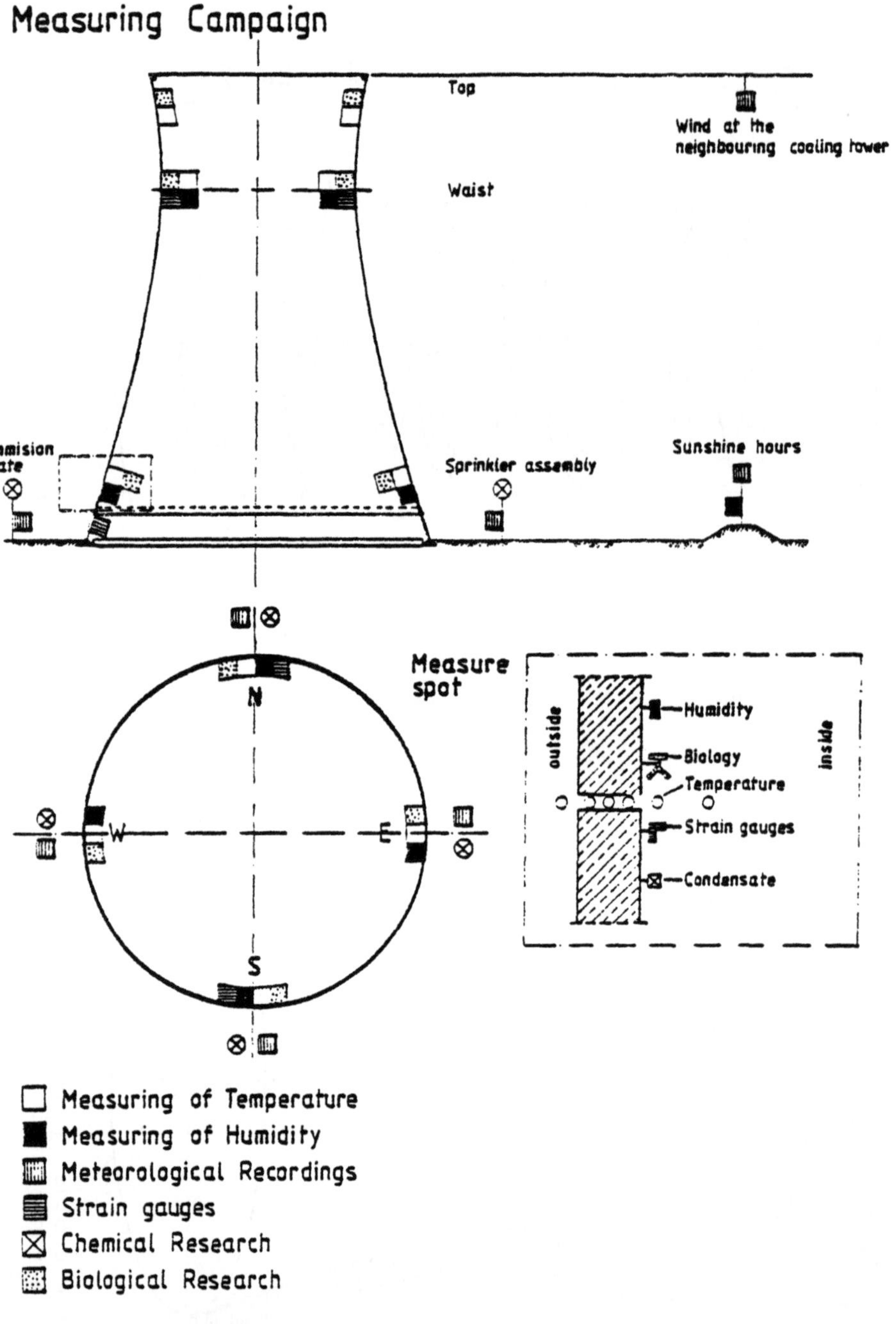

Fig. 2 : Graph of the performed measuring campaign and of the researched components.

Temperature Difference across the Shell Wall in Operation

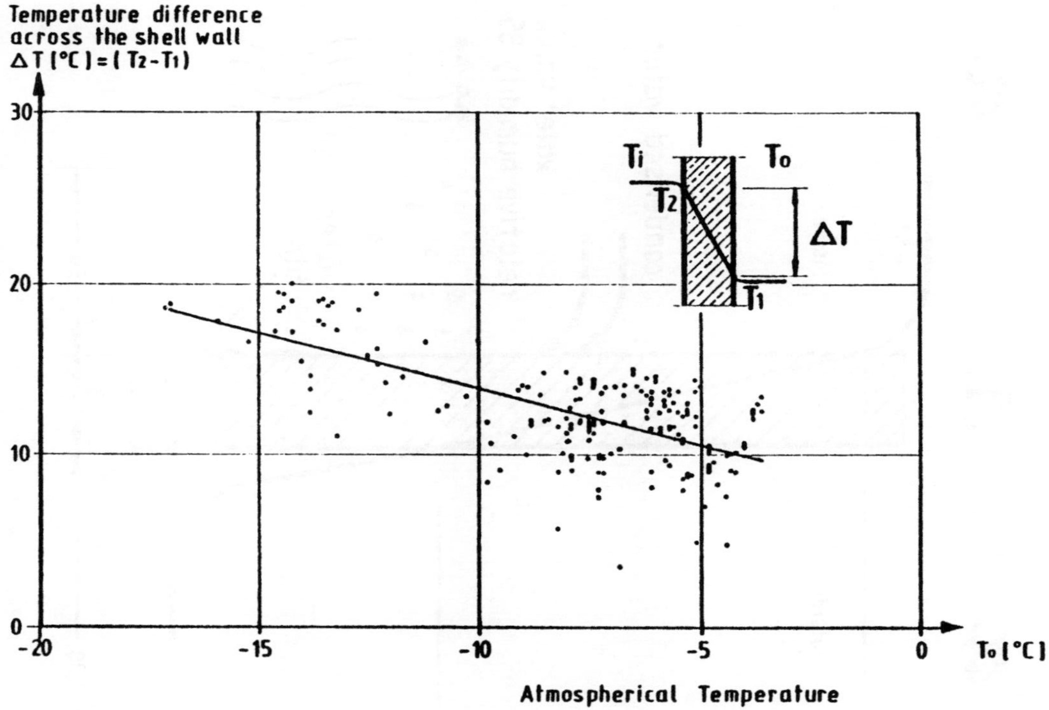

Fig. 3 : The temperature difference between the two surfaces of the shell wall tends to rise when atmospherical temperatures fall.

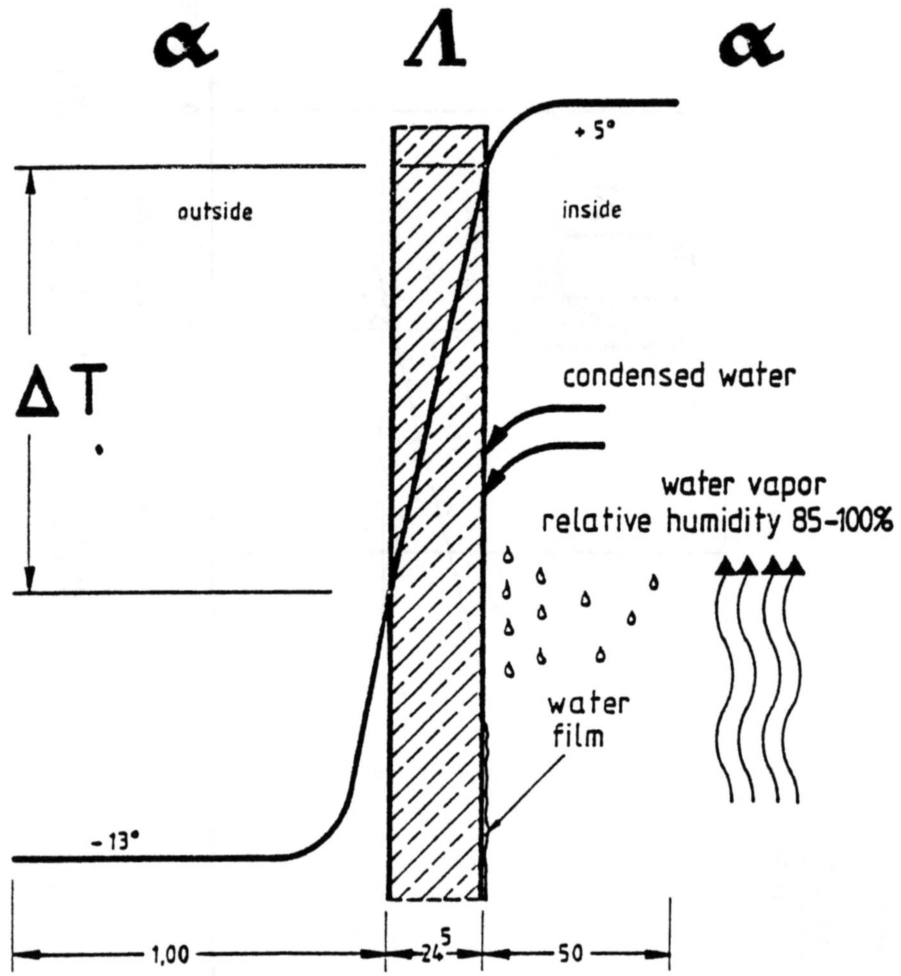

Fig. 4 : Formation of condensate at the cooling tower shell.

pH - Value
Circuit Water - Condensate

Period January - March 1983

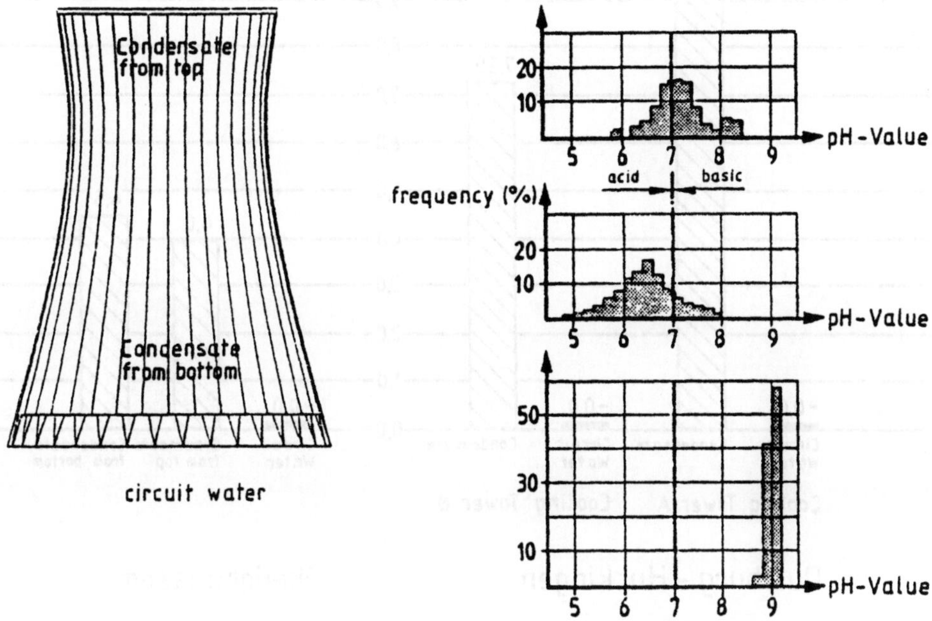

Fig. 5 : Comparison between the pH-values of the condensates and the pH-value of the circuit water.

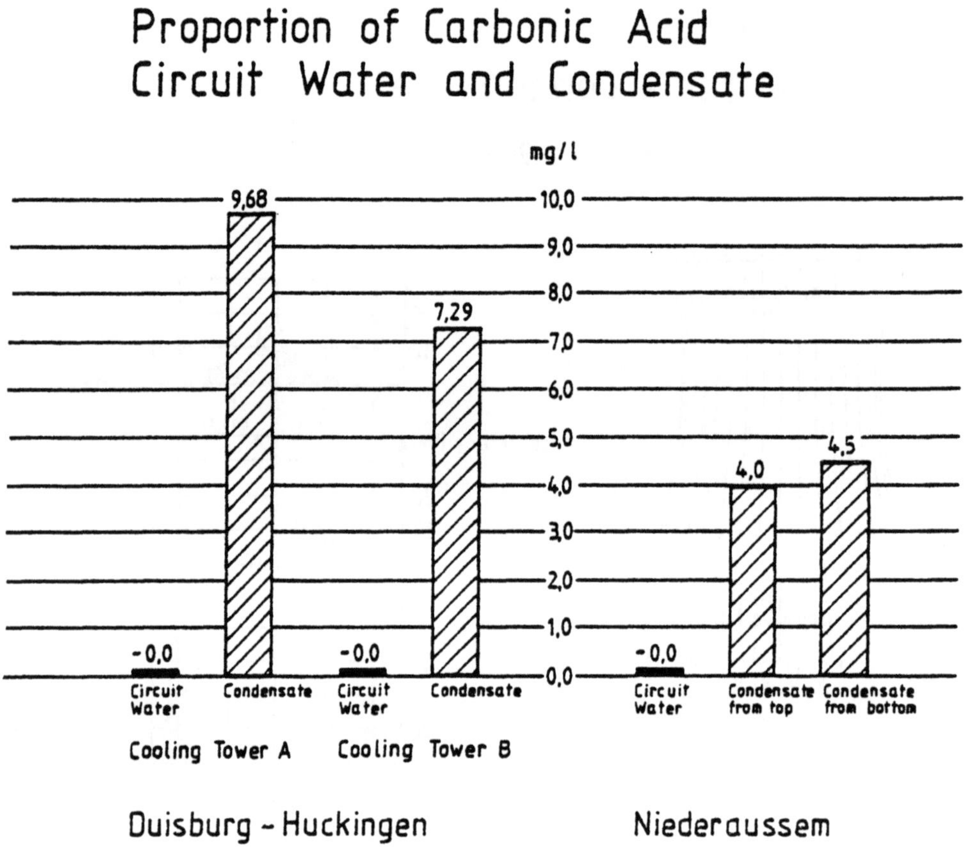

Fig. 6 : There is no lime-aggressive carbonic acid in the circuit water. But there are up to 10 mg/l lime-aggressive carbonic acid in the condensed water.

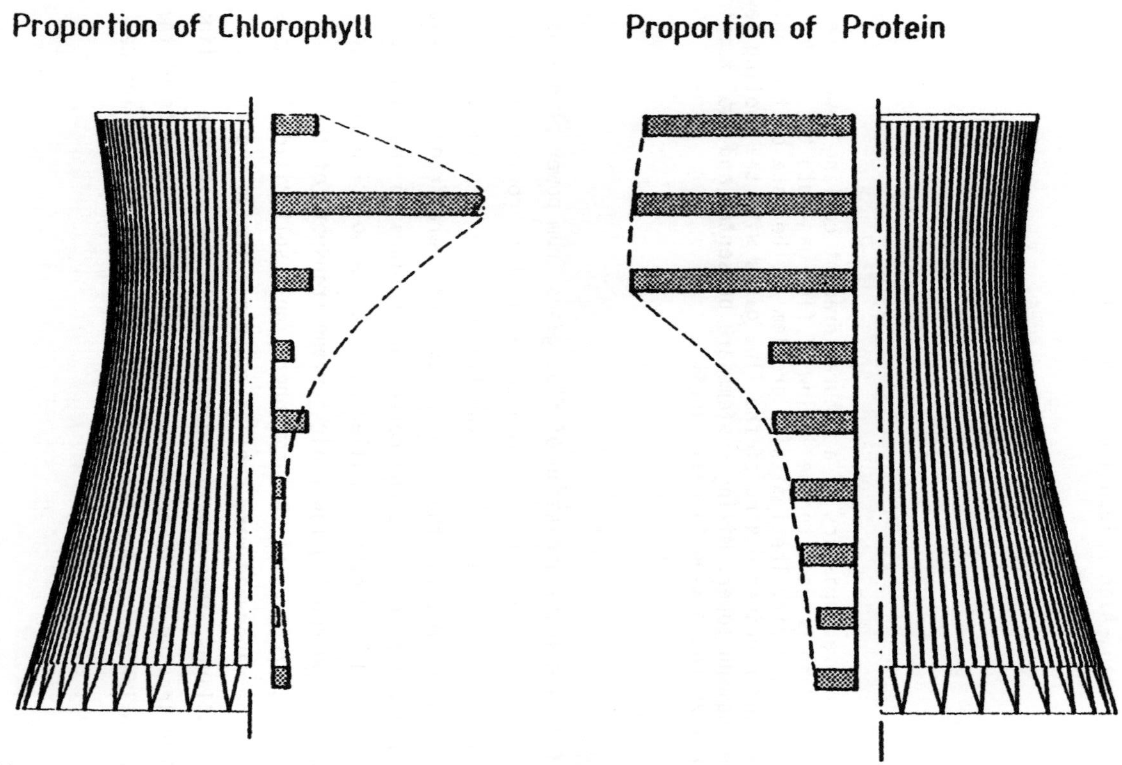

Fig. 7 : Proportion of chlorophyll as a scale for the extend of algae and proportion of protein as a scale for organic life.

Aerodynamic aspects of the Inflow Problem of Flue Gases into the Cooling Tower

H. Ruscheweyh

Institut für Leichtbau, Aachen Technical University

Summary

The internal flow characteristics of a natural draught cooling tower are shown and discussed in respect to the inflow of flue gas into the vapor and warm cooling tower air. The immission problem in the wake of the cooling tower at cross wind requires mixing of the flue gases with the cooling tower air within the cooling tower. Mixing systems are presented and two examples show how to realize the mixing in real cases.

1. Introduction

The problem of ground level immission of flue gases from power plants has become very important lately and the importance of this problem will increase in the future. Desulfurizing of the flue gas can help to solve some of the immission problems, but the desulfurized gases are frequently relatively cold and wet which increases the corrosion problems in the pipes and in the chimney stack and they have a smaller buoyancy. A solution of these problems is to use the cooling tower air for the transport of the flue gases into very high regions of the atmosphere.The desulfurized flue gases are passed into the vapor and warm cooling tower air inside the tower, i.e. just above the fillings. Then the flue gases have to mix with the cooling tower air before leaving the tower mouth. Some aerodynamic aspects of the flow condition inside the cooling tower and the mixing procedure are described here.

2. The flow condition inside the cooling tower

The flow condition inside the cooling tower is very complicated and is influenced by many parameters, as operating condition, heat distribution in the fillings, wind speed, inflow condition at the cooling tower inlet, the shape of the cooling tower top, ect. Only few investigations have been undertaken in the past to describe this flow. Akulowa /1/ published model test results and he showed that intensive recurrent and vortex regions exist inside the cooling tower at cross wind (see Fig. 1). These results are confirmed and extended by water tunnel tests of the author /2,3/ with a heated model in simulated atmospheric wind (=water) flow. It could be

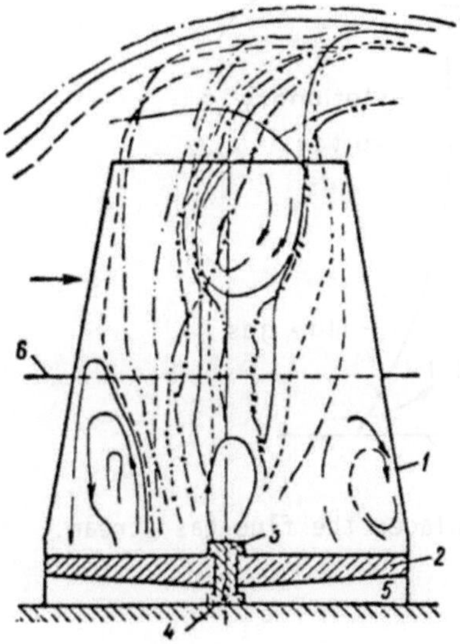

Fig. 1: Internal flow condition at cross wind /1/

Fig. 2: Internal flow condition without cross wind, heated model /3/.

shown that large vortices are established inside the cooling tower (Fig.2) which grow up and vanish in an irregular manner /3/.

Following a streamline starting from the center of the cooling tower base (Fig. 3) the streamline is displaced by the center vortex to an outer region of the cooling tower air stream, so that it touches the concrete wall. If this streamline is a flue gas stream, effects have to be considered for the ground level immission and the corrosion of the concrete wall of the cooling tower.

3. Immission conditions behind a large natural cooling tower

Investigations of the immission conditions behind large natural cooling towers show that the ground level concentration of flue gases emitted

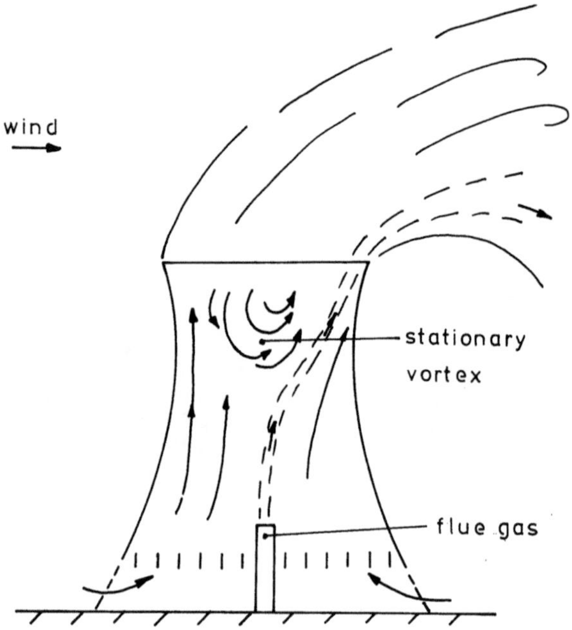

Fig. 3: The stationary center vortex displaces the flue-gas stream.

Fig. 4: Downdrift behind a full scale cooling tower.

from stacks nearby the cooling towers can increase very much if the flue gas stream touches the lateral top vortices of the cooling tower which induce a downdrift behind the cooling tower. By this downdrift the effective emission altitude is decreased so that the ground level concentration is increased respectively (Fig. 4 and 5). This effect is increased for groups of cooling towers by interference effects /4/.

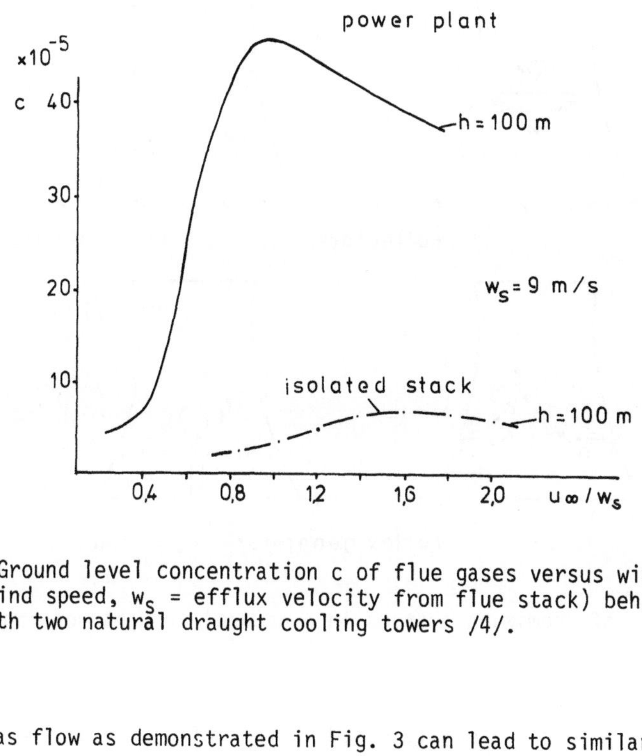

Fig. 5: Ground level concentration c of flue gases versus wind speed ratio (u_∞ = wind speed, w_S = efflux velocity from flue stack) behind a power plant with two natural draught cooling towers /4/.

A flue gas flow as demonstrated in Fig. 3 can lead to similar immission conditions as in Fig. 5. In order to reduce the immission concentration to an acceptable value the flue gases have to be mixed with the cooling tower air before leaving the cooling tower. This way the emission concentration can be decreased by a factor of 5 to 10.

4. The mixing process

The mixing process in a cooling tower has to be realized within an extremely short distance of approximalety 1.5 times the cooling tower diameter. (Mixing in a turbulent flow normally takes a distance of 20-30 x diameters). The mixing process therefore has to be supported by so called "premixing systems" which produce a good "distributive mixing" in the cross section /5/. Furthermore a high turbulence is generated, too, giving an intensive "turbulent mixing" following the distributive mixing.

Different premixing systems have been proposed in the past. Some examples are given in Fig. 6.

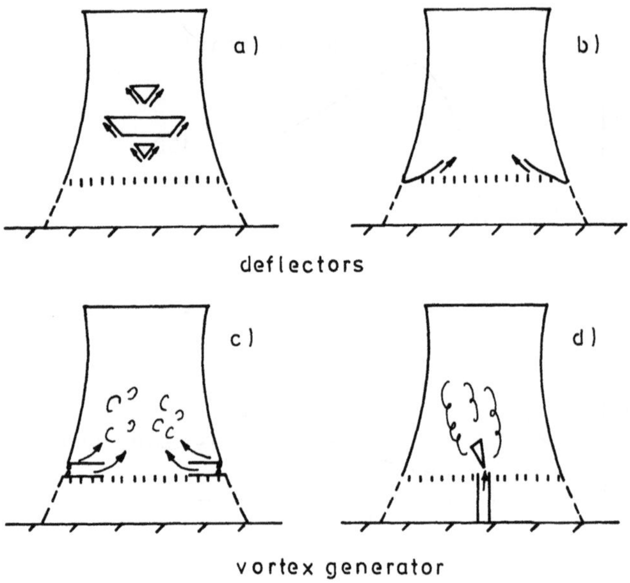

Fig. 6: Examples of premixing systems in a natural draught cooling tower.

The systems a.) and b.) are mainly based on the deflection of the stream while the systems c.) and d.) work with generated vortices.

For the application of these premixing systems the pressure drop, the constructive volume and the danger of contamination must be considered.

5. Examples

Two examples are presented to demonstrate practical solutions. While the first example was realized in full scale the second example is still in the state of design.

5.1 Desulfurizing plant inside of a cooling tower

A consequent solution in respect to short pipes for the desulfurized flue gases is shown in Fig. 7. The desulfurizing plant is integrated in the center of the cooling tower and the desulfurized flue gases are emitted immediately above the roof.

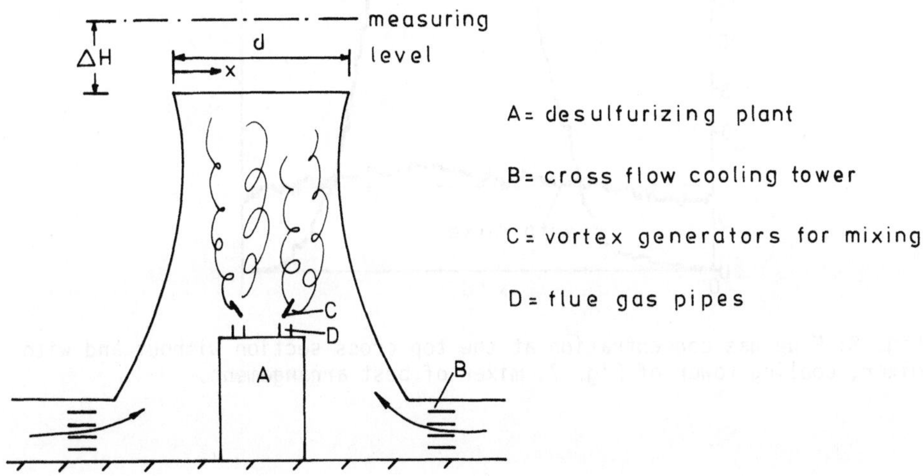

Fig. 7 : Sketch of a cross flow natural draught cooling tower with integrated desulfurizing plant and flue gas mixer.

Vortex generator plates of type d.) of Fig. 6 are installed directly over the flue gas outlets. The vortices generated by these plates give an intensive distributive mixing and simultaneously a high turbulence mixing. Fig.8 shows the flue gas concentration at the top cross section of the cooling tower without and with mixer.

Without any mixing system a peak exists in the flue gas concentration. With the premixing system of best arrangement a nearly constant flue gas concentration is attained for operation at full load. In order to include good mixing for the operation at partial load, too, the mixers have been arranged in an optimized position. For this arrangement the concentration distribution is modified as shown in Fig. 9. The two flat maxima are smoothed within a short distance above the cooling tower, so that an ideal mixing distribution is attained in this altitude.

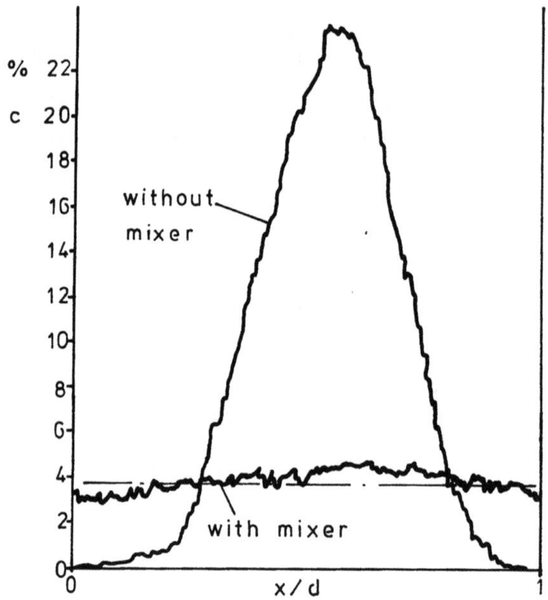

Fig. 8: Flue gas concentration at the top cross section without and with mixer, cooling tower of Fig. 7, mixer of best arrangement.

Measurements at the prototype confirmed the model test results.

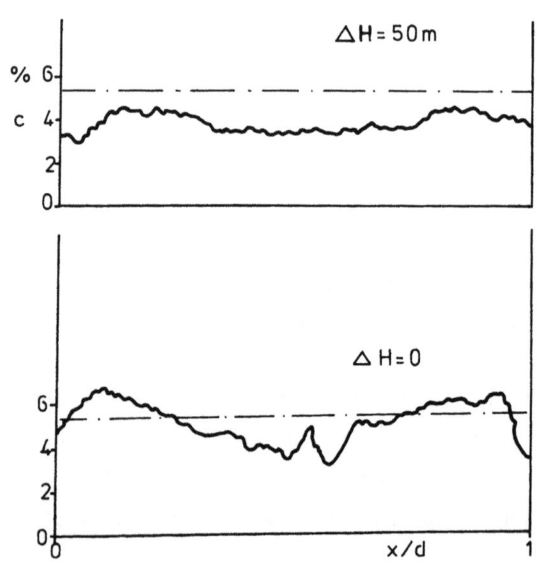

Fig. 9: Flue-gas concentration at the top and 50 m above the cooling tower with mixer in optimized arrangement, cooling tower of Fig. 7.

5.2 Flue gas inflow above the fillings

A model study was made to verify the inflow positions of the flue gas pipes above the fillings. Four flue gas pipes were used with vertical output direction (Fig. 10). A premixing system with vortex generating plates similar to the type d.) of Fig. 6 was installed above the outlet pipes. Fig. 11 shows the flue gas concentration in the outlet cross section of the cooling tower. Without the mixing system a very irregular distribution of the flue gas exits.

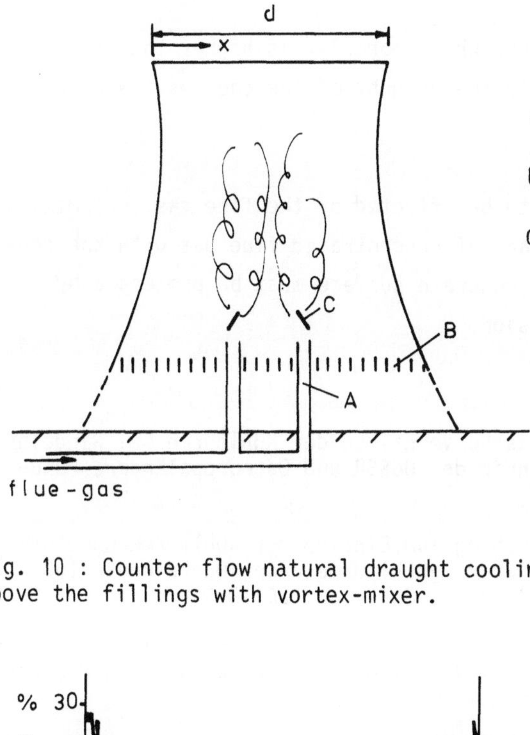

A = flue gas pipes

B = conterflow cooling tower

C = vortex generator

Fig. 10 : Counter flow natural draught cooling tower with flue gas inflow above the fillings with vortex-mixer.

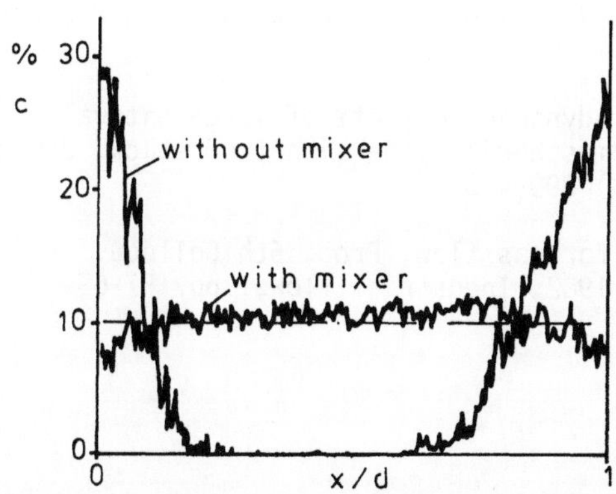

Fig. 11: Flue gas concentration c at the top of the cooling tower with and without vortex-mixer.

The concentration of the flue gas has its maximum at the concrete wall. With the mixing system a nearly constant distribution is attained. This mixing system works well for operation at full load and at partial load.

6. Additional remarks

Three points are worth mentioning:

a.) The installed mixing system stabilizes the cooling tower air stream because of its momentum exchange.

b.) Due to the flue gas temperature which normally is higher than the temperature of the cooling air the draught of the cooling tower increases.

c.) The concrete wall is likely to be affected by the flue gas. The mixing system should avoid the contact of concentrated flue gas with the concrete wall. Nevertheless the concrete surface must be protected by special covers against corrosion.

7. References

1. Akulowa, L.G.; Das aerodynamische Verhalten der Kühltürme bei Windeinwirkung, Notizen aus der Technik der UdSSR und Osteuropaländer, Puhu-Verlag, München, 1963

2. Ruscheweyh, H.; Modelluntersuchung zum Einfluß der Kühlturmkronenform auf das Zugverhalten großer Naturzugkühltürme unter Querwindeinfluß, Brennst.-Wärme-Kraft 34(1982) Nr. 7, Juli, pp. 361-369

3. Henning, H., Ruscheweyh, H.; Air flow conditions in a cooling tower and means of control, Proc.of the IAHR cooling tower workshop, San Francisco (1980) EPRI, Palo Alto, Calif.

4. Ruscheweyh, H., Fischer, K.; Aerodynamics effects of large natural-draught cooling towers on the atmospheric dispersion from stack, Journal of Industr. Aerodynamics, 4(1979) 399-413

5. Ruscheweyh, H.; A mixing system for gas flow, Proc. 5th Colloqu. on Industrial Aerodynamics, Aachen 1982, Industrial flows, pp. 57-67

FSC C105338

If you have any concerns about our products,
you can contact us on
ProductSafety@springernature.com

In case Publisher is established outside the EU,
the EU authorised representative is:
Springer Nature Customer Service Center GmbH
Europaplatz 3, 69115 Heidelberg, Germany

Printed by Libri Plureos GmbH
in Hamburg, Germany

MIX
Papier aus verantwortungsvollen Quellen
Paper from responsible sources
FSC® C105338

If you have any concerns about our products,
you can contact us on
ProductSafety@springernature.com

In case Publisher is established outside the EU,
the EU authorized representative is:
Springer Nature Customer Service Center GmbH
Europaplatz 3, 69115 Heidelberg, Germany

Printed by Libri Plureos GmbH
in Hamburg, Germany